Transient Analysis of Power Systems

Transient Analysis of Power Systems

A Practical Approach

Edited by

Juan A. Martinez-Velasco
Retired Professor
Polytechnic University of Catalonia
Barcelona
Spain

Registered Offices
John Wiley & Sons, Inc., 111 River Street, Hoboken, NJ 07030, USA
John Wiley & Sons Ltd, The Atrium, Southern Gate, Chichester, West Sussex, PO19 8SQ, UK

Editorial Office
The Atrium, Southern Gate, Chichester, West Sussex, PO19 8SQ, UK

For details of our global editorial offices, customer services, and more information about Wiley products visit us at www.wiley.com.

Wiley also publishes its books in a variety of electronic formats and by print-on-demand. Some content that appears in standard print versions of this book may not be available in other formats.

Library of Congress Cataloging-in-Publication Data

Names: Martinez-Velasco, Juan A., editor.
Title: Transient analysis of power systems : a practical approach / edited
    by Juan A. Martinez-Velasco, Retired Professor, Polytechnic University
    of Catalonia, Barcelona, Spain.
Description: Hoboken, NJ : Wiley-IEEE Press, 2020. | Includes
    bibliographical references and index.
Identifiers: LCCN 2019027811 (print) | LCCN 2019027812 (ebook) | ISBN
    9781119480532 (hardback) | ISBN 9781119480303 (adobe pdf) | ISBN
    9781119480495 (epub)
Subjects: LCSH: Transients (Electricity)–Simulation methods.
Classification: LCC TK3226 .T76 2020 (print) | LCC TK3226 (ebook) | DDC
    621.319/21–dc23
LC record available at https://lccn.loc.gov/2019027811
LC ebook record available at https://lccn.loc.gov/2019027812

Cover Design: Wiley
Cover Image: © kentoh/Shutterstock

Set in 10/12pt WarnockPro by SPi Global, Chennai, India
Printed and bound in Singapore by Markono Print Media Pte Ltd

10 9 8 7 6 5 4 3 2 1

Contents

About the Editor *xv*
List of Contributors *xvii*
Preface *xix*
About the Companion Website *xxi*

1   Introduction to Transients Analysis of Power Systems with ATP *1*
Juan A. Martinez-Velasco
1.1   Overview *1*
1.2   The ATP Package *3*
1.3   ATP Documentation *5*
1.4   Scope of the Book *6*
         References *8*

2   Modelling of Power Components for Transients Studies *11*
Juan A. Martinez-Velasco
2.1   Introduction *11*
2.2   Overhead Lines *12*
2.2.1   Overview *12*
2.2.2   Multi-conductor Transmission Line Equations and Models *13*
2.2.2.1 Transmission Line Equations *13*
2.2.2.2 Corona Effect *15*
2.2.2.3 Line Constants Routine *15*
2.2.3   Transmission Line Towers *16*
2.2.4   Transmission Line Grounding *17*
2.2.4.1 Introduction *17*
2.2.4.2 Low-Frequency Models *17*
2.2.4.3 High-Frequency Models *18*
2.2.4.4 Treatment of Soil Ionization *20*
2.2.5   Transmission Line Insulation *21*
2.2.5.1 Voltage-Time Curves *21*
2.2.5.2 Integration Methods *22*
2.2.5.3 Physical Models *22*
2.3   Insulated Cables *23*
2.3.1   Overview *23*
2.3.2   Insulated Cable Designs *24*
2.3.3   Bonding Techniques *25*
2.3.4   Material Properties *26*
2.3.5   Discussion *27*

2.3.6   Cable Constants/Parameters Routines   *27*
2.4   Transformers   *28*
2.4.1   Overview   *28*
2.4.2   Transformer Models for Low-Frequency Transients   *31*
2.4.2.1   Introduction to Low-Frequency Models   *31*
2.4.2.2   Single-Phase Transformer Models   *32*
2.4.2.3   Three-Phase Transformer Models   *36*
2.4.3   Transformer Modelling for High-Frequency Transients   *37*
2.4.3.1   Introduction to High-Frequency Models   *37*
2.4.3.2   Models for Internal Voltage Calculation   *39*
2.4.3.3   Terminal Models   *41*
2.5   Rotating Machines   *45*
2.5.1   Overview   *45*
2.5.2   Rotating Machine Models for Low-Frequency Transients   *46*
2.5.2.1   Introduction   *46*
2.5.2.2   Modelling of Induction Machines   *46*
2.5.2.3   Modelling of Synchronous Machines   *51*
2.5.3   High-Frequency Models for Rotating Machine Windings   *55*
2.5.3.1   Introduction   *55*
2.5.3.2   Internal Models   *56*
2.5.3.3   Terminal Models   *58*
2.6   Circuit Breakers   *58*
2.6.1   Overview   *58*
2.6.2   Circuit Breaker Models for Opening Operations   *59*
2.6.2.1   Current Interruption   *59*
2.6.2.2   Circuit Breaker Models   *60*
2.6.2.3   Gas-Filled Circuit Breaker Models   *61*
2.6.2.4   Vacuum Circuit Breaker Models   *62*
2.6.3   Circuit Breaker Models for Closing Operations   *64*
2.6.3.1   Introduction   *64*
2.6.3.2   Statistical Switches   *65*
2.6.3.3   Prestrike Models   *66*
           Acknowledgement   *66*
           References   *66*

3   **Solution Techniques for Electromagnetic Transient Analysis**   *75*
           Juan A. Martinez-Velasco
3.1   Introduction   *75*
3.2   Modelling of Power System Components for Transient Analysis   *76*
3.3   Solution Techniques for Electromagnetic Transients Analysis   *78*
3.3.1   Introduction   *78*
3.3.2   Solution Techniques for Linear Networks   *78*
3.3.2.1   The Trapezoidal Rule   *78*
3.3.2.2   Companion Circuits of Basic Circuit Elements   *79*
3.3.2.3   Computation of Transients in Linear Networks   *85*
3.3.2.4   Example: Transient Solution of a Linear Network   *86*
3.3.3   Networks with Nonlinear Elements   *87*
3.3.3.1   Introduction   *87*
3.3.3.2   Compensation Methods   *87*

3.3.3.3   Piecewise Linear Representation   *89*
3.3.4   Solution Methods for Networks with Switches   *90*
3.3.5   Numerical Oscillations   *91*
3.4   Transient Analysis of Control Systems   *96*
3.5   Initialization   *97*
3.5.1   Introduction   *97*
3.5.2   Initialization of the Power Network   *97*
3.5.2.1   Options for Steady-State Solution Without Harmonics   *97*
3.5.2.2   Steady-State Solution   *98*
3.5.3   Load Flow Solution   *99*
3.5.4   Initialization of Control Systems   *100*
3.6   Discussion   *100*
3.6.1   Solution Techniques Implemented in ATP   *101*
3.6.2   Other Solution Techniques   *101*
3.6.2.1   Transient Solution of Networks   *101*
3.6.2.2   Transient Analysis of Control Systems   *102*
3.6.2.3   Steady-State Initialization   *102*
           Acknowledgement   *103*
           References   *103*
           To Probe Further   *106*

4   The ATP Package: Capabilities and Applications   *107*
           Juan A. Martinez-Velasco and Jacinto Martin-Arnedo
4.1   Introduction   *107*
4.2   Capabilities of the ATP Package   *108*
4.2.1   Overview   *108*
4.2.2   The Simulation Module – TPBIG   *109*
4.2.2.1   Overview   *109*
4.2.2.2   Modelling Capabilities   *110*
4.2.2.3   Solution Techniques   *117*
4.2.3   The Graphical User Interface – ATPDraw   *120*
4.2.3.1   Overview   *120*
4.2.3.2   Main Functionalities   *120*
4.2.3.3   Supporting Modules for Power System Components   *123*
4.2.4   The Postprocessor – TOP   *125*
4.2.4.1   Data Management   *125*
4.2.4.2   Data Display   *126*
4.2.4.3   Data Processing   *127*
4.2.4.4   Data Formatting   *127*
4.2.4.5   Graphical Output   *127*
4.3   Applications   *128*
4.4   Illustrative Case Studies   *129*
4.4.1   Introduction   *129*
4.4.2   Case Study 1: Optimum Allocation of Capacitor Banks   *130*
4.4.3   Case Study 2: Parallel Resonance Between Transmission Lines   *132*
4.4.4   Case Study 3: Selection of Surge Arresters   *133*
4.5   Remarks   *136*
           References   *136*
           To Probe Further   *138*

5   **Introduction to the Simulation of Electromagnetic Transients Using ATP**   *139*
Juan A. Martinez-Velasco and Francisco González-Molin
5.1   Introduction   *139*
5.2   Input Data File Using ATP Formats   *140*
5.3   Some Important Issues   *142*
5.3.1   Before Simulating the Test Case   *142*
5.3.1.1   Setting Up a System Model   *142*
5.3.1.2   Topology Requirements   *142*
5.3.1.3   Selection of the Time-Step Size and the Simulation Time   *143*
5.3.1.4   Units   *143*
5.3.1.5   Output Selection   *144*
5.3.2   After Simulating the Test Case   *144*
5.3.2.1   Verifying the Results   *144*
5.3.2.2   Debugging Suggestions   *144*
5.4   Introductory Cases. Linear Circuits   *145*
5.4.1   The Series and Parallel RLC Circuits   *145*
5.4.2   The Series RLC Circuit: Energization Transient   *145*
5.4.2.1   Theoretical Analysis   *145*
5.4.2.2   ATP Implementation   *147*
5.4.2.3   Simulation Results   *148*
5.4.3   The Parallel RLC Circuit: De-energization Transient   *150*
5.4.3.1   Theoretical Analysis   *150*
5.4.3.2   ATP Implementation   *152*
5.4.3.3   Simulation Results   *153*
5.5   Switching of Capacitive Currents   *155*
5.5.1   Introduction   *155*
5.5.2   Switching Transients in Simple Capacitive Circuits – DC Supply   *155*
5.5.2.1   Energization of a Capacitor Bank   *155*
5.5.2.2   Energization of a Back-to-Back Capacitor Bank   *157*
5.5.3   Switching Transients in Simple Capacitive Circuits – AC Supply   *159*
5.5.3.1   Energization of a Capacitor Bank   *159*
5.5.3.2   Energization of a Back-to-Back Capacitor Bank   *160*
5.5.3.3   Reclosing into Trapped Charge   *162*
5.5.4   Discharge of a Capacitor Bank   *164*
5.6   Switching of Inductive Currents   *168*
5.6.1   Introduction   *168*
5.6.2   Switching of Inductive Currents in Linear Circuits   *168*
5.6.2.1   Interruption of Inductive Currents   *168*
5.6.2.2   Voltage Escalation During the Interruption of Inductive Currents   *170*
5.6.2.3   Current Chopping   *172*
5.6.2.4   Making of Inductive Currents   *175*
5.6.3   Switching of Inductive Currents in Nonlinear Circuits   *176*
5.6.4   Transients in Nonlinear Reactances   *178*
5.6.4.1   Interruption of an Inductive Current   *180*
5.6.4.2   Energization of a Nonlinear Reactance   *181*
5.6.5   Ferroresonance   *184*
5.7   Transient Analysis of Circuits with Distributed Parameters   *187*
5.7.1   Introduction   *187*
5.7.2   Transients in Linear Circuits with Distributed-Parameter Components   *187*

5.7.2.1   Energization of Lines and Cables   *187*
5.7.2.2   Transient Recovery Voltage During Fault Clearing   *191*
5.7.3   Transients in Nonlinear Circuits with Distributed-Parameter Components   *195*
5.7.3.1   Surge Arrester Protection   *195*
5.7.3.2   Protection Against Lightning Overvoltages Using Surge Arresters   *196*
           References   *201*
           Acknowledgement   *202*
           To Probe Further   *202*

6   Calculation of Power System Overvoltages   *203*
           Juan A. Martinez-Velasco and Ferley Castro-Aranda
6.1   Introduction   *203*
6.2   Power System Overvoltages: Causes and Characterization   *204*
6.3   Modelling for Simulation of Power System Overvoltages   *206*
6.3.1   Introduction   *206*
6.3.2   Modelling Guidelines for Temporary Overvoltages   *207*
6.3.3   Modelling Guidelines for Slow-Front Overvoltages   *208*
6.3.3.1   Lines and Cables   *208*
6.3.3.2   Transformers   *208*
6.3.3.3   Switchgear   *208*
6.3.3.4   Capacitors and Reactors   *209*
6.3.3.5   Surge Arresters   *209*
6.3.3.6   Loads   *210*
6.3.3.7   Power Supply   *210*
6.3.4   Modelling Guidelines for Fast-Front Overvoltages   *210*
6.3.4.1   Overhead Transmission Lines   *210*
6.3.4.2   Substations   *212*
6.3.4.3   Surge Arresters   *213*
6.3.4.4   Sources   *214*
6.3.5   Modelling Guidelines for Very Fast-Front Overvoltages in Gas Insulated
           Substations   *214*
6.4   ATP Capabilities for Power System Overvoltage Studies   *216*
6.5   Case Studies   *216*
6.5.1   Introduction   *216*
6.5.2   Low-Frequency Overvoltages   *216*
6.5.2.1   Case Study 1: Resonance Between Parallel Lines   *217*
6.5.2.2   Case Study 2: Ferroresonance in a Distribution System   *219*
6.5.3   Slow-Front Overvoltages   *225*
6.5.3.1   Case Study 3: Transmission Line Energization   *227*
6.5.3.2   Case Study 4: Capacitor Bank Switching   *238*
6.5.4   Fast-Front Overvoltages   *243*
6.5.4.1   Case Study 5: Lightning Performance of an Overhead Transmission Line   *244*
6.5.5   Very Fast-Front Overvoltages   *261*
6.5.5.1   Case Study 6: Origin of Very Fast-Front Transients in GIS   *262*
6.5.5.2   Case Study 7: Propagation of Very Fast-Front Transients in GIS   *263*
6.5.5.3   Case Study 8: Very Fast-Front Transients in a 765 kV GIS   *267*
           References   *270*
           To Probe Further   *274*

7   **Simulation of Rotating Machine Dynamics**   *275*
           Juan A. Martinez-Velasco
7.1   Introduction   *275*
7.2   Representation of Rotating Machines in Transients Studies   *275*
7.3   ATP Rotating Machines Models   *276*
7.3.1   Background   *276*
7.3.2   Built-in Rotating Machine Models   *276*
7.3.3   Rotating Machine Models for Fast Transients Simulation   *278*
7.4   Solution Methods   *278*
7.4.1   Introduction   *278*
7.4.2   Three-Phase Synchronous Machine Model   *278*
7.4.3   Universal Machine Module   *281*
7.4.4   WindSyn-Based Models   *284*
7.5   Procedure to Edit Machine Data Input   *284*
7.6   Capabilities of Rotating Machine Models   *285*
7.7   Case Studies: Three-Phase Synchronous Machine   *287*
7.7.1   Overview   *287*
7.7.2   Case Study 1: Stand-Alone Three-Phase Synchronous Generator   *288*
7.7.3   Case Study 2: Load Rejection   *288*
7.7.4   Case Study 3: Transient Stability   *298*
7.7.5   Case Study 4: Subsynchronous Resonance   *302*
7.8   Case Studies: Three-Phase Induction Machine   *309*
7.8.1   Overview   *309*
7.8.2   Case Study 5: Induction Machine Test   *310*
7.8.3   Case Study 6: Transient Response of the Induction Machine   *313*
7.8.3.1   First Case   *314*
7.8.3.2   Second Case   *314*
7.8.3.3   Third Case   *318*
7.8.4   Case Study 7: SCIM-Based Wind Power Generation   *323*
           References   *328*
           To Probe Further   *331*

8   **Power Electronics Applications**   *333*
           Juan A. Martinez-Velasco and Jacinto Martin-Arnedo
8.1   Introduction   *333*
8.2   Converter Models   *334*
8.2.1   Switching Models   *334*
8.2.2   Dynamic Average Models   *334*
8.3   Power Semiconductor Models   *335*
8.3.1   Introduction   *335*
8.3.2   Ideal Device Models   *335*
8.3.3   More Detailed Device Models   *335*
8.3.4   Approximate Models   *336*
8.4   Solution Methods for Power Electronics Studies   *337*
8.4.1   Introduction   *337*
8.4.2   Time-Domain Transient Solution   *337*
8.4.3   Initialization   *338*
8.5   ATP Simulation of Power Electronics Systems   *338*
8.5.1   Introduction   *338*
8.5.2   Switching Devices   *339*

8.5.2.1   Built-in Semiconductor Models   *339*
8.5.2.2   Custom-made Semiconductor Models   *340*
8.5.3   Power Electronics Systems   *342*
8.5.4   Power Systems   *343*
8.5.5   Control Systems   *343*
8.5.6   Rotating Machines   *344*
8.5.6.1   Built-in Rotating Machine Models   *344*
8.5.6.2   Custom-made Rotating Machine Models   *344*
8.5.7   Simulation Errors   *345*
8.6   Power Electronics Applications in Transmission, Distribution, Generation and Storage Systems   *345*
8.6.1   Overview   *345*
8.6.2   Transmission Systems   *346*
8.6.3   Distribution Systems   *346*
8.6.4   DER Systems   *347*
8.7   Introduction to the Simulation of Power Electronics Systems   *349*
8.7.1   Overview   *349*
8.7.2   One-Switch Case Studies   *350*
8.7.3   Two-Switches Case Studies   *351*
8.7.4   Application of the GIFU Request   *355*
8.7.5   Simulation of Power Electronics Converters   *361*
8.7.5.1   Single-phase Inverter   *361*
8.7.5.2   Three-phase Line-Commutated Diode Bridge Rectifier   *362*
8.7.6   Discussion   *365*
8.8   Case Studies   *367*
8.8.1   Introduction   *367*
8.8.2   Case Study 1: Three-phase Controlled Rectifier   *367*
8.8.3   Case Study 2: Three-phase Adjustable Speed AC Drive   *369*
8.8.4   Case Study 3: Digitally-controlled Static VAR Compensator   *373*
8.8.4.1   Test System   *375*
8.8.4.2   Control Strategy   *375*
8.8.5   Case Study 4: Unified Power Flow Controller   *382*
8.8.5.1   Configuration   *382*
8.8.5.2   Control   *382*
8.8.5.3   Modelling   *384*
8.8.5.4   ATPDraw Implementation   *385*
8.8.5.5   Simulation Results   *385*
8.8.6   Case Study 5: Solid State Transformer   *386*
8.8.6.1   Introduction   *386*
8.8.6.2   SST Configuration   *388*
8.8.6.3   Control Strategies   *388*
8.8.6.4   Test System and Modelling Guidelines   *393*
8.8.6.5   Case Studies   *396*
       Acknowledgement   *399*
       References   *399*
       To Probe Further   *404*

9   **Creation of Libraries**   *405*
          Juan A. Martinez Velasco and Jacinto Martin-Arnedo
9.1   Introduction   *405*
9.2   Creation of Custom-Made Modules   *406*
9.2.1   Introduction   *406*
9.2.2   Application of DATA BASE MODULE   *406*
9.2.3   Application of MODELS   *411*
9.2.4   The Group Option   *417*
9.3   Application of the ATP to Power Quality Studies   *419*
9.3.1   Introduction   *419*
9.3.2   Power Quality Issues   *419*
9.3.3   Simulation of Power Quality Problems   *422*
9.3.4   Power Quality Studies   *423*
9.4   Custom-Made Modules for Power Quality Studies   *426*
9.5   Case Studies   *426*
9.5.1   Overview   *426*
9.5.2   Harmonics Analysis   *426*
9.5.2.1   Case Study 1: Generation of Harmonic Waveforms   *428*
9.5.2.2   Case Study 2: Harmonic Resonance   *431*
9.5.2.3   Case Study 3: Harmonic Frequency Scan   *434*
9.5.2.4   Case Study 4: Compensation of Harmonic Currents   *441*
9.5.3   Voltage Dip Studies in Distribution Systems   *447*
9.5.3.1   Overview   *447*
9.5.3.2   Case Study 5: Voltage Dip Measurement   *449*
9.5.3.3   Case Study 6: Voltage Dip Characterization   *454*
9.5.3.4   Case Study 7: Voltage Dip Mitigation   *462*
          References   *466*
          To Probe Further   *470*

10   **Protection Systems**   *471*
          Juan A. Martinez-Velasco and Jacinto Martin-Arnedo
10.1   Introduction   *471*
10.2   Modelling Guidelines for Protection Studies   *472*
10.2.1   Line and Cable Models   *472*
10.2.1.1 Models for Steady-State Studies   *473*
10.2.1.2 Models for Transient Studies   *473*
10.2.2   Transformer Models   *473*
10.2.2.1 Low-frequency Transformer Models   *474*
10.2.2.2 High-frequency Transformer Models   *475*
10.2.3   Source Models   *475*
10.2.4   Circuit Breaker Models   *475*
10.3   Models of Instrument Transformers   *476*
10.3.1   Introduction   *476*
10.3.2   Current Transformers   *476*
10.3.3   Coupling Capacitor Voltage Transformers   *478*
10.3.4   Voltage Transformers   *479*
10.3.5   Case Studies   *480*
10.3.5.1 Case Study 1: Current Transformer Test   *480*
10.3.5.2 Case Study 2: Coupling Capacitor Voltage Transformer Test   *482*

10.3.6   Discussion   *484*
10.4   Relay Modelling   *484*
10.4.1   Introduction   *484*
10.4.2   Classification of Relay Models   *485*
10.4.3   Implementation of Relay Models   *486*
10.4.4   Applications of Relay Models   *488*
10.4.5   Testing and Validation of Relay Models   *488*
10.4.6   Accuracy and Limitations of Relay Models   *490*
10.4.7   Case Studies   *490*
10.4.7.1   Overview   *490*
10.4.7.2   Case Study 3: Simulation of an Electromechanical Distance Relay   *491*
10.4.7.3   Case Study 4: Simulation of a Numerical Distance Relay   *497*
10.5   Protection of Distribution Systems   *508*
10.5.1   Introduction   *508*
10.5.2   Protection of Distribution Systems with Distributed Generation   *508*
10.5.2.1   Distribution Feeder Protection   *508*
10.5.2.2   Interconnection Protection   *508*
10.5.3   Modelling of Distribution Feeder Protective Devices   *509*
10.5.3.1   Circuit Breakers – Overcurrent Relays   *509*
10.5.3.2   Reclosers   *511*
10.5.3.3   Fuses   *511*
10.5.3.4   Sectionalizers   *512*
10.5.4   Protection of the Interconnection of Distributed Generators   *513*
10.5.5   Case Studies   *514*
10.5.5.1   Case Study 5: Testing the Models   *514*
10.5.5.2   Case Study 6: Coordination Between Protective Devices   *524*
10.5.5.3   Case Study 7: Protection of Distributed Generation   *525*
10.6   Discussion   *531*
           Acknowledgement   *533*
           References   *533*
           To Probe Further   *537*

11   **ATP Applications Using a Parallel Computing Environment**   *539*
           Javier A. Corea-Araujo, Gerardo Guerra and Juan A. Martinez-Velasco
11.1   Introduction   *539*
11.2   Bifurcation Diagrams for Ferroresonance Characterization   *540*
11.2.1   Introduction   *540*
11.2.2   Characterization of Ferroresonance   *540*
11.2.3   Modelling Guidelines for Ferroresonance Analysis   *541*
11.2.4   Generation of Bifurcation Diagrams   *541*
11.2.5   Parametric Analysis Using a Multicore Environment   *542*
11.2.6   Case Studies   *544*
11.2.6.1   Case 1: An Illustrative Example   *544*
11.2.6.2   Case 2: Ferroresonant Behaviour of a Voltage Transformer   *545*
11.2.6.3   Case 3: Ferroresonance in a Five-Legged Core Transformer   *545*
11.2.7   Discussion   *550*
11.3   Lightning Performance Analysis of Transmission Lines   *550*
11.3.1   Introduction   *550*
11.3.2   Lightning Stroke Characterization   *551*

11.3.3   Modelling for Lightning Overvoltage Calculations   *552*

11.3.4   Implementation of the Monte Carlo Procedure Using Parallel Computing   *554*

11.3.5   Illustrative Example   *555*

11.3.5.1 Test Line   *555*

11.3.5.2 Line and Lightning Stroke Parameters   *555*

11.3.5.3 Simulation Results   *559*

11.3.6   Discussion   *562*

11.4   Optimum Design of a Hybrid HVDC Circuit Breaker   *563*

11.4.1   Introduction   *563*

11.4.2   Design and Operation of the Hybrid HVDC Circuit Breaker   *563*

11.4.3   ATP Implementation of the Hybrid HVDC Circuit Breaker   *565*

11.4.4   Test System   *566*

11.4.5   Transient Response of the Hybrid Circuit Breaker   *567*

11.4.6   Implementation of a Parallel Genetic Algorithm   *568*

11.4.7   Simulation Results   *570*

11.4.8   Discussion   *574*

          Acknowledgement   *575*

          References   *575*

A   Characteristics of the Multicore Installation   *579*

B   Test System Parameters for Ferroresonance Studies   *579*

          To Probe Further   *580*

          Index   *581*

About the Editor

Juan A. Martinez-Velasco was born in Barcelona, Spain. He received the Ingeniero Industrial and Doctor Ingeniero Industrial degrees from the Universitat Politècnica de Catalunya (UPC), Spain. He is retired and working as private consultant.

He has authored and co-authored more than 200 journal and conference papers, most of them on Transient Analysis of Power Systems. He has been involved in several EMTP (Electro-Magnetic Transients Program) courses and worked as consultant for some Spanish companies. His teaching and research areas cover Power Systems Analysis, Transmission and Distribution, Power Quality and Electromagnetic Transients. He has been an active member of several IEEE and CIGRE Working Groups.

He has been involved as editor or co-author in several books. He is also coeditor of the IEEE publication 'Modeling and Analysis of System Transients Using Digital Programs' (1999). In 2010, he was the coordinator of the Tutorial Course 'Transient Analysis of Power Systems. Solution Techniques, Tools, and Applications', given at the 2010 IEEE PES General Meeting, July 2010 and held in Minneapolis.

In 1999, he was given the '1999 PES Working Group Award for Technical Report', for his participation in the tasks performed by the IEEE Task Force on Modeling and Analysis of Slow Transients. In 2000, he was given the '2000 PES Working Group Award for Technical Report', for his participation in the edition of the special publication 'Modeling and Analysis of System Transients using Digital Programs'. In 2009 he was also given the 'Technical Committee Working Group Award' of the IEEE PES Transmission and Distribution Committee.

List of Contributors

Ferley Castro-Aranda
Universidad del Valle, Laboratorio de Alta
Tensión
Cali, Colombia

Javier A. Corea-Araujo
Applus+ IDIADA
Santa Oliva, Spain

Francisco González-Molina
Universitat Rovira i Virgili, Depto. de
Ingeniería Electrónica, Eléctrica y
Automática
Tarragona, Spain

Gerardo Guerra
DNV GL
Barcelona, Spain

Jacinto Martín-Arnedo
Estabanell y Pahisa Energia,
Granollers, Spain

Juan A. Martinez-Velasco
Retired – Formerly with Universitat
Politecnica de Catalunya
Barcelona, Spain

Preface

The transient analysis is an important task for power system analysis and design. Several simulation tools are currently available for this purpose. One of the most popular is the Alternative Transients Program (ATP).

The ATP is a royalty-free package integrated by at least three tools: (i) ATPDraw, a graphical user interface (GUI) for creating and/or editing input files; (ii) TPBIG, the main processor for transients and harmonics simulations; and (iii) one postprocessor for plotting simulation results. Actually, ATP users can also take advantage of several other tools, and create a custom-made environment with links to other packages.

The acronym ATP was originally used to name the tool for transients simulation. For many users, ATP is still the simulation tool; in this book, ATP is used to name indistinctly the package or the transients simulation tool, while TPBIG is used to name (when used) the simulation tool. ATPDraw is an interactive Windows-based GUI that can act as a shell for the whole package; that is, users can control the execution of all programs integrated in the package from ATPDraw. Several royalty-free tools with different capabilities are currently available for postproccesing simulation results.

ATP users can take advantage of several books for improving their knowledge on transient analysis and the application of this tool. Although this new book covers topics already covered by formerly released books, the overlapping with all of them is rather small; the main differences are in the organisation of the book and the case studies that illustrate potential ATP applications.

Actually, some readers might miss some equations and mathematical artefacts needed to detail and describe the performance of power components and systems. This aspect has been sacrificed to give room to more practical aspects. Readers are referred to other books that satisfactorily cover this part of the transient analysis.

It is important to emphasize that, although the contents of the book honour its title, the book cannot be used as a Reference Manual or Rule Book. In other words, readers will not learn how to use the package with this book; for that purpose, they should use the so-called ATP Rule Book, and the manuals of the complementary tools (e.g. ATPDraw and the selected plotting program).

The main goal of this book is to provide a clear scope of the studies that can be carried out with the ATP package. Although some complex studies and sophisticated custom-made simulation environments (with ATP as a simulation tool) are presented, the average level of the cases studied here is intermediate; a great majority of studies are related to small and medium test systems. However, the book could also be useful for beginners; Chapter 5 has been written with that purpose.

The chapters of this book can be classified into two groups; the first four chapters are dedicated to introduce the transient analysis of power systems and detail ATP capabilities; the rest of the book is dedicated to introduce some of the most common applications of the ATP package with a large enough number of case studies.

A very important aspect is the complementary collection of data files available to ATP users from the website of this book. For every case study presented in the book, readers will find one or several data files. When ATPDraw capabilities can satisfactorily create and edit a data file, this option has been used; however, there are a few cases for which those capabilities cannot be used, then the file has been manually edited using a simple text editor and taking into account the formats detailed in the Rule Book. This is a very important aspect about which ATP users should be aware: sooner or later some knowledge of ATP formats will be required, mainly for those interested in developing their own custom-made models.

In addition, it is worth mentioning that although this book uses the most important capabilities implemented in the simulation tool (either named as ATP or TPBIG), there are dozens of ATP options and requests that are not covered or applied here. The required length for illustrating those missed options could easily double or triple that of this book.

It is also worth mentioning that the tools of the package are continuously updated. This is important because, during the preparation of this book, models and capabilities were either added or modified. Not all of these models/capabilities have been covered in this book.

As for the applications, I am aware that some that are of concern for many ATP users have not been included in this book. For instance, very little is said about distributed energy resources. This is, without any doubt, a very important aspect; however, although more than one hundred data files are provided with the book, a similar reasoning could be made even for applications that are covered in the book: some topics could need more case studies for a better understanding. At the end, some selection had to be made.

Although all the date files used in this book have been implemented by the contributors, several case studies are based on models and parameters provided by other authors. In general, a reference to the original source is made in the chapter or in the Acknowledgement. Some of the case studies, even if they are now implemented in ATPDraw, were initially developed when this GUI was not yet released. This means that those case studies are rather old. Since the trace to the original source was neither clear nor available for all cases, I apologize if our gratitude to some authors is not mentioned in any part of this book.

Finally, I want to thank Dr. W. Scott Meyer and all those who became involved in the development of any of package tools for their work and effort, without which this book would have not been published.

Barcelona, Spain   *Juan A. Martinez-Velasco*
February 2019

About the Companion Website

The companion website for this book is at:

www.wiley.com/go/martinez/power_systems

The website includes:

- PCH and ACP files

Scan this QR code to visit the companion website.

1

Introduction to Transients Analysis of Power Systems with ATP

Juan A. Martinez-Velasco

1.1   Overview

Transient analysis has become a fundamental methodology for understanding the performance of power systems, determining power component ratings, explaining equipment failures, or testing protection devices. The study of transients is a mature field that can help to analyse and design modern power systems.

A significant effort has been dedicated to the development of new techniques and software tools adequate for transient analysis of power systems. Sophisticated models, complex solution techniques, and powerful simulation tools have been developed to perform studies that are of paramount importance in the analysis and design of modern power systems. Current tools for transient analysis can be applied into a myriad of studies (e.g. overvoltage calculation, flexible AC transmission systems (FACTS) and Custom Power applications, protective relay performance, power quality studies) for which detailed models and accurate solutions can be crucial.

Transient phenomena in power systems are associated with disturbances caused by faults, switching operations, lightning strikes, or load variations. These phenomena can stress and damage power equipment. The importance of their study is basically due to the effects they can have on the system performance or the failures they can cause to power equipment. Therefore, protection against these stresses is necessary. This protection can be provided by specialized equipment whose operation is aimed at either isolating the power system section where the disturbance has been originated (e.g. a power component failure that causes short-circuit) or limiting the stress across power equipment terminals (e.g. by installing a surge arrester that will mitigate voltage stresses). In addition, a better performance against stresses caused by transient phenomena can be also achieved with an adequate design of power equipment (e.g. by shielding overhead transmission lines to limit flashovers caused by direct lightning strokes). That is, although the power system operates most of the time under normal conditions, it must be designed to cope with the consequences associated to transient phenomena.

A rigorous and accurate analysis of transients in power systems is difficult due to the size of the system, the complexity of the interaction between power devices, and the physical phenomena that need to be analysed. Aspects that contribute to this complexity are the variety of causes, the nature of the physical phenomena, and the timescale of power system transients. In order to select an adequate protection against any type of stress, it is fundamental to know their origin, calculate their main characteristics, and estimate the most adverse conditions. Disturbances can be external (lightning strokes) or internal (faults, switching operations, load variations). Power system transients can be electromagnetic, when it is necessary to analyse the interaction between the (electric) energy stored in capacitors and the (magnetic) energy

Transient Analysis of Power Systems: A Practical Approach, First Edition. Edited by Juan A. Martinez-Velasco.
© 2020 John Wiley & Sons Ltd. Published 2020 by John Wiley & Sons Ltd.
Companion Website: www.wiley.com/go/martinez/power_systems

stored in inductors, or electromechanical, when the analysis involves the interaction between the energy supplied by sources, the electric energy stored in circuit elements, and the mechanical energy stored in rotating machines. To accurately analyse physical phenomena associated with transients, it is necessary to examine the power system for a time interval as short as a few nanoseconds or as long as several minutes. This is a challenge since the behaviour of power equipment is very dependent on the transient phenomena; namely, it depends on the range of frequencies associated to transients. Despite the powerful numerical techniques, simulation tools, and graphical user interfaces (GUIs) currently available, those involved in transients studies, sooner or later, face limitations of those models available in transients packages, the lack of reliable data and conversion procedures for parameter estimation, or insufficient studies aimed at validating models.

Figure 1.1 depicts the steps of a typical procedure when simulating transients in power systems [1].

1. *The selection of the study zone and the most adequate representation of each component involved in the transient.* The system zone is selected taking into account the frequency range of the transients to be simulated: the higher the frequencies, the smaller the zone modelled. In general, it is advisable to minimize the study zone since a larger number of components does not necessarily increase accuracy; instead it will increase the simulation time and there will be a higher probability of insufficient or incorrect modelling. Although a high number of works has been dedicated to provide guidelines on these aspects [2–4], some expertise is necessary to choose the study zone and the models.

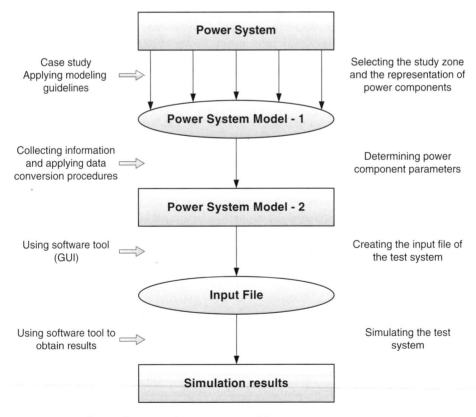

Figure 1.1 Simulation of transients in power systems [1].

2. *The estimation of parameters to be specified in the mathematical models.* Once the mathematical model has been selected, it is necessary to collect the information that could be useful to obtain the values of parameters to be specified. Details about parameter determination of some power components were presented in [5]. A sensitivity study should be carried out if one or several parameters cannot be accurately determined. Results derived from such study will show which parameters are of concern.
3. *The application of a simulation tool.* The steadily increasing capabilities of hardware and software tools have led to the development of powerful simulation tools that can cope with large and complex power systems. Modern software for transient analysis incorporates friendly GUIs that can be very useful when creating the input file of the test system model.
4. *The analysis of results.* Simulation of electromagnetic transients can be used, among others, for determining component ratings (e.g. insulation levels or energy absorption capabilities), testing control and protection systems, validating power component representations or understanding equipment failures. A deep analysis of simulation results is an important aspect of the entire procedure since each of these studies may involve an iterative procedure in which models and parameters values must be adjusted.

Readers interested in transients analysis can consult specialized literature [3, 6–18].

1.2 The ATP Package

ATP is an acronym that stands for Alternative Transients Program [19]. The ATP package is integrated by at least three tools: (i) ATPDraw, a GUI for creating/editing input files [20, 21]; (ii) TPBIG, the main processor for transients and harmonics simulations; (iii) one postprocessor for plotting simulation results. Actually, ATP users can also take advantage of other tools (e.g. ATP Control Center and ATPDesigner [22] which can be used as a control center for the entire package) or add other tools that can be useful for some specific tasks.

ATPDraw is an interactive Windows-based program that can act as a shell for the whole package; that is, users can control the execution of all modules integrated in the package from ATPDraw. As for the postprocessor, several tools have been developed to obtain graphical results (e.g. PCPlot, TPPLOT, GTPPLOT, TOP, PlotXY, ATP Analyzer), and it is possible to run most of them from ATPDraw. The most popular postprocessor among ATP users is PlotXY, developed by Maximo Ceraolo (University of Pisa, Italy) [23]. TOP (The Output Program), a royalty-free tool created by Electrotek Concepts, is the postprocessor used with most of the case studies presented in this book [24].

The acronym ATP was initially used to denote the transients simulation tool here named TPBIG. Presently, many users use ATP to indistinctly name either the transients simulation tool or the entire package.

ATP was originally developed for simulation of electromagnetic transients in power systems. However, the package can also be used to perform AC steady-state calculations and simulate electromechanical transients (e.g. subsynchronous resonance, AC drives). Solution methods to solve systems with nonlinear components have also been implemented [9, 19]. ATP can represent control systems and interface them with an electric network. Finally, several non-simulation supporting routines are also available to create model data files; these supporting routines can be used for computing parameters and creating models of lines, cables, and transformers. Figure 1.2 shows a schematic diagram of the connectivity between simulation capabilities, supporting routines, external programs, and all types of files. A more detailed description of ATPDraw and TPBIG capabilities is provided in Chapter 4.

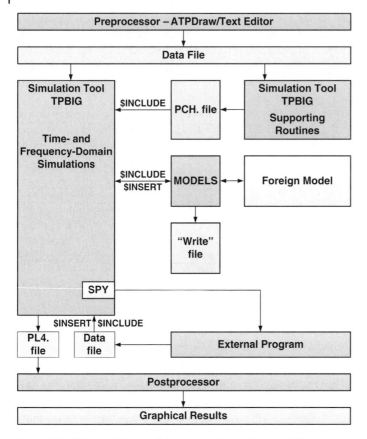

Figure 1.2 ATP simulation modules, supporting routines, and files.

The applications that can be covered by the ATP can be classified as follows:

- *Time-domain simulations.* They are generally used for simulation of transients, such as switching or lightning overvoltages; however, they can also be used for analysing harmonic distortion created by power electronics devices.
- *Frequency-domain simulations.* ATP capabilities can also be used to obtain the driving point impedance at a particular node versus frequency, detect resonance conditions, design filter banks, or analyse harmonic propagation.
- *Parametric studies.* They are usually carried out to evaluate the relationship between variables and parameters. When one or more parameter cannot be accurately specified, this analysis will determine the range of values which may be of concern.
- *Statistical studies.* Several ATP capabilities can be applied to perform these studies (e.g. studies based on the Monte Carlo method). Their results can be of paramount importance in some insulation coordination and power quality studies.

ATP capabilities can also be used to expand the fields of applications; with this tool a data case can be simulated several times before deactivating the program, parameters of the system under simulation can be changed according to a given law, some components can be either disconnected or activated, and some calculations can be carried out by external programs. In addition, it is possible, if required, to modify the simulation time on line or the number of runs in a parametric study.

The following concepts can be of paramount importance for expanding the applications of ATP:

- *Multiple run option.* A data case can be simulated as many times as necessary, while changes are introduced into the system model at every run. This option is known as POCKET CALCULATOR VARIES PARAMETERS (PCVP); see the ATP Rule Book [19]. Such an option can be used to perform statistical and parametric studies. However, it can also be used in many other applications. For instance, once the target of the study has been set, PCVP can be used to run the case as many times as required while one or several parameters are gradually adjusted or the system topology is modified until the target is reached.
- *Open system.* A link to external tools can be established before, during and after a simulation to take advantage of the capabilities of these tools and to add or test new capabilities. This option can be used, for instance, to link the ATP to MATLAB and take advantage of its features, or to run a custom-made program that can derive the parameters of a power component using a data conversion procedure not yet implemented in the package.
- *Data symbol replacement.* $PARAMETER is a declaration that can be used to replace data symbols of arbitrary length prior to a simulation [19]. Up to three replacement modes can be used: simple character replacement (one string is replaced by another with the same length), mathematical replacement (string is replaced by a number deduced from a mathematical formula), integer serialization (used to encode strings within a DO loop). Conditional branching (IF-THEN-ENDIF) is a built-in feature that can be used to select between two or more choices.
- *Data module.* ATP-coded templates have been used in the past for the development of data modules that could facilitate the use of the tool by beginners, or to simplify the use of power components and extend modelling capabilities to more complex equipment [5]. Presently, custom-made models are represented by a module and its associated ATPDraw icon. Although the development of new modules generally relies on the routine DATA BASE MODULE, other ATP capabilities can be used to perform simple calculations with module arguments, to decide what parts of a module can be activated at a given run, or what parts should remain sleeping. The so-called Type 94 component could be the best solution for developing some nonlinear components.
- *Interactivity.* Several simulation modules will usually be involved in a general procedure. Interactivity between them is critical as calculations will be performed in several modules. The connectivity between a power system and a control section to pass variables in both senses has been a feature since the earliest development of control capabilities. However, it has been the possibility of passing also parameters what has added flexibility to some of the capabilities described above and increased the type of applications.

Actually, the type of tasks that the ATP package can carry out is practically unlimited. For instance, by using some simple rules and taking advantage of some capabilities (e.g. TO SUPPORTING PROGRAM feature to run supporting routines, DO loops to serialize power components, string replacement), it is possible to develop a data section aimed at creating the code of a component taking into account the transient process to be simulated and the information available.

1.3 ATP Documentation

The following material will be of help when using any of the basic tools that constitute the package.

- The so-called *EMTP Theory Book*, written by H.W. Dommel [9], should be used by those interested in models and solution techniques implemented in TPBIG. Although the book needs to be updated, it is still a very valuable source of information for users of any transients tool.
- The rules to be followed for creating an input data file are presented in the *ATP Rule Book* [19]. Although the average user will create and run input data files by relying only on the GUI ATPDraw, there are many situations for which the Rule Book will be a necessary resource. Examples where a user will need to consult the Rule Book are those case studies in which the control section is based on MODELS language (see Chapter 4) or those applications for which custom-made models based on DATA BASE MODULE might be required (see Chapter 9).
- ATPDraw is currently the gate used by most ATP users for creating and running case studies. Although based on a friendly environment and easy to use, ATPDraw has many capabilities and not all of them are obvious; the *ATPDraw Reference Manual* can be then a valuable source for consultation [20, 21].

In general, postprocessing tools (e.g. PlotXY, TOP) are easy to use and the help capabilities available in most of them are clear enough.

1.4   Scope of the Book

This book provides a basic background on the main aspects to be considered when performing transients studies with ATP and a scope of the package applications. The chapters are dedicated to

- summarizing modelling guidelines in transient simulations and the most basic solution techniques implemented in ATP;
- covering the main application fields of ATP (overvoltage calculation, rotating machine dynamics, protection of power systems, power electronics applications, power quality studies);
- describing the procedure and select the ATP capabilities that can be used for building custom-made models;
- providing some insights about the construction of simulation environments in which ATP is the tool dedicated to carry out transient calculations.

The man topics covered by each chapter are summarized in the following paragraphs.

Modelling of power components. The representation of power components for transients studies depends of the phenomena to be analysed (i.e. the cause, the range of frequencies with which the transient occurs, the components involved in the transient). Chapter 2 provides a summary of the guidelines to be followed when representing some of the most important power components in a transient study. Modelling guidelines will also be provided in other chapters, and summarized in most of the case studies presented in this book.

Solution techniques. ATP is an off-line circuit-oriented simulation tool that can be used to simulate transients in power systems using a time-domain solution technique [9]. However, the steady-state of the system under study, prior to the calculation of a transient process, is usually required, and its calculation is performed in the frequency-domain. Chapter 3 details the solution techniques implemented in ATP for steady-state and transient solutions of power and control systems.

ATP capabilities and applications. ATP capabilities can be used to simulate power systems, develop custom-made models, or create new simulation environments. As already mentioned, ATP can be applied not only for simulation of electromagnetic transients but also to an extensive range of studies. Chapter 4 details the capabilities and built-in models available in ATP. The chapter will include a few examples to illustrate the range of applications of the package.

Introduction to the simulation of power systems transients. Chapter 5 presents the simulation of some simple case studies using the ATP package. The chapter summarizes the procedure to be followed with each module of the ATP package; discusses the modelling guidelines to be applied, and analyses the results obtained from each case study. The selected examples will illustrate the usage of elementary linear and nonlinear components, with either lumped- and/or distributed-parameter components.

Calculation of overvoltages in power systems. An overvoltage is a voltage having a crest value exceeding the corresponding crest of the maximum system voltage. Overvoltages can occur with a very wide range of waveshapes and durations. Causes and main characteristics of overvoltages are well known, and they are classified in standards (IEC, IEEE). For instance, the magnitude of external lightning overvoltages remains essentially independent of the system design, whereas that of internal switching overvoltages increases with the operating voltage of the system. The estimation of overvoltage magnitudes and shapes is fundamental for the insulation design of power components, and for the selection of protective devices [25, 26]. Chapter 6 analyses the different types of overvoltages and their causes, summarizes modelling guidelines for overvoltage calculation with ATP, and presents some illustrative cases of overvoltages.

Simulation of rotating machine dynamics. Two options are available in ATP for representing conventional rotating machines: the Three-Phase Synchronous Machine model and the so-called Universal Machine module [19]. To date, these capabilities have mostly been used to simulate only three-phase synchronous and induction machines. However, the applications are endless and a significant experience is already available. Chapter 7 provides a summary of the features available in the two options, summarizes the methods implemented for interfacing the rotating machine models to the power system, and includes several illustrative examples that will cover some of the most important applications of ATP in the study of rotating machine dynamics.

Simulation of power electronics devices. Power electronics applications have quickly spread to all voltage levels, from high-voltage transmission to low-voltage circuits in end-user facilities [27–29]. Modelling and simulation of power electronics devices are important tasks for concept validation and design of new devices. Chapter 8 provides general modelling guidelines and procedures for simulation of power electronics devices using ATP capabilities, and presents several case studies that will cover some important applications (FACTS and Custom Power devices, drives, solid state transformer) of the package in this field.

Development of custom-made models and libraries. Several component models needed in some studies (e.g. protection, power quality) are not available in the ATP. However, many capabilities of the package can be used to develop custom-made models aimed at representing missed components. Chapter 9 shows how ATP capabilities can be used for the development of a library of component modules that can be called from ATPDraw as built-in models [30]. The chapter details the steps to be made for the development of a library of models aimed at carrying out power quality studies. Power quality is a multidisciplinary area related to the assessment, analysis, characterization, and quantification of the mutual interaction between the utility and its customers (i.e. the interaction equipment and the power system). The concept can be considered as a combination of voltage and current quality, so it is, therefore, concerned with deviations of voltage and current from the ideal single-frequency sine waves of constant amplitude and frequency. Power quality disturbances can be generally classified in two categories: variations (small deviations of voltage and/or current characteristics from their nominal or declared value/waveform) and events (large deviations of voltage or current characteristics from their nominal or declared values/waveforms) [31, 32]. The chapter shows how the ATP capabilities can be used to analyse the effect of voltage dips, assess distortion caused by harmonic sources, and simulate techniques for mitigating voltage dips and harmonic currents.

Protection systems. Protection systems are critical components and their behaviour is an important part of the power system response to a transient event. A system aimed at protecting against overcurrents consists of three major parts: instrument transformers (current, wound electromagnetic voltage, and capacitor voltage transformers), protective relays, and circuit breakers [33, 34]. Chapter 10 summarizes modelling guidelines for representing the power system, instrument transformers and the different types of relays (electromechanical, static/electronic, microprocessor-based) at both transmission and distribution levels using ATP capabilities. The chapter includes some case studies that illustrate the potential of ATP in this field, and the application of custom-made models developed for representing distribution-level protective devices following the procedure proposed in Chapter 9.

Advanced applications. ATP users can develop simulation environments in which ATP capabilities are combined with capabilities from other simulation tools. Such combinations can allow users to create powerful tools that are able to significantly expand ATP applications. Chapter 11 proposes a general procedure based on a MATLAB-ATP link and the usage of a multicore environment to expand ATP applications and reduce simulation times. The chapter details three different case studies that show how ATP can also be used as a design tool that could be applied to studies that require additional capabilities.

References

1 Martinez-Velasco, J.A. (ed.) (2015). *Transient Analysis of Power Systems. Solution Techniques, Tools and Applications.* Chichester (United Kingdom): Wiley/IEEE Press.
2 CIGRE WG 33.02 (1990). *Guidelines for Representation of Network Elements when Calculating Transients.* CIGRE Brochure 39.
3 Gole, A., Martinez-Velasco, J.A. and Keri, A. (eds.) (1998). *Modeling and Analysis of Power System Transients Using Digital Programs.* IEEE Special Publication TP133, 1–187.
4 IEC TR 60071-4 (2004). *Insulation Co-ordination - Part 4: Computational Guide to Insulation Co-ordination and Modeling of Electrical Networks.*
5 Martinez-Velasco, J.A. (ed.) (2009). *Power System Transients. Parameter Determination.* Boca Raton (FL, USA): CRC Press.
6 Phadke, A.G. (1980). *Digital Simulation of Electrical Transient Phenomena.* IEEE Publication 81 EHO173-5-PWR.
7 Tziouvaras, D.A. (1999). *Electromagnetic Transient Program Applications to Power System Protection.* IEEE Special Publication TP150, 1–101.
8 Martinez-Velasco, J.A. (2011). *Transient Analysis of Power Systems: Solutions Techniques, Tools and Applications.* IEEE Special Publication 11TP255E, 100–194.
9 Dommel, H.W. (1986). *Electromagnetic Transients Program Reference Manual (EMTP Theory Book).* Portland (OR, USA): Bonneville Power Administration.
10 Greenwood, A. (1991). *Electrical Transients in Power Systems,* 2e. New York (NY, USA): John Wiley.
11 van der Sluis, L. (2001). *Transients in Power Systems.* Chichester (United Kingdom): Wiley.
12 Chowdhuri, P. (2003). *Electromagnetic Transients in Power Systems,* 2e. Taunton (United Kingdom): RS Press/Wiley.

13 Watson, N. and Arrillaga, J. (2003). *Power Systems Electromagnetic Transients Simulation*. Stevenage (United Kingdom): The Institution of Electrical Engineers.

14 Shenkman, A.L. (2005). *Transient Analysis of Electric Power Circuits Handbook*. Dordrecht (The Netherlands): Springer.

15 Das, J.C. (2010). *Transients in Electrical Systems. Analysis, Recognition, and Mitigation*. New York (NY, USA): McGraw-Hill.

16 Ametani, A., Nagaoka, N., Baba, Y. et al. (2017). *Power System Transients: Theory and Applications*, 2e. Boca Raton (FL, USA): CRC Press.

17 Ametani, A. (ed.) (2015). *Numerical Analysis of Power System Transients and Dynamics*. Stevenage (United Kingdom): The Institution of Engineering and Technology.

18 Haginomori, E., Koshiduka, T., Arai, J., and Ikeda, H. (2016). *Power System Transient Analysis. Theory and Practice Using Simulation Programs (ATP-EMTP)*. Chichester (United Kingdom): Wiley.

19 Can/Am Users Group (2000). *ATP Rule Book*. Can/Am Users Group.

20 Høidalen, H.K., Prikler, L., and Hall, J.L. (1999). ATPDraw – Graphical preprocessor to ATP, Windows version. Presented at the IPST'99 in Budapest, Hungary (June 1999).

21 Prikler, L. and Høidalen, H.K. (2002). *ATPDraw for Windows User's Manual*, SEFAS TR F5680.

22 Kizilcay, M. and Hoidalen, H.K. (2015). EMTP-ATP. In: *Chapter 2 of Numerical Analysis of Power System Transients and Dynamics* (ed. A. Ametani). Stevenage (United Kingdom): The Institution of Engineering and Technology.

23 Ceraolo, M. (2018). Experiences in creating a software tool to analyze and postprocess simulated and measured data. *Software: Practice and Experience* 48 (12): 2380–2388. Visit the following link http://ceraolo-plotxy.ing.unipi.it/default.htm.

24 Grebe, T. and Smith, S. (1999). Visualize system simulation and measurement data. *IEEE Computer Applications in Power* 12 (3): 46–51.

25 Ragaller, K. (ed.) (1980). *Surges in High-Voltage Networks*. New York (NY, USA): Plenum Press.

26 Hileman, A.R. (1999). *Insulation Coordination for Power Systems*. New York (NY, USA): Marcel Dekker.

27 Sen, K.K. and Sen, M.L. (2009). *Introduction to FACTS Controller: Theory, Modeling, and Applications*. Hoboken (NJ, USA): Wiley/IEEE Press.

28 Ghosh, A. and Ledwich, G. (2002). *Power Quality Enhancement using Custom Power Devices*. Norwell (MA, USA): Kluwer Academic Publishers.

29 Yazdani, A. and Iravani, R. (2010). *Voltage-Sourced Converters: Modeling, Control, and Applications*. Hoboken (NJ, USA): Wiley.

30 Martinez-Velasco, J.A. and Martin-Arnedo, J. (1999). EMTP modular library for power quality analysis. Presented at IEEE Budapest Power Tech in Budapest, Hungary (29 August–2 September).

31 Bollen, M.H.J. (2000). *Understanding Power Quality Problems. Voltage Dips and Interruptions*. New York (NY, USA): IEEE Press.

32 Dugan, R.C., McGranaghan, M.F., Santoso, S., and Wayne Beaty, H. (2012). *Electrical Power Systems Quality*, 3e. New York (NY, USA): Mc-Graw Hill.

33 Anderson, P.M. (1998). *Power System Protection*. New York (NY, USA): Mc-Graw Hill/IEEE Press.

34 Phadke, A.G. and Thorp, J.S. (2009). *Computer Relaying for Power Systems*, 2e. Chichester (United Kingdom): Wiley.

2

Modelling of Power Components for Transients Studies

Juan A. Martinez-Velasco

2.1 Introduction

The accurate simulation of transient phenomena requires a representation of network components valid for a frequency range that varies from direct current (dc) to dozens of MHz. On the other hand, such simulation implies not only the selection of component models but also the selection of the system area that must be represented. The rules to be considered when selecting models and the system area for the simulation of electromagnetic transients can be summarized as follows [1]:

1) Select the system zone taking into account the frequency range of the transients; the higher the frequencies, the smaller the zone to be modelled.
2) Minimize the part of the system to be represented. An increased number of components do not necessarily mean increased accuracy, as there could be a higher probability of insufficient or incorrect modelling. In addition, a very detailed representation of a system will usually require longer simulation time.
3) Implement an adequate representation of losses. Since their effect on maximum voltages and oscillation frequencies is limited, they do not play a critical role in many cases. However, there are some cases (e.g. ferroresonance or capacitor bank switching) for which losses are critical to defining the magnitude of overvoltages.
4) Consider an idealized representation of some components if the system to be simulated is too complex. Such representation will facilitate the edition of the data file and simplify the analysis of simulation results.
5) Perform a parametric study if one or several parameters cannot accurately be determined. Results derived from such a study will show which parameters are of concern.

Modelling of power components taking into account the frequency-dependence of parameters can be achieved currently through the use of mathematical models which are accurate enough for a specific range of frequencies. Each range of frequencies usually corresponds to some particular transient phenomena. One of the most accepted classifications divides frequency ranges into four groups [2, 3]: low-frequency transients, from 0.1 Hz to 3 kHz, slow-front transients, from 50/60 Hz to 20 kHz, fast-front transients, from 10 kHz to 3 MHz, very fast-front transients, from 100 kHz to 50 MHz. Note that there is overlap between frequency ranges.

If a representation is already available for each frequency range, the selection of the model may suppose an iterative procedure: the model must be selected based on the frequency range of the transients to be simulated; however, the frequency ranges of the case study are not usually known before performing the simulation. This task can be alleviated by looking into

Transient Analysis of Power Systems: A Practical Approach, First Edition. Edited by Juan A. Martinez-Velasco.
© 2020 John Wiley & Sons Ltd. Published 2020 by John Wiley & Sons Ltd.
Companion Website: www.wiley.com/go/martinez/power_systems

widely accepted classification tables; see [1]. Much effort has been dedicated to clarify the main aspects to be considered when representing power components in transient simulations. Nowadays users of electromagnetic transients tools can obtain information on this field from several sources [3–5].

This chapter provides a short summary of the modelling guidelines suggested for representing power system components involved in the generation and delivery of electric energy.

2.2 Overhead Lines

2.2.1 Overview

The selection of an adequate line model is crucial in most transient studies. Voltage stresses to be considered in overhead line design can also be classified into groups, each one having a different frequency range [6–8]: (i) power-frequency voltages in the presence of contamination; (ii) temporary (low-frequency) overvoltages produced by faults, load rejection or ferroresonance; (iii) slow-front overvoltages produced by switching operations; (iv) fast-front overvoltages, generally caused by lightning flashes.

Two types of time-domain models have been developed for overhead lines: lumped- and distributed-parameter models. The appropriate selection of a model depends on the highest frequency involved in the phenomenon under study and, to less a lesser extent, on the line length.

Lumped-parameter line models represent transmission systems by lumped R, L, G, and C elements whose values are calculated at a single frequency. These models, known as PI models, are adequate for steady-state calculations, although they can also be used for transient simulations in the neighbourhood of the frequency at which parameters were evaluated. The most accurate models for transient calculations are those that take into account the distributed nature of the line parameters [3–5]. Two categories can be distinguished for these models: constant parameters and frequency-dependent parameters. The number of spans and the different hardware of a transmission line, as well as the models required to represent each part (conductors and shield wires, towers, grounding, insulation), depend on the cause of the voltage stress. The following rules summarize the modelling guidelines to be followed in each case [9].

1. In power-frequency and temporary overvoltage calculations, the whole transmission line length must be included in the model, but only the representation of phase conductors is usually needed. A multi-phase model with lumped and constant parameters, including conductor asymmetry, will generally suffice. For transients with a frequency range above 1 kHz, a frequency-dependent model could be needed to account for the ground propagation mode. Corona effect can also be important if phase-conductor voltages exceed the corona inception voltage.
2. In switching overvoltage calculations, a multi-phase distributed-parameter model of the whole transmission line length, including conductor asymmetry, is generally required. As for temporary overvoltages, frequency-dependence of parameters is important for the ground propagation mode, and only phase conductors need to be represented.
3. The calculation of lightning-caused overvoltages requires a more detailed model, in which towers, footing impedances, insulators and tower clearances, in addition to phase conductors and shield wires, are represented. However, only a few spans at both sides of the point of impact must be considered in the line model. Since lightning is a fast-front transient phenomenon, a multi-phase model with distributed parameters, including conductor asymmetry and corona effect, is required for the representation of each span.

The length extent of an overhead line that must be included in a model depends on the type of transient to be analysed. As a rule of thumb, the lower the frequencies, the more length of line to be represented. For low- and mid-frequency transients, the whole line length is included in the model. For fast-front and very fast-front transients, a few line spans will usually suffice.

The following subsections are respectively dedicated to present the equations needed to model multi-conductor lines, and the models to be considered for representing towers, grounding impedances and insulators. See reference [9] for more details on every subject.

2.2.2 Multi-conductor Transmission Line Equations and Models

2.2.2.1 Transmission Line Equations

Figure 2.1 depicts a differential section of a multi-conductor overhead line illustrating the couplings among series inductances and amongst shunt capacitances. The behaviour of this line is described in the frequency domain by two matrix equations:

$$-\frac{d\mathbf{V}_x(\omega)}{dx} = \mathbf{Z}(\omega)\,\mathbf{I}_x(\omega) \tag{2.1a}$$

$$-\frac{d\mathbf{I}_x(\omega)}{dx} = \mathbf{Y}(\omega)\,\mathbf{V}_x(\omega) \tag{2.1b}$$

where $\mathbf{Z}(\omega)$ and $\mathbf{Y}(\omega)$ are respectively the series impedance and the shunt admittance matrices per unit length.

The series impedance matrix of an overhead line can be decomposed as follows:

$$\mathbf{Z}(\omega) = \mathbf{R}(\omega) + j\omega\mathbf{L}(\omega) \tag{2.2}$$

where $\mathbf{Z}$ is a complex and symmetric matrix, whose elements are frequency-dependent. For transient analysis, elements of $\mathbf{R}$ and $\mathbf{L}$ must be calculated taking into account the skin effect in conductors and ground. This is achieved by using either Carson's ground impedance [10], or Schelkunoff's surface impedance formulae for cylindrical conductors [11].

The shunt admittance can be expressed as follows:

$$\mathbf{Y}(\omega) = \mathbf{G} + j\omega\mathbf{C} \tag{2.3}$$

where $\mathbf{Y}$ is also a complex and symmetric matrix, with frequency-dependent elements. In general, elements of $\mathbf{G}$ can be associated with currents leaking to ground through insulator strings, which can mainly occur with contaminated insulators. Their values can usually be neglected for most studies; however, under corona effect conductance values can be significant. That is, under non-corona conditions, with clean insulators and dry weather, conductances can be neglected.

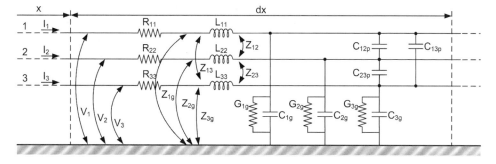

Figure 2.1 Differential section of a three-phase overhead line.

As for **C** elements, their frequency dependence can be neglected within the frequency range that is of concern for overhead line design.

If matrices **R**, **L**, **G**, and **C** are assumed constant (i.e. independent of frequency), Eq. (2.1) can be stated as follows:

$$-\frac{\partial \mathbf{v}(x,t)}{\partial x} = \mathbf{R}\mathbf{i}(x,t) + \mathbf{L}\frac{\partial \mathbf{i}(x,t)}{\partial t} \tag{2.4a}$$

$$-\frac{\partial \mathbf{i}(x,t)}{\partial x} = \mathbf{G}\mathbf{v}(x,t) + \mathbf{C}\frac{\partial \mathbf{v}(x,t)}{\partial t} \tag{2.4b}$$

where $\mathbf{v}(x,t)$ and $\mathbf{i}(x,t)$ are the voltage and current vectors respectively.

Advanced models can consider an additional distance-dependence of the line parameters (non-uniform line), the effect of induced voltages due to distributed sources caused by nearby lightning (illuminated line), and the dependence of the line capacitance with respect to the voltage (nonlinear line, due to corona effect). Given the frequency dependence of the series parameters, the approach to the solution of the line equations, even in transient calculations, is performed in the frequency domain.

The general solution of the line equations with uniformly-distributed parameters in the frequency domain can be expressed as follows:

$$\mathbf{I}_x(\omega) = e^{-\Gamma(\omega)x}\mathbf{I}_f(\omega) + e^{+\Gamma(\omega)x}\mathbf{I}_b(\omega) \tag{2.5a}$$

$$\mathbf{V}_x(\omega) = \mathbf{Y}_c^{-1}(\omega)[e^{-\Gamma(\omega)x}\mathbf{I}_f(\omega) - e^{+\Gamma(\omega)x}\mathbf{I}_b(\omega)] \tag{2.5b}$$

where $\mathbf{I}_f(\omega)$ and $\mathbf{I}_b(\omega)$ are the vectors of forward and backward travelling wave currents at $x = 0$, $\Gamma(\omega)$ is the propagation constant matrix, and $\mathbf{Y}_c(\omega)$ is the characteristic admittance matrix:

$$\Gamma(\omega) = \sqrt{\mathbf{YZ}} \tag{2.6a}$$

$$\mathbf{Y}_c(\omega) = \sqrt{\mathbf{Z}^{-1}\mathbf{Y}} \tag{2.6b}$$

$\mathbf{I}_f(\omega)$ and $\mathbf{I}_b(\omega)$ can be deduced from the boundary conditions of the line. Considering the frame shown in Figure 2.2, the solution at line ends can be formulated as follows:

$$\mathbf{I}_k(\omega) = \mathbf{Y}_c(\omega)\mathbf{V}_k(\omega) - \mathbf{H}(\omega)[\mathbf{Y}_c(\omega)\mathbf{V}_m(\omega) + \mathbf{I}_m(\omega)] \tag{2.7a}$$

$$\mathbf{I}_m(\omega) = \mathbf{Y}_c(\omega)\mathbf{V}_m(\omega) - \mathbf{H}(\omega)[\mathbf{Y}_c(\omega)\mathbf{V}_k(\omega) + \mathbf{I}_k(\omega)] \tag{2.7b}$$

where $\mathbf{H} = \exp(-\Gamma\ell)$, being ℓ the length of the line.

Transforming Eq. (2.7) into the time domain gives:

$$\mathbf{i}_k(t) = \mathbf{y}_c(t) * \mathbf{v}_k(t) - \mathbf{h}(t) * \{\mathbf{y}_c(t) * \mathbf{v}_m(t) + \mathbf{i}_m(t)\} \tag{2.8a}$$

$$\mathbf{i}_m(t) = \mathbf{y}_c(t) * \mathbf{v}_m(t) - \mathbf{h}(t) * \{\mathbf{y}_c(t) * \mathbf{v}_k(t) + \mathbf{i}_k(t)\} \tag{2.8b}$$

where symbol * indicates convolution and $\mathbf{x}(t) = \mathrm{F}^{-1}\{\mathbf{X}(\omega)\}$ is the inverse Fourier transform.

These equations suggest that an overhead line can be represented at each end by a multi-terminal admittance paralleled by a multi-terminal current source, as shown in Figure 2.3. The implementation of this equivalent circuit requires the synthesis of an electrical network

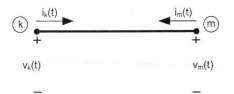

Figure 2.2 Line model – reference frame.

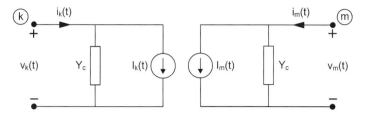

Figure 2.3 Equivalent circuit for time-domain simulations.

to represent the multi-terminal admittance. In addition, the current source values have to be updated at every time step during the time-domain calculation. Both tasks are not straight-forward, and many approaches have been developed to cope with this problem. The techniques developed to solve the equations of a multi-conductor frequency-dependent overhead line can be classified into two main categories: modal-domain techniques and phase-domain techniques. For an overview of the main approaches, see [9].

2.2.2.2 Corona Effect
When the voltage of a conductor reaches a critical value (v_c) and the electrical field in its neigh-bourhood is higher than the dielectric strength of the air, ionization is produced around the conductor. As a consequence, there will be storage and movement of charges in the ionized region, which can be viewed as an increase of the conductor radius and consequently of the capacitance to ground [8]. This phenomenon is known as corona effect. The increase in capac-itance results in both a decrease in the velocity of propagation and a decrease in the surge impedance. The decrease in velocity causes distortion of the surge voltage during propagation; that is, the wave front is pushed back and the steepness of the surge is decreased. Depending on the tail of the initial surge, the crest voltage is also decreased. A model based on the physi-cal processes is very complicated and impractical for transient simulation using a time-domain tool. In propagation analysis, it is common to use models based on a macroscopic description, specifically models based on charge-voltage curves (*q-v* curves). The models proposed in the literature can be classified into two groups: (i) static models, in which the corona capacitance is only a function of the voltage, and (ii) dynamic models, in which the capacitance is a function of the voltage and its derivatives. In general, all methods need to calculate the corona inception voltage beforehand.

2.2.2.3 Line Constants Routine
When only phase conductors and shield wires are to be included in the overhead line model, the line parameters can be calculated from the line geometry and physical properties of phase conductors, shield wires and ground. Great accuracy is not usually required when specifying input values if the goal is to duplicate low-frequency and slow-front transients, but more care is needed, mainly with the ground resistivity value, if the goal is to simulate fast transients [9, 12]. Alternative Transients Program (ATP) users obtain overhead line parameters and models by means of a dedicated supporting routine known as LINE CONSTANTS [12, 13]: users enter the geometry of the line towers, as well as the physical parameters of the line conductors, and select the desired type of line model. This routine can create the following models: lumped-parameter equivalent or nominal PI circuits, at the specified frequency; constant distributed-parameter model, at the specified frequency; frequency-dependent distributed-parameter model, fitted for a given frequency range. The following information can be provided by the routine: the capacitance or the susceptance matrix; the series impedance matrix; resistance, inductance and capacitance per unit length for zero and positive sequences, at a given frequency or for a

specified frequency range; surge impedance, attenuation, propagation velocity and wavelength for zero and positive sequences, at a given frequency or for a specified frequency range. Line matrices can be provided for the system of physical conductors, the system of equivalent phase conductors, or symmetrical components of the equivalent phase conductors.

2.2.3 Transmission Line Towers

The representation of a tower is usually made in circuit terms; that is, the tower is represented by means of several line sections and circuit elements that are assembled taking into account the tower structure [14–23]. Due to the fast-front times associated to lightning stroke currents, most tower models assume that the tower response is dominated by the transverse electromagnetic mode (TEM) wave and neglects other types of radiation. More sophisticated models based on non-uniform transmission lines or on a combination of lumped- and distributed-parameter circuit elements have been developed [18, 19]. In many cases, it is important to obtain lightning overvoltages across insulators located at different heights above ground; this is particularly important when two or more transmission lines with different voltage levels share the same tower.

Models based on a constant-parameter circuit representation may be classified as follows [19]:

a) *Single-conductor line models.* The first models represented the tower by means of simple geometric forms [24]. The surge propagation velocity along tower elements can be assumed that of the light. However, the multiple paths of the lattice structure and the crossarms introduce some time delays; consequently, the time for a complete reflection from ground is longer than that obtained from a travel time whose value is the tower height divided by the speed of light. Therefore, the propagation velocity in some models was reduced to include this effect in the tower response. Crossarms behave as short-stub lines with open-circuit ends [14]. Experimental results have shown that travel times in crossarms are longer than those derived by assuming a propagation velocity equal to that of light. Figure 2.4 shows two of the tower configurations used in the IEEE FLASH program, and the expressions to obtain the parameters of the lossless line models [20, 21].

b) *Multi-conductor line models.* Each segment of the tower between crossarms is represented as a multi-conductor vertical line, which can be reduced to a single conductor. The tower

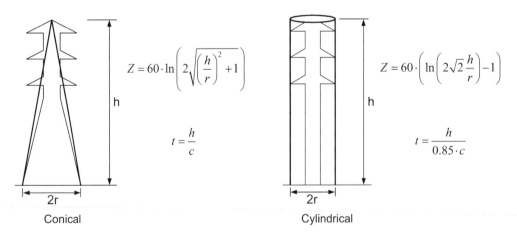

$$Z = 60 \cdot \ln\left(2\sqrt{\left(\frac{h}{r}\right)^2 + 1} \right)$$

$$t = \frac{h}{c}$$

Conical

$$Z = 60 \cdot \left(\ln\left(2\sqrt{2}\frac{h}{r} \right) - 1 \right)$$

$$t = \frac{h}{0.85 \cdot c}$$

Cylindrical

Figure 2.4 Tower configurations.

model is then a single-phase line whose section increases from top to ground. The model presented in [16] included the effect of bracings (represented by lossless lines in parallel to the main legs) and crossarms (represented as lossless line branched at junction points). The multi-storey model is a special case composed of four sections that represent the tower sections between crossarms [15, 23]. The approach was originally developed for representing towers of Ultra High Voltage (UHV) transmission lines. Each section consists of a lossless line in series with a parallel RL circuit, included for attenuation of the travelling waves. A study presented in [25] concluded that the multi-storey model with surge impedance values originally proposed in [15] was not adequate for representing towers of lower voltage transmission lines: the tower model for shorter towers can be simpler than that proposed in [15], and a single lossless line for each tower section will suffice.

2.2.4 Transmission Line Grounding

2.2.4.1 Introduction
High voltages can be generated on grounded parts of an overhead line support when either a ground wire or a phase conductor is struck by lightning. If lightning strikes a tower or a ground wire, the discharge should then be safely led to the earth and dissipated there. The purpose of grounding for protection against lightning is to bypass the energy of the lightning discharge safely to the ground (i.e. most of the energy of the lightning discharge should be dissipated into the ground without raising the voltage of the protected system). When current is discharged into the soil through a ground electrode, potential gradients are set up as a result of the conduction of current through the soil. The grounding impedance depends on the area of the tower steel (or grounding conductor) in contact with the earth, and on the resistivity of the earth. The latter is not constant, fluctuates over time, and is a function of soil type, moisture content, temperature, current magnitude, and waveshape. Another factor is soil ionization which leads to an additional decay in the electrode resistance when high currents are discharged into the soil. Under lightning surge conditions and some power-frequency fault conditions, the high current density in the soil increases the electric field strength up to values that cause electrical discharges in the soil that surrounds the electrode. The threshold level and intensity of the ionization are especially high when the soil is dry or when it has a high resistivity.

The representation of the grounding impedance depends on the frequency range of the discharged current. Grounding models can be classified into two groups: low- and high-frequency models. In practice, they correspond respectively to power-frequency and lightning currents. Grounding systems of transmission lines can be broadly classified as compact (concentrated) and extended (distributed) [9, 26, 27]. A compact grounding can be represented by lumped-parameter circuit elements. In an extended grounding system, the travel time of the electromagnetic fields along the electrodes is comparable with that along the support itself; this generally applies to grounding systems with physical dimensions exceeding 20 m.

2.2.4.2 Low-Frequency Models
The power-frequency impedance of a grounding system can be expressed as follows [28]:

$$Z = (R_L + R_E + R_C) + j(X_E + X_L) \tag{2.9}$$

where R_L and X_L are respectively the resistance and the reactance of the ground electrode leads, R_C is the contact resistance between the surface of the electrode and the surrounding soil, due to the imperfect contact between the soil and the surface of the electrode, R_E is the dissipation resistance of the soil that surrounds the electrode (i.e. the resistance of the earth between the electrode surface and a remote earth), X_E is the reactance of the current paths in the soil.

The series resistance of the metallic electrode conductors and leads is typically much lower than the contact resistance and the resistance of the surrounding soil; the reactance of the metallic conductors and leads is much higher than the reactance of the current dissipated in the soil; the contact resistance can typically be neglected when the soil has settled around the electrode conductors; finally, at power frequencies, the reactance of the grounding conductors and leads becomes negligible compared to the dissipation resistance. Therefore, at low frequencies, the grounding impedance can only be represented by the dissipation resistance.

Compact grounding systems. The rod is the most common type of ground electrode. Ground rods are generally made of galvanized steel and driven vertically down from the earth's surface. When the length of the ground rod is much greater than its radius, the low-current low-frequency impedance of a single rod behaves as a resistance, whose value decreases as either the buried length or the radius of the rod increase, but it does not decrease directly with length, so an increase in length above certain limit will not significantly reduce the resistance. Ground resistance can be reduced by connecting several rods in parallel. The resistance is inversely proportional to the number of parallel rods, provided the spacing between rods is large compared to their length. That is, when the rods are closely spaced compared with their length, the whole ground arrangement behaves as one rod with a larger apparent diameter and a small reduction in resistance. As the rod spacing increases, the combined resistance decreases. The proximity effect between the ground rods tends to increase the combined resistance, thus diminishing the advantage of multiple rods. For other electrode geometries, see [9, 28].

Extended grounding systems. The contact area of a grounding system with the earth can be increased by installing a counterpoise, which is a conductor buried in the ground at a depth of about 1 m. Common arrangements include one or more radial wires extending out from each tower base, single or multiple continuous wires from tower to tower, or combinations of radial and continuous wires. Neither the conductor radius nor the burial depth has a significant impact on the resistance. Several short wires, arranged radially, may be more effective than a single long wire even if the total length and contact resistance of both arrangements are the same.

Grounding resistance in non-homogeneous soils. Several methods have been developed to deal with non-homogeneous soils. One of the methods consists of stratifying the soil in two layers: the first layer reaches a depth d and is characterized by a resistivity ρ_1, the second one has an infinite depth and a resistivity ρ_2. In general, it is possible to determine a two-layer soil structure that can represent typical soils for grounding purposes; if it is not possible, then a three-layer soil model should be considered. A simpler two-layer soil treatment is appropriate when $\rho_2 \gg \rho_1$. Approximations to obtain the ground resistance in non-homogeneous soils were proposed in [29].

2.2.4.3 High-Frequency Models

For high-frequency phenomena, such as lightning, the effective impedance of a buried horizontal wire is not constant. Such behaviour can be explained by considering surge propagation along the buried conductor, and can be summarized as follows [30]:

- The effective impedance is initially equal to the surge impedance of the buried wire and it reduces in a few μs to a level that corresponds to the leakage resistance.
- The transition from the initial surge impedance to the final leakage resistance is accomplished in a few round-trip travel times along the wire. Although multi-velocity waves will exist, the only one of importance is slow and travels at approximately 30% of the velocity of light.

- The surge impedance rises abruptly in less than 1 μs, and increases at a slow rate thereafter; the leakage resistance is initially a very high value, but it decreases as the reflection of the travelling waves builds up the voltage along the conductor, being its final value equal to the low-frequency resistance. Resistance quickly dominates as a current wave propagates along the conductor.

Distributed-parameter models. The electrode is represented as the equivalent circuit shown in Figure 2.5, where R is the parallel resistance, L is the series inductance, and C the parallel capacitance to ground per unit length. To take into account soil ionization a nonlinear resistance might be included in the model instead of a constant one. However, soil ionization is not instantaneous and soil resistivity decreases with a time constant of about 2 μs. This value is rather large compared with the front times associated with fast-front lightning strokes; e.g. most negative subsequent strokes. Moreover, if soil ionization occurs, it will always cause a reduction of the ground potential rise. Therefore, ignoring this phenomenon always gives conservative results. For the calculation of parameters R, L, and C, see [9].

Lumped-parameter models. The lightning behaviour of counterpoise electrodes could be represented by the simple equivalent circuit presented in Figure 2.6a [30], where R_c is the counterpoise leakage resistance, R_s is a resistor selected so that the high-frequency impedance of the circuit corresponds to the surge impedance of the counterpoise (Z_c), and L_c is the inductor responsible for the transition from the surge impedance to the low-frequency impedance; its value depends on the length of the counterpoise. Experimental studies of reflections from tower bases showed that the initial reflection differed from that predicted by assuming a lumped-circuit representation of the grounding impedance; this can be justified by including transient ground-plane impedance, and may be incorporated as an additional series inductance in the grounding impedance. This equivalent inductance is valid for the normal case, in which the front time is much greater than the tower travel time. Longer front times give higher values of average inductance, but the voltage rise from the inductance at the crest current is actually lower. Figure 2.6b shows the high-frequency lumped-parameter model of a ground electrode proposed in [31].

Figure 2.5 Equivalent circuit of a ground electrode at high frequencies.

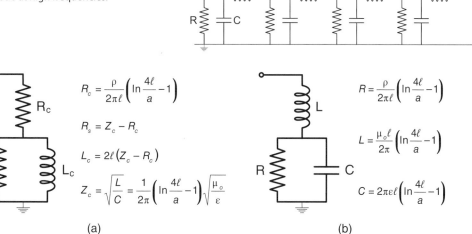

In (a):
$$R_c = \frac{\rho}{2\pi\ell}\left(\ln\frac{4\ell}{a} - 1\right)$$

$$R_s = Z_c - R_c$$

$$L_c = 2\ell\left(Z_c - R_c\right)$$

$$Z_c = \sqrt{\frac{L}{C}} = \frac{1}{2\pi}\left(\ln\frac{4\ell}{a} - 1\right)\sqrt{\frac{\mu_o}{\varepsilon}}$$

In (b):
$$R = \frac{\rho}{2\pi\ell}\left(\ln\frac{4\ell}{a} - 1\right)$$

$$L = \frac{\mu_o\ell}{2\pi}\left(\ln\frac{4\ell}{a} - 1\right)$$

$$C = 2\pi\varepsilon\ell\left(\ln\frac{4\ell}{a} - 1\right)$$

(a)   (b)

Figure 2.6 High-frequency lumped equivalent circuits. (a) Counterpoise, (b) Ground electrode.

The following aspects should be taken into account when using the above grounding models:

- The application of the lumped-parameter circuit model should be limited to cases where the length of the rod is less than one-tenth the wavelength in earth, which limits the frequency range of the validity of this model to low frequencies. This approach can be used for a preliminary analysis, but keeping in mind that it greatly overestimates the ground rod impedance at high frequencies.
- The approximate distributed-parameter circuit reduces the overestimation of the ground rod impedance at high frequencies in comparison with the *RLC* circuit.
- The best fit to ground rod impedance is achieved by means of a model based on antenna theory, with a non-uniform distributed-parameter model, whose parameters can be deduced by curve matching [31].

2.2.4.4 Treatment of Soil Ionization

The resistance of a ground electrode may decrease due to ionization of the soil. When a current is injected to the electrode, ionization will occur around the electrode if a critical field gradient is exceeded. In those regions, low ohmic discharge channels are formed, being the resistance of the ionized zone reduced to a negligible value. The ground resistance of an electrode remains at the value determined by the electrode geometry and the soil resistivity until ionization breakdown is reached; after breakdown, the resistance varies. This soil breakdown can be viewed as an increase of the geometry of the electrode. The transient electric fields needed to ionize small volumes of soil, or to flashover across the soil surface are typically between 100 and 1000 kV/m [26]. For a vertical rod, this process may be represented with the simplified model of Figure 2.7, which also shows the final area. As the ionization progresses, the shape of the zone becomes more spherical; that is, when the gradient exceeds a critical value E_0, breakdown of soil occurs and the ground rod can be modelled as a hemisphere electrode. The ionization zone is described by the critical field strength E_0 at which the radius is equal to r.

The impulse resistance is inversely proportional to the reciprocal of the square root of the current, but on a log-to-log plot it is a straight line, as shown in Figure 2.8a [32]. The low-current resistance value R_0 is maintained until the current exceeds the threshold current, I_g, that produces the critical field strength E_0; beyond this current the resistance starts decreasing. A simplification for rod electrodes should account for some aspects (i.e. they have the low-current resistance for currents close to zero, they approach the square root dependence for very high current values, and they approximate the log dependence between these two extremes). As the current increases, a point is reached where the ionized zone is approximately spherical and R_i decreases as illustrated in Figure 2.8b.

This simplification is valid for ground electrodes with small extension. In unfavourable soil conditions in which counterpoises with large extensions are required, the simplification is not

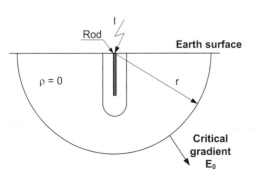

Figure 2.7 After soil ionization a rod becomes a hemisphere.

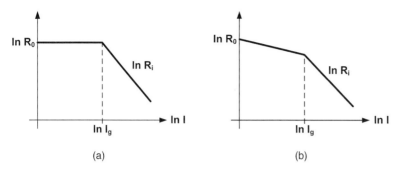

Figure 2.8 Impulse resistance including soil ionization. (a) Hemisphere electrode. (b) Rod electrode.

accurate enough. In addition, the critical breakdown voltage gradient of soil needs to be known accurately, but E_0 is difficult to define.

2.2.5 Transmission Line Insulation

Power line insulation is of external type (i.e. exposed to atmospheric conditions) and self-restoring (i.e. it recovers its insulating properties after a disruptive discharge). Insulation strength is expressed in terms of withstand voltage, a quantity determined by tests conducted under specified conditions with specified waveshapes. The same insulation may have different withstand voltages for different voltage waveshapes; that is, insulation withstand strength depends greatly on the waveshape of the applied voltage.

In general, the strength characteristic of self-restoring insulation may be represented by a cumulative normal o Gaussian distribution [6–8], whose mean is known in IEEE standards as the critical flashover voltage (CFO) and in IEC standards as the U_{50}, or 50% flashover voltage. The insulation exhibits a 50% probability of flashover when the U_{50} is applied; i.e. half the impulses flashover. Assuming a normal distribution, insulation strength is fully specified by providing the values of U_{50} and the standard deviation, σ_f (or σ_f/U_{50}).

The Gaussian distribution is unbounded to right and left (i.e. it is defined between plus and minus infinity). A limit of minus infinity indicates that there exists a probability of flashover for a voltage equal to zero, which is physically impossible. For this reason, IEC has adopted the Weibull distribution for representing the strength characteristic of self-restoring insulation [6].

The wide variety of lightning stroke characteristics, together with the modification effects that the line components introduce, stresses line insulation with a diversity of impulse voltage waveshapes, so it is important to be able to evaluate insulation performance when stressed by nonstandard lightning impulses. The most popular insulation models do not describe physical phenomena, instead they are based on the characteristics of the various phases of the discharge mechanism. A thorough review of these models was presented in [33].

The following paragraphs summarize the main characteristics of the models proposed for analysing the dielectric strength of air gaps under lightning overvoltages.

2.2.5.1 Voltage-Time Curves

They give the dependence of the specific impulse shape peak voltage on the time-to-breakdown, see Figure 2.9. Voltage-time curves are determined experimentally for a specific gap or for an insulator string, and may be represented with empirical equations, applicable only within the range of parameters covered experimentally [34]. In practice, measurements can be affected by several factors: impulse front shape, front times of the applied standard lightning impulse, gap distance and geometry, polarity, internal impedance of the impulse generator. Voltage-time

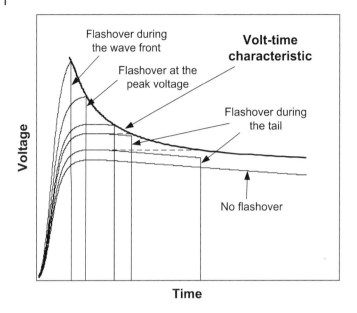

Figure 2.9 Volt-time characteristic.

curves do not apply to multiple flashover studies and their accuracy might be poor for long times to breakdown or low probability flashovers.

2.2.5.2 Integration Methods

Insulation performance is a function of one or more significant parameters of the nonstandard voltage waveshape [35–37]. Their common basic assumptions are: (i) there is a minimum voltage V_0 that must be exceeded before any breakdown can start or continue; (ii) the time to breakdown is a function of both the magnitude and the time duration of the applied voltage above the minimum voltage V_0; (iii) there exists a unique set of constants associated with breakdown for each insulator configuration. The dielectric breakdown of the insulation is obtained then from the following equation:

$$D = \int_{t_0}^{t_c} (v(t) - V_0)^n dt \qquad (2.10)$$

where t_0 is the time after which the voltage $v(t)$ is higher than the required minimum voltage V_0 (also known as reference voltage) and t_c is the time-to-breakdown. The constant D is known as disruptive effect constant. If $n = 1$, the method is known as the equal-area law. These methods can be applied to specific geometries and voltage shapes only.

2.2.5.3 Physical Models

They consider the different phases of the discharge mechanism and their dependence on the applied voltage, and compute the time necessary for completion of all the phases of the discharge process [18, 38–40]. When the applied voltage exceeds the corona inception voltage, streamers propagate and cross the gap after a certain time if the voltage remains high enough. Only when the streamers have crossed the gap, the leaders can start their development. Usually the leader velocity increases exponentially. Both the streamer and the leader phases can develop from only one or both electrodes. When the leader has crossed the gap, or when the two leaders meet, the breakdown occurs.

The time-to-breakdown is the sum of three components: the time to the corona inception voltage, the time the streamers need to cross the gap or to meet the streamers from the opposite electrode, and the leader propagation time. The time the streamers need to cross the gap decreases as voltage is raised and is almost independent of voltage polarity, electrode configuration and gap clearance, but it is strongly dependent on the ratio E/E_{50}, where E is the average field in the gap at the applied voltage and E_{50} is the average field at CFO. These models assume that the streamer phase is completed when the applied voltage has reached a value that gives an average field in the gap equal to E_{50}. The leader propagation time can be calculated from its velocity, which depends on the applied voltage and the leader length. Various expressions have been proposed for the leader velocity.

The following approach for calculating the leader velocity was proposed in [40]:

$$\frac{d\ell}{dt} = kv(t) \left(\frac{v(t)}{g - \ell} - E_0 \right) \tag{2.11}$$

where g is the gap length and ℓ is the leader length. $v(t)$ is the actual (absolute value) voltage in the gap. The constants k and E_0 depend on the gap configuration and insulator type.

The *leader progression model* (LPM) shown in Eq. (2.11) is valid for a large variety of impulse shapes and for the evaluation of the dielectric strength of a variety of geometries. The integration methods have comparable accuracies but more restricted application in relation to waveshapes. The empirical methods can give a good accuracy when they are used within their validity limits (i.e. when specific data are used for a specific insulator or gap, together with a careful application of the model); the use of volt-time curves works well in the short time-to-breakdown domain (2÷6 µs). No single approach alone can be recommended for all applications.

2.3 Insulated Cables

2.3.1 Overview

The behaviour of an insulated cable can be described by the following equations [12, 41, 42]:

$$\mathbf{Z}(\omega) = \mathbf{R}(\omega) + j\omega\mathbf{L}(\omega) \tag{2.12a}$$

$$\mathbf{Y}(\omega) = \mathbf{G}(\omega) + j\omega\mathbf{C}(\omega) \tag{2.12b}$$

where $\mathbf{R}$, $\mathbf{L}$, $\mathbf{G}$, and $\mathbf{C}$ are the cable parameter matrices expressed in per unit length. These quantities are $(n \times n)$ matrices, being n the number of (parallel) conductors of the cable system. The variable ω stresses the fact that these quantities depend on frequency.

ATP has two supporting routines for the calculation of cable parameters known as CABLE CONSTANTS (CC) and CABLE PARAMETERS (CP) [13].

Guidelines for representing insulated cables in electromagnetic transients studies are similar to those proposed for overhead lines. In addition, the solution of cable equations can be carried out following the same techniques proposed in the previous section. However, the large variety of cable designs makes very difficult the development of a single computer routine for calculating the parameter of each design. The calculation of matrices $\mathbf{Z}$ and $\mathbf{Y}$ uses cable geometry and material properties as input parameters. In general, CC/CP users must specify:

1) *Geometry*. Location of each conductor ($x - y$ coordinates); inner and outer radii of each conductor; burial depth of the cable system.
2) *Material properties*. Resistivity, ρ, and relative permeability, μ_r, of all conductors (μ_r is unity for all non-magnetic materials); resistivity and relative permeability of the surrounding medium, ρ, μ_r; relative permittivity of each insulating material, ε_r.

Accurate input data are in general more difficult to obtain for cable systems than for overhead lines as the small geometrical distances make the CP highly sensitive to errors in the specified geometry. In addition, it is not straightforward to represent certain features such as wire screens, semiconducting screens, armours, and lossy insulation materials. It is worth noting that CC/CP routines take the skin effect into account but neglect proximity effects. Besides these routines have some shortcomings in representing certain cable features. A previous conversion procedure may be required in order to bring the available cable data into a form which can be used as input to a CC/CP routine. This conversion is frequently needed because input cable data can have alternative representations, while CC/CP routines only support one representation and they do not consider certain cable features, such as semiconducting screens and wire screens.

An important aspect when calculating electromagnetic transients in insulated cables is the bonding technique used with the cable under study.

The rest of this section is dedicated to introduce the main cable designs for high-voltage applications, summarize the bonding techniques used with high-voltage cable designs, and discuss how to prepare input data of a cable whose design cannot be directly specified in the CC/CP routines. For more details on the calculation of cable parameters, see references [12] and [43–45].

2.3.2 Insulated Cable Designs

Single core self-contained cables. They are coaxial in nature, see Figure 2.10. The insulation system can be based on extruded insulation (e.g. Cross Linked Polyethylene (XLPE)) or oil-impregnated paper (fluid-filled or mass-impregnated). The core conductor can be hollow in the case of fluid-filled cables. Self-contained (SC) cables for high-voltage applications are always designed with a metallic sheath conductor, which can be made of lead, corrugated aluminium, or copper wires. Such cables are also designed with an inner and an outer semi-conducting screen, which are in contact with the core conductor and the sheath conductor, respectively.

Three-phase self-contained cables. They consist of three SC cables which are contained in a common shell. The insulation system of each SC cable can be based on extruded insulation or on paper-oil. These cables can be differentiated into the two designs shown in Figure 2.11:

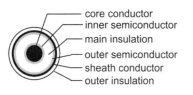

Figure 2.10 SC XLPE cable, with and without armour.

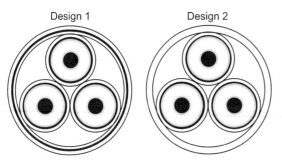

Figure 2.11 Three-phase cable designs.

Figure 2.12 Pipe type cable.

- *Design 1*. One metallic sheath for each SC cable, with cables enclosed within metallic pipe (sheath/armour). This design can be directly modelled using the pipe-type representation available in some CC routines.
- *Design 2*. One metallic sheath for each SC cable, with cables enclosed within insulating pipe. None of the present CC routines can directly deal with this type of design due to the common *insulating* enclosure. This limitation can be overcome in one of the following ways: (i) place a very thin conductive conductor on the inside of the insulating pipe, the cable can then be represented as a pipe-type cable in a CC routine; (ii) place the three SC cables directly in earth (and ignore the insulating pipe). Both options should give reasonably accurate results when the sheath conductors are grounded at both ends. However, these approaches are not valid when calculating induced sheath overvoltages.
  The space between the SC cables and the enclosing pipe is for both designs filled by a composition of insulating materials; however, CC routines only permit to specify a homogenous material between sheaths and the metallic pipe.
Pipe-type cables. They consist of three SC paper cables that are laid asymmetrically within a steel pipe, which is filled with pressurized low viscosity oil or gas, see Figure 2.12. Each SC cable is fitted with a metallic sheath. The sheaths may be touching each other.

2.3.3 Bonding Techniques

The screens (sheaths and armours) of three-phase cables are normally installed in one of the following bonding configurations [45, 46]:

1. *Single-end bonding*. Grounding of the screens at one end only: This option virtually reduces the currents circulating in the screen to zero, since there is no closed loop. However, it can cause an increase of the screen voltage: one end of the screen is grounded, while a voltage appears at the other end, being the voltage magnitude proportional to the length of the cable.
2. *Both-ends bonding*. Grounding of the screens at both ends: This option reduces the voltage in the screen, which is now close to zero at both ends. However, it provides a closed path for the current in the screen, so both currents and losses are larger than with the previous option. For submarine cables, it is common to adopt solid bonding due to the difficulty in constructing joints offshore. To reduce the loss caused by higher screen currents, the screens of submarine cables are usually designed with a large cross-section (i.e. low resistance).
3. *Cross-bonding*. Grounding of the screens at both ends with transposition of the screens: The cable is divided into three minor sections. The screens are transposed between minor sections and grounded at every third minor section, forming a major section. The minor sections should have a similar length in order to keep the system as balanced as possible, and cable may have as many major sections as necessary. The transposition of the screens

assures that each screen is exposed to the magnetic field generated by each phase. Assuming a balance system (i.e. same magnitude and 120° phase difference between phases) installed in trefoil configuration and that each minor section had exactly the same length, the induced voltages would cancel and the circulating current would be null. Since HV cables are rarely installed in a balanced configuration, it is difficult to achieve such conditions; this option cannot eliminate neither circulating currents nor induced voltages, but it reduces both to low values.

At the distribution level it is normal to use the both-ends bonding technique. Cables longer than 3 km normally adopt cross-bonding to reduce screen currents, suppress screen voltages and increase the thermal capacity of the cable. The relationships between currents and voltages in a cross-bonded cable are not the same that for cables in which other bonding option has been implemented. That is, the per unit length series impedance and shunt admittance matrices of a cross-bonded cable are different from those presented above. A procedure for obtaining the series impedance and shunt admittance matrices of a homogeneously cross-bonded cable is presented in [45]. Single-end bonding is limited to short cables.

2.3.4 Material Properties

Table 2.1 shows appropriate values for materials used in insulated cable designs [43, 44].

Conductors. Stranded conductors need to be modelled as massive conductors. The resistivity should be increased with the inverse of the fill factor of the conductor surface so as to give the correct resistance of the conductor. The resistivity of the surrounding ground depends strongly on the soil characteristics, ranging from about $1\,\Omega\,m$ (wet soil) to about $10\,k\Omega\,m$ (rock). The resistivity of sea water lies between 0.1 and $1\,\Omega\,m$.

Insulations. The relative permittivity of the main insulation is usually obtained from the manufacturer. The values shown in Table 2.1 were measured at power frequency. Most extruded insulations, including XLPE and Polyethylene (PE), are practically lossless up to $1\,MHz$, whereas paper-oil type insulations exhibit significant losses also at lower frequencies. The losses are associated with a permittivity that is complex and frequency-dependent:

$$\varepsilon_r(\omega) = \varepsilon_r'(\omega) - j\varepsilon_r''(\omega) \qquad \tan \delta(\omega) = \frac{\varepsilon_r''}{\varepsilon_r'} \tag{2.13}$$

where $\tan \delta$ is the insulation loss factor. ATP CC routine does not allow entering a frequency-dependent loss factor, so a constant value has to be specified. However, this could lead to

Table 2.1 Resistivity of conductive materials.

Cable section	Property	Material and values	
Conductors	Resistivity ($\Omega\,m$)	Copper	1.72E-8
		Aluminium	2.83E-8
		Lead	22E-8
		Steel	18E-8
Insulation layers	Relative permittivity	XLPE	2.3
		Mass-impregnated	4.2
		Fluid-filled	3.5
Semiconducting layers	Resistivity ($\Omega\,m$)	<1E-3	
	Relative permittivity	>1000	

non-physical frequency responses which cannot be accurately fitted by frequency-dependent transmission line models. Therefore, the loss-angle should instead be specified as zero. Breien and Johansen fitted the measured frequency response of insulation samples of a low-pressure fluid-filled cable in the frequency range 10 kHz–100 MHz; according to their results, the frequency-dependent permittivity causes additional attenuation of pulses shorter than 5 μs [47].

Semiconducting materials. The main insulation of high-voltage cables is frequently sandwiched between two semiconducting layers. The electric parameters of semiconducting screens can vary between wide limits. The values shown in Table 2.1 are indicative for extruded insulation. The resistivity is required by norm to be smaller than $1E-3\,\Omega$ m. Semiconducting layers can in most cases be taken into account by using a simplistic approach [43, 44].

2.3.5   Discussion

High-frequency cable transients essentially propagate as decoupled coaxial waves between cores and sheaths [42, 48], so the transient behaviour of the cable is sensitive to the modelling of the core, main insulation, semiconductors, and metallic sheath. The sensitivity of coaxial waves can be summarized as follows [43, 44]:

1) Increasing the core resistivity increases the attenuation and slightly decreases propagation velocity.
2) Increasing the sheath resistivity (or decreasing the sheath thickness) increases the attenuation.
3) Increasing the insulation permittivity increases the cable capacitance, and this decreases velocity and surge impedance.
4) With a fixed insulation thickness, adding semiconducting screens increases the inductance of the core-sheath loop without changing the capacitance. This decreases velocity and increases surge impedance.
5) Cable sheaths are normally grounded at both ends, so the potential along these conductors is low compared to that of the core conductors, even in transient conditions. Consequently, the transients on phase conductors are insensitive to the specified properties of insulating materials external to the sheath.
6) The magnetic flux external to the sheath is small at frequencies above which the penetration depth is smaller than the sheath thickness. High-frequency transients are not very sensitive to ground and to the conductors external to the sheaths. The shielding effect increases with a decreasing sheath resistance.
7) Some care is needed when modelling armoured SC cables at low and intermediate frequencies as the return path of each coaxial mode divides between the sheath and the armour. This makes the propagation characteristics sensitive to the armour model (and to the separation distance between the sheath and armour). The armour permeability is then an important parameter. In addition, the armour/pipe must be accurately represented during ground faults since it can strongly affect the zero-sequence impedance of the cable and thus the magnitude of the fault current.

2.3.6   Cable Constants/Parameters Routines

As mentioned above, ATP users obtain insulated cable parameters and models by means of two supporting routines (CC versus CP) [13]. Users enter geometrical and physical parameters, as well as the grounding conditions, of the cable and select the desired model. The following information can be provided by the routines: the shunt admittance matrix and the series impedance matrices per unit length at a given frequency; resistance, inductance, conductance, and capacitance matrices per unit length at a given frequency; the characteristic impedance matrix in phase variables; voltage and current transformation matrices; modal quantities (i.e. attenuation constant, propagation velocity, series impedance and shunt admittance per unit length, characteristic impedance, and characteristic admittance for each mode). Cable matrices can be provided only for the system of physical conductors.

ATP routines allow users to request the following models [13]: (i) lumped-parameter PI circuit at the specified frequency; (ii) frequency-dependent distributed-parameter model, fitted for a given frequency range. Remember that the user can specify either a cross-bonded or a non-cross-bonded design. Additionally, the routines can also be used to obtain overhead line parameters and equivalent circuits.

However, actual cable designs can be different from those assumed by CC/CP routines:

- CC/CP routines assume that the relative permittivity of each insulating layer is real and frequency independent; therefore, any relaxation phenomenon in the insulation is neglected.
- CC/CP routines do not allow users to directly specify the semiconducting layers, which must be introduced by a modification of the input data. Semiconducting layers, which are in contact with a metallic conductor, can be taken into account by replacing the semiconductors with the insulating material of the main insulation, and increasing the permittivity of the total insulation so that the electric capacitance between the core and the sheath remains unchanged. The validity of this approach has been verified by measurements up to at least 1 MHz [43]. For a rigorous treatment of semiconducting layers see [49].
- Cable routines take into account the skin effect in the conductors, but neglect the proximity effect between parallel cables; i.e. a cylindrically symmetrical current distribution is assumed in all conductors and the helical winding effect of the wire screen is not taken into account.

Users have to decide how to represent the core stranding, the inner semiconducting screen, the outer semiconducting screen or the wire screen (sheath). Guidelines aimed at adapting actual cable data to the requirements of CC/CP routines were provided in [43, 44].

When preparing cable data, users should care about the following aspects:

1) Supporting routines do not directly apply to SC cables with semiconducting screens, so a conversion procedure is needed before entering the cable data.
2) In addition, these routines do not take into account any additional attenuation at very high frequencies resulting from the semiconducting screens. A correct modelling of semiconducting screens of SC coaxial type cables is important; a careless model can have too low surge impedance and a too high propagation velocity. The effect is strongly dependent on the transient.
3) The nominal thickness of the various cable screens can be smaller than those found in actual cables. This can result in an error for the propagation characteristics of the cable model.

2.4 Transformers

2.4.1 Overview

There are many transients in power systems (e.g. connection and disconnection of transformers, resonance and ferroresonance, lightning or switching surges travelling towards a transformer) for which the transformer response is dominant and an accurate transformer model is crucial. Some of these transients require proper modelling of the nonlinear behaviour of the transformer core caused by magnetization and hysteresis. Others require adequate representation of the frequency dependence of the leakage and/or magnetizing parameters. Very high-frequency transients require the accurate representation of all capacitances: to ground, between windings, and inter-turn [50].

Although there is enough theoretical background for the satisfactory modelling of the transformer subjected to all kind of transients, very often it is not possible to obtain the necessary data to compute the model parameters: some knowledge of internal construction details is required, but geometrical data is only available to manufacturers. Several factors make transformer modelling a difficult task: transformer's behaviour is nonlinear and frequency-dependent, many topological variations on core and coil construction are possible (see Figures 2.13 and 2.14), and there are many physical attributes whose behaviour may need to be correctly represented (self- and mutual-inductances between coils, leakage fluxes, skin effect and proximity effect in coils, magnetic core saturation, hysteresis and eddy current losses

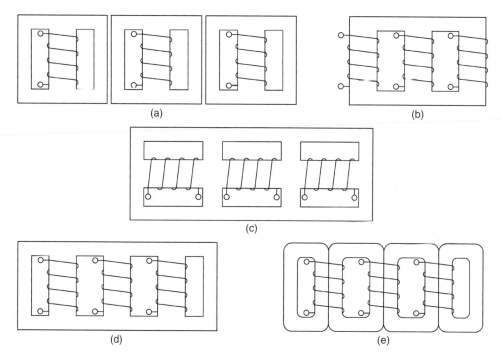

Figure 2.13 Three-phase core designs. (a) Triplex core. (b) Three-legged stacked-core. (c) Shell core. (d) Five-legged stacked-core. (e) Five-legged wound-core.

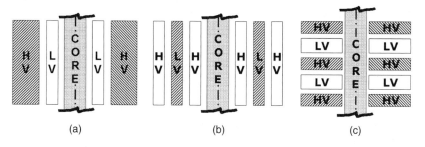

Figure 2.14 Winding designs. (a) Concentric design. (b) Interleaved design. (c) 'Pancake' design.

in core, capacitive effects) [50]. Transformer energization is a good example that illustrates difficulties related to transformer modelling [51]: before the flux penetrates the ferromagnetic core, the inductance is basically that of an air core and losses are basically originated in conductors and dielectric; after the flux has penetrated the core completely, the inductance becomes that of an iron core, and losses occur in conductors, core, dielectric and transformer tank.

ATP transformer models are suitable only for low- and mid-frequency transients; i.e. phenomena below the first winding resonance (i.e. few kHz). The transients for which these models can be applied include most line switching, capacitor switching, harmonic interactions, ferroresonance, and controller interactions. When subjected to fast-front or very fast-front surges, the behaviour of a transformer is very different from that under low- or mid-frequency transient surges, since magnetic flux no longer penetrates in the core, skin effect takes place, and proximity losses cannot be neglected. Simulation of fast-front transients (e.g. those caused by lightning) should be based on a detailed winding model that is usually not available as a standard model. This type of model can be developed as a standalone model using the transmission line theory (for a strict distributed-parameter representation) or a state space representation (using a ladder connection of lumped-parameter segments). However, the assembly of such a model requires a high degree of expertise by the user, as well as substantial knowledge of the transformer construction details. The latter is not typically available except to the transformer manufacturer. Another concern is on the internal resonance which occurs when one or more frequencies of the input surge are equal to some of the resonance frequencies of the transformer. The inter-turn insulation is particularly vulnerable to high-frequency oscillation, so estimating the distribution of inter-turn overvoltages is of essential interest.

Transformer models can be classified into three categories [52]. A summary of the scopes, limitations, and advantages of the models that belong to each category is given below.

A. *Black-box models* take advantage of mathematical identification methods to fit the terminal behaviour of transformers with respect to field measurements (either time domain and/or frequency-domain data) [53, 54]. These models accurately replicate the terminal response of the transformer under study. However, this approach does not provide information regarding the internal behaviour of the transformer, the models cannot be used for all operating conditions, and different black-box models need to be produced for different situations. These models have been generally applied for mid- and high-frequency applications.

B. *White-box models* are usually built from basic electrical components that correspond to physical parts of the transformer structure (i.e. internal geometry) and have a physical meaning [55–58]. Their main drawback is the need of the actual transformer dimensions and design details. This information is proprietary and not usually available; therefore, their construction outside the laboratory or the manufacturing firm is not usually possible. They are mainly suitable for manufacturers or transformer designers. Their major advantage is that they are useful for studies in where a detailed analysis of internal overvoltages, the estimation of the magnetic flux distribution, or the prediction of eddy currents and losses in different regions of the transformer geometry are crucial. These models can give an accurate understanding of the transformer behaviour at different operating conditions.

C. *Grey-box models* are a compromise between black- and white-box models [59–62]; they may be fairly accurate and retain some physicality. The topology and the structure components are derived physically as for white-box models; however, the model parameters are estimated from terminal measurement data, such as saturation inductance measurements, as for black-box models. The grey-box modelling overcomes the limitations associated with access to the transformer's construction and material information; the main challenge is to estimate the parameters from terminal measurements.

Table 2.2 Modelling guidelines for transformers.

Parameter/effect	Low-frequency transients	Slow-front transients	Fast-front transients	Very fast-front transients
Short-circuit impedance	Very important[a]	Very important	Important	Negligible
Saturation	Very important[b]	Important	Negligible	Negligible
Iron losses	Important[c]	Negligible	Negligible	Negligible
Eddy currents	Very important	Important	Negligible	Negligible
Capacitive coupling	Negligible[d]	Important	Very important	Very important

a) Unimportant for ferroresonance. However, it may have an effect on the frequency of the upstream network, especially critical under 300 Hz (harmonics range).
b) Unimportant for most control interaction cases, harmonic conditions not caused by saturation, and other non-saturation cases.
c) Only for resonance phenomena.
d) Capacitances can be very important for some ferroresonance cases.

Since not every characteristic of the transformer plays a role in all transients, the modelling guidelines given in Table 2.2 are usually recommended. The table, which shows the importance of some parameters and effects when modelling a transformer for a specific frequency range, is a modified version of that proposed in [3]. Details on modelling guidelines for transformers have also been presented in [3–5, 50, 52, 63–66]. This section provides an overview of transformer models, and has been divided into two subsections dedicated to summarize transformer modelling for analysis of low- and high-frequency transients.

2.4.2 Transformer Models for Low-Frequency Transients

2.4.2.1 Introduction to Low-Frequency Models

A unified and generally accepted model for transformers capable of reproducing their behaviour under all low-frequency operating conditions does not exist. For this range of frequencies, transformer models can be classified into several groups [1, 63, 64].

Matrix representation: The steady state equations of a single-phase multi-winding transformer can be expressed by an impedance equation [67]:

$$\mathbf{V} = \mathbf{ZI} \tag{2.14}$$

The formulation can be extended to three-phase transformers by replacing any element of $\mathbf{Z}$ by a (3×3) submatrix.

For transient calculations, the equation must be rewritten in the following form:

$$\mathbf{v} = \mathbf{Ri} + \mathbf{L}\frac{d\mathbf{i}}{dt} \tag{2.15}$$

where $\mathbf{R}$ and $j\omega\mathbf{L}$ are respectively the real and the imaginary part of the impedance matrix.

In case of a very low excitation current, the transformer should be described by an admittance formulation:

$$\mathbf{I} = \mathbf{YV} \tag{2.16}$$

For transients simulations this expression becomes:

$$\frac{d\mathbf{i}}{dt} = \mathbf{L}^{-1}\mathbf{v} - \mathbf{L}^{-1}\mathbf{Ri} \tag{2.17}$$

Both approaches include phase-to-phase couplings and terminal characteristics, but they do not consider differences in core or winding topology. These models are linear and theoretically valid only for the frequency at which the nameplate data was obtained, although they are reasonably accurate for frequencies below 1 kHz. For simulation of saturable cores, excitation is externally attached at the model terminals in the form of non-linear elements; such core model is not always topologically correct, but good enough in many cases.

Saturable transformer component (STC). A single-phase multi-winding transformer model can be based on a star-circuit representation, see Figure 2.15 [12, 13]. Its application to three-phase core units can be made through the addition of a zero-sequence magnetizing inductance. This model is of limited application, even for single-phase units.

An important aspect is the node to which the magnetizing inductance is connected. It is unimportant if the inductance is not saturated, because of its high value; but it can be important if the inductance is saturated. The star point is not always the correct topological connecting point.

Topology-based models. They can accurately represent any type of core design and can be derived using two different approaches: (i) models based on a formulation similar to that of Eq. (2.15) [68]; (ii) models based on the principle of duality. Duality models include the effects of saturation in each individual leg of the core, inter-phase magnetic coupling, and leakage effects [63, 64, 69–72]; in the equivalent magnetic circuit, windings appear as magnetomotive force (MMF) sources, leakage paths appear as linear reluctances, and magnetic cores appear as saturable reluctances. The mesh and node equations of the magnetic circuit are duals of the electrical equivalent node and mesh equations, respectively. Winding resistances, core losses, and capacitive coupling effects are not obtained directly from the transformation, but can be added to the equivalent circuit. Reference [65] classifies these models into two groups: models derived from the magnetic circuit (i.e. the principle of duality is applied once the magnetic circuit has been built) and models derived from the direct application of the principle of duality (i.e. the principle of duality is applied from the geometry of the transformer without drawing the reluctance circuit). The main difference between topology-based models and matrix-based models is the way in which core and winding models are assembled: in the matrix formulation the location of the core model in the equivalent circuit is arbitrary, which may lead to wrong results for applications where an accurate representation of the saturation is mandatory; in duality-based models it is clear where the core should be connected, and each inductor can be related one-to-one with a section of the core, so the nonlinear characteristics of the core are properly considered.

Hybrid models. They are a combination of duality-based and matrix-based models, and incorporate the capacitive effects and the frequency dependence of resistances [61]. These models are topologically correct and include leakages between core and coils. That is, they represent the actual internal core and coil arrangements, and can be universally applied, regardless of the winding connections.

2.4.2.2 Single-Phase Transformer Models

Figure 2.16 shows the two-winding single-phase transformer T-equivalent circuit. In this circuit R_H and R_L are the series resistances, which include the Joule losses and eddy current losses in the windings (when data is available); L_H and L_L represent the leakage inductance (or series inductance), which has been divided among the two windings; R_m and L_m describe the core behaviour. Parameters to be specified in this model are usually derived from nameplate or test data. Excitation and short-circuit currents, voltages and losses, must be provided from both

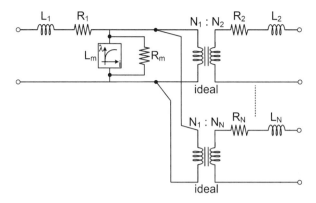

Figure 2.15 Star-circuit representation of single-phase N-winding transformers.

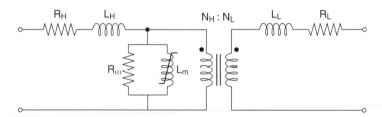

Figure 2.16 Traditional equivalent circuit of a single-phase core.

positive- and zero-sequence measurements. Standards recommend procedures for measuring the above values and provide specifications and requirements for conducting tests.

Although most textbooks use this model to describe the behaviour of a single-phase transformer, it is not topologically correct: except for the winding resistances, the other circuit elements cannot be related in a topological sense to specific physical regions of the transformer. The T-model is relatively easy to implement, but its main disadvantage is the quasi-arbitrary division of the leakage inductance, since it cannot be divided into two (winding) inductances while keeping a physical meaning. The division of the leakage inductance may also result in a wrong location of the core model in the equivalent circuit.

A topologically-correct model of a two-winding single-phase transformer for low- and mid-frequency transients is the so-called PI model shown in Figure 2.17 [52, 73, 74]. In this model the internal elements represent physically the magnetic circuit. There is only one leakage inductance in the middle (L_{sh}) and two magnetizing branches, which can have different values. The leakage impedance L_{sh} is linear while the magnetizing branches are nonlinear; all of them are frequency dependent. As with any duality-derived model, the parameters can be accurately estimated when the design dimensions are known.

Winding model

The series parameters of the above circuits represent the transformer winding and may be computed from the short-circuit test. For concentric winding designs, the inner winding has smaller reactance than the outer winding; most often the inner winding is the lower-voltage winding; see Figure 2.16. In Figure 2.17, the leakage inductance is divided among windings: when saturation is neglected, the division between primary and secondary inductance is arbitrary; when the core saturates, the PI model should be selected since an arbitrary division can provide wrong results. To obtain more accuracy, winding design details are required.

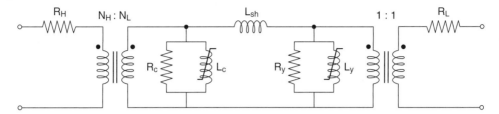

Figure 2.17 PI-shaped model for a single-phase transformer.

Core model

A usual representation is a parallel combination of a nonlinear inductance L_m, representing magnetic core saturation, and a constant resistance R_m, representing core losses. Although this model is accurate enough for many studies, more sophisticated models are mandatory for some transients. Transformer saturation is an important component of many low-frequency electromagnetic transient phenomena (e.g. ferroresonance, transformer energization leading to inrush currents); in general, saturation needs to be included in transients involving high flux. The total flux in the iron core during an energization is the sum of two fluxes: the residual and the forced flux. For most phenomena, the critical transformer saturation parameters are the slope (air-core inductance), the zero-current intercept of the saturation curve, the nominal flux and the corresponding excitation current, and the location of the saturation representation in the transformer model topology, see Figure 2.18.

Hysteresis is a nonlinear, history- and frequency-dependent phenomenon; see Figure 2.19. Iron core losses in grain-oriented steel, the predominant core material, can be divided into three general categories [52]: static hysteresis (which account for about 40% of the total), classical eddy current (about 20%), and excess losses (about 40%) [75, 76]. The term excess loss reflects the fact that it is anomalous from the viewpoint of the classical loss theory [77], which is mainly applied to non-oriented steels used in generators and motors. Based on how the iron core models account for the loss components, two types of hysteresis models exist: static and dynamic [52]. The hysteretic curve may be built using the following additional parameters: (i) the iron core losses, including hysteresis losses described by the area of the cycle, (ii) the coercitive current, which may be given by simple formulas, and (iii) the residual flux at the intercept of zero current.

Saturation with or without hysteresis can be incorporated into a transformer model using test data curves or estimating the key parameters from transformer geometry. It is important to take into account that the exciting current includes core loss and magnetizing components, and manufacturers usually provide rms currents, not peak. Remember that

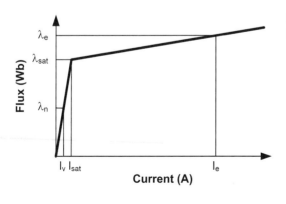

Figure 2.18 Description of the saturation curve of a transformer (main parameters).

Figure 2.19 Hysteresis curve.

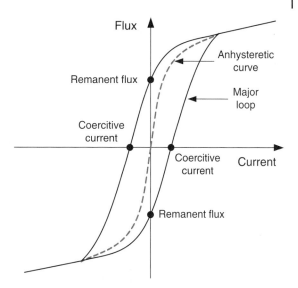

winding capacitances can affect low-current data and hysteresis biases saturation curve (i.e. cobra-shaped curves [50]). The anhysteretic saturation characteristic can usually be modelled by a piecewise linear inductance with two slopes, since increasing the number of slopes does not always improve the accuracy. However, there are some cases (e.g. ferroresonance), for which a more detailed representation may be required (i.e. the inductance should be represented by means of a characteristic curve with more slopes or by assuming hysteresis). Core losses can also be critically important in some phenomena involving saturation (e.g. ferroresonance).

Eddy currents

Eddy currents are induced in windings and core when the magnetic flux changes with time. In steady state, the eddy currents are undesirable because they produce losses. However, when a transformer is subjected to a transient the induced eddy currents are beneficial because they add damping, so the proper modelling of this damping can be important.

Eddy currents in windings. Winding losses are frequency dependent. Winding resistances must incorporate an ac component due to eddy currents in windings, skin effect, and stray losses. Figure 2.20 shows a series-parallel *RL* circuit (Foster equivalent) that can be used for an accurate representation of the winding resistance and the leakage inductance in low-frequency and switching transients. To obtain parameters of such circuit, a frequency response test must be performed and a fitting procedure applied [57, 78]. If such information is not available, then effective resistance approximations presented in the literature could be considered. A correction of the resistance value to account for temperature effect should be also considered. The series Foster equivalent circuit is only a terminal model. The elements of the circuit are not related to a section or region of the winding in a dual way. The elements of a

Figure 2.20 Simplified frequency-dependent representation of winding.

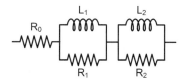

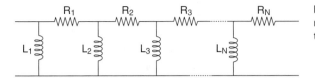

Cauer equivalent circuit to represent the eddy current effects can be related to currents and magnetic fluxes in the duality sense [79].

Eddy currents in core. The magnetization curves presented above are only valid for slowly varying phenomena since it has been assumed that the magnetic field can penetrate the core completely. For high frequency, this is not true. A change in the magnetic field induces eddy currents in the iron. Consequently, the flux density will be lower than that given by the normal magnetization curve. For high frequencies, the flux will be confined to a thin layer close to the lamination surface, whose thickness decreases as the frequency increases. This indicates that inductances representing iron path magnetization and resistances representing eddy current losses are frequency dependent [79]. The circulation of these eddy currents introduces additional losses. To limit their influence, a transformer core is built up from a large number of parallel laminations. Eddy current models intended for simulation of the frequency dependence of the magnetizing inductance, as well as losses, can be classified into two categories obtained, respectively, by the realization of the analytical expression for the magnetizing impedance as a function of frequency, and by subdivision of the lamination into sub-laminations and the generation of their electrical equivalents. Computationally efficient models have been derived by synthesizing a Foster or a dual Cauer equivalent circuit to match the equivalent impedance of either a single lamination or a coil wound around a laminated iron core leg [78]. The accuracy of the standard Cauer representation (see Figure 2.21) over a defined frequency range depends on the number of terms retained in a partial fraction expansion and, therefore, on the number of sections. To represent the frequency range up to 200 kHz with an error of less than 5%, only four terms are required [80]. The first section governs its characteristics at frequencies up to a few kilohertz; each subsequent section comes into play as the frequency range of concern increases.

Tank model

Often a significant amount of flux will find a path through the tank and/or magnetic tank shunts under core saturation conditions. The presence of the tank changes the value of the saturation inductance. The transformer model does not have to be changed when the tank exists, but the saturation needs to be properly computed [52]. The saturation inductance increases with the tank when it is not saturated: the magnetic flux closes through the tank with higher permeability than air; the flux paths change in the presence of the tank (i.e. the mean flux path lengths decreases), so if the tank saturates, the magnetic flux passes through the walls as if it was air. Therefore, the terminal saturation inductance when the tank is saturated is equivalent to the saturation inductance without tank. These circumstances are valid when transformers tanks have shields to prevent eddy current effects: if the magnetic effects of the induced eddy currents are taken into account, the final slope of the terminal saturation inductance may be even lower than the air-core inductance because the equivalent inductance includes now (one or more) short-circuited inductors in series [52].

2.4.2.3 Three-Phase Transformer Models

Three-phase core configurations have been separated into two subsections, assuming that the transformer model to be used in transient simulations can be linear or must incorporate nonlinear behaviour, respectively; see [12, 50, 52, 63–65].

Matrix representation. This representation is usually derived from standardized tests, (i.e. short-circuit and no-load tests). In order to get model parameters directly from those tests a supporting routine, BCTRAN, can be used [13, 67]; it can be applied to both single- and multi-phase multi-winding transformers. This model is based on measurements made at steady state (50/60 Hz), so it may only give accurate information for low-frequency transients. The elements of the impedance matrix **Z**, see (2.14), can be derived from the open- and short-circuit tests. Excitation losses are not included in **Z**, but they can be added as shunt resistances across one or more windings. Winding resistances are obtained from the losses computed during the short-circuit tests at power frequency. There could be some accuracy problems with the above calculations when exciting currents are low or ignored because in such cases the matrix can become singular. To solve those problems an admittance matrix representation should be used [12, 67]. A matrix representation has a very simple usage, as it is based on nameplate data; however, the models are linear and theoretically valid only for the frequency at which the nameplate data was obtained. Nonlinear behaviour can be incorporated by attaching the core representation at the model terminals. The real part of the elements of the matrix can be negative in certain cases [50]. Other aspects discussed in the original reference are the modifications to be used with delta-connected transformers [67].

Topology-based models. Several difficulties arise when modelling and simulating three-phase core configurations: three-phase transformers have magnetic coupling between phases; four- and five-legged transformers, as well as shell-type transformers, have low reluctance for zero sequence fluxes, as they can circulate directly through the core. Three-legged core type transformers have a high reluctance path for zero sequence fluxes, which close through the air and the transformer tank. Different approaches have been proposed to date for representation of the same core configurations. Some differences exist between models for representing the same core configuration, even when the same principle (e.g. duality) is used. As a consequence, different procedures for parameter estimation have been proposed for models of the same core configuration. Tests to obtain some characteristics values to be specified in several models are not covered by present standards. Capacitances between terminals and ground (core and tank), between windings and between phases could be added to the equivalent circuit. Dividing the flux paths into two parts results in a more symmetric and convenient connection for the core equivalent. In the case of unbalanced excitation that includes zero sequence, the total fluxes linked by the three sets of coils on the phase legs will not add to zero, and the zero-sequence flux will circulate through the surrounding oil and air and through fittings and tank walls. Following the assumptions made in the development of the equivalent circuit, this flux is distributed across the top of the coils on the three legs [64]. Since the tank affects the behaviour of three-phase transformers under the existence of zero-sequence currents and over-excitation, an accurate tank model needs to represent the correct behaviour for balanced and unbalanced over excitation conditions; the model should properly represent the induced currents from nearby conductors and from conductors that cross the structure. For more details on this effect, see [52].

Hybrid models. They are a combination of duality-based and matrix-based models [61]. This approach is topologically correct and can be applied regardless of the winding connections.

2.4.3 Transformer Modelling for High-Frequency Transients

2.4.3.1 Introduction to High-Frequency Models

Transformer windings may be subjected to high-frequency waves arising from switching operations, lightning discharges, and other changes in the operating conditions of the system

[50, 51, 66]. Experience shows that transient overvoltages are not only dangerous because of their amplitude, but also because of their rate of rise, so frequent overvoltages with low amplitude and higher rate of rise can be as dangerous as overvoltages with higher amplitude.

Electromagnetic transients in transformers due to high-frequency waves (i.e. steep-fronted waves) are commonly studied using *internal* models, which consider the propagation and distribution of the incident impulse along the transformer windings, and *terminal* models, which consider the response of the transformer from its terminals and may also permit the calculation of transferred voltages [50].

A terminal model can be represented by a circuit that interconnects the different terminals of the transformer with as many nodes as terminals, plus one to represent ground. The equations of such circuit can be expressed in the frequency domain in terms of either its admittance or impedance matrix as follows:

$$\mathbf{I}(\omega) = \mathbf{Y}(\omega)\mathbf{V}(\omega) \quad \Rightarrow \quad \mathbf{V}(\omega) = \mathbf{Z}(\omega)\mathbf{I}(\omega) \qquad (\mathbf{Z}(\omega) = [\mathbf{Y}(\omega)]^{-1}) \tag{2.18}$$

where $\mathbf{Y}(\omega)$ and $\mathbf{Z}(\omega)$ are respectively the admittance and the impedance matrices, $\mathbf{I}(\omega)$ is the vector of currents injected into the nodes, and $\mathbf{V}(\omega)$ is the vector of voltages between nodes and ground.

For system studies, it is sufficient to model the component as a black-box model: its terminal impedance characteristic is matched within the frequency range of concern. However, when the internal transient response is required, it is necessary to use a much more detailed model in which all regions of critical dielectric stress can be identified. Internal transient response is a result of the distributed electrostatic and electromagnetic characteristics of the windings. The transient waves propagate into the winding with a certain velocity, the winding has a certain wave transit time, and the wavefront of the transient can be regarded as being distributed along a length of the winding. For a steep-fronted voltage surge, most of the wave front will reside across the first few turns, which can be overstressed. The wave front slopes off and the amplitude is attenuated as the wave penetrates along the winding because of damping due to the eddy currents. For all practical winding structures, this phenomenon is quite complex and can only be investigated by constructing a detailed model and carrying out a numerical solution for the transient response and frequency characteristics in the regions of concern.

Internal models for high-frequency transients can be described either by a distributed-parameter representation or as a ladder connection of lumped-parameter segments [50, 66]. Proper choice of the segment length for lumped-parameter modelling is fundamental. Analysis of steep-fronted transients (in the order of dozens or hundreds of kHz) using one segment per coil of the winding can be sufficient, whereas very fast front transients (in the order of MHz) may require considering one segment per turn.

Surges affecting one of the transformer windings can give rise to overvoltages in the other windings. The analysis of this transference phenomenon can also be of importance at the design stage of the winding insulation. An equivalent network for a multi-winding transformer, in which the conventional ladder network used for a single winding is extended for multiple windings, permits the analysis of the transferred voltage to other windings to which the impulse is not directly applied. However, surge voltage transfer can also be analysed by means of terminal (black-box) models. Black-box models have been used for calculating the interaction between a transformer and the system or the transferred voltages to other transformer windings [53, 54, 66, 81, 82].

In general, at high frequencies, the core acts as a flux barrier, the flux does not penetrate in the core, and the iron core losses can be neglected accordingly; that is, the core inductance is considered to behave as a linear element. However, the iron core losses influence the frequency transients up to 1 MHz, so the flux penetration dynamics in the core should be taken into account for fast-front transients, mainly those due to switching with frequencies below 100 kHz [3].

The frequency dependence of winding parameters has to be considered. Skin and proximity effects produce frequency dependence of winding and core impedances because of the reduced flux penetration. At very high frequencies, the conductance representing the capacitive loss in the winding's dielectric also depends on frequency.

2.4.3.2 Models for Internal Voltage Calculation

A transformer winding behaves as a distributed-parameter multi-conductor transmission line (MTL) when it is subjected to high-frequency surges [83]. Although it would be ideal to compute voltages between turns by representing each turn as a separate line, transformers are normally manufactured with a great number of turns, so such approach can be time consuming. On the other hand, a very detailed representation of every turn is not required for many practical cases. A much simpler representation can be obtained by lumping elements. An alternative approach when all turns and coils of the transformer winding must be included may be based on the application of a single-phase transmission line (STL) model in which each coil is considered as a single-phase distributed-parameter line.

Distributed-parameter models. Figure 2.22 shows the diagram of a coil with n turns and the schematic representation of a differential length segment for this representation. This model

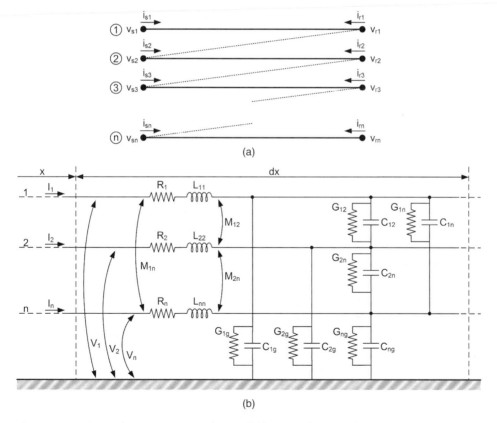

Figure 2.22 Multi-conductor transmission line model for a transformer coil. (a) Multi-conductor transmission line model. (b) Differential section of a multi-conductor transmission line.

takes into account the distributed nature of winding parameters and the coupling between turns, and includes resistances and conductances required to represent the various types of losses. The formulation of an MTL-based model in the Laplace domain can be expressed by means of the telegrapher's equations as follows:

$$\frac{d\mathbf{V}(x,s)}{dx} = -\mathbf{Z}(s)\mathbf{I}(x,s) \tag{2.19a}$$

$$\frac{d\mathbf{I}(x,s)}{dx} = -\mathbf{Y}(s)\mathbf{V}(x,s) \tag{2.19b}$$

where $\mathbf{Z}$ and $\mathbf{Y}$ are the n x n matrices of series impedances and shunt admittances per unit length, n is the number of conductors (discs or turns), $\mathbf{V}(x, s)$ and $\mathbf{I}(x, s)$ are the vectors of voltages and currents at point x of the winding. According to Figure 2.22a, the end of each turn is connected to the beginning of the next turn, resulting in a zig-zag connection. An impedance connected at the end of the n-th element in Figure 2.22a can be used to represent either the neutral impedance or the remaining part of the winding, when only a section of the winding is modelled in detail.

Lumped-parameter models. The model shown in Figure 2.22b can be reduced by lumping series elements within a turn and shunt elements between turns as shown in Figure 2.23a, in which

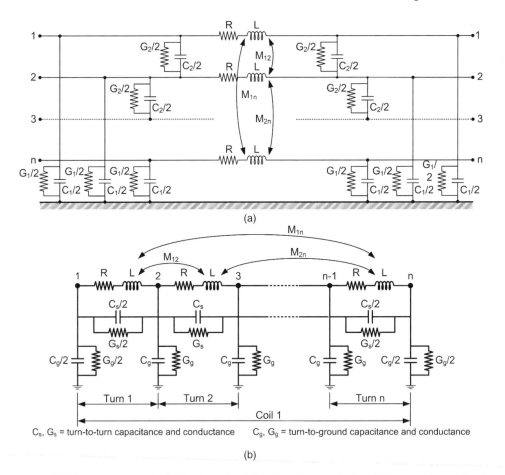

Figure 2.23 Lumped-parameter ladder-type circuit for a transformer winding. (a) Lumped-parameter PI model of a coil section. (b) Equivalent lumped-parameter circuit of a coil or winding.

Figure 2.24 Equivalent circuit per unit length of a transformer winding.

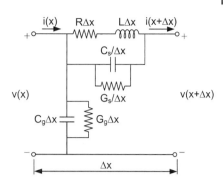

only capacitances between adjacent turns are considered. Taking into account the connection between turns indicated in Figure 2.22a, a single turn can be represented by a series inductance with mutual inductances between turns, and parallel and series capacitances arranged as in Figure 2.23b. Note that the turn-to-turn capacitance has been lumped in parallel with the inductance, while the ground capacitance has been lumped and halved at each end of a turn. The coil can be then represented by as many circuit blocks as there are turns. Further order reduction can be achieved by lumping parameters within a coil model [84]. The resulting model is a series of circuits with mutual magnetic couplings similar to that displayed in Figure 2.23b, in which each segment represents several turns or even a complete coil. Proper choice of the segment length for lumped-parameter modelling is fundamental. Analysis of fast-front transients (in the order of hundreds of kHz) using one segment per coil of the winding can be sufficient, whereas very fast-front transient analysis (in the order of MHz) can require considering one segment per turn.

STL theory. A representation that can be derived from the circuit shown in Figure 2.23b may assume that parameters are distributed. If the coupling between circuit segments is neglected or included in the inductance elements, the resulting representation for a differential segment could be that shown in Figure 2.24 [50, 66, 85]. This circuit model is similar to that of a STL in which parameters per unit length are defined as follows: L is the series inductance of the winding, R is the loss component of L, C_s is the series (turn-to-turn) capacitance of the winding, G_s is the loss component of C_s, C_g is the turn-to-ground capacitance of the winding, G_g is the loss component of C_g.

Combined STL and MTL models. If all turns and coils of the winding might need to be considered in the study, a MTL-based model would result in very large model and a significant computational effort. This problem can be addressed by combining the STL and MTL-based models described above [84, 86]: (i) each coil is represented by a STL model, so that voltages at the coil ends can be obtained; (ii) each coil is represented by a MTL model to compute the distribution of the inter-turn overvoltages independently from the other coils.

Voltage transfer analysis. An impulse propagating along a winding is transferred to the other windings due to capacitive and inductive couplings. If a distributed-parameter model is chosen, the equivalent circuit for a differential segment of a two-winding transformer could be that shown in Figure 2.25. A lumped-parameter model of a two-winding transformer for analysis of internal voltage distribution and voltage transfer can be obtained by extending the ladder-type model to the second winding and adding inductive and capacitive coupling between elements of both windings [87, 88].

2.4.3.3 Terminal Models

A terminal model of a transformer is a representation seen from its terminals that can be used to analyse the performance of the system to which the transformer is connected, obtain the voltage

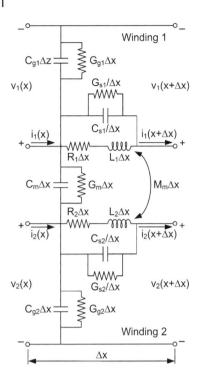

Figure 2.25 Equivalent circuit per unit length of a single-phase two-winding transformer.

developed between terminals in front of external stimulus, or estimate the voltage transfer to other phases or other windings.

Table 2.3 shows the models proposed by CIGRE for representing single-phase windings in fast-front and very fast-front transients [3]. The first row displays the models that can be used when only the interaction with the system is of concern, while the second row shows the models of a two-winding transformer when surge transfer is of concern. Note that all models are linear (i.e. saturation effects are neglected), which might not be entirely correct for the lowest frequency ranges, and they suggest that capacitances become crucial for steep-fronted stimuli. Since a winding is seen as a frequency-dependent impedance, it may be represented as a black-box model whose terminal impedance or admittance characteristic is matched within the frequency range of concern to the measured frequency response. Therefore, the terminal impedance and the corresponding circuit representation can be derived from the frequency response of the winding (which will provide a representation from all terminals), or from a circuit representation like that used to obtain the internal voltage distribution.

The models recommended in the table show that for very fast-front transients rather simple models can suffice. In general, when voltage transfer to other phase of the same winding or to other windings is not of concern, the model may consist of the surge impedance of the winding paralleled by a capacitance-to-ground. Most full frequency-dependent models are based on the fitting of the elements of the nodal admittance or impedance matrix, which represent the winding as seen from its terminals. These matrices depend on the internal connection of the windings (e.g. wye or delta). The relative complexity, accuracy, and numerical stability of a model is related to the choice of network representation. Several high-frequency transformer models have been derived from measurements [53, 82, 89].

Black-box models

A terminal model can be described in the frequency domain in terms of its admittance or its impedance matrices; see Eq. (2.17). Figure 2.26 schematizes the way in which the elements of

Table 2.3 Terminal models for transformers.

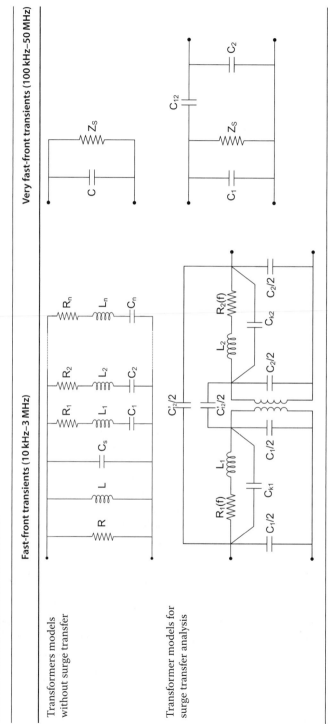

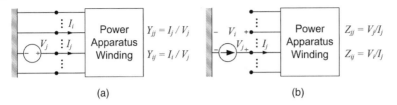

Figure 2.26 Measurement procedures for obtaining admittance and impedance matrices. (a) Admittance measurements. (b) Impedance measurements.

the two matrices can be derived. Consider, for instance, the admittance formulation. If a 1-pu voltage is applied at node j of the apparatus while the remaining terminals are short-circuited, the jth column of $\mathbf{Y}(\omega)$ will be equivalent to the currents measured from ground to each terminal. Applying this procedure, as presented in Figure 2.26, direct measurement of all elements of $\mathbf{Y}(\omega)$ can be completed. Alternatively, when a 1-pu current is injected at node j while the remaining terminals are open, the jth column of $\mathbf{Z}(\omega)$ will be equivalent to the voltages measured from each terminal to ground.

Two models are considered: the simplest one is based on a single input (i.e. voltage, current) and a single output (i.e. current or voltage), while the most complex approach can be also used for voltage transfer analysis between phases and between windings.

1. *Single-input single-output model.* The equivalent circuit of a single-phase winding can be derived from either the impedance-frequency plot or the admittance-frequency plot. If the admittance function is available, it can be reduced to a rational function with the following form [90]:

$$y(s) = \sum_{m=1}^{N} \frac{c_m}{s - a_m} + d + se \tag{2.20}$$

   For the realization of this function as a passive circuit, see [90].

2. *Multiple-input multiple-output model.* A complete model of a component can be obtained, irrespective of the number of phases and windings, as long as its frequency response is known from either measurements or calculations based on the physical layout. The idea originally proposed in [82] is to obtain an equivalent circuit whose nodal admittance matrix matches the nodal admittance matrix of the original component over the frequency range of interest. For a description of numerical methods for curve fitting and calculation of poles and residues of a rational function that matches a frequency response, see [77, 82].

Topological models
The classical power-frequency T-circuit for single-phase transformers can be used to model the interaction among windings belonging to the same phase (i.e. mounted on the same leg of the core): the model of the frequency-dependent series branch (short-circuit impedance) can be separated from the model of the shunt branches (constant stray capacitances and magnetizing branch), so the representation is independent of the external connection among windings (i.e. wye or delta) [91]. Stray capacitances are connected at the terminals of the circuit, so the internal part has exactly the same form as the conventional power-frequency model. The series impedances include the equivalent series resistance (current dependent losses) and the equivalent leakage inductance (self and mutual leakage flux) of the windings, in combination with part of winding-to-winding capacitance. The frequency-dependent response of this branch can be obtained from short-circuit test measurements, and the measured positive and zero sequence impedances can be synthesized with a network of constant *RLC* elements, following

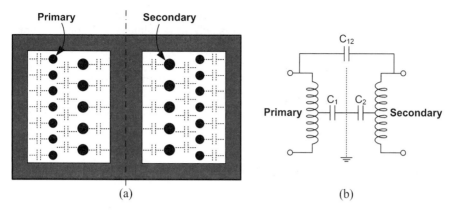

Figure 2.27 Capacitive coupling between primary and secondary windings of a transformer. (a) Schematic representation. (b) Equivalent circuit.

a procedure similar to those mentioned above. The value of the stray capacitances can be determined from experimental measurements. The model was developed under the assumption that the interaction between phases in a single-core transformer unit is symmetrical; this assumption neglects the effect of uneven capacitive and leakage-inductance coupling among phases mounted in the outer and central limbs of a three-phase core-type transformer, permits the use of a balanced-system transformation matrix that can be taken as real and constant to decouple the system at all frequencies, and facilitates the design of experiments to measure the parameters without having to open the internal connections among windings. The model was validated for a frequency range up to 100 kHz.

Terminal models for very fast-front transients
Under a steep-fronted voltage stimulus, a winding initially behaves like a capacitance. The capacitive elements also dominate the voltage transferred to other windings when a fast-rising surge impinges in a terminal. Field tests show that the value of the surge impedance of the transformer winding is about few thousands ohms; therefore, for practical purposes, the winding termination may be also represented as an open circuit [26]. These approaches (i.e. a capacitance or an open circuit) are commonly used in some transient studies, mostly those in which lightning surges are involved [8].

Other representations have been proposed [51]. Figure 2.27a is a schematic representation of a two-winding transformer in which capacitances between windings and capacitances to the core are indicated. As a first approximation, this forms the capacitive circuit shown in Figure 2.27b. The capacitances can be determined from the geometry of the coil and core structure. The windings and the core may be seen as a potential divider so that the voltage at the secondary winding can represent a fraction of the surge voltage, with no relationship to the turns ratio of the transformer.

2.5 Rotating Machines

2.5.1 Overview

The models available in ATP are adequate only for representing rotating machines in low-frequency transient studies (e.g. transient stability, subsynchronous resonance, load rejection, generator tripping, generator synchronization, inadvertent energization). However, coils and windings of rotating machines may be subjected to abnormal steep-fronted (high-frequency) voltage surges that may damage the windings. Under steady-state voltage conditions, the voltage distribution is uniformly distributed in the winding and the interturn voltages are low; under steep-fronted transient conditions, the voltage distribution is non-homogeneous and some interturn voltages can reach very high values. As for transformer windings, an accurate evaluation of these transients is required. This section has been divided into two main parts dedicated to summarize the rotating machine models that need to be considered for low- and high-frequency transients, respectively.

2.5.2 Rotating Machine Models for Low-Frequency Transients

2.5.2.1 Introduction

The behaviour of a rotating machine in low-frequency transients can be represented by the equations of its electrical and mechanical subsystems. Models available in ATP can be used for representing conventional rotating machines in low-frequency transients. The capabilities of the so-called Universal Machine can be used to simulate up to 12 different types of rotating machines [13]. Since ATP models have been traditionally used for representing three-phase induction (asynchronous) and synchronous machines only, this chapter summarizes exclusively the modelling approaches that can be considered for representing these machines [92]. Additional details about the capabilities of the rotating machine models available in ATP and the techniques implemented for interfacing machine models to the power system model will be provided in Chapters 4 and 7.

2.5.2.2 Modelling of Induction Machines

The induction machine equations

A simplified cross-sectional view of a conventional three-phase induction machine is shown in Figure 2.28, where the positive direction of the magnetic axes corresponds to the direction of flux linkages induced by the positive phase currents. The stator magnetic axes *as*, *bs*, and *cs* are stationary, symmetric, and 120 electrical degrees apart from each other. The rotor magnetic axes *ar*, *br*, and *cr* are also symmetric, 120 electrical degrees apart from each other, and can rotate with respect to the stator. The rotor position and speed are denoted by θ_r and ω_r, respectively.

A three-phase wye-connected induction machine may be represented by the coupled circuits shown in Figure 2.29. Although the rotor circuit may be short-circuited to represent a squirrel-cage induction machine, it is left open for representing machines with accessible rotor terminals (e.g. a doubly-fed induction machine) as a more general case. Using the motor

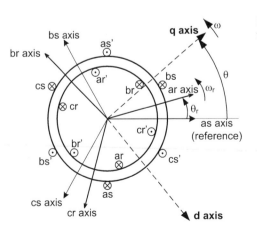

Figure 2.28 Simplified cross-sectional view of a three-phase symmetrical induction machine.

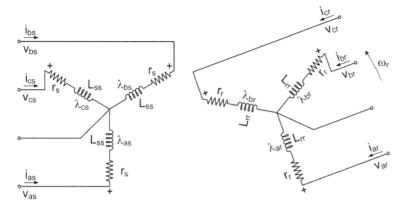

Figure 2.29 Stator and rotor coupled-circuits of a three-phase induction machine.

convention, stator and rotor currents flow into the machine when positive-polarity phase voltages are applied. For convenience, all rotor variables and parameters are assumed to be referred to the stator side using appropriate turns ratio. The stator and rotor windings are assumed to be magnetically-coupled, symmetric, and sinusoidally distributed. The figures also show the rotating reference frame quadrature and direct axes (i.e. q and d axes). Without loss of generality, the q-axis leads the d-axis by 90 electrical degrees, and the position of the reference frame is measured between the as and the q axes, and denoted by θ.

The voltage equations of the coupled circuit shown in Figure 2.29 may be expressed as [93]:

$$\begin{bmatrix} \mathbf{v}_{abcs} \\ \mathbf{v}_{abcr} \end{bmatrix} = \begin{bmatrix} \mathbf{R}_s & \mathbf{0} \\ \mathbf{0} & \mathbf{R}_r \end{bmatrix} \begin{bmatrix} \mathbf{i}_{abcs} \\ \mathbf{i}_{abcr} \end{bmatrix} + \frac{d}{dt} \begin{bmatrix} \boldsymbol{\lambda}_{abcs} \\ \boldsymbol{\lambda}_{abcr} \end{bmatrix} \tag{2.21}$$

where $\mathbf{R}_s$ ($= \mathrm{diag}(r_s\ r_s\ r_s)$) and $\mathbf{R}_r$ ($= \mathrm{diag}(r_r\ r_r\ r_r)$) are the diagonal 3×3 matrices of stator and rotor resistances, $\mathbf{v}_{abcs} = [v_{as}\ v_{bs}\ v_{cs}]^T$ is the vector of stator voltages, $\mathbf{v}_{abcr} = [v_{ar}\ v_{br}\ v_{cr}]^T$ is the vector of rotor voltages, $\mathbf{i}_{abcs} = [i_{as}\ i_{bs}\ i_{cs}]^T$ is the vector of stator currents, and $\mathbf{i}_{abcr} = [i_{ar}\ i_{br}\ i_{cr}]^T$ is the vector of rotor currents. For a conventional squirrel-cage short-circuited rotor, $\mathbf{v}_{abcr} = 0$.

The corresponding flux linkage equations are:

$$\begin{bmatrix} \boldsymbol{\lambda}_{abcs} \\ \boldsymbol{\lambda}_{abcr} \end{bmatrix} = \begin{bmatrix} \mathbf{L}_s & \mathbf{L}_{sr} \\ \mathbf{L}_{sr}^T & \mathbf{L}_r \end{bmatrix} \begin{bmatrix} \mathbf{i}_{abcs} \\ \mathbf{i}_{abcr} \end{bmatrix} \tag{2.22}$$

where $\mathbf{L}_s$ and $\mathbf{L}_r$ are respectively the stator and rotor self-inductance matrices, while $\mathbf{L}_{sr}$ is the mutual inductance matrix. $\mathbf{L}_s$ and $\mathbf{L}_r$ contain the self and mutual inductances of stator and rotor windings, and their coefficients are constant [92].

The qd0 induction machine model
The induction machine is round and symmetric with respect to all its windings. The stator equations and variables can be transformed using the following transformation matrix:

$$\mathbf{K}_s = \frac{2}{3} \begin{bmatrix} \cos(\theta) & \cos\left(\theta - \frac{2\pi}{3}\right) & \cos\left(\theta + \frac{2\pi}{3}\right) \\ \sin(\theta) & \sin\left(\theta - \frac{2\pi}{3}\right) & \sin\left(\theta + \frac{2\pi}{3}\right) \\ \frac{1}{2} & \frac{1}{2} & \frac{1}{2} \end{bmatrix} \tag{2.23}$$

The rotor magnetic axes *ar*, *br*, and *cr* are rotating. The relative angle between the *ar*-axis and the arbitrary reference frame *q*-axis is $(\theta - \theta_r)$. The rotor equations and variables are transformed using a matrix that uses the angle $(\theta - \theta_r)$:

$$\mathbf{K}_r = \frac{2}{3} \begin{bmatrix} \cos(\theta - \theta_r) & \cos\left(\theta - \theta_r - \frac{2\pi}{3}\right) & \cos\left(\theta - \theta_r + \frac{2\pi}{3}\right) \\ \sin(\theta - \theta_r) & \sin\left(\theta - \theta_r - \frac{2\pi}{3}\right) & \sin\left(\theta - \theta_r + \frac{2\pi}{3}\right) \\ \frac{1}{2} & \frac{1}{2} & \frac{1}{2} \end{bmatrix} \tag{2.24}$$

Upon application of the following transformations:

$$\begin{bmatrix} \mathbf{i}_{qd0s} \\ \mathbf{i}_{qd0r} \end{bmatrix} = \begin{bmatrix} \mathbf{K}_s & \mathbf{0} \\ \mathbf{0} & \mathbf{K}_r \end{bmatrix} \begin{bmatrix} \mathbf{i}_{abcs} \\ \mathbf{i}_{abcr} \end{bmatrix} \tag{2.25a}$$

$$\begin{bmatrix} \mathbf{\lambda}_{qd0s} \\ \mathbf{\lambda}_{qd0r} \end{bmatrix} = \begin{bmatrix} \mathbf{K}_s & \mathbf{0} \\ \mathbf{0} & \mathbf{K}_r \end{bmatrix} \begin{bmatrix} \mathbf{\lambda}_{abcs} \\ \mathbf{\lambda}_{abcr} \end{bmatrix} = \begin{bmatrix} \mathbf{K}_s & \mathbf{0} \\ \mathbf{0} & \mathbf{K}_r \end{bmatrix} \begin{bmatrix} \mathbf{L}_s & \mathbf{L}_{sr} \\ \mathbf{L}_{sr}^T & \mathbf{L}_r \end{bmatrix} \begin{bmatrix} \mathbf{K}_s & \mathbf{0} \\ \mathbf{0} & \mathbf{K}_r \end{bmatrix}^{-1} \begin{bmatrix} \mathbf{i}_{qd0} \\ \mathbf{i}_{qd0} \end{bmatrix} \tag{2.25b}$$

$$\begin{bmatrix} \mathbf{v}_{qd0s} \\ \mathbf{v}_{qd0r} \end{bmatrix} = \begin{bmatrix} \mathbf{K}_s & \mathbf{0} \\ \mathbf{0} & \mathbf{K}_r \end{bmatrix} \begin{bmatrix} \mathbf{v}_{abcs} \\ \mathbf{v}_{abcr} \end{bmatrix} \tag{2.25c}$$

the matrices for the flux equations in terms of transformed *qd0* variables are constant and the transformed equations result in the equivalent circuits depicted in Figure 2.30.

Modelling higher-order rotor effects

The conventional *qd0* model of Figure 2.30 generally assumes a single equivalent rotor winding per magnetic axis and implies a low-order model. However, such model may not be sufficient to represent the behaviour of the rotor design with deep bar conductors and double-layer designs. Additionally, it is not possible to have a model accurate enough over a wide range of frequencies without increasing its order to account for the rotor dynamics with a distributed-parameter

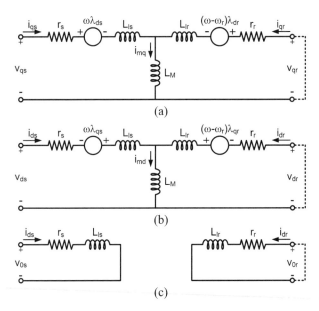

Figure 2.30 Equivalent circuits for conventional three-phase induction machine. (a) *q*-axis equivalent circuit. (b) *d*-axis equivalent circuit. (c) Zero-sequence equivalent circuit.

nature. The equivalent rotor resistance r_r changes with the frequency of rotor currents due to the deep-rotor-bar effect [93]. During transients over a wide range of speeds (from stall to nominal), the effective changes in r_r can be quite significant. The equivalent resistance increases with the frequency due to the skin effect in the rotor bars. To represent the double-cage rotor effect with better accuracy, several simple models have been proposed in the literature [94, 95]. The common approach is based on representing the rotor circuit using several *RL* branches: the order of the circuit is increased by one additional branch with the goal of capturing the changes in equivalent rotor resistance and reactance. To improve the modelling of the deep-bar effects, the rotor can be represented as a high-order admittance (impedance) transfer function: the admittance transfer function is identified by fitting the measured frequency response in the desired range [96].

Modelling of saturation

The magnetic saturation of the main flux path and that of the leakage flux path are treated independently [97]. For simplicity and without loss of generality, only the main flux saturation is discussed here. The stator and rotor leakage flux saturation can be easily included using the same general approach.

The main flux linkage λ_m of a round isotropic rotor can be assumed to be aligned with the magnetizing current vector i_m. The flux linkages λ_{mq} and λ_{md} are the projections of the main flux linkage λ_m onto the q and d axes, respectively, as shown in Figure 2.31. Therefore, the main flux λ_m and the total magnetizing current i_m may be expressed in terms of their axes components as:

$$\lambda_m = \sqrt{\lambda_{mq}^2 + \lambda_{md}^2} \qquad i_m = \sqrt{i_{mq}^2 + i_{md}^2} \qquad (2.26)$$

This approach captures the so-called cross saturation effect [97], a magnetic coupling between q and d axes due to the presence of magnetic saturation. The magnetic saturation can be implemented using either a piecewise-linear representation of the nonlinear magnetic characteristic with relatively small number of linear segments, as depicted in Figure 2.32a, or by a nonlinear monotonic function (which may be composed on many linear segments) in terms of the total magnetizing current, $\lambda_m = \Lambda(i_m)$.

The dynamic or incremental inductance can be then obtained as [98]:

$$L = \frac{\partial \lambda_m}{\partial i_m} \qquad (2.27)$$

This nonlinear magnetizing inductance L can be approximated by L_D within a very small range Δi_m, see Figure 2.32b. Upon discretization, this expression becomes

$$L_D = \frac{\Delta \lambda_m}{\Delta i_m} \qquad (2.28)$$

Figure 2.31 Main magnetizing flux for induction machine with round and isotropic rotor: *q*- and *d*-axes flux and current components.

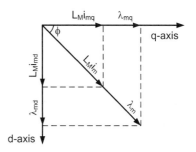

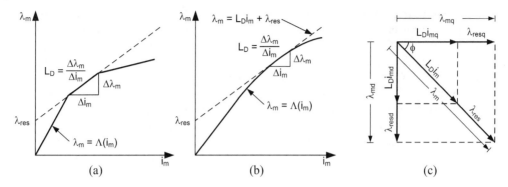

Figure 2.32 Saturation representation for induction machines: (a) Piecewise-linear saturation function; (b) Smooth saturation function; and (c) Projections of residual fluxes onto *q* and *d* axes.

The saturation function is then represented as follows:

$$\lambda_m(t) = L_D i_m(t) + \lambda_{res} \tag{2.29}$$

where λ_{res} is the residual flux.

For round-rotor induction machine, λ_{mq} and λ_{md} may be expressed in terms of incremental magnetizing inductances and residual fluxes as follows (see Figure 2.32c,):

$$\lambda_{mq} = L_D i_{mq} + \lambda_{resq} \qquad \lambda_{md} = L_D i_{md} + \lambda_{resd} \tag{2.30}$$

where

$$\lambda_{resq} = \lambda_{res} \cos \varphi \qquad \lambda_{resd} = \lambda_{res} \sin \varphi \tag{2.31}$$

Electromagnetic torque

The electromagnetic torque may be expressed in direct physical machine variables as follows:

$$T_e = \left(\frac{P}{2}\right) [\mathbf{i}_{abcs}]^T \frac{\partial}{\partial \theta_r} [\mathbf{L}_{sr}][\mathbf{i}_{abcr}] \tag{2.32}$$

In terms of transformed variables, it could be written as follows [92]:

$$T_e = \left(\frac{3}{2}\right) \left(\frac{P}{2}\right) (\lambda_{md} i_{qs} - \lambda_{mq} i_{ds}) \tag{2.33}$$

where *P* is the number of poles.

Mechanical system equations

In general, the mechanical subsystem of an induction machine is represented as a single-mass. Without loss of generality, if motor convention is used, the equations may be expressed as follows [92]:

$$\frac{d\theta_r}{dt} = \omega_r \tag{2.34a}$$

$$J\frac{d\omega_r}{dt} = \frac{P}{2}(T_e - T_m) \tag{2.34b}$$

where θ_r is the rotor position in electrical degrees, ω_r is the angular electrical speed, *J* is the moment of inertia, T_m and T_e are respectively the mechanical torque and electromagnetic torque.

The relationship between the rotor position in mechanical degrees and its position in electrical degrees (and between the rotor speed in electrical and mechanical degrees) is as follows:

$$\theta_{rm} = \frac{2}{P}\theta_r \qquad \omega_{rm} = \frac{2}{P}\omega_r \tag{2.35}$$

For simplicity of notation, θ_r and ω_r are used in equations and figures; they correspond to an equivalent 2-pole machine.

2.5.2.3 Modelling of Synchronous Machines

Synchronous machine equations

A simplified view of a three-phase synchronous machine cross-section is shown in Figure 2.33, where a salient-pole rotor is assumed. Motor convention is used again in the voltage equations so that the stator currents flowing into the machine have a positive sign in the voltage equations. The flux linkage of each winding is assumed to have the same sign as the current flowing in that winding. The rotor position and speed with respect to the reference as-axis are denoted as usual by θ_r and ω_r, respectively. The machine may be represented as the coupled circuit depicted in Figure 2.34. The stator circuit consists of three windings, as, bs, and cs, while the rotor circuit

Figure 2.33 Simplified cross-sectional view of a three-phase synchronous machine coupled circuit model.

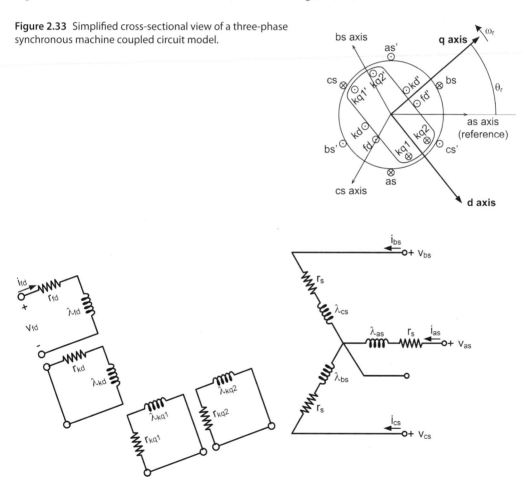

Figure 2.34 Stator and rotor coupled-circuit model of a three-phase synchronous machine.

includes one field winding *fd*, one damper winding *kd* in *d* axis, and two damper windings *kq*1 and *kq*2 in *q* axis. The number of damper windings in the model and their parameters are generally determined to match the machine's frequency response in each axis. Additional damper windings will be usually required for high-frequency transients.

A three-phase synchronous machine can be modelled by a lumped-parameter coupled circuit whose voltage equations may be expressed in matrix form as [93]:

$$
\begin{bmatrix} \mathbf{v}_{abcs} \\ \mathbf{v}_{abcr} \end{bmatrix} = \begin{bmatrix} \mathbf{R}_s & \mathbf{0} \\ \mathbf{0} & \mathbf{R}_r \end{bmatrix} \begin{bmatrix} \mathbf{i}_{abcs} \\ \mathbf{i}_{abcr} \end{bmatrix} + \frac{d}{dt} \begin{bmatrix} \boldsymbol{\lambda}_{abcs} \\ \boldsymbol{\lambda}_{abcr} \end{bmatrix}
\tag{2.36}
$$

where $\mathbf{R}_s$ ($= \mathrm{diag}(r_s \, r_s \, r_s)$) and $\mathbf{R}_r$ ($= \mathrm{diag}(r_{kq1} \, r_{kq2} \, r_{fd} \, r_{kd})$) are the diagonal matrices of stator and rotor resistances, $\mathbf{v}_{abcs} = [v_{as} \, v_{bs} \, v_{cs}]^T$ is the vector of stator voltages, $\mathbf{v}_{qdr} = [0 \, 0 \, v_{fd} \, 0]^T$ is the vector of rotor voltages, $\mathbf{i}_{abcs} = [i_{as} \, i_{bs} \, i_{cs}]^T$ is the vector of stator currents, and $\mathbf{i}_{qdr} = [i_{kq1} \, i_{kq2} \, i_{fd} \, i_{kd}]^T$ is the vector of rotor currents.

The corresponding flux linkage equations are:

$$
\begin{bmatrix} \boldsymbol{\lambda}_{abcs} \\ \boldsymbol{\lambda}_{qdr} \end{bmatrix} = \begin{bmatrix} \mathbf{L}_s & \mathbf{L}_{sr} \\ \mathbf{L}_{rs} & \mathbf{L}_r \end{bmatrix} \begin{bmatrix} \mathbf{i}_{abcs} \\ \mathbf{i}_{qdr} \end{bmatrix}
\tag{2.37}
$$

The coefficients of the stator self-inductance and mutual inductance matrices depend on the rotor position, while the coefficients of the rotor inductance matrix are constant [92].

The qd0 synchronous machine model

Voltage and flux linkage equations of the synchronous machine are expressed in the rotor reference frame using the Park's transformation, which is given by the transformation (2.23). Since the rotor variables are already expressed in *qd*0 coordinates, the Park's transformation is applied only to the stator circuit and variables:

$$
\begin{bmatrix} \mathbf{i}_{qd0s} \\ \mathbf{i}_{qd0r} \end{bmatrix} = \begin{bmatrix} \mathbf{K}_s & \mathbf{0} \\ \mathbf{0} & \mathbf{U} \end{bmatrix} \begin{bmatrix} \mathbf{i}_{abcs} \\ \mathbf{i}_{abcr} \end{bmatrix}.
\tag{2.38a}
$$

$$
\begin{bmatrix} \boldsymbol{\lambda}_{qd0s} \\ \boldsymbol{\lambda}_{qd0r} \end{bmatrix} = \begin{bmatrix} \mathbf{K}_s & \mathbf{0} \\ \mathbf{0} & \mathbf{U} \end{bmatrix} \begin{bmatrix} \boldsymbol{\lambda}_{abcs} \\ \boldsymbol{\lambda}_{abcr} \end{bmatrix} = \begin{bmatrix} \mathbf{K}_s & \mathbf{0} \\ \mathbf{0} & \mathbf{U} \end{bmatrix} \begin{bmatrix} \mathbf{L}_s & \mathbf{L}_{sr} \\ \mathbf{L}_{sr}^T & \mathbf{L}_r \end{bmatrix} \begin{bmatrix} \mathbf{K}_s & \mathbf{0} \\ \mathbf{0} & \mathbf{U} \end{bmatrix}^{-1} \begin{bmatrix} \mathbf{i}_{qd0} \\ \mathbf{i}_{qd0} \end{bmatrix}
\tag{2.38b}
$$

$$
\begin{bmatrix} \mathbf{v}_{qd0s} \\ \mathbf{v}_{qd0r} \end{bmatrix} = \begin{bmatrix} \mathbf{K}_s & \mathbf{0} \\ \mathbf{0} & \mathbf{U} \end{bmatrix} \begin{bmatrix} \mathbf{v}_{abcs} \\ \mathbf{v}_{abcr} \end{bmatrix}.
\tag{2.38c}
$$

where $\mathbf{U}$ is the unit matrix.

The transformed equations result in the equivalent circuits displayed in Figure 2.35.

Modelling of saturation

The anisotropic rotor structure difficulties the determination of the saturation characteristic for salient-pole synchronous machines: the saturation characteristic along the *q* axis is generally not measurable through simple experiments [99–101]; it is instead approximated. Saturation modelling for synchronous machines can be based on the saliency factor defined as follows [102, 103]:

$$
S_F = \sqrt{\frac{L_{mqu}}{L_{mdu}}} = \sqrt{\frac{L_{mqs}}{L_{mds}}}
\tag{2.39}
$$

where L_{mqu}, L_{mdu}, L_{mqs}, and L_{mds} are respectively the unsaturated and saturated *q* and *d* axes magnetizing inductances.

Assuming that the saliency factor remains constant at all saturation levels allows the anisotropic salient-pole machine to be converted into an equivalent isotropic machine with

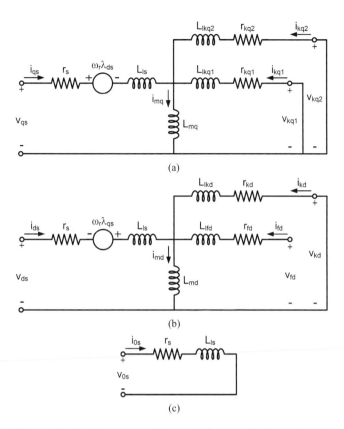

Figure 2.35 Synchronous machine equivalent circuits. (a) *q*-axis equivalent circuit. (b) *d*-axis equivalent circuit. (c) 0-sequence equivalent circuit.

Figure 2.36 Saliency factor approach for representing the magnetizing flux and current of an equivalent isotropic machine.

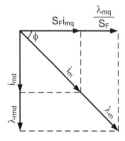

appropriately rescaled q axis. Thereby, the main flux vector λ_m and the magnetizing current i_m of the equivalent isotropic machine are defined as follows (see Figure 2.36):

$$\lambda_m = \sqrt{\left(\frac{\lambda_{mq}}{S_F}\right)^2 + \lambda_{md}^2} \qquad\qquad i_m = \sqrt{(S_F i_{mq})^2 + i_{md}^2} \qquad\qquad (2.40)$$

It is assumed that the d axis magnetic saturation characteristic is represented by a nonlinear monotonic saturation function Λ_d. Based on the saliency factor, the main-flux magnetic saturation characteristic of an equivalent isotropic machine may also be represented by the d-axis characteristic, $\lambda_m = \Lambda(i_m)$. This approach becomes similar to that for induction machines: upon calculation of the partial derivative of $\Lambda(i_m)$ with respect to i_m, the incremental inductance is

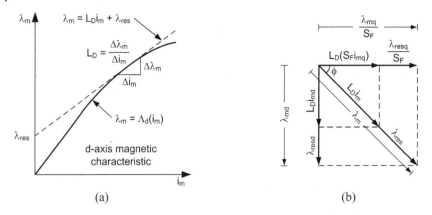

(a)   (b)

Figure 2.37 Saturation representation for synchronous machines: (a) Piecewise linear representation of magnetic saturation using *d*-axis characteristic; (b) Projections of the magnetizing and residual fluxes onto *qd* axes considering the saliency factor.

obtained using (2.27) and (2.28). The residual flux λ_{res} and the local slope incremental inductance L_D are shown in Figure 2.37a. As the machine equations are expressed in *qd* coordinates, the linear relationship (2.36) is projected into the *qd* axes taking into account the saliency factor (see Figure 2.37b):

$$\frac{\lambda_{mq}}{S_F} = L_D(S_F i_{mq}) + \frac{\lambda_{resq}}{S_F} \qquad \lambda_{md} = L_D i_{md} + \lambda_{resd} \tag{2.41}$$

where residual fluxes are also found as:

$$\lambda_{resq} = \lambda_{res} \cos \varphi \cdot S_F \qquad \lambda_{resd} = \lambda_{res} \sin \varphi \tag{2.42}$$

The angle φ takes into account the position of the main flux within the rotor reference frame, *qd* coordinates. The magnetizing fluxes can be then expressed as follows:

$$\lambda_{mq} = L_D S_F^2 (i_{qs} + i_{kq1} + i_{kq2}) + \lambda_{resq} \qquad \lambda_{md} = L_D (i_{ds} + i_{fd} + i_{kd}) + \lambda_{resd} \tag{2.43}$$

Electromagnetic torque
The expression of the electromagnetic torque using machine variables might be as follows:

$$T_e = \left(\frac{P}{2}\right) [\mathbf{i}_{abcs}]^T \left(\frac{1}{2} \frac{\partial [\mathbf{L}_s]}{\partial \theta_r} [\mathbf{i}_{abcs}] + \frac{\partial [\mathbf{L}_{sr}]}{\partial \theta_r} [\mathbf{i}_{qdr}]\right) \tag{2.44}$$

In terms of transformed variables, the expression would be as follows [92]:

$$T_e = \left(\frac{3}{2}\right) \left(\frac{P}{2}\right) (\lambda_{md} i_{qs} - \lambda_{mq} i_{ds}) \tag{2.45}$$

Mechanical system equations
The rotor of a turbine-generator setup has a very complex mechanical structure. A single-mass representation of the mechanical system of such setup is adequate when analysing power system dynamic performance that accounts for the oscillation of the entire turbine-generator rotor with respect to other generators. However, a continuum (distributed-parameter) model might be considered to account for the complete range of torsional oscillations. In general, a simple lumped multi-mass model is adequate for studying problems related to torsional oscillations [104]: each rotor element is modelled as a rigid mass connected to adjacent elements by massless shafts with a single equivalent torsional stiffness constant. Figure 2.38 illustrates the

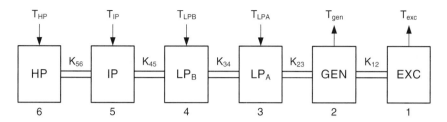

Figure 2.38 Structure of a lumped multi-mass shaft system of a synchronous machine.

structure of a typical lumped multi-mass model of a steam turbine-generator setup [105]. The six torsional masses represent the rotors of the exciter, the generator, two low pressure (LP) turbine sections, the intermediate pressure (IP) turbine section and the high pressure (HP) turbine section, respectively. Such shaft system can be considered as a mass-spring-damper system and its characteristics are defined by three set of parameters: the moment of inertia of each individual mass, the torsional stiffness of each shaft section, and the damping coefficients associated with each mass.

There are a number of sources contributing to mechanical damping: (i) steam forces on turbine blades, which are represented by damping torques proportional to the speed deviations of individual masses from synchronous mechanical speed; (ii) bearing friction and windage on shaft elements, which are represented by damping torques proportional to the angular speed of individual masses; (iii) shaft material hysteresis, which are represented by damping torques proportional to the angular speed difference between the adjacent masses.

If generator convention is used, the equation for the ith mass connected to masses $(i-1)$ and $(i+1)$ of a two-pole machine can be expressed as follows:

$$J_i \frac{d\omega_i}{dt} + D_{asi}\omega_i + D_{ssi}(\omega_i - \omega_0) + D_{mi,i-1}(\omega_i - \omega_{i-1}) + D_{mi,i+1}(\omega_i - \omega_{i+1})$$
$$+ K_{i,i-1}(\theta_i - \theta_{i-1}) + K_{i,i+1}(\theta_i - \theta_{i+1}) = T_{mi} - T_{ei} \tag{2.46}$$

where

J_i is the moment of inertia of the ith mass, in kg-m^2,
D_{asi} is the absolute speed self-damping coefficient of the ith mass, in N-m-s,
D_{ssi} is the speed deviation self-damping coefficient of the ith mass, in N-m-s,
$D_{mi,i-1}$ is the mutual-damping coefficient between the ith and the $(i-1)$th mass, in N-m-s,
$K_{i,i-1}$ is the torsional stiffness of shaft section between the ith and the $(i-1)$th mass, in N-m,
T_{mi} is the mechanical torque developed by the ith turbine section (T_{HP}, T_{IP}, T_{LPB}, T_{LPA}), in N-m,
T_{ei} is the electromagnetic torques of either the generator or the exciter (T_{gen}, T_{ex}), in N-m,
ω_0 is the synchronously rotating reference, in rad/sec,
ω_i is the angular speed of ith mass, in rad/sec,
θ_i is the angular position of ith mass, in radians.

2.5.3 High-Frequency Models for Rotating Machine Windings

2.5.3.1 Introduction

As for transformers, two types of models can be considered for analysing the behaviour of rotating machine windings under high-frequency (step-fronted) stresses: (i) *internal* models, aimed at analysing the voltage distribution within the machine winding; (ii) *terminal* models, used to analyse the interaction of the rotating machine with the system [66]. A short summary about both modelling approaches is provided in the subsequent sections.

2.5.3.2 Internal Models

The machine winding consists of a chain of series-connected coils that are distributed around the machine stator. Under steep-fronted transient conditions, the effective self-inductance of a coil differs considerably from the power-frequency value; initially, the self-inductance arises from flux that is confined mainly to paths outside the high-permeability iron core by eddy currents that are set up in the core by the incident surge. The reluctance of the flux paths changes as the flux penetrates into the core. For calculation purposes it may be necessary to consider self-inductance as a time-varying parameter. Similar considerations apply to the mutual coupling between coils. However, due to the limited extent of flux penetration into the core, the flux linkage between coils in a neighbouring slot is very small, so the mutual coupling between coils under surge voltage conditions is very small too. Unlike transformer windings, the capacitance between coils is very low because each coil is embedded in a slot that acts as a grounded boundary. The inter-coil capacitance is usually limited to that in the line-end coil and is very small too; however, due to the fact that the coil is embedded in the slot, the coil-to-ground capacitance is large [106, 107].

The equivalent circuit of a machine winding is based on the following assumptions:

- The behaviour of the core iron is like that of a grounded sheath, and the slot iron boundary may be replaced by a grounded sheath, which is impenetrable to high-frequency waves. The series inductance and resistance of the coils are also frequency dependent due to the eddy currents in the core and to the skin effect in conductors.

- The two opposite overhang parts of the stator core are considered uncoupled because eddy currents in the core provide effective shielding at high frequencies. Overhang and slot parts are also uncoupled because of the eddy current in the core. The two parts of the coil at the coil entry are uncoupled since they are nearly perpendicular to each other over most of their length and are further shielded from each other by eddy currents in adjacent coils. Insulation between the lamination permits magnetic coupling to the coils inside adjacent slots. However, the two slot parts of the coil are not coupled because of the eddy current in the neighbouring coils. Coupling between adjacent coils of different layers in the same slot is a lower effect that the close coupling between adjacent turns.

- The capacitive couplings between coils of one phase winding and between coils of different phase windings are very small and are usually neglected. The capacitance between turns in a coil and between the coil and the core are important, and should be taken into account. The dielectric losses must also be represented.

- Only TEM propagation mode has to be considered, so the MTL theory can thus be applied to the slot sections [108]. In addition, it can be assumed that the effects of coil insulation in the line-end coil sections dominate over those of the air spaces and waves propagate through these sections with the same velocity as through the slots.

- The basic unit in a winding equivalent circuit is a coil. A stator coil occupies two distinct regions of the machine (see Figure 2.39): the slot region, in which the active coil sides are placed inside the slots in the magnetic core structure, and the overhang region, in which the end turns are positioned in air. The two slot regions are electromagnetically remote, as are the two end-winding regions (at either end of the stator core). A uniform untransposed MTL model with a number of conductors equal to that of the coil turns is used for each region. The multi-conductor lines have different electrical characteristics in each region. The coil has five transmission lines with discontinuities at the junctions: four discontinuities are due to iron-air interfaces and the fifth due to interruption of an end-winding section by the coil terminals.

- The time duration for a transient study can be limited to that corresponding to the period of time of propagation of the surge voltage through these coils. The effect of adding more

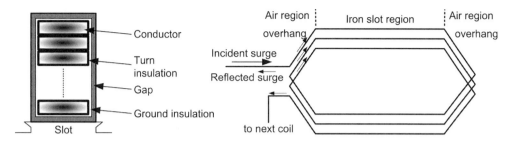

Figure 2.39 Scheme of a form wound coil and its subdivision.

coils to the model on the voltage distribution in the line end coil diminishes as the number of coils increases. The number of coils needed in a winding to enable the line end coil voltage distribution to be accurately predicted increases as the number of turns per coil is reduced.

- When a surge impinges on the line-end coil of a machine winding, see Figure 2.39, it breaks up into transmitted surges moving away from the overhang region and a reflected surge travelling back into the surge source network. The transmitted surges propagate along the overhang section until they reach the slot entry, where they encounter a change in surge impedance; this change causes further reflections and refractions. Each turn of the coil may be regarded as a single conductor transmission line coupled to its neighbour turns, being the end of one turn the start of the next turn, and the coil can be regarded as an MTL in that all the turns run in parallel. One phase of the armature winding consists of series connection of the separate coils of the phase. The complete model of a machine consists of the models of the winding phases, which can be either delta or star connected.

Distributed-parameter models
The models presented for transformers are also valid for rotating machine coils and windings [108–111]. However, rotating machine coils are non-uniform. Consecutive sections of the coil model can be then connected using their two-port constant matrices to give a single-section model of the coil. Since the penetration of magnetic flux into conductors, cores and tank walls, and of the eddy currents induced in them depend on frequency, winding parameters must be frequency-dependent. In most practical examples, transient currents across the branches of the equivalent circuit have several frequencies. Since a single passive element cannot characterize the properties of frequency-dependent impedances, a solution is to represent any circuit branch by a circuit block whose impedance matches the actual winding behaviour at a number of frequencies. A hybrid approach may be considered in which surge propagation is represented by a constant distributed-parameter MTL and the frequency dependence of some parameters is represented by lumped-parameter circuits whose frequency response matches the measured characteristic.

Lumped-parameter models
Lumped-parameter circuits can be derived by representing each section of a turn as a multi-phase PI circuit: the complete coil model consists of five sections (see Figure 2.39) for each of which the equivalent circuit is that shown in Figure 2.22b [112]. Consecutive sections of the coil model are connected in cascade to obtain the transmission matrix of a single coil, which corresponds to a PI circuit in which each phase represents a turn and whose terminals are connected to the adjacent turns [113].

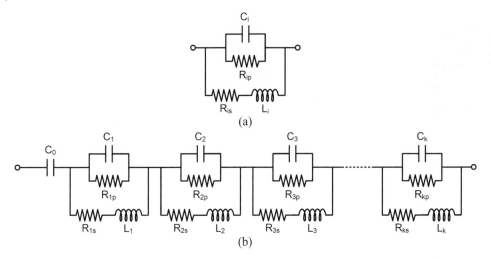

Figure 2.40 Synthesis of motor winding model. (a) Second order circuit for synthesis of motor model. (b) Four stage representation.

2.5.3.3 Terminal Models

For rotating machines, the terminal impedance and the corresponding circuit representation can be also derived from the frequency response of the winding or from a circuit representation like that used to obtain the internal voltage distribution. So, the basic theory presented for Transformers to obtain terminals models can be also applied to rotating machines. A single-input single-output model can be derived by factoring the impedance function as follows [51]:

$$Z(s) = \frac{K_0}{s} + \sum_i \frac{2K_i(s + a_i)}{(s + a_i)^2 + \omega_i^2} \tag{2.47}$$

where ω_i values correspond to the frequencies at which the impedance function peaks. By adjusting the values of a_i, ω_i, and K_i it is possible to select circuit parameters and provide a response that closely matches any measured peak in the machine. As shown in Figure 2.40, the entire equivalent circuit is obtained by connecting a number of series circuit blocks, with capacitor C_0 to provide the pole at zero frequency (i.e. under a very fast-front voltage stimulus, a winding initially behaves as a capacitance).

Reference [66] provides a short review of models proposed for representing motor and generator windings under fast- and very fast-front voltages.

2.6 Circuit Breakers

2.6.1 Overview

A circuit breaker is a mechanical switching device, capable of making, carrying and breaking currents under normal circuit conditions, and also making, carrying for a specified time and breaking currents under specified abnormal circuit conditions such as those of short-circuits [114–119]. The performance of an ideal circuit breaker may be summarized as follows:

1. When closed, it is a good conductor and withstands thermally and mechanically any current below or equal to the rated short-circuit current.

2. When opened, it is a good insulator and withstands the voltage between contacts, the voltage to ground or to the other phases.
3. When closed, it can quickly and safely interrupt any current below or equal to the rated short circuit current.
4. When opened, it can quickly and safely close a shorted circuit.

Circuit breaker models are needed to analyse both closing and opening operations. Unless an ideal representation was assumed, no single model can be used in either type of operation.

In normal operating conditions, a circuit breaker is in the closed position and some current flows through the closed contacts. The circuit breaker opens its contacts when receives a tripping signal. An electric arc is formed in a circuit breaker after contacts start separating; it changes from a conducting to a non-conducting state in a very short period of time. The arc is normally extinguished as the current reaches a natural zero in the alternating current cycle; this mechanism is assisted by drawing the arc out to maximum length, increasing its resistance and limiting its current. Various techniques have been adopted to extend the arc; they differ according to size, rating and application [119]. In high-voltage circuit breakers (>1 kV), the current interruption is performed by cooling the arc. Power circuit breakers are categorized according to the extinguishing medium in which the arc is formed. Although a large number of arc models have been proposed, and some of them have been successfully applied, there is no model that takes into account all physical phenomena in a circuit breaker. Approaches that can be used to reproduce the arc interruption phenomenon are listed below [120]:

a) *Physical arc models*, which include the physical process in detail; that is, the overall arc behaviour is calculated from conservation laws, gas and plasma properties, and more or less detailed models of exchange mechanisms (radiation, heat conduction, turbulence).
b) *Black box models*, which consider the arc as a two-pole, determine the transfer function using a chosen mathematical form, and fit the remaining free parameters to measured voltage and current traces [121, 122]. They describe the interaction of an arc and an electrical circuit during an interruption process; rather than physical processes, it is the electrical behaviour of the arc which is of importance.
c) *Formulae and diagrams*, which give parameter dependencies for special cases and scaling laws. They can be derived from either tests or calculations with the two previous models.

Several models with different level of complexity can also be used to represent a circuit breaker in closing operations. When using a black-box type model, the simplest approach assumes that the breaker behaves as an ideal switch whose impedance passes instantaneously from an infinite value to a zero value at the closing time. A more sophisticated approach assumes that there is a closing time from the moment at which the contacts start to close to the moment that they finally make, and its withstand voltage decreases as the separation distance between contacts decreases. Finally, a dynamic arc representation can also be included in case of breaker prestrike.

The rest of this section is dedicated to discuss the different approaches proposed for representing circuit breakers during both opening and closing operations using black-box models.

2.6.2 Circuit Breaker Models for Opening Operations

2.6.2.1 Current Interruption

Circuit breakers accomplish the task of interrupting an electrical current by using some interrupting medium for dissipating the energy input. For most breaking technologies the interrupting process is well understood and can be described with some accuracy.

The electric arc is a self-sustained discharge capable of supporting large currents with a relatively low voltage drop. The arc acts like a nonlinear resistor: after current interruption, the resistance value does not change rapidly from a low to an infinite value. For several microseconds after interruption, a post-arc current flows. The interruption process can be separated into three periods: the arcing period, the current-zero period, the dielectric recovery period. During arcing period, the time constant of the dc component in a short circuit current becomes smaller because of the arc resistance. As the arc current approaches a current-zero point, the ratio of arc heat loss to electrical input energy increases and the arc voltage rises abruptly.

After the arc current is extinguished, the recovery voltage appears. Since the space between contacts does not change to a completely insulating state, a small current, the post-arc current, flows through the breaker as the recovery voltage builds up and soon disappears. If the extinguishing ability of the circuit breaker is small, the post-arc current does not decrease, and an interruption failure can occur; it is known as *thermal failure* or *reignition*. The recovery voltage has two components: a transient recovery voltage (TRV) and a power-frequency recovery voltage (PFRV). The TRV has a direct effect on the interrupting ability of a circuit breaker, but the PFRV is also important because it determines the centre of the TRV oscillation. The waveform and magnitude of the TRV vary according to many factors, such as system voltages, equipment parameters and fault types. The waveform determined by the system parameters alone is the inherent TRV. All regions of the TRV have an effect on the breaking interrupting ability. For some kinds of circuit breakers, the initial TRV is the most critical period. The circuit interruption is complete only when the circuit breaker contacts have recovered sufficient dielectric strength after arc extinction near current zero. Here, the dielectric strength recovery characteristics play an important role. Depending on the type of circuit breaker, this period is also dangerous. An interruption failure during this period is known as *dielectric failure* or *restrike*.

2.6.2.2 Circuit Breaker Models

The main objectives of a circuit breaker model are [119, 120]: (i) from the system viewpoint, to determine all voltages and currents that are produced within the system as a result of the breaker action; (ii) from the breaker viewpoint, to determine whether the breaker will be successful when operating within a given system under a given set of conditions.

The aim of a black-box model is to describe the interaction of the switching device and the corresponding electrical circuit during an interruption process. Rather than internal processes, it is the electrical behaviour of the circuit breaker which is of importance. Black-box models are aimed at obtaining a quantitatively correct performance of the circuit breaker. Several levels of model complexity are possible [120]:

1) The breaker is represented as an ideal switch that opens at first current zero crossing after the tripping signal is given. This model can be used to obtain the voltage across the breaker; this voltage is to be compared with a pre-specified TRV withstand capability for the breaker. This model cannot reproduce any interaction between the arc and the system.

2) The arc is modelled as a time-varying resistance or conductance. The time variation is determined ahead of time based on the breaker characteristic and perhaps upon the knowledge of the initial interrupting current. This model can represent the effect of the arc on the system, but requires advanced knowledge of the effect of the system on the arc. Arc parameters are not always easy to obtain and the model still requires the use of precomputed TRV curves to determine the adequacy of the breaker.

3) The breaker is represented as a dynamically-varying resistance or conductance, whose value depends on the past history of arc voltage and current. This model can represent both, the

effect of the arc on the system and the effect of the system on the arc. No precomputed TRV curves are required. These models are generally developed to determine initial arc quenching; that is, to study the thermal period only, although some models can be used to determine arc reignition due to insufficient voltage withstand capability of the dielectric between breaker contacts. Their most important application cases are short-line fault interruption and switching of small inductive currents.

Many models for circuit breakers, represented as a dynamic resistance/conductance, have been proposed. A survey on black-box models of gas (air, SF_6) circuit breakers was presented in [119]. ATP implementation of dynamic arc models, adequate for gas and oil circuit breakers, was presented in [123]. All those models are useful to represent a circuit breaker during the thermal period; models for representation of SF_6 breakers during thermal and dielectric periods were discussed in [124, 125]. Vacuum circuit breakers lack an extinguishing medium and their representation has to consider statistical properties; their models may be used for studying the mean chopping current value with known di/dt, dielectric breakdown voltage characteristic, contacts separation dynamics, probability of high frequency arc quenching capability, or probability of high-frequency zero current passing. For a bibliography on circuit breaker models, see [119].

The next subsections summarize the most common approaches proposed for representing both gas-filled and vacuum circuit breakers.

2.6.2.3 Gas-Filled Circuit Breaker Models

Ideal switch models
The breaker has as a zero-impedance when closed and opens at first current-zero crossing after the tripping signal is given. The breaking action is therefore completely independent of the arc. Depending on the goal, two options can be considered:

a) The model is used to obtain the voltage across the breaker; this voltage may be later compared with the TRV withstand capability for the breaker. With this approach, the model does not reproduce any interaction between the breaker and the system.
b) The model incorporates the specified TRV curve, allowing in this manner the possibility of reproducing a breakdown and the interaction of the breaker and the system.

Dynamic arc representation
Arc models are expressed as formulae for the time varying resistance or conductance as a function of arc current, arc voltage and several time varying para-meters representing arc properties. Several arc models are available in the literature; most models keep the basic idea of describing arc behaviour using parameters with different physical interpretation. Basic descriptions of arc behaviour were first described by Cassie and Mayr [126, 127]:

- The Cassie model is given by the following equation:

$$\frac{1}{g_c}\frac{dg_c}{dt} = \frac{1}{\tau_c}\left(\left(\frac{v}{v_0}\right)^2 - 1\right) = \frac{1}{\tau_c}\left(\left(\frac{i}{v_0 g_c}\right)^2 - 1\right) \tag{2.48}$$

This model assumes an arc channel with constant temperature, current density and electric field strength. Changes of the arc conductance result from changes of arc cross section; energy removal is obtained by convection.

- The Mayr model is given by the following equation:

$$\frac{1}{g_m}\frac{dg_m}{dt} = \frac{1}{\tau_m}\left(\frac{vi}{P_0} - 1\right) = \frac{1}{\tau_m}\left(\frac{i^2}{P_0 g_m} - 1\right) \tag{2.49}$$

This model assumes that changes of arc temperature are dominant, and size and profile of the arc column are constant. Thermal conduction is the main mechanism of energy removal.

In these equations, g is the arc conductance, v is the arc voltage, i is the arc current, τ is the arc time constant, P_0 is the steady-state power loss, and v_0 is the constant part of the arc voltage. g_c is in the region of 1 μs (SF_6) and g_m is in a region between 0.1 and 0.5 μs (SF_6). Although these parameters are not strictly constant, their variation is slow enough to assume them constant.

Note that both models can be defined by only two parameters: (i) τ_c and v_0 for Cassie model; (ii) τ_m and P_0 for Mayr model. They give a qualitative description of arc behaviour, and should be carefully used for quantitative representations. A great number of modifications have been formulated; they introduce more parameters into the equation or define the equation in a more general form.

For more details on arc models, see references [119–122, 128].

2.6.2.4   Vacuum Circuit Breaker Models

Introduction

The vacuum arc can be divided into three regions: the cathode spot region, the inter-electrode region, and the space charge sheath in front of the anode. The cathode spot region and the anode region have a constant thickness, and both are very small compared to the inter-electrode space [129–131]. The vacuum arc ceases when the cathode spots disappear. At current zero, the inter-electrode space still contains a certain amount of conductive charge. As the current reverses polarity, the old anode becomes the new cathode, but in the absence of cathode spots, the overall breaker conductance drops, which allows a rise of the recovery voltage. The combination of the residual plasma conductance and this recovery voltage gives rise to a post-arc current, which contains a considerable scatter that disturbs the relationship between the arcing conditions and the post-arc current. This has to do with the final position of the last cathode spot: when its final position is near the edge of the cathode, a significant amount of charge is ejected away from the contacts, and less charge is returned to the external electric circuit, compared to the situation in which the final spot position is close to the centre of the cathode. Since the cathode spots move randomly across the cathode surface, the final position is unknown, and the post-arc conditions are different for each measurement, which gives the post-arc current a random nature.

It is generally accepted that the post-arc current is divided into three phases [132]:

1. Before current zero, ions are launched from the cathode towards the anode. At current zero, the ions already produced continue to move towards the anode due to their inertia.
2. Immediately after current zero, the electrons reduce their velocity, and the flux of positive charge arrives to the post-arc cathode. This process continues until the electrons reverse their direction. With no charge, the voltage across the gap remains zero. As soon as the electrons reverse their direction and move away from the cathode, the post-arc current enters its second phase, leaving an ionic space charge sheath behind. The gap between electrodes is not neutral, with a TRV that mostly stands across the sheath.
3. The sheath continues to expand until it reaches the new anode. At that moment, the post-arc current starts its third phase. The electric current drops, since all electrons have been removed from the gap. The electric field between the contacts moves the remaining ions towards the cathode, but the current that results from this process is negligible.

Restrikes and reignitions

The arc current in vacuum is formed from electrons which are withdrawn from the metal surface by the electric field; the resulting current density increases locally the contact temperature, which may eventually lead to the formation of a cathode spot, and to a subsequent *dielectric restrike*. For perfectly smooth contact surfaces, the electric field under normal operating conditions is generally too low to cause dielectric restrike; however, irregularities that increase the local electric field can be present on the contact surface. This concentrates the current density to smaller surface areas, increasing the heating and creating conditions for a restrike. The restrike is not only caused by irregularities on the contact surface, but also by particles in the vacuum, which is in general inevitable; they may originate from protrusions on the contact surface, which are drawn under the influence of an electric field. *Thermal reignition* occurs when a breaker fails in the period immediately following current zero: the gap contains charge and vapour from the arc, the contacts are still hot, and there can also be pools of hot liquid metal on their surface. It takes several microseconds for the charge to remove, but it takes several milliseconds for the vapour to diffuse, and the pools to cool. A failure that occurs when vapour is still present is called *dielectric reignition*. With the increased contact temperature, the conditions for a failure are improved. When an electron, accelerated in the electric field, hits a neutral vapour particle with sufficient momentum, it knocks out an electron from the neutral. This process reduces the kinetic energy of the first electron; both electrons accelerate in the electric field, hit other neutral particles and cause an avalanche of electrons in the gap, which eventually causes reignition. This process enhances when the probability of an electron hitting a vapour particle increases. This can be prevented by either increasing the vapour pressure or the gap length. Both methods reduce the reignition voltage, but at some point, electrons collide with particles before reaching the appropriate ionization energy. As a result, the reignition voltage eventually rises with increasing vapour pressure or gap length.

Models of vacuum circuit breakers

The arc extinction in gas-filled circuit breakers involves the balance between the internal energy of the arc and the cooling power of the surrounding gas; after current zero, the cooling power has to be stronger than the arc internal energy to successfully extinguish the arc. Vacuum circuit breakers lack an extinguishing medium and their arc internal energy is almost completely concentrated in the cathode spots. The reignition process is quite different, since the source of the new arc (i.e. cathode spots) arises at a different location (the new cathode) before than current zero. For this reason, post-arc modelling has been mainly focused on simulating the anode temperature during arcing and the behaviour of the cathodic space charge sheath after current zero, because both involve the conditions at the location of potentially new cathode spots. The representation of vacuum circuit breakers during both interruption of low currents and the post-zero current is summarized below.

Low-current models. The models have to include the cold withstand voltage characteristic, the high-frequency quenching capability, and the chopping current level [133]. The cold withstand voltage characteristic of a vacuum circuit breaker is a function of the contact distance. A crucial parameter is the speed of contact separation [134, 135]. Reference [136] represented the withstand voltage characteristic with an exponential expression, while [134] showed that the failure can occur at short gaps, and it is sufficient to use a straight line. The high frequency quenching capability is defined by the slope of the reignited current at current zero; this characteristic depends also on the reignited voltage [137], and exhibits a time dependent behaviour. The chopping current depends mainly on the contact material [138]. The characteristics describing whether or not reignition occurs are [134]:

$$V_b = A(t - t_{open}) + B \tag{2.50a}$$

$$\frac{di}{dt} = C(t - t_{open}) + D \tag{2.50b}$$

where t_{open} is the moment of contact opening. The quantities V_b and di/dt represent the dielectric and arc quenching capability of the circuit breaker, respectively.

Post-arc current models. The post-arc current has been usually modelled with the aid of Child's law [139], modified by Andrews and Varey [140]. The model is based on the kinetic motion of charged particles: the direction in which the charged particles move is controlled by the electric field, which implies that during the sheath growth, only ions arrive at the cathode. The model assumes that the sheath is completely free of electrons, and the electric field at the plasma-sheath edge is zero. However, since the sheath grows into the plasma, stationary ions move through the sheath edge and an additional ion current flows through the sheath. According to the original model [139], ions start at rest, but they always reach an initial velocity as long as the sheath moves. The model lacks a proper theory for the voltage-zero phase, which can be explained when the thermal velocity of the particles is taken into account; see reference [128] for more details.

2.6.3 Circuit Breaker Models for Closing Operations

2.6.3.1 Introduction
A closing operation can produce transient overvoltages whose maximum peaks depend on several factors (e.g. the network representation on the source side of the breaker, the charge trapped on transmission lines in a reclosing operation). A factor that has some influence on the maximum peak is the instant of closing, which can be different for every pole of a three-phase breaker: a circuit breaker has a voltage-withstand capability between contacts that depends on the separation between them; as the contacts close and the gap between them gets shorter, breakdown will occur if the voltage across the gap exceeds the dielectric strength. In other words, electrical closing can happen before mechanical closing. Figure 2.41 shows the prestrike phenomenon at the time the stress exceeds the strength [120]. An effect of prestrikes is that the probability distribution of closing instants will not be uniformly distributed, as shown in Figure 2.42 [120]. For a multi-phase breaker, the probability distribution of closing times is more complicated to determine since the order in which breaker poles close will be random in a cycle. ATP allows users to analyse the influence of the closing time and obtain a statistical distribution of switching overvoltages, usually in the form of an accumulative frequency distribution.

For some studies it may be of interest to represent also the prestrike arc conductance. Since an oscillatory arc current at a very high frequency may be caused during prestrike, such current may result in multiple quenchings and prestrikes prior to the ultimate contact.

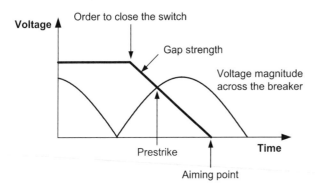

Figure 2.41 Prestrike phenomenon.

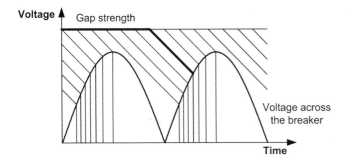

Figure 2.42 Distribution of the closing time during prestrike.

Several models can be used to represent a circuit breaker in closing operations [4, 120]:

- The simplest model assumes that the breaker behaves as an ideal switch whose impedance passes instantaneously from an infinite value, when open, to a zero value at the closing time. This performance can be represented at any part of a power cycle. The effect of the operation depends on the closing time. In multi-phase systems an additional factor is added since all poles do not close simultaneously. A further improvement assumes that the closing instant for either single-phase or multi-phase breakers is randomly determined.
- A more advanced approach assumes that there is a closing time from the moment at which the contacts start to close to the moment that they finally make. The withstand voltage decreases as the separation distance between contacts decreases, an arc will strike before the contacts have completely closed if the voltage across them exceeds the withstand voltage of the dielectric medium. Modelling of the pre-strike effect and its influence on the switching overvoltages produced during line energization has been analysed in [141].
- A third approach includes the prestrike dynamic arc conductance.

The two first approaches are presented below.

2.6.3.2 Statistical Switches

The calculation of switching overvoltages is fundamental for the insulation design of transmission level components. However, equipment insulation design is not usually based on the highest overvoltage because this particular event may have a low probability of occurrence, and the design would not be economical. The insulation design of power system equipment is based on the concepts *stress* and *strength*. The probability distribution of switching overvoltages (stress) is compared to equipment insulation (strength) that is also described statistically; the goal is to choose the insulation level so as to achieve a failure rate criterion. See reference [8].

The *statistical switches* implemented in ATP can facilitate the calculation of the statistical distribution of switching overvoltages. Two approaches can be considered [12]:

- The closing time is systematically varied from a minimum to a maximum instant in equal increments of time; this type of switch is known as *systematic* switch. An accurate evaluation of the switching overvoltage probability distribution using this type of switches can be very laborious. For instance, if the closing operation is performed over the entire range of a cycle and a time increment of 0.2 ms is chosen, then the number of energizations per phase in a 50 Hz system is 100, and the total number is (100*100*100 =) 1E + 6.
- The closing time is randomly varied according to a given probability distribution; this type of switch is known as *statistics* switch. Data required to represent these switches are the mean closing time and the standard deviation. The closing time of a statistics switch is randomly

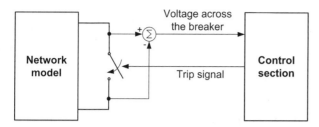

Figure 2.43 Modelling of a circuit breaker prestrike.

selected according to a probability distribution, which is usually uniform or normal (Gaussian). In general, it is assumed that the closing of an independent breaker pole may occur at any point of the power-frequency cycle with equal likelihood.

A three-phase breaker can be represented as three single breakers each with an independent probability distribution. However, the three poles can be mechanically linked so that each pole attempt to close at the same instant, the aiming point. In reality, there will be a finite time or pole span between the first and the last pole to close. If pre-insertion resistors are used to mitigate switching overvoltages, the closing times of both main and auxiliary contacts are statistically determined using a dependency similar to that described above. A way to obtain the closing times is to consider that the main contacts are the latest to close [142].

2.6.3.3 Prestrike Models

For most purposes the prestrike phenomenon can be represented as an ideal switch for which the only information of concern is the closing time of each pole. This model can be easily represented using ATP capabilities. Figure 2.43 shows a possible representation in which the circuit breaker is modelled as an ideal switch whose operation is controlled by the voltage across terminals and taking into account the time variation of the gap strength [120]. A more sophisticated model may include the arc conductance and the possibility of multiple quenchings and restrikes prior to the ultimate metallic contact if during prestrike a very high frequency oscillatory current is induced.

Acknowledgement

The material of this chapter comes from more than a dozen of chapters previously published in other books and special publications. I thank all colleagues who were involved in those chapters, somehow a little of their work is here.

References

1 Martinez-Velasco, J.A. (ed.) (2009). *Power System Transients. Parameter Determination.* CRC Press.
2 IEC 60071-1. (2010). *Insulation Co-ordination, Part 1: Definitions, Principles and Rules.*
3 CIGRE WG 33.02 (1990). *Guidelines for Representation of Network Elements when Calculating Transients.* CIGRE Brochure 39.
4 Gole, A., Martinez-Velasco, J.A. and Keri, A. (eds.) (1998). *Modeling and Analysis of Power System Transients Using Digital Programs.* IEEE Special Publication TP-133-0.

5 IEC TR 60071-4 (2004). *Insulation Co-ordination - Part 4: Computational Guide to Insulation Co-ordination and Modeling of Electrical Networks.*

6 IEC 60071-2 (1996). *Insulation Co-ordination, Part 2: Application Guide.*

7 IEEE Std 1313.2 (1999). *IEEE Guide for the Application of Insulation Coordination.*

8 Hileman, A.R. (1999). *Insulation Coordination for Power Systems.* Marcel Dekker.

9 Martinez-Velasco, J.A., Ramirez, A.I., and Dávila, M. (2009). Overhead lines. In: *Power System Transients. Parameter Determination* (ed. J.A. Martinez-Velasco), 17–135. CRC Press.

10 Carson, J.R. (1926). Wave propagation in overhead wires with ground return. *Bell System Technical Journal* 5: 539–554.

11 Schelkunoff, S.A. (1934). The electromagnetic theory of coaxial transmission lines and cylindrical shields. *Bell System Technical Journal* 13: 532–579.

12 Dommel, H.W. (1986). *Electromagnetic Transients Program. Reference Manual (EMTP Theory Book).* Portland: Bonneville Power Administration.

13 Canadian/American EMTP User Group (2014). *ATP Rule Book.* Portland: Can/Am EMTP User Group.

14 Chisholm, W.A., Chow, Y.L., and Srivastava, K.D. (1985). Travel time of transmission towers. *IEEE Transactions on Power Apparatus and Systems* 104 (10): 2922–2928.

15 Ishii, M., Kawamura, T., Kouno, T. et al. (1991). Multistory transmission tower model for lightning surge analysis. *IEEE Transactions on Power Delivery* 6 (3): 1327–1335.

16 Hara, T. and Yamamoto, O. (1996). Modelling of a transmission tower for lightning surge analysis. *IEE Proceedings - Generation, Transmission and Distribution* 143 (3): 283–289.

17 Gutiérrez, J.A., Moreno, P., Naredo, J.L. et al. (2004). Nonuniform transmission tower model for lightning transient studies. *IEEE Transactions on Power Delivery* 19 (2): 490–496.

18 CIGRE WG 33-01. (1991). Guide to Procedures for Estimating the Lightning Performance of Transmission Lines. CIGRE Brochure 63.

19 Martinez, J.A. and Castro-Aranda, F. (2005). Tower modeling for lightning analysis of overhead transmission lines. IEEE PES General Meeting in San Francisco, USA (June 2005).

20 IEEE WG on Lightning Performance of Transmission Lines (1993). Estimating lightning performance of transmission lines II: updates to analytical models. *IEEE Transactions on Power Delivery* 8 (3): 1254–1267.

21 IEEE WG on Lightning Performance of Transmission Lines (1985). A simplified method for estimating lightning performance of transmission lines. *IEEE Transactions on Power Apparatus and Systems* 104 (4): 919–932.

22 IEEE Std 1243. (1997). IEEE Guide for Improving the Lightning Performance of Transmission Lines.

23 Yamada, T., Mochizuki, A., Sawada, J. et al. (1995). Experimental evaluation of a UHV tower model for lightning surge analysis. *IEEE Transactions on Power Delivery* 10 (1): 393–402.

24 Chisholm, W.A., Chow, Y.L., and Srivastava, K.D. (1983). Lightning surge response of transmission towers. *IEEE Transactions on Power Apparatus and Systems* 102 (9): 3232–3242.

25 Ito, T., Ueda, T., Watanabe, H. et al. (2003). Lightning flashover on 77-kV systems: observed voltage bias effects and analysis. *IEEE Transactions on Power Delivery* 18 (2): 545–550.

26 Chowdhuri, P. (2003). *Electromagnetic Transients in Power Systems*, 2e. Taunton, UK: RS Press/Wiley.

27 Chisholm, W.A. (2007). Transmission line transients – grounding. In: *Power Systems* (ed. L.L. Grigsby), 10–93. CRC Press.

28 Phillips, A. (2006). *Guide for Transmission Line Grounding: A Roadmap for Design, Testing and Remediation: Part I – Theory Book*. EPRI Report 1013594.

29 Loyka, S.L. (1999). A simple formula for the ground resistance calculation. *IEEE Transactions on Electromagnetic Compatibility* 41 (2): 152–154.

30 Bewley, L.V. (1951). *Traveling Waves on Transmission Systems*, 2e. New York: Wiley.

31 Grcev, L. and Popov, M. (2005). On high-frequency circuit equivalents of a vertical ground rod. *IEEE Transactions on Power Delivery* 20 (2): 1598–1603.

32 Weck, K.H. (1988). Remarks to the current dependence of tower footing resistances. CIGRE Session, Paper 33–85.

33 CIGRE WG 33.01. (1992). *Guide for the Evaluation of the Dielectric Strength of External Insulation*. CIGRE Brochure 72.

34 IEC 60060-1. (1989). *High-voltage Test Techniques – Part 1: General Definitions and Test Requirements*.

35 Jones, A.R. (1954). Evaluation of the integration method for analysis of non-standard surge voltages. *AIEE Transmission* 73: 984–990.

36 Suzuki, T. and Miyake, K. (1977). Experimental study of the breakdown voltage time characteristics of large air-gaps with lightning impulses. *IEEE Transactions on Power Apparatus and Systems* 96 (1): 227–233.

37 Darveniza, M. and Vlastos, A.E. (1988). The generalized integration method for predicting impulse volt-time characteristics for non-standard wave shapes – a theoretical basis. *IEEE Transactions on Electrical Insulation* 23 (3): 373–381.

38 Baldo, G., Hutzler, B., Pigini, A. and Garbagnati, E. (1992). Dielectric strength under fast front overvoltages in reference ambient conditions. Guide for the Evaluation of the Dielectric Strength of External Insulations. CIGRE Brochure 72.

39 Pigini, A., Rizzi, G., Garbagnati, E. et al. (1989). Performance of large air gaps under lightning overvoltages: experimental study and analysis of accuracy of predetermination methods. *IEEE Transactions on Power Delivery* 4 (2): 1379–1392.

40 Weck, K.H. (1981). Lightning performance of substations. CIGRE Conference in Rio de Janeiro, Brazil.

41 Wedepohl, L.M. and Wilcox, D.J. (1973). Transient analysis of underground power-transmission systems. System-model and wave-propagation characteristics. *Proceedings of the IEE* 120 (2): 253–260.

42 Ametani, A. (1980). A general formulation of impedance and admittance of cables. *IEEE Transactions on Power Apparatus and Systems* 99 (3): 902–909.

43 Gustavsen, B., Martinez, J.A., and Durbak, D. (2005). Parameter determination for modeling systems transients. Part II: insulated cables. *IEEE Transactions on Power Delivery* 20 (3): 2045–2050.

44 Gustavsen, B., Noda, T., Naredo, J.L. et al. (2009). Insulated cables. In: *Power System Transients. Parameter Determination* (ed. J.A. Martinez-Velasco), 137–175. CRC Press.

45 Ametani, A., Nagaoka, N., Baba, Y. et al. (2017). *Power System Transients. Theory and Applications*, 2e. CRC Press.

46 Faria da Silva, F.F. and Bak, C.L. (2013). *Electromagnetic Transients in Power Cables*. Springer.

47 Breien, O. and Johansen, I. (1971). Attenuation of traveling waves in single-phase high-voltage cables. *Proceedings of the IEE* 118 (6): 787–793.

48 Noualy, J.P. and Le Roy, G. (1977). Wave-propagation modes on high-voltage cables. *IEEE Transactions on Power Apparatus and Systems* 96 (1): 158–165.

49 Ametani, A., Miyamoto, Y., and Nagaoka, N. (2004). Semiconducting layer impedance and its effect on cable wave-propagation and transient characteristics. *IEEE Transactions on Power Delivery* 19 (4): 1523–1531.

50 de León, F., Gómez, P., Martinez-Velasco, J.A., and Rioual, M. (2009). Transformers. In: *Power System Transients. Parameter Determination* (ed. J.A. Martinez-Velasco), 177–249. CRC Press.

51 Greenwood, A. (1991). *Electrical Transients in Power Systems.* Wiley.

52 Jazebi, S., Zirka, S.E., Lambert, M. et al. (2016). Duality derived transformer models for low-frequency electromagnetic transients – Part II: complementary modeling guidelines. *IEEE Transactions on Power Delivery* 31 (5): 2420–2430.

53 Gustavsen, B. (2004). Wide band modeling of power transformers. *IEEE Transactions on Power Delivery* 19 (1): 414–422.

54 Gustavsen, B. (2016). Wide-band transformer modeling including core nonlinear effects. *IEEE Transactions on Power Delivery* 31 (1): 219–227.

55 Álvarez-Mariño, C., de León, F., and López-Fernández, X.M. (2012). Equivalent circuit for the leakage inductance of multiwinding transformers: unification of terminal and duality models. *IEEE Transactions on Power Delivery* 27 (1): 353–361.

56 Jazebi, S., de León, F., and Vahidi, B. (2013). Duality-synthesized circuit for eddy current effects in transformer windings. *IEEE Transactions on Power Delivery* 28 (2): 1063–1072.

57 Tarasiewicz, E.J., Morched, A.S., Narang, A., and Dick, E.P. (1993). Frequency dependent eddy current models for nonlinear iron cores. *IEEE Transactions on Power Systems* 8 (2): 588–597.

58 de León, F. and Semlyen, A. (1994). Complete transformer model for electromagnetic transients. *IEEE Transactions on Power Delivery* 9 (1): 231–239.

59 Jazebi, S. and de León, F. (2015). Duality-based transformer model including eddy current effects in the windings. *IEEE Transactions on Power Delivery* 30 (5): 2312–2320.

60 Jazebi, S., de León, F., Farazamand, A., and Deswal, D. (2013). Dual reversible transformer model for the calculation of low-frequency transients. *IEEE Transactions on Power Delivery* 28 (4): 2509–2517.

61 Mork, B.A., Gonzalez, F., Ishchenko, D. et al. (2007). Hybrid transformer model for transient simulation – Part I: development and parameters. *IEEE Transactions on Power Delivery* 22 (1): 248–255.

62 Rezaei-Zare, A. (2015). Enhanced transformer model for low- and mid-frequency transients – Part I: model development. *IEEE Transactions on Power Delivery* 30 (1): 307–315.

63 Martinez, J.A. and Mork, B. (2005). Transformer modeling for low- and mid-frequency transients – a review. *IEEE Transactions on Power Delivery* 20 (2): 1625–1632.

64 Martinez, J.A., Walling, R., Mork, B. et al. (2005). Parameter determination for modeling systems transients. Part III: transformers. *IEEE Transactions on Power Delivery* 20 (3): 2051–2062.

65 Jazebi, S., Zirka, S.E., Lambert, M. et al. (2016). Duality derived transformer models for low-frequency electromagnetic transients – Part I: topological models. *IEEE Transactions on Power Delivery* 31 (5): 2410–2419.

66 Martinez-Velasco, J.A. (2013). Basic methods for analysis of high frequency transients in power apparatus windings. In: *Electromagnetic Transients in Transformer and Rotating Machine Windings* (ed. C.Q. Su), 45–110. IGI Global.

67 Brandwajn, V., Dommel, H.W., and Dommel, I.I. (1982). Matrix representation of three-phase n-winding transformers for steady-state and transient studies. *IEEE Transactions on Power Apparatus and Systems* 101 (6): 1369–1378.

68 Hatziargyriou, N.D., Prousalidis, J.M., and Papadias, B.C. (1993). Generalised transformer model based on the analysis of its magnetic core circuit. *IEE Proceedings C* 140 (4): 269–278.

69 Slemon, G.R. (1953). Equivalent circuits for transformers and machines including non-linear effects. *Proceedings of the IEE* 100: 129–143.

70 Arturi, C.M. (1991). Transient simulation and analysis of a three phase five-limb step up transformer following an out-of-phase synchronization. *IEEE Transactions on Power Delivery* 6 (1): 196–207.

71 Narang, A. and Brierley, R.H. (1994). Topology based magnetic model for steady state and transient studies for three phase core type transformers. *IEEE Transactions on Power Systems* 9 (3): 1337–1349.

72 Mork, B.A. (1999). Five-legged wound-core transformer model: Derivation, parameters, implementation, and evaluation. *IEEE Transactions on Power Delivery* 14 (4): 1519–1526.

73 de León, F., Farazmand, A., and Joseph, P. (2012). Comparing the T and π equivalent circuits for the calculation of transformer inrush currents. *IEEE Transactions on Power Delivery* 27 (4): 2390–2398.

74 Azebi, S., Farazmand, A., Murali, B.P., and de León, F. (2013). A comparative study on π and T equivalent models for the analysis of transformer ferroresonance. *IEEE Transactions on Power Delivery* 28 (1): 526–528.

75 Foster, K., Werner, F.E., and Del Vecchio, R.M. (1982). Loss separation measurements for several electrical steels. *Journal of Applied Physics* 53: 8308–8310.

76 Zirka, S.E., Moroz, Y.I., Moses, A.J., and Arturi, C.M. (2011). Static and dynamic hysteresis models for studying transformer transients. *IEEE Transactions on Power Delivery* 26 (4): 2352–2362.

77 Gustavsen, B. and Semlyen, A. (1998). Calculation of transmission line transients using polar decomposition. *IEEE Transactions on Power Delivery* 13 (3): 855–862.

78 de León, F. and Semlyen, A. (1993). Time domain modeling of eddy current effects for transformer transients. *IEEE Transactions on Power Delivery* 8 (1): 271–280.

79 Avila-Rosales, J. and Alvarado, F.L. (1982). Nonlinear frequency dependent transformer model for electromagnetic transient studies in power systems. *IEEE Transactions on Power Apparatus and Systems* 101 (11): 4281–4288.

80 Fuchs, E.E., Yildirim, D., and Grady, W.M. (2000). Measurement of eddy-current loss coefficient PEC-R, derating of single-phase transformers, and comparison with K-factor approach. *IEEE Transactions on Power Delivery* 15 (1): 148–154.

81 Degeneff, R.C. (1977). A general method for determining resonances in transformer windings. *IEEE Transactions on Power Apparatus and Systems* 96 (2): 423–430.

82 Morched, A., Marti, L., and Ottevangers, J. (1993). A high frequency transformer model for the EMTP. *IEEE Transactions on Power Delivery* 8 (3): 1615–1626.

83 Shibuya, Y., Fujita, S., and Tamaki, E. (2001). Analysis of very fast transients in transformer. *IEE Proceedings C Generation, Transmission and Distribution* 148 (5): 377–383.

84 McNutt, W.J., Blalock, T.J., and Hinton, R.A. (1974). Response of transformer windings to system transient voltages. *IEEE Transactions on Power Apparatus and Systems* 93 (2): 457–466.

85 AlFuhaid, A.S. (2001). Frequency characteristics of single phase two winding transformer using distributed parameter modeling. *IEEE Transactions on Power Delivery* 16 (4): 637–642.

86 Popov, M., Sluis, L.V., Paap, G.C., and De Herdt, H. (2003). Computation of very fast transient overvoltages in transformer windings. *IEEE Transactions on Power Delivery* 18 (4): 1268–1274.

87 Ragavan, K. and Satish, L. (2005). An efficient method to compute transfer function of a transformer from its equivalent circuit. *IEEE Transactions on Power Delivery* 20 (2): 780–788.

88 Abeywickrama, K.G.N.B., Serdyuk, Y.V., and Gubanski, S.M. (2006). Exploring possibilities for characterization of power transformer insulation by frequency response analysis (FRA). *IEEE Transactions on Power Delivery* 21 (3): 1375–1382.

89 Gustavsen, B. (2010). A hybrid measurement approach for wideband characterization and modeling of power transformers. *IEEE Transactions on Power Delivery* 25 (3): 1932–1939.

90 Gustavsen, B. (2002). Computer code for rational approximation of frequency dependent admittance matrices. *IEEE Transactions on Power Delivery* 17 (4): 1093–1098.

91 Chimklai, S. and Marti, J.R. (1995). Simplified three-phase transformer model for electromagnetic transient studies. *IEEE Transactions on Power Delivery* 10 (3): 1316–1325.

92 Martinez-Velasco, J.A., Jatskevich, J., Filizadeh, S. et al. (2012). Modelling of power components for transient analysis. In: *Power System Transients*, 1–84. EOLSS.

93 Krause, P.C., Wasynczuk, O., and Sudhoff, S.D. (2002). *Analysis of Electric Machinery and Drive Systems*, 2e. Piscataway, NJ: IEEE Press.

94 Smith, A.C., Healey, R.C., and Williamson, S. (1996). A transient induction motor model including saturation and deep-rotor-bar effect. *IEEE Transactions on Energy Conversion* 11 (1): 8–15.

95 Ikeda, M. and Hiyama, T. (2007). Simulation studies of the transients of squirrel-cage induction motors. *IEEE Transactions on Energy Conversion* 22 (2): 233–239.

96 Sudhoff, S.D., Aliprantis, D.C., Kuhn, B.T., and Chapman, P.L. (2002). An induction machine model for predicting inverter-machine interaction. *IEEE Transactions on Energy Conversion* 17 (2): 203–210.

97 Levi, E. (1995). A unified approach to main flux saturation modeling in D-Q axis models of induction machines. *IEEE Transactions on Energy Conversion* 10 (3): 455–461.

98 Pekarek, S.D., Wasynczuk, O., and Hegner, H.J. (1998). An efficient and accurate model for the simulation and analysis of synchronous machine/converter systems. *IEEE Transactions on Energy Conversion* 13 (1): 42–48.

99 Brandwajn, V. (1980). Representation of magnetic saturation in the synchronous machine model in an electromagnetic transients program. *IEEE Transactions on Power Apparatus and Systems* 99 (5): 1996–2002.

100 IEEE Std. 115. (1995). IEEE Guide: Test Procedures for Synchronous Machines.

101 IEC 34–4. (1985). Rotating electrical machines – Part 4: Methods for determining synchronous machine quantities from tests.

102 Levi, E. (1998). State-space d-q axis models of saturated salient pole synchronous machines. *IEE Proceedings - Electric Power Applications* 145 (3): 206–216.

103 Levi, E. (1999). Saturation modelling in d-q axis models of salient pole synchronous machines. *IEEE Transactions on Energy Conversion* 14 (1): 44–50.

104 IEEE Task Force on Slow Transients (Chairman: M.R. Iravani) (1995). Modeling and analysis guidelines for slow transients. Part I: torsional oscillations; transient torques; turbine blade vibrations; fast bus transfer. *IEEE Transactions on Power Delivery* 10 (4): 1950–1955.

105 Karaagac, U., Mahseredjian, J., and Martinez-Velasco, J.A. (2009). Synchronous machines. In: *Power System Transients. Parameter Determination* (ed. J.A. Martinez-Velasco), 251–350. CRC Press.

106 Adjaye, R.E. and Cornick, K.J. (1979). Distribution of switching surges in the line-end coils of cable-connected motors. *Electric Power Applications* 2 (1): 11–21.

107 Wright, M.T., Yang, S.J., and McLeay, K. (1983). General theory of fast-fronted interturn voltage distribution in electrical machine windings. *Proceedings of the IEE* 130, 4: 245–256.

108 McLaren, P.G. and Oraee, H. (1985). Multiconductor transmission line model for the line end coil of large AC machines. *Proceedings of the IEE* 132, 3: 149–156.

109 Guardado, J.L. and Cornick, K.J. (1989). A computer model for calculating steep-fronted surge distribution in machine windings. *IEEE Transactions on Energy Conversion* 4 (1): 95–101.

110 Guardado, J.L., Carrillo, V., and Cornick, K.J. (1995). Calculation of interturn voltages in machine windings during switching transients measured on terminals. *IEEE Transactions on Energy Conversion* 10 (1): 87–94.

111 Lupo, G., Petrarca, C., Vitelli, M., and Tucci, V. (2002). Multiconductor transmission line analysis of steep-front surges in machine windings. *IEEE Transactions on Dielectrics and Electrical Insulation* 9 (3): 467–478.

112 Bacvarov, D.C. and Sarma, D.K. (1986). Risk of winding insulation breakdown in large ac motors caused by steep switching surges. Part I: computed switching surges. *IEEE Transactions on Energy Conversion* (1, 1): 130–139.

113 Rhudy, R.G., Owen, E.L., and Sharma, D.K. (1986). Voltage distribution among the coils and turns of a form wound ac rotating machine exposed to impulse voltage. *IEEE Transactions on Energy Conversion* 1 (2): 50–60.

114 van der Sluis, L. (2001). *Transients in Power Systems*. Wiley.

115 Browne, T.E. (ed.) (1984). *Circuit Interruption. Theory and Techniques*. Marcel Dekker.

116 Flurscheim, C.H. (ed.) (1985). *Power Circuit Breaker Theory and Design*. Peter Peregrinus.

117 Nakanishi, K. (ed.) (1991). *Switching Phenomena in High-Voltage Circuit Breakers*. Marcel Dekker.

118 Garzon, R.D. (ed.) (2002). *High Voltage Circuit Breakers. Design and Applications*, 2e. Marcel Dekker.

119 Martinez-Velasco, J.A. and Popov, M. (2009). Circuit breakers. In: *Power System Transients. Parameter Determination* (ed. J.A. Martinez-Velasco), 447–555. Boca Raton, FL: CRC Press.

120 EPRI Report EL-4651. (1987). EMTP Workbook II.

121 CIGRE Working Group 13.01 (1988). Practical application of arc physics in circuit breakers. Survey of calculation methods and application guide. *Electra* 118: 64–79.

122 CIGRE Working Group 13.01 (1993). Applications of black box modelling to circuit breakers. *Electra* 149: 40–71.

123 Kizilcay, M. (1985). Dynamic arc modelling in EMTP. *EMTP News* 5 (3): 15–26.

124 van der Sluis, L., Rutgers, W.R., and Koreman, C.G.A. (1992). A physical arc model for the simulation of current zero behaviour of high-voltage circuit breakers. *IEEE Transactions on Power Delivery* 7 (2): 1016–1022.

125 van der Sluis, L. and Rutgers, W.R. (1992). Comparison of test circuits for high-voltage circuit breakers by numerical calculations with arc models. *IEEE Transactions on Power Delivery* 7 (4): 2037–2045.

126 Cassie, A.M. (1939). Arc rupture and circuit severity: A new theory. CIGRE Session, Report 102.

127 Mayr, O. (1943). Beitraege zur theorie des statischen und dynamischen lichtbogens. *Archiv für Elektrotechnik* 37 (12): 588–608.

128 Martinez, J.A., Mahseredjian, J., and Khodabakhchian, B. (2005). Parameter determination for modeling systems transients. Part VI: circuit breakers. *IEEE Transactions on Power Delivery* 20 (3): 2079–2085.

129 Slade, P.G. (2008). *The Vacuum Interrupter. Theory, Design and Application*. CRC Press.

130 Glinkowski, M.T. and Greenwood, A. (1989). Computer simulation of post-arc plasma behavior at short contact separation in vacuum. *IEEE Transactions on Plasma Science* 17 (1): 45–50.

131 Yanabu, S., Homma, M., Kaneko, E., and Tamagawa, T. (1985). Post arc current of vacuum interrupters. *IEEE Transactions on Power Apparatus and Systems* 104 (1): 166–172.

132 van Lanen, E.P.A., Smeets, R.P.P., Popov, M., and van der Sluis, L. (2007). Vacuum circuit breaker postarc current modelling based on the theory of Langmuir probes. *IEEE Transactions on Plasma Science* 35 (4): 925–932.

133 Popov, M., van der Sluis, L., and Paap, G.C. (2001). Investigation of the circuit breaker reignition overvoltages caused by no-load transformer switching surges. *European Transactions on Electrical Power* 11 (6): 413–422.

134 Glinkowski, M.T., Gutierrez, M.R., and Braun, D. (1997). Voltage escalation and reignition behavior of vacuum circuit breakers during load shedding. *IEEE Transactions on Power Delivery* 12 (1): 219–226.

135 Roguski, T.A. (1989). Experimental investigation of the dielectric recovery strength between the separating contacts of vacuum circuit breaker. *IEEE Transactions on Power Delivery* 4 (2): 1063–1069.

136 Smeets, R.P.P., Funahashi, T., Kaneko, E., and Ohshima, I. (1993). Types of reignition following high-frequency current zero in vacuum interrupters with two types of contact material. *IEEE Transactions on Plasma Science* 21 (5): 478–483.

137 Helmer, J. and Lindmayer, M. (1996). Mathematical modelling of the high frequency behaviour of vacuum interrupters and comparison with measured transients in power systems. XVIIth International Symposium on Discharges and Electrical Insulation in Vacuum in Berkeley, USA.

138 Kosmac, J. and Zunko, P. (1995). A statistical vacuum circuit breaker model for simulation of transients overvoltages. *IEEE Transactions on Power Delivery* 10 (1): 294–300.

139 Child, C.D. (1911). Discharge from hot CaO. *Physical Review* 32 (5): 492–511.

140 Andrews, J.G. and Varey, R.H. (1971). Sheath growth in a low pressure plasma. *Physics of Fluids* 14 (2): 339–343.

141 Woodford, D.A. and Wedepohl, L.M. (1997). Impact of circuit breaker pre-strike on transmission line energization transients. International Conference on Power Systems Transients in Seattle, USA (June 1997).

142 EPRI Report EL-4202. (1985). Electromagnetic Transients Program (EMTP) Primer.

3

Solution Techniques for Electromagnetic Transient Analysis

Juan A. Martinez-Velasco

3.1 Introduction

A transient phenomenon in any type of system can be caused by a change of the operating conditions or the system configuration. Transient phenomena in power systems are caused by faults, switching operations, lightning strokes, or load variations. The importance of their study is mainly due to the effects these disturbances can have on the system performance or the failures they can cause to power equipment [1–9].

Stresses that can damage power equipment are of two types: overcurrents and overvoltages. Overcurrents may damage some power components due to excessive heat dissipation; overvoltages may cause insulation breakdowns (failure through solid or liquid insulation) or flashovers (insulation failure through air). Protection against overcurrents is performed by specialized equipment whose operation is aimed at disconnecting the faulted position from the rest of the system by separating the minimum number of power components from the unfaulted sections. Protection against overvoltages can be achieved by selecting an adequate insulation level of power equipment or by installing devices aimed at mitigating voltage stresses. In order to select an adequate protection against both types of stresses, it is fundamental to know their origin, estimate the most adverse conditions, and calculate the transients they can produce.

Modern power systems are complex and require advanced solution methods for analysis and design. The solution of most transients in power systems is not easy by hand calculation. For some case studies, one can drastically reduce the size of the equivalent circuit and obtain an equation whose solution can be found in textbooks. For the majority of transients, an accurate, or even an approximate, solution can be only obtained by means of a computer.

Power system transients can be classified according to the nature of the physical phenomena into electromechanical and electromagnetic. Electromechanical transients can be solved using some approximation methods: the power system is assumed to remain in quasi steady-state, whereas the response of the generating units (e.g. synchronous or asynchronous machines) is solved using differential equations in time-domain; such an approach can be used efficiently to simulate very large scale systems for rotor angle stability problems [10]. Electromagnetic transients correspond to the widest range of frequencies: power system variables are visualized in time-domain at the waveform level and can contain frequencies from dc to dozens of MHz. Such a wideband of frequencies requires detailed models and sophisticated solution methods.

Several tools have been used over the years to analyse electromagnetic transients. At early stages, miniature power system models, known as Transient Network Analyzers (TNAs), were used. At present, the digital computer is the most popular tool, although TNAs are still used. Since the release of the first digital computers, a significant effort has been dedicated to the development of numerical techniques and simulation tools aimed at solving transients in

Transient Analysis of Power Systems: A Practical Approach, First Edition. Edited by Juan A. Martinez-Velasco.
© 2020 John Wiley & Sons Ltd. Published 2020 by John Wiley & Sons Ltd.
Companion Website: www.wiley.com/go/martinez/power_systems

power systems. Hardware and software advances have also motivated the development of more powerful techniques and simulation tools. Computer transients programs have significantly reduced time and cost of simulations. Software packages used for computing electromagnetic transients can be classified into two main categories: off-line and on-line (real-time) [11]. The purpose of an off-line simulation tool is to conduct simulations on a generic computer. Off-line tools use powerful graphical user interfaces (GUIs), numerical methods, and programming techniques; they do not have any computing time constraints and can be made as precise as desired within the available data, models, and related mathematics. Real-time simulation tools are capable of generating results in synchronism with a real-time clock. Such tools are capable of interfacing with physical devices and maintaining data exchanges within the real-time clock [12]. The capability to compute and interface within real-time, imposes restrictions on the design of such tools. As this book is aimed at illustrating the scope of applications that can be covered with the Alternative Transients Program (ATP), this chapter targets only off-line solution methods.

The off-line techniques developed to date for computation of electromagnetic transients in power systems can be classified into two groups: time-domain and frequency-domain. Some hybrid approaches (i.e. a combination of both techniques) have also been proposed. Among the time-domain solution methods, the most popular one is the algorithm developed by H.W. Dommel [13–15], which is a combination of the trapezoidal rule and the method of characteristics, also known as Bergeron method. This algorithm was the origin of the ElectroMagnetic Transients Program (EMTP) [13, 16].

The model of an actual power system can include linear and nonlinear power delivery components, sources and control systems, as well as protection components. Techniques aimed at duplicating the behaviour of a power system during a transient phenomenon must include techniques for estimating the steady-state solution prior to the beginning of the transient, techniques to solve the system equations during the time loop and capable of coping with both linear and nonlinear elements, and techniques for solving the transient response of control systems [17].

The next sections provide an introduction to the most basic techniques implemented in the ATP for simulating electromagnetic transients in power systems and the dynamic behaviour of control systems [18]. A tool for simulating the behaviour of a power system during a transient phenomenon must also include capabilities for calculating the steady-state solution prior to the beginning of the transient. The chapter summarizes solution techniques for the computation of electromagnetic transients using single-phase representations, solutions to the numerical oscillations produced by the trapezoidal rule, methods to obtain initial steady-state solutions, and procedures to solve control systems and the interface between power networks and control systems. The last section presents a short overview of solution methods implemented in other transients tools.

3.2   Modelling of Power System Components for Transient Analysis

The goal of a power system is to satisfy the energy demand of a variety of users by generating, transmitting, and distributing the electric energy. These functions are performed by components whose design and behaviour are very complex; as a consequence, the analysis of transient phenomena in power systems is a difficult task due to the complexity of power components and the interactions that can occur among them.

The performance of power system components during a transient phenomenon depends on their design and dimensions, as well as on the nature of the transient phenomenon. Voltages and currents propagate along conductors with finite velocity; so, by default, certain models for electromagnetic transient analysis should consider that electrical parameters are distributed. Only when physical dimensions of those parts of a component affected by a transient are small compared with the wavelength of the main frequencies, a representation based on lumped parameters could be used. The selection of the most adequate representation of a power component in transient simulations is not an easy task due to the frequency ranges of the transients that can appear in power systems and to the different behaviour that a component can have for each frequency range.

However, a representation valid throughout a wide range of frequencies is not practically possible for most components. Consider, for instance, the behaviour of a transformer during electromagnetic transients. A transformer is a device whose behaviour is dominated by magnetic coupling between windings and core saturation when transients are of low or mid frequency. However, when the transients are caused by high-frequency disturbances, such as lightning strokes, then the behaviour of the transformer is dominated by stray capacitances as well as capacitances among windings. Modelling of power components, taking into account the frequency dependence of parameters, can be practically made by developing mathematical models that are accurate enough for a specific range of frequencies. Each range of frequencies usually corresponds to some particular transient phenomena. See Chapter 2.

Guidelines for representing power components in time-domain simulations have been the main subject of several publications [19–21]. The aspects to be considered in digital simulations of electromagnetic transients are the target of the study, the study zone of the test system to be represented in model, or the estimation of power component parameters [22].

For an introduction to the simulation of transients in power systems it is advisable to proceed with simplified models of the components involved in the phenomena to be analysed. Only single-phase models are used in this chapter for representing power components in transient phenomena, and only transients of electromagnetic nature are analysed; that is, the representation of mechanical parts is omitted.

A simplified model of many power components for electromagnetic transient analysis can be constructed by using basic circuit elements that can be classified into the following three categories:

1. *Sources.* They are used to represent power generators and external disturbances that can be the origin of some transients (e.g. lightning strokes). Two types of sources can be distinguished: a voltage source (Thevenin representation) and a current source (Norton representation). The equivalent scheme of a voltage source includes a series impedance, while the equivalent scheme of a current source incorporates a parallel admittance. An ideal behaviour for each type of source can be assumed by decreasing to zero the series impedance of a voltage source and the parallel admittance of a current source.

2. *Passive elements.* Depending on the transient phenomenon, the behaviour of some components can be either linear or nonlinear. The transformer is an example for which either a linear or a saturable model can be required. The representation of linear components will be based on lumped-parameter elements (resistance, inductance, capacitance) and distributed-parameter elements (single-phase lossless line). This list can be expanded by adding other basic elements such as the magnetic coupling or the ideal transformer. If the behaviour of a component is nonlinear, then its representation can include nonlinear resistances and/or saturable inductances.

3. *Switches.* They modify the configuration of a network by connecting or disconnecting components, although they will also be used to represent faults or short-circuits. The behaviour of an ideal switch can be summarized as follows: its impedance is infinite when it is opened

and zero when it is closed. It can close at any instant regardless of the voltage value at the source side; however, it will open only when the current goes through zero. The possibility of opening an ideal switch with non-zero currents can be needed to analyse some phenomena, e.g. the *current chopping* phenomenon. Several types of switches can be modelled by using different criteria to determine when they should open or close.

3.3 Solution Techniques for Electromagnetic Transients Analysis

3.3.1 Introduction

The transient analysis of large power systems is usually carried out by means of a time-domain numerical technique. This type of technique is based on the integration of the differential equations that represent the behaviour of a system. Numerical integration is used to transform the differential equations of a circuit element into algebraic equations that involve voltages, current, and past values. These algebraic equations can be interpreted as representing a resistive companion circuit of that circuit element. The equations of the whole companion network are assembled using a network technique (e.g. the nodal admittance equations), and solved as a function of time at discrete instants.

The studies to solve travelling wave problems by means of a digital computer were started in the early 1960's to solve small networks, with linear and nonlinear lumped-parameter, as well as distributed-parameter elements. This section presents the principles of the scheme developed by H.W. Dommel; it combines the Bergeron method and the trapezoidal rule into an algorithm capable of solving transients in single- and multi-phase networks with lumped and distributed parameters [15]. The scheme was initially implemented for solving linear systems, and later expanded to cope with nonlinear networks.

This section shows how to apply the Dommel scheme to the solution of circuits with linear and nonlinear elements, how to obtain the transient solution in circuits with switches, and how to solve the numerical oscillations caused by the trapezoidal rule.

3.3.2 Solution Techniques for Linear Networks

3.3.2.1 The Trapezoidal Rule
Assume that the differential equation that represents the behaviour of a circuit element is the following one:

$$\frac{dy(t)}{dt} = x(t) \tag{3.1}$$

This equation can be also written as follows:

$$y(t) = y(0) + \int_0^t x(z)dz \tag{3.2}$$

Figure 3.1 shows the principle of the numerical integration. The area corresponding to a given interval (t_{n-1}, t_n) is approximated by a trapezoid; that is, $x(t)$ is assumed to vary linearly inside the interval. Then, if the value of y has been computed at time t_{n-1}, the value at time t_n will be approximated by means of the following expression:

$$y_n = y_{n-1} + \frac{x_{n-1} + x_n}{2}(t_n - t_{n-1}) \tag{3.3}$$

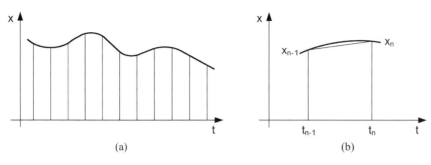

Figure 3.1 Application of the trapezoidal rule: (a) Primitive function. (b) Trapezoidal integration.

If the computation is carried out using a constant time interval Δt, also known as *integration time step*, the procedure can be expressed as follows:

$$y(t) = y(t - \Delta t) + \frac{\Delta t}{2}[x(t) + x(t - \Delta t)] \tag{3.4}$$

3.3.2.2 Companion Circuits of Basic Circuit Elements

The derivation of the companion equivalents of basic circuit elements is presented in the following paragraphs [15].

1. *Resistance.* Since the behaviour of a resistance is represented by an algebraic equation:

$$v_k(t) - v_m(t) = v_{km}(t) = R i_{km}(t) \tag{3.5}$$

the application of the trapezoidal rule is not needed and the companion circuit is the resistance itself.

2. *Inductance.* The behaviour of an inductance can be represented by a differential equation:

$$v_k(t) - v_m(t) = v_{km}(t) = L \frac{d i_{km}(t)}{dt} \tag{3.6}$$

The following expression is deduced from the application of the trapezoidal rule:

$$i_{km}(t) = i_{km}(t - \Delta t) + \frac{\Delta t}{2L}[v_{km}(t) + v_{km}(t - \Delta t)] \tag{3.7}$$

which can be rewritten as follows:

$$i_{km}(t) = \frac{\Delta t}{2L} v_{km}(t) + \left[\frac{\Delta t}{2L} v_{km}(t - \Delta t) + i_{km}(t - \Delta t) \right] \tag{3.8}$$

The second term at the right side is known as the *history term*. Using the notation:

$$I_{km}(t) = \left[\frac{\Delta t}{2L} v_{km}(t - \Delta t) + i_{km}(t - \Delta t) \right] \tag{3.9}$$

Equation (3.8) can be written as follows:

$$i_{km}(t) = \frac{\Delta t}{2L} v_{km}(t) + I_{km}(t) \tag{3.10}$$

Figure 3.2 shows the Norton companion circuit of an inductance, in which the history term is represented as an auxiliary current source.

3. *Capacitance.* The behaviour of a capacitance can be represented by a differential equation:

$$i_{km}(t) = C \frac{d}{dt}[v_k(t) - v_m(t)] = C \frac{d v_{km}(t)}{dt} \tag{3.11}$$

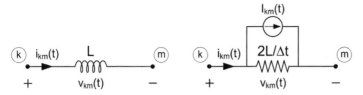

Figure 3.2 Companion circuit of an inductance.

The following expression is deduced from the application of the trapezoidal rule:

$$v_{km}(t) = v_{km}(t - \Delta t) + \frac{\Delta t}{2C}[i_{km}(t) + i_{km}(t - \Delta t)] \tag{3.12}$$

Solving for the current yields:

$$i_{km}(t) = \frac{2C}{\Delta t}v_{km}(t) - \left[\frac{2C}{\Delta t}v_{km}(t - \Delta t) + i_{km}(t - \Delta t)\right] \tag{3.13}$$

As for the inductance, the second term at the right side of (3.13) is regarded as a history-current term. Using the notation:

$$I_{km}(t) = -\left[\frac{2C}{\Delta t}v_{km}(t - \Delta t) + i_{km}(t - \Delta t)\right] \tag{3.14}$$

Equation (3.13) can be written as follows:

$$i_{km}(t) = \frac{2C}{\Delta t}v_{km}(t) + I_{km}(t) \tag{3.15}$$

Figure 3.3 shows the Norton companion circuit of a capacitance, in which the history–current term is also represented as an auxiliary current source.

Note: The term $I_{km}(t)$, known as *history current*, has been traditionally written as $hist(t - \Delta t)$. The notation used here is different: the instant specified in these terms is t because it is used in the companion model at time t, although all the current values are calculated at $(t - \Delta t)$; see [15].

4. *Ideal line.* The equations of a single-phase lossless line, see Figure 3.4, can be expressed as follows:

$$\frac{\partial v(x, t)}{\partial x} = -L\frac{\partial i(x, t)}{\partial t}$$
$$\frac{\partial i(x, t)}{\partial x} = -C\frac{\partial v(x, t)}{\partial t} \tag{3.16}$$

where L y C are respectively the inductance and capacitance per unit length, and x is the distance with respect to the sending end of the line.

After differentiating with respect to the variable x, these equations become:

$$\frac{\partial^2 v(x, t)}{\partial x^2} = LC\frac{\partial^2 v(x, t)}{\partial t^2}$$
$$\frac{\partial^2 i(x, t)}{\partial x^2} = LC\frac{\partial^2 i(x, t)}{\partial t^2} \tag{3.17}$$

The general solution of the voltage equation has the following form:

$$v(x, t) = f_1(x - vt) + f_2(x + vt) \tag{3.18}$$

where

$$v = \frac{1}{\sqrt{LC}} \tag{3.19}$$

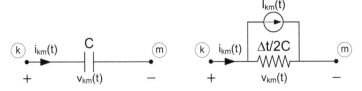

Figure 3.3 Companion circuit of a capacitance.

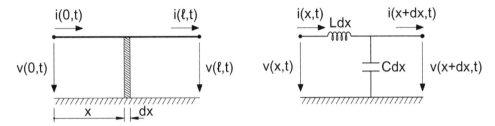

Figure 3.4 Scheme of a single-phase lossless line.

Both f_1 and f_2 are voltage functions, and υ is the so-called *propagation velocity*. Since the current equation in (3.17) has the same form that the voltage equation, its solution will also have a similar expression. It is, however, possible to obtain a general solution of the current equation based on that for the voltage. This solution could be expressed as follows:

$$i(x, t) = \frac{f_1(x - \upsilon t) - f_2(x + \upsilon t)}{Z_c} \tag{3.20}$$

where Z_c is the *surge impedance* of the line

$$Z_c = \sqrt{\frac{L}{C}} \tag{3.21}$$

Note that $f_1(x - \upsilon t)$ remains constant if the value of the quantity $(x - \upsilon t)$ is also constant. Thus, the function $f_1(x - \upsilon t)$ represents a voltage wave that propagates with velocity υ towards increasing x, while $f_2(x + \upsilon t)$ represents a voltage travelling wave towards decreasing x. Both waves are neither distorted nor damped while propagating along the line. The general solution of the voltage and the current at any point along an ideal line is, therefore, constructed by superposition of waves that travel in both directions. The expressions for f_1 and f_2 are determined from the boundary and initial conditions.

From (3.18) and (3.20) one can deduce that the ratio between voltage and current waves is always the surge impedance of the line, Z_c. However, this ratio can be positive or negative depending on the direction of propagation.

Upon manipulation of Eqs. (3.18) and (3.20), the following form is derived:

$$v(x, t) + Z_c i(x, t) = 2f_1(x - \upsilon t) \tag{3.22}$$

Assume ℓ is the line length and τ ($= \ell/\upsilon$) is the line travel time, then the value of $(v + Z_c i)$ at end m of the line in Figure 3.4 will be the same as at end k an interval of time τ before; thus:

$$v(m, t) + Z_c i(m, t) = v(k, t - \tau) + Z_c i(k, t - \tau) \tag{3.23}$$

Using the change of variables indicated in Figure 3.5:

$$v_k(t) = v(k, t) \qquad i_{km}(t) = +i(k, t)$$
$$v_m(t) = v(m, t) \qquad i_{mk}(t) = -i(m, t) \tag{3.24}$$

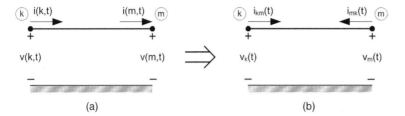

Figure 3.5 (a) Line boundary conditions. (b) Change of notation in the line variables.

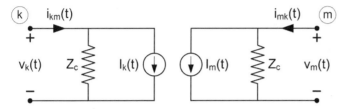

Figure 3.6 Companion circuit of a single-phase lossless line.

the above relationship becomes:

$$v_m(t) - Z_c i_{mk}(t) = v_k(t - \tau) + Z_c i_{km}(t - \tau) \tag{3.25}$$

Following a similar reasoning for waves travelling in the opposite direction, the relationship would be:

$$v_k(t) - Z_c i_{km}(t) = v_m(t - \tau) + Z_c i_{mk}(t - \tau) \tag{3.26}$$

Solving for currents at both line ends yields:

$$i_{km}(t) = \frac{v_k(t)}{Z_c} - \left[\frac{v_m(t - \tau)}{Z_c} + i_{mk}(t - \tau) \right]$$
$$i_{mk}(t) = \frac{v_m(t)}{Z_c} - \left[\frac{v_k(t - \tau)}{Z_c} + i_{km}(t - \tau) \right] \tag{3.27}$$

Using the notation:

$$I_k(t) = - \left[\frac{v_m(t - \tau)}{Z_c} + i_{mk}(t - \tau) \right]$$
$$I_m(t) = - \left[\frac{v_k(t - \tau)}{Z_c} + i_{km}(t - \tau) \right] \tag{3.28}$$

the equations of the line can be written as follows:

$$i_{km}(t) = \frac{v_k(t)}{Z_c} + I_k(t)$$
$$i_{mk}(t) = \frac{v_m(t)}{Z_c} + I_m(t) \tag{3.29}$$

These equations can be interpreted as the Norton companion circuit shown in Figure 3.6.

Table 3.1 shows a summary of the companion circuits derived from applying the trapezoidal rule to the circuits of lumped-parameter components and the Bergeron method to the circuit of a distributed-parameter lossless line.

Table 3.1 Companion circuits of basic circuit elements.

Element	Time-domain circuit	Companion circuit

Resistance

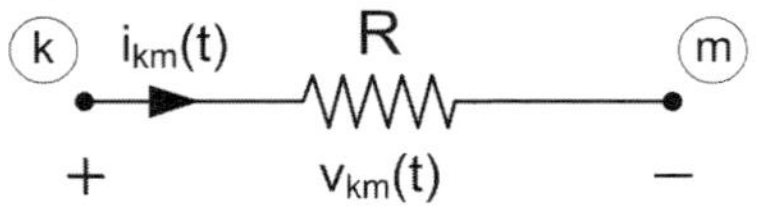

$$v_{km}(t) = Ri_{km}(t)$$

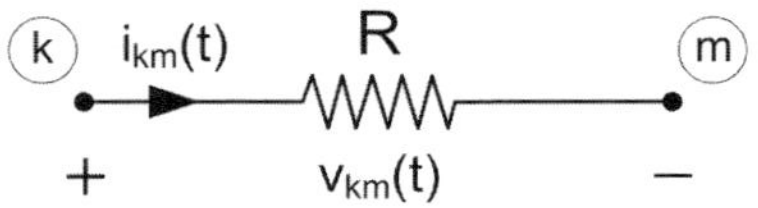

$$v_{km}(t) = Ri_{km}(t)$$

Inductance

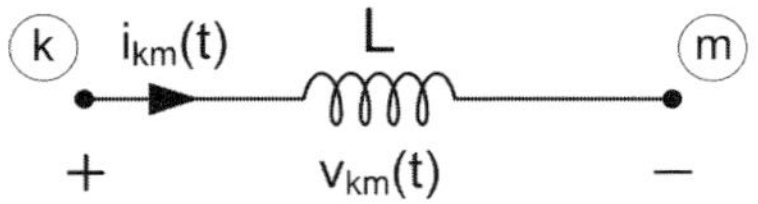

$$v_{km}(t) = L\frac{di_{km}(t)}{dt}$$

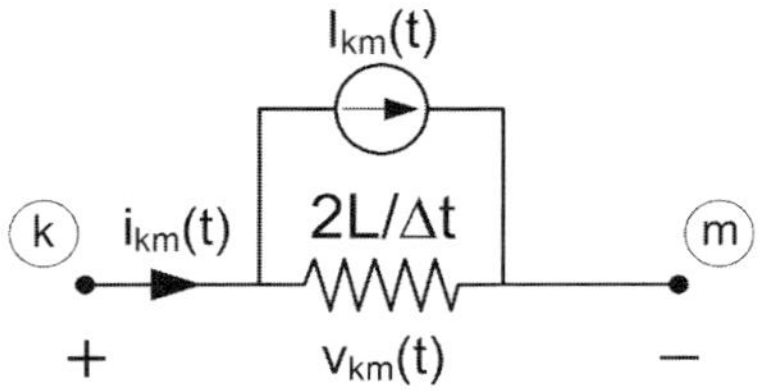

$$i_{km}(t) = \frac{\Delta t}{2L}v_{km}(t) + I_{km}(t)$$

$$I_{km}(t) = \left[\frac{\Delta t}{2L}v_{km}(t - \Delta t) + i_{km}(t - \Delta t)\right]$$

(Continued)

Table 3.1 (Continued)

Element	Time-domain circuit	Companion circuit
Capacitance		
Lossless line		

Capacitance, time-domain:

$$i_{km}(t) = C\frac{dv_{km}(t)}{dt}$$

Capacitance, companion circuit:

$$i_{km}(t) = \frac{2C}{\Delta t}v_{km}(t) + I_{km}(t)$$

$$I_{km}(t) = -\left[\frac{2C}{\Delta t}v_{km}(t-\Delta t) + i_{km}(t-\Delta t)\right]$$

Lossless line, time-domain:

$$\frac{\partial v(x,t)}{\partial x} = -L\frac{\partial i(x,t)}{\partial t}$$

$$\frac{\partial i(x,t)}{\partial x} = -C\frac{\partial v(x,t)}{\partial t}$$

Lossless line, companion circuit:

$$i_{km}(t) = \frac{v_k(t)}{Z_c} + I_k(t) \quad I_k(t) = -\left[\frac{v_m(t-\tau)}{Z_c} + i_{mk}(t-\tau)\right]$$

$$i_{mk}(t) = \frac{v_m(t)}{Z_c} + I_m(t) \quad I_m(t) = -\left[\frac{v_k(t-\tau)}{Z_c} + i_{km}(t-\tau)\right]$$

3.3.2.3 Computation of Transients in Linear Networks

Several frameworks can be used to solve the equations of linear resistive networks; the approach implemented in ATP is based on the nodal admittance equations detailed below. The derivation of the algebraic nodal equations for a linear network is illustrated by means of Figure 3.7, which shows several circuit elements connected to a given node [15].

The following equations are deduced from the companion circuit of each branch connected to this node:

$$i_{12}(t) = \frac{1}{R} v_{12}(t)$$

$$i_{13}(t) = \frac{\Delta t}{2L} v_{13}(t) + I_{13}(t) \qquad I_{13}(t) = \frac{\Delta t}{2L} v_{13}(t - \Delta t) + i_{13}(t - \Delta t)$$

$$i_{14}(t) = \frac{2C}{\Delta t} v_{14}(t) + I_{14}(t) \qquad I_{14}(t) = -\frac{2C}{\Delta t} v_{14}(t - \Delta t) - i_{14}(t - \Delta t)$$

$$i_{15}(t) = \frac{1}{Z_c} v_1(t) + I_{\ell 1}(t) \qquad I_{\ell 1}(t) = -\frac{1}{Z_c} v_5(t - \tau) - i_{51}(t - \tau) \qquad (3.30)$$

The application of the Kirchhoff's current law to node 1 in the figure gives:

$$i_{12}(t) + i_{13}(t) + i_{14}(t) + i_{15}(t) = i_1(t) \qquad (3.31)$$

The following equation is obtained upon substitution of (3.30) into (3.31):

$$\left[\frac{1}{R} + \frac{\Delta t}{2L} + \frac{2C}{\Delta t} + \frac{1}{Z_c}\right] v_1(t) - \frac{1}{R} v_2(t) - \frac{\Delta t}{2L} v_3(t) - \frac{2C}{\Delta t} v_4(t)$$

$$= i_1(t) - I_{13}(t) - I_{14}(t) - I_{\ell 1}(t) \qquad (3.32)$$

Using the previous procedure with all nodes, the equations of a network of any size can be assembled and written as follows [15]:

$$[G][v(t)] = [i(t)] - [I(t)] \qquad (3.33)$$

where G is the nodal conductance matrix, $v(t)$ is the vector of node voltages, $i(t)$ is the vector of current sources, and $I(t)$ is the vector of history terms.

If Δt is constant, only the right-hand side of Eq. (3.33) must be updated at each time step as elements of the nodal conductance matrix remain constant. In addition, the resulting conductance matrix is symmetrical. The solution of the transient process is then obtained using

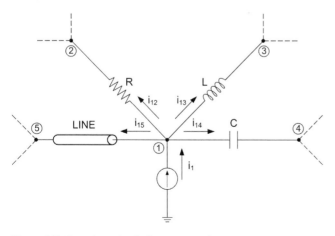

Figure 3.7 Generic node of a linear network.

triangular factorization. One of the main advantages of this procedure is that it can be applied to networks of arbitrary size in a very simple fashion.

In general, some nodes of a power system have known voltages because voltage sources are connected to them, so the order of the vector of unknown node voltages can be reduced. Assume that A and B are used to denote the sets of unknown and known node voltages, respectively. Nodal admittance equations can then be rewritten in a partitioned matrix form [15]:

$$\begin{bmatrix} G_{AA} & G_{AB} \\ G_{BA} & G_{BB} \end{bmatrix} \begin{bmatrix} v_A(t) \\ v_B(t) \end{bmatrix} = \begin{bmatrix} i_A(t) \\ i_B(t) \end{bmatrix} - \begin{bmatrix} I_A(t) \\ I_B(t) \end{bmatrix} \tag{3.34}$$

Upon performing the partitioned matrix product in (3.34) and discarding part B, the following set of equations are obtained [15]:

$$[G_{AA}][v_A(t)] = [i_A(t)] - [I_A(t)] - [G_{AB}][v_B(t)] \tag{3.35}$$

The application of the above technique, based on the companion circuit equivalents, is illustrated with the following example.

3.3.2.4 Example: Transient Solution of a Linear Network

Figure 3.8 depicts the system to be analysed. A single-phase lossless overhead line, characterized by its surge impedance, Z_c, and its travel time, τ, feeds a capacitor at its receiving end. According to the figure, the supply side is represented by a dc voltage source in series with its equivalent inductance. The goal of this study is to establish the equations that should be used to obtain the voltages at each end of the overhead line.

Assume the switch is already closed. From the companion circuit of each element of the test circuit, the companion circuit shown in Figure 3.9 is obtained.

The current equations for the two circuit nodes in Figure 3.9 take the following form:

$$\frac{e(t) - v_1(t)}{\dfrac{2L}{\Delta t}} + I_L(t) = \frac{v_1(t)}{Z_c} + I_{l1}(t)$$

$$\frac{v_2(t)}{Z_c} + I_{l2}(t) + \frac{v_2(t)}{\dfrac{\Delta t}{2C}} + I_C(t) = 0$$

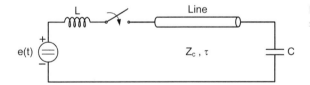

Figure 3.8 Example: diagram of the test system.

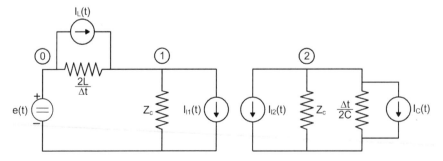

Figure 3.9 Example: companion circuit of the test system.

with (see Table 3.1)

$$I_L(t) = \frac{\Delta t}{2L} v_L(t - \Delta t) + i_L(t - \Delta t) \qquad v_L(t) = e(t) - v_1(t)$$

$$I_C(t) = -\left[\frac{2C}{\Delta t} v_C(t - \Delta t) + i_C(t - \Delta t)\right] \qquad v_C(t) = v_2(t)$$

$$I_{l1}(t) = -\left[\frac{v_2(t - \tau)}{Z_c} + i_{21}(t - \tau)\right] \qquad I_{l2}(t) = -\left[\frac{v_1(t - \tau)}{Z_c} + i_{12}(t - \tau)\right]$$

The nodal admittance equations can be written as follows:

$$e(t)\frac{\Delta t}{2L} + I_L(t) - I_{l1}(t) = v_1(t)\left[\frac{\Delta t}{2L} + \frac{1}{Z_c}\right]$$

$$-I_C(t) - I_{l2}(t) = v_2(t)\left[\frac{1}{Z_c} + \frac{2C}{\Delta t}\right]$$

The following form can be then derived:

$$\begin{bmatrix} \frac{\Delta t}{2L} + \frac{1}{Z_c} & 0 \\ 0 & \frac{1}{Z_c} + \frac{2C}{\Delta t} \end{bmatrix} \begin{bmatrix} v_1(t) \\ v_2(t) \end{bmatrix} = \begin{bmatrix} e(t)\frac{\Delta t}{2L} + I_L(t) - I_{l1}(t) \\ -I_C(t) - I_{l2}(t) \end{bmatrix}$$

Upon solving voltage nodes, the final solution is as follows:

$$\begin{bmatrix} v_1(t) \\ v_2(t) \end{bmatrix} = \begin{bmatrix} \dfrac{2LZ_c}{2L + Z_c\Delta t} & 0 \\ 0 & \dfrac{Z_c\Delta t}{2Z_c C + \Delta t} \end{bmatrix} \begin{bmatrix} e(t)\frac{\Delta t}{2L} + I_L(t) - I_{l1}(t) \\ -I_C(t) - I_{l2}(t) \end{bmatrix}$$

3.3.3 Networks with Nonlinear Elements

3.3.3.1 Introduction

The scheme detailed above can be used to solve linear networks. However, many power components (e.g. transformers, surge arresters) exhibit nonlinear behaviour. Several modifications to the basic method have been proposed to cope with nonlinear and time-varying elements [15, 23]. Two different schemes are used with ATP: (i) compensation methods; (ii) piecewise linear representations. Both approaches are detailed below for power systems components that can be represented as single-phase nonlinear elements.

3.3.3.2 Compensation Methods

In a compensation-based method, a nonlinear element is simulated as a current injection that is super-imposed on the linear network after a solution without the nonlinear element has been found [24]. These methods are first presented for solving systems with one nonlinear element and then for systems with two or more nonlinear elements [15].

One nonlinear element: Assume that the network contains only one nonlinear element between nodes k and m, as indicated in Figure 3.10. The compensation method states that the nonlinear branch is initially excluded from the network solution and later simulated as a current source that leaves node k and enters node m.

The current i_{km} must fulfil two equations: the network equation of the linear part (represented by the instantaneous Thevenin equivalent circuit between nodes k and m)

$$i_{km} = v_{km-0} - R_{Thev} i_{km} \qquad (v_{km-0} = v_{k-0} - v_{m-0}) \tag{3.36}$$

(subscript '0' indicates solution without the nonlinear branch, $v_{km} = v_k - v_m$) and the relationship of the nonlinear branch itself

$$v_{km} = f(i_{km}, di_{km}/dt, t, \cdots) \tag{3.37}$$

The value of the Thevenin resistance R_{Thev} in Eq. (3.36) is pre-computed once before entering the time step loop, and re-computed whenever switches open and close.

The network Eq. (3.35) can be rewritten as

$$[G_{AA}][v_A] = [k_A] \tag{3.38}$$

with $[k_A]$ being the known right-hand side from (3.35).

The solution with compensation proceeds as follows in each time step:

1) Compute the node voltages $[v_{A-0}]$ without the nonlinear branch. From this solution and the other known voltages $[v_B]$, extract the open circuit voltage $v_{km-0} = v_{k-0} - v_{m-0}$.
2) Solve Eqs. (3.36) and (3.37) simultaneously for i_{km}. If Eq. (3.37) is given analytically, then the Newton-Raphson can be used. If Eq. (3.36) is defined point-by-point as a piecewise linear curve, then the intersection of the two curves must be found through a search procedure, as indicated in Figure 3.10.
3) Find the final solution by superimposing the response of the current i_{km}.

$$[v_A] = [v_{A-0}] - [r_{Thev}]i_{km} \tag{3.39}$$

Remember that superposition is permissible as long as the rest of the system is linear.

The simultaneous solution of the network equation with the nonlinear equation, as illustrated in Figure 3.10, is straightforward if the nonlinear branch is a nonlinear resistance defined by $v_{km} = f(i_{km})$, or if it is a time-varying resistance with $v_{km} = R(t)i_{km}$.

Two or more nonlinear elements: Assume the system model has M nonlinear branches. Since the compensation method models each branch as a current source, then, M vectors $[r_{Thev-1}]$, ... $[r_{Thev-M}]$ must be pre-computed (and re-computed whenever switches change position). The Thevenin equivalent resistance becomes an $M \times M$-matrix $[R_{Thev}]$ in this case. The first column of this matrix is created by calculating the differences $r_{Thev-ki} - r_{Thev-mi}$ for all M nonlinear elements $i = 1, \ldots M$ from $[r_{Thev-1}]$, the second column by doing the same from $[r_{Thev-2}]$, etc. In the solution process, step (1) remains identical as with one element, but step (2) now requires the solution of M nonlinear equations. Step (3) uses M vectors $[r_{Thev-1}]$, ... $[r_{Thev-M}]$ in place of one vector.

Lines with distributed parameters decouple the network equations for the two ends; see the companion circuit shown in Figure 3.6: phenomena at one line end are seen at the other end one travel time τ later. Nonlinear elements decoupled by distributed-parameter lines

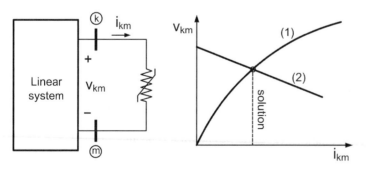

Figure 3.10 Principles of the compensation method with a nonlinear resistance.

can therefore be solved independently from each other, because each has its own Thevenin equivalent equation decoupled from the others. The $[r_{Thev}]$-vector of a particular nonlinearity will only have nonzero entries for the nodes of its own area. Therefore, all $[r_{Thev}]$-vectors can be merged into a single vector, at the expense of another vector which contains the area number for each component [13].

If an M-phase Thevenin equivalent circuit must be used, in cases where the M nonlinear elements are not separated by travel time, then a system of nonlinear equations must be solved. The Newton-Raphson iteration method is an efficient approach for solving systems of nonlinear equations [5].

Since ATP works with a fixed time step size Δt, numerical problems can arise with nonlinear elements. If Δt is too large, artificial negative damping or hysteresis can occur and cause numerical instability.

3.3.3.3 Piecewise Linear Representation

The solution for nonlinear elements represented by means of a piecewise linear representation is summarized separately for inductances and resistances.

- *Piecewise linear inductance*. A saturable inductance can be accurately represented as a piecewise linear inductance with two or more slopes (Figure 3.11). In the ATP, the conductance matrix is changed and re-triangularized whenever the solution moves from one straight-line segment to another segment. This approach essentially permits any number of piecewise linear segments. The approach for inductances with hysteretic behaviour, see Figure 3.12, is similar. Moving along any linear segment is still described by the same differential equation $v = L(di/dt)$ used for linear inductance. Therefore, the representation in the transient solution part is the equivalent resistance $2L/\Delta t$ in parallel with a current source known from the history in the preceding time step (Figure 3.2). However, the equivalent resistance must be changed whenever the simulation moves from one segment into another. The fact that the linear segment does not (in general) pass through the origin is automatically accounted for by the history terms. Starting from a residual flux is permitted. This representation with hysteresis can be tricky to use; see [25] for more details.
- *Piecewise linear resistance*. History terms are not needed in this case. Each linear segment with a slope of $R = dv/di$ is represented as a voltage source v_{KNEE} in series with a resistance R, or a current source v_{KNEE}/R in parallel with a resistance R (Figure 3.13).

Multi-slope piecewise linear elements can create special problems [15]. For instance, if the piecewise linear resistance is used to model a silicon-carbide surge arrester with a spark gap, then the ATP does not automatically know which segment it should jump to after sparkover. Therefore, the user must specify the segment number as part of the input data; this may require a trial run.

Figure 3.11 Piecewise linear inductance – Characteristic curve.

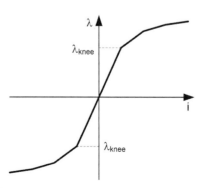

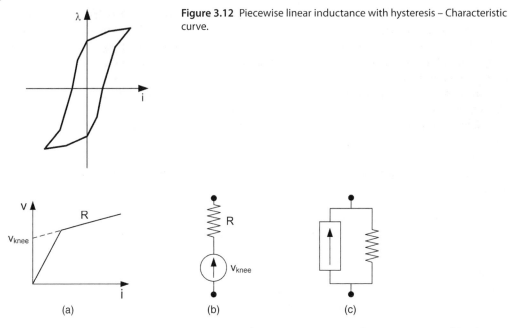

Figure 3.12 Piecewise linear inductance with hysteresis – Characteristic curve.

Figure 3.13 Piecewise linear resistance. (a) Piecewise linear representation – Characteristic curve. (b) Voltage source representation. (c) Current source representation.

All piecewise linear representations cause overshoots, because the need for changing to the next segment is only recognized after the last point has gone outside its proper range. Caution is therefore needed in the choice of Δt to keep the overshoot small. The overshoot is usually less severe on piecewise linear inductances because the flux, being the integral over the voltage, cannot change abruptly, while voltages across nonlinear resistances can change very quickly. A cure for the overshoot problem would be an interpolation method which moved the solution backwards by a fraction of Δt, and then restarted the solution again at that point with Δt; the points along the time axis would then no longer be spaced at equal distances; see [15].

Both the piecewise linear representation and the compensation method suffer from the fact that nonlinear inductances are approximated as linear inductances in the ac phasor solution. Since transformers and shunt reactors do saturate in normal steady-state operation, a jump from the linear to the nonlinear characteristic will therefore occur at $t = 0$. Whether the jump occurs in λ or i, or in both depends to some extent on the type of representation [15].

3.3.4 Solution Methods for Networks with Switches

A very simple approach for simulating transients in a linear network is to use ideal switches for modifying the configuration of the network and for connecting/disconnecting elements. There is more than one way of handling changing switch positions in the transient solution loop. Different schemes can be used to solve networks with switches: (i) a switch is represented as a resistance; (ii) the compensation method; (iii) reconfiguration of nodal admittance matrix.

The switch can be represented as a resistance with a very large value if the switch is open and a very small value if the switch is closed. Very large values of R do not cause numerical problems in solution methods based on nodal equations, but very small values can cause

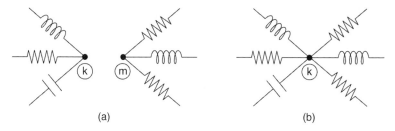

Figure 3.14 Representation of switches. (a) Open. (b) Closed.

numerical problems. For a switch located between nodes k and m, the calculation of the switch current under any circumstance is trivial

$$i_{km} = \frac{v_{km}}{R} = \frac{v_k - v_m}{R} \tag{3.40}$$

Using compensation to represent M switches, an M-phase Thevenin equivalent circuit would be pre-computed with an equation of the form (3.36). The switch currents, which are needed for the superposition calculation (see Section 3.3.3), are

$$[i_{km}] = 0 \text{ if all switches are open} \tag{3.41a}$$

$$[i_{km}] = [R_{Thev}]^{-1}[v_{km-0}] \text{ if all switches are closed} \tag{3.41b}$$

If only some switches are closed, the $[R_{Thev}]$ is a submatrix obtained from the full matrix after deleting the rows and columns for the open switches. The switch currents are automatically obtained in this approach, and there should not be any numerical problems.

A third approach is to change the network connections whenever a switch position changes. As indicated in Figure 3.14, there are two nodes whenever the switch is open, and only a single node whenever the switch is closed. This approach can be implemented in at least two different ways [15]:

- *Network reduction to switch nodes.* Before entering the time step loop, normal Gaussian elimination is used on those nodes with unknown voltages (subset A) which do not have switches connected to them. For the rest of the nodes of subset A with switches, all switches are assumed to be open. Whenever a switch position changes in the time step loop, this reduced matrix is first modified to reflect the actual switch positions. After this modification, the triangularization is completed for the entire matrix of subset A.
- *Complete re-triangularization.* The matrix is built and triangularized completely again whenever switch positions change, or when the slope of piecewise linear elements changes. The current is calculated from the original row of either node k or node m, with all switches open, with the proper right-hand side. With this scheme, any number of switches can be connected to any node, as long as the current in each switch is uniquely defined.

3.3.5 Numerical Oscillations

In some situations, such as switching operations or transitions between segments in piecewise-linear inductances, the trapezoidal rule acts as a differentiator, and introduces sustained numerical oscillations. In addition, the trapezoidal rule filters out high-frequency voltages for given current injections and amplifies high-frequency currents for given voltages across capacitances [15].

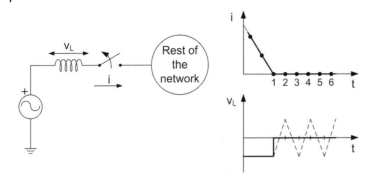

Figure 3.15 Voltage after current interruption.

Consider the case of a current interrupted in a circuit breaker (Figure 3.15). The exact solution for v_L is shown as a solid line, with a sudden jump to zero at the instant of current interruption, whereas the ATP solution is shown as a dotted line. Since

$$v_L(t) = \frac{2L}{\Delta t}(i(t) - i(t - \Delta t)) - v_L(t - \Delta t) \qquad (3.42)$$

and assuming that the voltage solution was correct prior to current interruption, it follows that $v_L(t) = -v_L(t - \Delta t)$ in points 2, 3, 4, … as soon as the current at $(t - \Delta t)$ and t become zero; therefore, the solution for v_L will oscillate around zero with the amplitude of the pre-interruption value.

In some cases, a sudden jump could be an unacceptable answer that would indicate improper modelling of the real system. An example would be the calculation of transient recovery voltages: since any circuit breaker would reignite if the voltage was to rise with an infinite rate of rise immediately after current interruption, the cure could consist on including the proper stray capacitance between the circuit breaker terminals.

A similar problem can arise when the circuit configuration has a capacitance loop: the trapezoidal can cause numerical oscillations in the capacitance current.

Since these numerical oscillations always oscillate around the correct answer (i.e. around zero in Figure 3.15), the smooth option available in ATP can be used to get rid of them and obtain a better response. However, it is crucial to avoid this problem if these oscillations can cause numerical problems in other parts of the network.

Several techniques have been proposed to control or reduce these numerical oscillations. These techniques can be classified into two groups: techniques that use an external solution and techniques based on a modification of the solution technique. Numerical oscillations with ATP can be solved by means of an external solution. One of these techniques uses additional damping to force oscillations to decay [26]. This damping is externally provided by adding fictitious resistances in parallel with inductances, and in series with capacitors; see Figure 3.16. This solution was suggested in [27, 28].

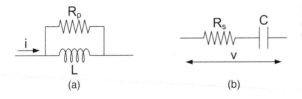

(a)   (b)

Figure 3.16 Additional damping. (a) Parallel damping for inductances. (b) Series damping for capacitances.

From the application of the trapezoidal solution to the parallel circuit of Figure 3.16a the following form is derived

$$v(t) = \frac{1}{\frac{\Delta t}{2L} + \frac{1}{R_p}}(i(t) - i(t - \Delta t)) - \frac{R_p - \frac{2L}{\Delta t}}{R_p + \frac{2L}{\Delta t}}v(t - \Delta t) \tag{3.43}$$

If a current impulse is injected into this circuit (e.g. with i rising linearly to I_{max} between 0 and Δt, dropping linearly back to zero between Δt and $2\Delta t$, and staying at zero thereafter), then, after the impulse has dropped back to zero, the first term in Eq. (3.43) will disappear, and the second term will begin oscillating

$$v(t) = -\alpha \cdot v(t - \Delta t) \tag{3.44}$$

with

$$\alpha = \frac{R_p - \frac{2L}{\Delta t}}{R_p + \frac{2L}{\Delta t}} \tag{3.45}$$

being the reciprocal of the damping factor.

A expression to obtain the value of R_p from the value of α can be derived

$$R_p = k_p \frac{2L}{\Delta t} \tag{3.46}$$

where

$$k_p = \frac{1 + \alpha}{1 - \alpha} \tag{3.47}$$

This oscillation term will be damped if $\alpha < 1$, and will disappear in one time step for $R_p = (2L/\Delta t)$ or $\alpha = 0$ (critically damped case). If R_p is too large, the damping effect is too small; if R_p is reduced until it approaches the value $(2L/\Delta t)$, then an unacceptable error can be introduced into the inductance representation.

The numerical oscillations analysed above can appear in capacitance currents when there is an abrupt change in dv/dt; however, these situations are less common and will seldom appear. In any case, a solution to these oscillations might be based on the damping provided by series resistances, as shown in Figure 3.16b. Following a similar analysis with the series damping resistance to be used in series with capacitances, the following expressions are derived

$$i(t) = -\alpha \cdot i(t - \Delta t) \tag{3.48}$$

with

$$\alpha = \frac{\frac{\Delta t}{2C} - R_s}{\frac{\Delta t}{2C} + R_s} \tag{3.49}$$

being the reciprocal of the damping factor.

A expression to obtain the value of R_s from the value of α can be derived

$$R_s = k_s \frac{\Delta t}{2C} \tag{3.50}$$

where

$$k_s = \frac{1-\alpha}{1+\alpha} \tag{3.51}$$

Typical range for k_p is 5–10; typical range for k_s is 0.1–0.2 [15].

This method can have an important effect on the accuracy of the solution, although the external resistance can be physically justified (e.g. losses of an inductor can be represented by means of a parallel resistor).

Another technique is based on the application of *snubber circuits* in parallel with switches/valves. A snubber circuit consists of a capacitor in series with a resistance, see Figure 3.17. This option is particularly interesting in power electronics applications as snubber circuits are very often placed in parallel with semiconductors to limit overvoltages across them. No matter whether they have to be installed or not, parallel snubber circuits can be used as a solution for numerical oscillations in applications where an inductive current is going to be interrupted. This solution was proposed by H. K. Law in 1980 [29], and is usually applied in power electronics simulations when semiconductors are in series with inductances.

Some basic rules have been proposed for estimating the parameters of the snubber circuit [29]: they will depend on the semiconductor peak current and the series parameters (e.g. load parameters). A basic rule to be followed in all applications is that the value of the snubber impedance should be much larger than the equivalent impedance of the load; that is, the maximum current that will flow across the snubber circuit should be a small percentage of the maximum current that will flow across the parallel valve or switch.

Figure 3.18 shows the results derived from the disconnection of an ideal inductor. The parameters of this circuit are: peak value of the voltage source, $V_{peak} = 100$ V; inductance, $L = 0.1$ mH; frequency, $f = 50$ Hz. The time step is 20 μs. The first plot (i.e. Figure 3.18b) shows

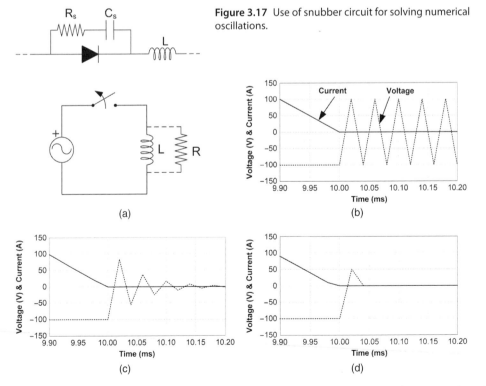

Figure 3.17 Use of snubber circuit for solving numerical oscillations.

(a)

(b)

(c)

(d)

Figure 3.18 Numerical oscillations when opening an inductive current. (a) Scheme of the circuit. (b) Without damping resistor. (c) With damping resistor ($k = 10$). (d) With damping resistor ($k = 1$).

the oscillations that occur once the inductor current reaches the zero value: the voltage across the inductor begins to oscillate. It can easily be checked whether the peak of the inductor voltage oscillates around zero with the amplitude of the pre-interruption value (i.e. 100 V). When external damping is added, oscillations disappear in a number of time steps that will depend on parameter k_p: with $k_p = 5$, the oscillating term fades away in about 0.20 ms; in a critically damped case ($k_p = 1$, $R_p = 2L/\Delta t$), the oscillations disappear in 40 µs, but the simulation error is not negligible as one can observe by comparing the voltage values with those obtained in the previous case ($k_p = 10$). Remember that the time required to fully damp oscillations depends on the time step used in the simulation.

Figure 3.19 shows a very simple case of single-phase rectifier simulation. Plots with and without snubber circuit in parallel with the diode illustrates this problem and its solution. The sinusoidal 50-Hz voltage source has a peak value of 100 V; the parameters of the circuit load are $R = 10\,\Omega$, $L = 4$ mH. The time step used in all simulations is 20 µs. It is evident from the plot shown in Figure 3.19b that without a parallel snubber circuit, the voltage across the RL load is not correct: this plot shows how numerical oscillations can occur across the RL load voltage during the period in which the diode current is zero. Figures 3.19c and d illustrate the effect of a parallel snubber circuit. The impedance values at power frequency (i.e. 50 Hz) and the time constants values are as follows:

- Figure 3.19c $Z_{load} = 11.26\,\Omega$ $Z_{snubber} \approx 319\,000\,\Omega$ $R_s C_s = 10$ µs
- Figure 3.19d $Z_{load} = 11.26\,\Omega$ $Z_{snubber} \approx 3190\,\Omega$ $R_s C_s = 100$ µs.

One can observe that the oscillations in the first case with a parallel snubber circuit need several time steps to get fully damped, while they disappear in the second case after two time steps; this is due to a much smaller impedance of the snubber circuit in parallel with the diode. In both cases $Z_{load} \ll Z_{snubber}$. However, the time constants $R_s C_s$ are respectively smaller and greater than the simulation time step.

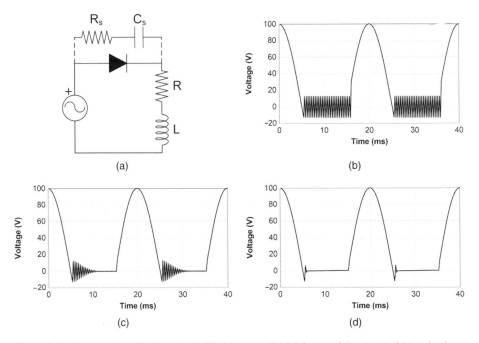

Figure 3.19 Numerical oscillations in a half-bridge circuit. (a) Scheme of the circuit. (b) Load voltage without snubber. (c) Load voltage with snubber ($R_s = 1000\,\Omega$, $C_s = 0.01$ µF). (d) Load voltage with snubber ($R_s = 100\,\Omega$, $C_s = 1$ µF).

3.4   Transient Analysis of Control Systems

The development of a section for representation of control systems in transients programs was initially motivated by studies of high-voltage direct current (HVDC) links. The Transient Analysis of Control Systems (TACS) option was implemented in the Bonneville Power Administration (BPA) EMTP in 1976 [30]. Although the main goal was the simulation of HVDC converters, it soon became obvious that TACS capabilities had many other applications, such as the representation of excitation of synchronous generators, dynamic arcs in circuit breakers, or protective relays.

Control systems are represented in TACS by block diagrams with interconnection between system elements. Control elements can be transfer functions, FORTRAN algebraic functions, logical expressions and some special devices (e.g. frequency sensor, min/max tracking, transport delay). The solution method used by TACS is also based on the trapezoidal rule.

A control block in the s-domain can be described by the following relationship [15]:

$$X(s) = G(s)U(s) \tag{3.52}$$

where $U(s)$ and $X(s)$ are respectively the input and the output in the Laplace domain, and $G(s)$ is a rational transfer function

$$G(s) = K\frac{N_0 + N_1 s + \cdots + N_m s^m}{D_0 + D_1 s + \cdots + D_n s^n} \qquad (m \leq n) \tag{3.53}$$

Transfer functions are converted into algebraic equations in the time-domain [15]:

$$cx(t) = Kdu(t) + hist(t - \Delta t) \tag{3.54}$$

where K is the gain, while constants c and d are obtained from the coefficients of $G(s)$ [15, 30].

A control system with linear blocks result in a system of equations with the following general form:

$$[A_{xx}][x(t)] + [A_{xu}][u(t)] = hist(t - \Delta t) \tag{3.55}$$

The resulting algebraic equations of a control system are by nature asymmetrical. Due to this fact, the electric network and the control system are solved separately. First, the network solution is advanced, next, network variables are passed to the control section, and then control equations are solved. Finally, the network receives control commands. The whole procedure introduces a time-step delay; see Figure 3.20.

Components other than transfer functions can be included in a TACS section, but they are seen as nonlinear blocks and not directly added into the simultaneous solution of transfer

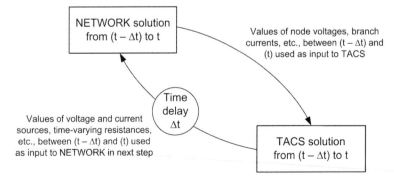

Figure 3.20 Interface between a network and a control system.

functions. When a nonlinear block is inside a closed-loop configuration, a true simultaneous solution is not possible. The procedure implemented in the TACS solution is simultaneous only for linear blocks, (i.e. *s*-transfer functions), and sequential for nonlinear blocks. When these blocks are present, the loop is broken and the system is solved by inserting a time delay. These delays inside control loops, as well as the delay between the network and the control system, are the sources of different effects. Instabilities, inaccuracies, and numerical oscillations produced by delays have been reported.

TACS capabilities in ATP have been expanded over the years; users can now implement supplemental devices by means of the TACS 69 device and use more complex structures, such as conditional branching (IF-THEN-ELSE-ENDIF). A more detailed list of ATP TACS capabilities is provided in Chapter 4.

Although the first release of TACS was a powerful and flexible tool, new applications have been demanding capabilities other than those implemented in the original version. One example is the digital controls used in HVDC converters and other flexible ac transmission systems (FACTS) devices. The execution of tasks only when needed, the simulation of conditional branching (IF-THEN-ELSE), or the manipulation of vector arrays are capabilities not available in the first TACS releases. Some of these limitations can be overcome by a second option for representation of control systems. The language MODELS [31], initially known as 'New TACS', was developed to substitute the TACS program. However, it became obvious that both options provided complementary capabilities; therefore, TACS was preserved in ATP. The interface MODELS-Network is the same as used by TACS, so there is also a one time-step delay between the network solution and the MODELS solution. The MODELS option has a capability to avoid a time-step delay when the control section combines linear and nonlinear 'blocks'. From the user's point of view, the selection of one of these options can be based on several aspects (e.g. type of application, CPU time, complexity of the model to be developed, etc.).

3.5 Initialization

3.5.1 Introduction

A general test system configuration can consist of an electrical network that may contain linear and nonlinear components, circuit breakers, semiconductor converters, protective relays, rotating machines and control systems. An automatic initialization step prior to entering the time loop is not possible in the most general situation. In such cases, the user may consider a simple *brute force* approach: the simulation is started without performing any initial calculation and carried out long enough to allow the transients to settle down to steady-state conditions. If the system is linear, with or without rotating machines, but without control systems, then some ATP initialization procedures can be used. Some illustrative examples are presented in Chapters 7 and 8. The principles of the ATP capabilities to be used for automatic steady-state initialization are summarized below, together with some initialization options applied to networks with control systems.

3.5.2 Initialization of the Power Network

3.5.2.1 Options for Steady-State Solution Without Harmonics

The solution of a transient phenomenon is dependent on the initial conditions with which the transient begins. Although some simulations can be performed with zero initial conditions, there are many instances for which the simulation must begin from power-frequency steady-state operating conditions. Transient studies are usually conducted when the test power system is in steady-state. This means that the simulator must be able to initialize the test system in order to obtain the steady state prior to any time-domain transient solution. At least two alternatives can be used with ATP for steady-state initialization: (i) ac steady-state solution; (ii) load flow solution. While the ac steady-state solution begins with the knowledge of all voltage sources, the load flow solution is based on the knowledge of some voltage and power constraints at certain system nodes, to which either system sources or system loads are connected. Load flow calculations are followed by the solution of the ac steady-state solution, once the voltages of all nodes to which sources are connected are known.

3.5.2.2 Steady-State Solution

The steady-state solution of linear networks at a single frequency is a rather simple task, and can be obtained using nodal admittance equations [15]:

$$[Y][V] = [I] \tag{3.56}$$

where $[Y]$ is the nodal complex admittance matrix, $[V]$ is the vector of node voltages, $[I]$ is the vector of current sources. Elements of both $[V]$ and $[I]$ are also complex phasor values.

As for the transient solution, this equation can be partitioned by separating nodes to ground for which the voltage is unknown, part A, from nodes for which the voltage is known, part B (e.g. voltage sources to ground)

$$\begin{bmatrix} Y_{AA} & Y_{AB} \\ Y_{BA} & Y_{BB} \end{bmatrix} \begin{bmatrix} V_A \\ V_B \end{bmatrix} = \begin{bmatrix} I_A \\ I_B \end{bmatrix} \tag{3.57}$$

From this equation, a form for obtaining $[V_A]$ can easily be deduced [15]:

$$[Y_{AA}][V_A] = [I_A] - [Y_{AB}][V_B] \tag{3.58}$$

The ac steady-state solution of a network with switches is very simple: the network model uses two nodes for open switches, and one node for closed switches. Therefore, the system of equations of one network depends on the initial status of all system switches.

The ATP routine for automatic steady-state initialization can be used with linear systems. This routine is, however, a useful tool on its own since there are many useful studies for which only a steady-state solution is required. Some application examples will be presented in subsequent chapters. Some of these applications are summarized below.

1) *Frequency scan.* ATP has a routine that can automatically vary the input frequency between two limits specified by the user. With this option, the solution is obtained at each frequency in the way detailed above. The output that can be derived from its application is the variation of voltages, currents, or impedances as a function of the frequency. This option can be very useful for finding the frequency-dependent impedance of a network seen from a particular location. Some of its applications are presented in Chapters 6 and 9.

2) *Different frequencies in disconnected parts.* ATP can find the steady-state solution in networks with sources having different frequencies, provided the network is disconnected into subnetworks, with each subnetwork only containing sources with the same frequency. This option is used for initializing the universal machine (see Chapter 7), and can be used to analyse trapped charge effects or systems with HVDC links. If the system under study has two disconnected subnetworks with frequencies f_1 and f_2, respectively, the nodal admittance equations can be expressed as follows [15]:

$$\begin{bmatrix} [Y_1] & 0 \\ 0 & [Y_2] \end{bmatrix} \begin{bmatrix} [V_1] \\ [V_2] \end{bmatrix} = \begin{bmatrix} [I_1] \\ [I_2] \end{bmatrix} \tag{3.59}$$

in which f_1 has been used in forming $[Y_1]$, and f_2 in forming $[Y_2]$.

3) *Steady-state initialization of dc networks.* For dc solutions, some program modifications are required because under a dc condition an inductance becomes short-circuit. Although ATP capabilities include dc source models, to obtain the steady-state initialization of a dc network, the dc sources are represented as ac sources with a very low frequency.
4) *Steady-state initialization with nonlinear elements.* The approach used by ATP is very simple: a nonlinear inductance is approximated as a linear inductance for initialization. However, since nonlinear elements are approximated as linear elements in the ac phasor solution, a sudden jump can occur at $t = 0$ between the linear and nonlinear representations. For nonlinear inductances, the problem can be minimized through proper voltage source rotations [15].

3.5.3 Load Flow Solution

ATP has a multiphase Load Flow module that can be used to obtain the initial conditions of the power system under study. A load-flow solution is based on constraints: the sources are replaced by PQ, PV, or slack bus constraints; the loads are replaced by PQ constraints. Upon convergence of the load-flow solution, all steady-state phasors become available. To apply this capability, network components must provide a load-flow solution model which is usually that used in steady-state calculations.

The load (power) flow option presently implemented in ATP was added to the BPA EMTP in 1983 by F. Rasmussen (Elkraft, Denmark) [15]. It adjusts the magnitude and angles of sinusoidal sources iteratively in a sequence of steady-state solutions, until specified voltage and power constraints (i.e. active and reactive power, active power and voltage magnitude, or some other specified criteria) are achieved.

Nodes at which the user specifies active power P and reactive power Q (or some other combination of P, Q, voltage magnitude, and voltage angle) are treated as voltage sources in the direct solution of the system of linear Eq. (3.58). After this solution, the current at the P, Q-nodes is calculated from the equations of part B which have been left out in (3.58):

$$I_k = \sum_{m=1}^{n} Y_{km} V_m \tag{3.60}$$

for all nodes k of part B, except for slack node and then the power from

$$P_k - jQ_k = I_k V_k^* \tag{3.61}$$

The calculated values of P_k, Q_k are then compared with the values specified by the user. Based on these differences, corrections are made to the angle θ_k and magnitude $|V_k|$ of each voltage V_k [15]:

$$\Delta\theta_k = \frac{P_k - P_{k-specified}}{\frac{1}{2}(|P|_k + |P_{k-specified}|)} \cdot 2.5F \tag{3.62a}$$

$$\Delta|V_k| = \frac{Q_k - Q_{k-specified}}{\frac{1}{2}(|Q|_k + |Q_{k-specified}|)} \cdot 100F \tag{3.62b}$$

F is a deceleration factor which decreases from 1.0 to 0.25 in 500 iterations, with the formula

$$F = \left(\frac{1000 - k}{1000}\right)^2 \tag{3.63}$$

where k is the iteration step.

Once the voltages have been corrected, another direct solution of Eq. (3.58) is obtained. This cycle of calculations is repeated until $\Delta\theta_k$ and $\Delta|V_k|$ become sufficiently small.

This method is comparable to the Gauss-Seidel load-flow solution method applied to the reduced system (i.e. part *B*).

The ATP load-flow solution method may be sensitive to the type of network being studied, and its convergence may be slow even if the reduced system is small.

3.5.4   Initialization of Control Systems

Control systems can be very complicated and so is their initialization.

- The input and output signals of a transfer function are usually dc quantities in steady state; if the control system consists of transfer function blocks only, a system of equations similar to (3.55) can be formed and solved; this is basically what ATP can do automatically with TACS initialization.
- A control system may have integrators; their steady-state initialization must be supplied by the user, but the initial values cannot be found easily. See Chapter 13 of [15] for a discussion about this aspect.
- TACS signal sources are not restricted to dc quantities in steady state; they can be pulse trains, sinusoidal functions, and other periodic functions. This complexity makes the implementation of an initialization procedure that could cope with all options extremely difficult.
- The ac steady-state solution of the electric network is found first; therefore, before TACS variables are initialized, NETWORK variables are available for automatic initialization in control systems. This may cause problems if a TACS output is used to define a sinusoidal voltage source in NETWORK whose initial amplitude and angle, as supplied by the user, could differ from the values coming from the TACS initialization.

The automatic initialization of a control system in ATP is restricted to only linear control systems defined by linear control blocks with dc and ac inputs. As an alternative, the user may specify an initial value for every TACS variable, but this option becomes impractical when the control system is complex and can include any type of control devices/blocks.

3.6   Discussion

Currently, most tools for transients simulation are based on the Dommel scheme. Time-domain simulation of electromagnetic transients using digital computers began in the early 1960's; since then a significant effort has been dedicated to solve some of the main drawbacks and limitations of the original scheme.

An ideal software tool for simulation of transients in power systems should provide users with capabilities for modelling any type of component for any type of transient phenomena, and solution techniques that would cope with all type of studies that the user could require (transient solution for linear and nonlinear systems, transient solution of control systems, steady-state solution regardless of the system model complexity). In reality, all tools have some limitations that affect modelling and simulation tasks (e.g. not all required models are available, there is no solution technique that can obtain the steady-state solution of a power system whose model includes control strategies).

Current software packages offer high levels of modelling and simulation capabilities for studying electromagnetic and electromechanical transients. This section summarizes the techniques implemented in ATP and other transients tools for solving transient phenomena in power systems.

3.6.1 Solution Techniques Implemented in ATP

The list of aspects ATP users should account for when constructing and testing the system model might include the following items:

- Numerical oscillations can occur as a consequence of some switching operations. Although they are usually an indication of improper modelling, there are some situations in which a simple model is desired. Some cures for this problem have been presented in Section 3.3.5; they will also be used in other chapters of this book (see Chapters 6 and 8).
- Some options are available in ATP for steady-state initialization; however, they cannot be efficiently used in many situations (e.g. systems whose model combines complex electrical networks and control systems, systems with nonlinear elements). A brute force approach cannot always be used (e.g. for systems with several synchronous generators).
- The Load Flow option implemented in ATP, known as FIX SOURCE, can be used efficiently with small power systems. For large or huge systems, the convergence could be very slow.
- Some cures for steady-state initialization problems caused by the limitations of the solution techniques currently implemented in ATP are suggested with some case studies presented in other chapters (see, for instance, Chapters 7 and 8).

3.6.2 Other Solution Techniques

A significant effort has been dedicated to overcome the limitations listed above and to expand capabilities in transient simulation tools. This section provides a list of techniques proposed to cope with the transient solution and steady state initialization for application in EMTP-like tools.

3.6.2.1 Transient Solution of Networks

The trapezoidal rule is an A-stable method which does not produce run-off instability [32]. However, this rule suffers from some drawbacks: it uses a fixed time-step size and can originate sustained numerical oscillations.

The step size determines the maximum frequency that can be simulated, so users have to know the frequency range of the transient simulation to be performed in advance; on the other hand, both slow and fast transients can occur at the same time in different system sections. A procedure by which two or more time-step sizes can be used in the trapezoidal integration was presented in [33].

Some efficient techniques developed to avoid numerical oscillations are based on the temporary modification of the solution method, when only numerical oscillations can occur, without affecting the rest of the simulation. One of these techniques is based on the CDA (Critical Damping Adjustment) procedure [34, 35]: during a switching operation, CDA uses a backward Euler rule and two half-size integration steps. This method does not require recalculation of the admittance matrix. Other techniques are based on interpolation [36]. The procedure presented in [37] uses two time-step sizes and represents switching devices (power electronics components) by means of characteristic curves. A modified interpolation is the approach proposed in some works, see [38–40]. A new formulation based on a modified-augmented-nodal analysis (MANA) was introduced in [41]. Later improvements provided a high ability for accommodating device equations, an empowered solution of nonlinearities and control systems, and a very flexible steady-state initialization, including power flow solution, see [42]. See also [22].

3.6.2.2 Transient Analysis of Control Systems

Some effort has also been dedicated to overcome limitations and solve problems originated by TACS:

- Improvements to solve internal time delays, initialization problems and some FORTRAN code limitations were presented in [43].
- Limitations in FORTRAN code capabilities were solved by developing an interface between TACS and FORTRAN subroutines. The interface presented in [44] maintains full TACS capabilities and takes advantage of the FORTRAN flexibility to represent digital controls.
- Several techniques can be used to simultaneously solve power network and control system equations and avoid problems related to the interface delay. Two procedures using compensation were presented in [45] and [46].
- A different and simple solution using filter interposition to solve inaccuracies caused by the interface time delay was proposed in [47].
- A new fixed-point approach where both the power system and control system are solved sequentially was presented in [48].

3.6.2.3 Steady-State Initialization

Steady-state initialization becomes a very difficult task when the system model includes nonlinear elements, power electronics converters and control systems. Automatic methods to obtain the initialization when the system model includes control systems do not yet exist. An initial solution with harmonics can be obtained using some simple approaches.

The simplest one is *brute force* (see Section 3.5.1). This approach can be very accurate, but its convergence can be very slow if the network has components with light damping. In the presence of synchronous and/or asynchronous machines, the simulation may reach abnormal operating modes if no proper initialization is applied. Even if the system model includes frequency-dependent line models (with increased damping over constant parameter models), the transients without initialization can require a very long period (e.g. several seconds) for attaining the actual steady-state response. This can demand very long computing time for large systems.

A more efficient method is to perform an approximate linear ac steady-state solution with nonlinear branches disconnected or represented by linearized models.

Some tools have a *snapshot* feature: the state of the system is saved after a run, so later runs can be started at this point. Using a brute-force initialization, the system is started from standstill and, once the steady-state has been reached, a snapshot is taken and saved for later studies.

Another option is to manually specify initial conditions (e.g. trapped charge). This can be useful for reproducing complex conditions such as those that can occur in a ferroresonance study; however, it becomes impractical in cases of large and complex systems.

An interesting method, known as *Initialization with Harmonics* (IwH), was presented in [49]. This procedure uses an iterative solution based on the frequency-domain representation of nonlinear inductances such as voltage-dependent harmonic current sources. An improved version of the IwH method, using a harmonic Norton modelling of nonlinear branches and a Newton-Raphson steady-state calculation was presented in [50].

Several procedures have been proposed to calculate initial conditions using time-domain techniques, such as gradient and shooting methods [51], or the waveform relaxation technique [52].

Hybrid approaches to calculate initial conditions in nonlinear networks using both frequency- and time-domain techniques have also been developed [53].

An improved initialization procedure, the Multiphase Harmonic Load Flow (MHLF), based on branch equations, was presented in [54, 55]. In this method, static compensators and other nonlinear elements, under balanced or unbalanced conditions, are represented by harmonic Norton equivalent circuits. Further improvements of this procedure incorporated a synchronous machine model into the initialization procedure [56].

Power-flow solution methods for application with transients tools have also been developed. A multiphase power-flow solution using a Newton-type method was presented in [57]. A different approach that takes advantage of a MANA formulation was presented in [58]; see also [22].

Acknowledgement

A high percentage of this chapter is based on material originally presented in the book written by H.W. Dommel under contract with Bonneville Power Administration (BPA); see reference [15]. During more than three decades that book has been a very valuable source of information for the author who is in debt to H.W. Dommel and BPA for the edition of the reference manual.

References

1 Bewley, L.V. (1951). *Traveling Waves on Transmission Systems*. New York (NY, USA): Wiley.

2 Peterson, H.A. (1951). *Transients in Power Systems*. New York (NY, USA): Wiley.

3 Rudenberg, R. (1968). *Electrical Shock Waves in Power Systems*. Cambridge (MA, USA): Harvard University Press.

4 Ragaller, K. (ed.) (1977). *Current Interruption in High-Voltage Networks*. New York (NY, USA): Plenum Press.

5 Ragaller, K. (ed.) (1979). *Surges in High-Voltage Networks*. New York (NY, USA): Plenum Press.

6 Greenwood, A. (1991). *Electrical Transients in Power Systems*, 2e. New York (NY, USA): Wiley.

7 Chowdhuri, P. (2003). *Electromagnetic Transients in Power Systems*, 2e. Taunton (UK): RS Press-Wiley.

8 van der Sluis, L. (2001). *Transients in Power Systems*. Chichester (UK): Wiley.

9 Das, J.C. (2010). *Transients in Electrical Systems. Analysis, Recognition, and Mitigation*. New York (NY, USA): McGraw-Hill.

10 Kundur, P. (1994). *Power System Stability and Control*. New York: McGraw-Hill.

11 Dufour, C. and Bélanger, J. (2015). Real-time simulation technologies in engineering. In: *Chapter 4 of Transient Analysis of Power Systems. Solution Techniques, Tools and Applications* (ed. J.A. Martinez-Velasco). Chichester (UK): Wiley – IEEE Press.

12 Mahseredjian, J., Dinavahi, V., and Martinez, J.A. (2009). Simulation tools for electromagnetic transients in power systems: overview and challenges. *IEEE Transactions on Power Delivery* 24 (3): 1657–1669.

13 Dommel, H.W. (1969). Digital computer solution of electromagnetic transients in single- and multi-phase networks. *IEEE Transactions on Power Apparatus and Systems* 88 (2): 734–741.

14 Dommel, H.W. (1997). Techniques for analyzing electromagnetic transients. *IEEE Computer Applications in Power* 10 (3): 18–21.

15 Dommel, H.W. (1986). *Electromagnetic Transients Program. Reference Manual (EMTP Theory Book)*. Portland (OR, USA): Bonneville Power Administration.

16 Dommel, H.W. and Scott Meyer, W. (1974). Computation of electromagnetic transients. *Proceedings of the IEEE* 62 (7): 983–993.

17 Mahseredjian, J., Kocar, I., and Karaagac, U. (2015). Solution techniques for electromagnetic transients in power systems. In: *Chapter 2 of Transient Analysis of Power Systems. Solution Techniques, Tools and Applications* (ed. J.A. Martinez-Velasco). Chichester (UK): Wiley/IEEE Press.

18 Kizilcay, M. and Hoidalen, H.K. (2015). EMTP-ATP. In: *Chapter 2 of Numerical Analysis of Power System Transients and Dynamics* (ed. A. Ametani). Stevenage (UK): The Institution of Engineering and Technology.

19 IEC TR 60071-4. (2004). *Insulation co-ordination - Part 4: Computational Guide to Insulation Co-ordination and Modelling of Electrical Networks*.

20 CIGRE WG 33.02. (1990). Guidelines for Representation of Network Elements when Calculating Transients, CIGRE Brochure no. 39.

21 Gole, A.M., Martinez-Velasco, J.A., and Keri, A.J.F. (eds.) (1999). *Modeling and Analysis of System Transients Using Digital Programs*. IEEE PES Special Publication, TP-133-0.

22 Martinez-Velasco, J.A. (2010). Parameter determination for electromagnetic transient analysis in power systems. In: *Chapter 1 of Power System Transients. Parameter Determination* (ed. J.A. Martinez-Velasco). Boca Raton (FL, USA): CRC Press.

23 Dommel, H.W. (1971). Nonlinear and time-varying elements in digital simulation of electromagnetic transients. *IEEE Transactions on Power Apparatus and Systems* 90 (6): 2561–2567.

24 Tinney, W.F. (1972). Compensation methods for network solutions by optimally ordered triangular factorization. *IEEE Transactions on Power Apparatus and Systems* 91 (1): 123–127.

25 Frame, J.G., Mohan, N., and Liu, T.H. (1982). Hysteresis modeling in an electro-magnetic transients program. *IEEE Transactions on Power Apparatus and Systems* 101 (9): 3403–3412.

26 Alvarado, F.L., Lasseter, R.H., and Sanchez, J.J. (1983). Testing of trapezoidal integration with damping for the solution of power transient studies. *IEEE Transactions on Power Apparatus and Systems* 102 (12): 3783–3790.

27 Brandwajn, V. (1982). Damping of numerical noise in the EMTP solution. *EMTP Newsletter* 2 (3): 10–19.

28 Alvarado, F. (1982). Eliminating numerical oscillations in trapezoidal integration. *EMTP Newsletter* 2 (3): 20–32.

29 Lauw, H.K. (1980). Design recommendations for numerical stability of power electronic converters. *EMTP Newsletter* 1 (4): 2–6.

30 Dube, L. and Dommel, H.W. (1977). Simulation of control systems in an electromagnetic transients program with TACS. In: *Proceedings of IEEE, PICA*, 266–271.

31 Dubé, L. and Bonfanti, I. (1992). MODELS: a new simulation tool in the EMTP. *European Transactions on Electrical Power Engineering* 2 (1): 45–50.

32 Tripathy, S.C., Rao, N.D., and Elangovan, S. (1978). Comparison of stability properties of numerical integration methods for switching surges. *IEEE Transactions on Power Apparatus and Systems* 97 (6): 2318–2326.

33 Semlyen, A. and de León, F. (1993). Computation of electromagnetic transients using dual or multiple time steps. *IEEE Transactions on Power Systems* 8 (3): 1274–1281.

34 Marti, J.R. and Lin, J. (1989). Suppression of numerical oscillations in the EMTP. *IEEE Transactions on Power Systems* 4 (2): 739–747.

35 Lin, J. and Marti, J.R. (1990). Implementation of the CDA procedure in the EMTP. *IEEE Transactions on Power Systems* 5 (2): 394–402.

36 Kulicke, B. (1981). Simulation program NETOMAC: difference conductance method for continuous and discontinuous systems. *Siemens Research and Development Reports* 10 (5): 299–302.

37 Maguire, T.L. and Gole, A.M. (1991). Digital simulation of flexible topology power electronic apparatus in power systems. *IEEE Transactions on Power Delivery* 6 (4): 1831–1840.

38 Sana, A.R., Mahseredjian, J., Dai-Do, X., and Dommel, H. (1995). Treatment of discontinuities in time-domain simulation of switched networks. *Mathematics and Computers in Simulation* 38: 377–387.

39 Kuffel, P., Kent, K., and Irwin, G. (1997). The implementation and effectiveness of linear interpolation within digital simulation. *Electrical Power and Energy System* 19 (4): 221–228.

40 Zoua, M., Mahseredjian, J., Joos, G. et al. (2006). Interpolation and reinitialization in time-domain simulation of power electronic circuits. *Electrical Power and Systems Research* 76: 688–694.

41 Mahseredjian, J. and Alvarado, F. (1997). Creating an electromagnetic transients program in MATLAB: MatEMTP. *IEEE Transactions on Power Delivery* 12 (1): 380–388.

42 Mahseredjian, J., Dennetière, S., Dubé, L. et al. (2007). On a new approach for the simulation of transients in power systems. *Electrical Power and Systems Research* 77: 1514–1520.

43 Lasseter, R. and Zhou, J. (1994). TACS enhancements for the electromagnetic transient program. *IEEE Transactions on Power Systems* 9 (2): 736–742.

44 Bui, L.X., Casoria, S., Morin, G., and Reeve, J. (1992). EMTP TACS-FORTRAN interface development for digital controls modelling. *IEEE Transactions on Power Systems* 7 (1): 314–319.

45 Araujo, A.E.A., Dommel, H.W., and Marti, J.R. (1993). Simultaneous solution of power and control equations. *IEEE Transactions on Power Systems* 8 (4): 1483–1489.

46 Lefebvre, S. and Mahseredjian, J. (1995). Improved control systems simulation in the EMTP through compensation. *IEEE Transactions on Power Delivery* 10 (4): 1654–1662.

47 Cao, X., Kurita, A., Yamanaka, T. et al. (1996). Suppression of numerical oscillation caused by the EMTP-TACS interface using filter interposition. *IEEE Transactions on Power Delivery* 11 (4): 2049–2055.

48 Mahseredjian, J., Dubé, L., Zou, M. et al. (2006). Simultaneous solution of control system equations in EMTP. *IEEE Transactions on Power Systems* 21 (1): 117–124.

49 Dommel, H.W., Yan, A., and Wei, S. (1986). Harmonics from transformer saturation. *IEEE Transactions on Power Systems* 1 (2): 209–215.

50 Lombard, X., Masheredjian, J., Lefebvre, S., and Kieny, C. (1995). Implementation of a new harmonic initialization method in EMTP. *IEEE Transactions on Power Delivery* 10 (3): 1343–1352.

51 Perkins, B.K., Marti, J.R., and Dommel, H.W. (1995). Nonlinear elements in the EMTP: steady-state initialization. *IEEE Transactions on Power Systems* 10 (2): 593–601.

52 Wang, Q. and Marti, J.R. (1996). A waveform relaxation technique for steady state initialization of circuits with nonlinear elements and ideal diodes. *IEEE Transactions on Power Delivery* 11 (3): 1437–1443.

53 Murere, G., Lefebvre, S., and Do, X.D. (1995). A generalized harmonic balance method for EMTP initialization. *IEEE Transactions on Power Delivery* 10 (3): 1353–1359.

54 Xu, W., Marti, J.R., and Dommel, H.W. (1991). A multiphase harmonic load flow solution technique. *IEEE Transactions on Power Systems* 6 (1): 174–182.

55 Xu, W., Marti, J.R., and Dommel, H.W. (1991). Harmonic analysis of systems with static compensators. *IEEE Transactions on Power Systems* 6 (1): 183–190.

56 Xu, W., Marti, J.R., and Dommel, H.W. (1991). A synchronous machine model for three-phase harmonic analysis and EMTP initialization. *IEEE Transactions on Power Systems* 6 (4): 1530–1538.

57 Allemong, J.J., Bennon, R.J., and Selent, P.W. (1993). Multiphase power flow solutions using EMTP and Newton's method. *IEEE Transactions on Power Systems* 8 (4): 1455–1462.

58 Peralta, J.A., de León, F., and Mahseredjian, J. (2008). Unbalanced multi-phase load-flow using a positive-sequence load-flow program. *IEEE Transactions on Power Systems* 23 (2): 469–476.

To Probe Further

Several ATP data files are provided in the website of this book. The cases deal with very simple problems related to some of the solution techniques summarized in this chapter. ATP beginners are encouraged to run the cases and check the results.

4

The ATP Package: Capabilities and Applications

Juan A. Martinez-Velasco and Jacinto Martin-Arnedo

4.1 Introduction

ATP is an acronym that stands for Alternative Transients Program [1, 2]. It is a non-commercial simulation program based on the ElectroMagnetic Transients Program (EMTP) developed by Bonneville Power Administration [3, 4]. ATP was originally developed for simulation of transients in power systems; current ATP capabilities allow users to apply it to the analysis of electromagnetic and electromechanical transients (i.e. overvoltages, subsynchronous resonance, wind power generation), the development of models for power electronics devices (FACTS, Custom Power), or the design of protective relays.

Since the first ATP version was released, several simulation tools have been designed to help ATP users. These tools can be integrated in a single package that users can customize for specific applications. Presently, the acronym ATP is used to denote a package that consists of at least three tools: (i) the graphical user interface (GUI) ATPDraw [5]; (ii) the simulation tool TPBIG; (iii) a postprocessor for graphical output.

The acronym ATP was originally used to name the simulation tool. For many users, ATP is still the simulation tool; in this book ATP is used interchangeably to name indistinctly the package or the tool, while TPBIG is used to name the simulation tool.

ATPDraw is an interactive Windows-based GUI that can act as a shell for the whole package; that is, users can control the execution of all programs integrated in the package from ATPDraw. Actually, the ATP package can include many tools depending on the case studies and results in which a user is interested. ATP standard components (i.e. built-in models available in TPBIG) are supported by this tool. Users can also create custom-made objects based on ATP capabilities such as DATA BASE MODULE and MODELS language [6]. By creating new objects and adding programs that can be controlled from ATPDraw, users can design a library of custom-made modules to support specific studies; see Chapter 9.

Several tools are currently available to ATP users for postprocessing the simulation results generated by TPBIG (e.g. PCPlot, TPPLOT, GTPPLOT, PlotXY, ATP Analyzer, TOP). PloxtXY, developed by M. Ceraolo (University of Pisa, Italy), is probably the most popular postprocessor among ATP users [7]. Most results presented in this book have been derived by using TOP (The Output Processor) [8].

Other tools that can be used as 'control centre' are ATPDesigner and ATP Control Center. They have not been used in the preparation of this book and are not presented in this chapter; for more details on their capabilities, see references [2, 9].

Transient Analysis of Power Systems: A Practical Approach, First Edition. Edited by Juan A. Martinez-Velasco.
© 2020 John Wiley & Sons Ltd. Published 2020 by John Wiley & Sons Ltd.
Companion Website: www.wiley.com/go/martinez/power_systems

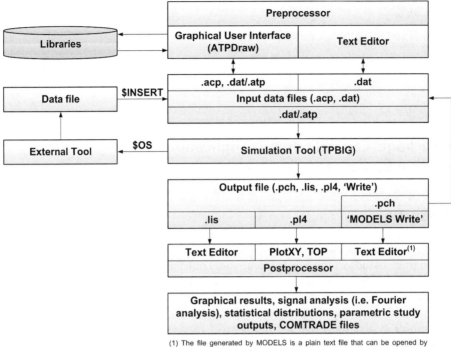

Figure 4.1 Tools and tasks of the ATP package.

Figure 4.1 depicts the most common tasks that can be carried out in typical studies by the tools usually integrated in the ATP package; see also Figure 1.2. Although other files can be generated and manipulated by the package tools, and more interactions between programs and files can be activated, the figure shows the most important connections between tools and files. Note, for instance, that MODELS, a general-purpose simulation language embedded in ATP for the representation of control systems and user-defined components, can be used to generate ASCII-code files and output information that could not be generated by any other tool.

This chapter summarizes ATP capabilities, provides an overview of its applications and presents some examples that illustrate the type of studies that can be carried out with this package. The subsequent sections detail the main features of each tool, define the meaning of the various types of files shown in the figure, and describe the connections between them.

4.2   Capabilities of the ATP Package

4.2.1   Overview

The capabilities of the ATP package are the combination of capabilities provided by each tool. Given that users can expand the package by adding more tools or creating custom-made environments, the ATP applications are almost unlimited. This section presents a summary of the main capabilities of the three basic tools of the package (i.e. pre-processor/GUI, simulation tool, postprocessor). As the scope of applications is basically that of the simulation tool, the section begins with capabilities of TPBIG.

4.2.2 The Simulation Module – TPBIG

4.2.2.1 Overview

TPBIG is a tool for simulation of transients in power systems. In addition to the modelling capabilities, several non-simulation supporting routines are available to calculate the parameters of lines, cables, and transformers, or create their models. The program can also represent control systems and interface them with an electric network. Although this program is mainly intended for transients simulation, it can also be used to perform ac steady-state calculations, to obtain system impedance as a function of the frequency, and to calculate harmonic power flows. In addition, some capabilities allow users to expand ATP applications beyond time- and frequency-domain simulations, or to develop custom-made models, as illustrated in all subsequent chapters. A schematic overview of available simulation modules and supporting routines, as well as their interactions, is shown in Figure 4.2. The subsequent subsections summarize the TPBIG capabilities for representing power systems components, calculating parameters and creating models for some components (e.g. lines, cables, transformers), performing time- and frequency-domain simulations, and developing custom-made models. Basic solution methods

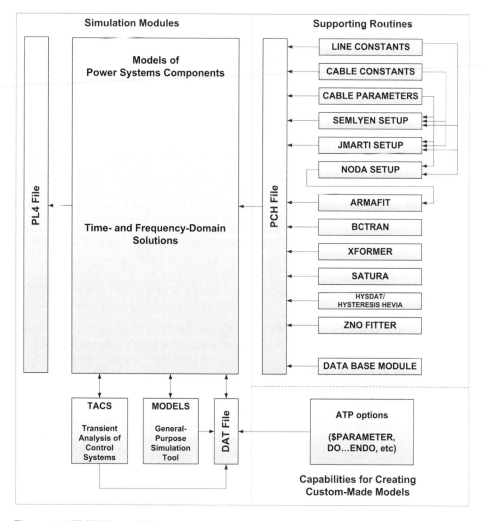

Figure 4.2 ATP-TPBIG capabilities.

for time- and frequency-domain simulations were presented in Chapter 3; this section provides a short summary.

4.2.2.2 Modelling Capabilities

Built-in electrical components

ATP can represent a high number of electrical network components, from simple passive linear and non-linear branches, switches, and static sources to sophisticated models of transmission lines, power transformers, and rotating machines. Basic capabilities of ATP include built-in models that cover multi-phase line and cable models with either lumped or distributed parameters, ideal and saturable multi-phase transformer models, rotating machine models, ideal time- and voltage-controlled switches, ideal current and voltage sources of various shapes. Table 4.1 provides an overview of built-in components, including control system devices. A summary of the ATP capabilities for modelling power system components is presented below.

Overhead lines and insulated cables The formulation of the equations and their solution is similar for both overhead lines and insulated cables, although the large variety of cable designs makes it very difficult for the development of a single model to represent every type of cable. Models for overhead lines and insulated cables can be classified into two groups: (i) lumped-parameter models that represent transmission systems by lumped elements whose values are calculated at a single frequency; (ii) distributed-parameter models, which can be classified into two categories, constant-parameter and frequency-dependent parameter models. The first type of model is adequate for steady-state calculations, although it can also be used for transient simulations in the neighbourhood of the frequency at which parameters were evaluated. The most accurate models for transient calculations take into account the distributed nature of parameters and consider their frequency-dependence.

Line and cable parameters are computed using a supporting routine (LINE CONSTANTS for lines, CABLE CONSTANTS and CABLE PARAMETERS for cables) [1]. The general formulation of impedances and admittances of single-core coaxial and pipe-type cables was presented in [10].

Several distributed-parameter models for overhead lines and insulated cables are available in ATP [2]:

1. Constant distributed-parameter line model (Bergeron model).
2. Second-order recursive-convolution line model (SEMLYEN SETUP) [11].
3. Frequency-dependent line model (JMARTI SETUP) [12].
4. Frequency-dependent ARMA (Auto-Regressive Moving-Average) line model (NODA SETUP) [13, 14].

The parameters of the Bergeron model are computed at a particular frequency; the other models consider the frequency dependence of line parameters. Line equations of models (2) and (3) are solved using a time-domain solution based on the modal theory: multiphase line/cable equations are decoupled through modal transformation matrices, so that each mode can be studied separately as a single-phase line/cable. The main drawback of these models is that modal transformation matrices are assumed constant and real; this assumption may introduce inaccurate results and, in some cases, lead to instability because of violation of the passivity conditions. Model (4) is a direct phase-domain model based on an Interpolated Auto-Regressive Moving Average (IARMA) convolution method [13, 14]. The modelling of either a line or a cable requires two steps: (i) calculation of parameters as a function of frequency using a supporting routine (e.g. CABLE PARAMETERS or LINE CONSTANTS); (ii) fitting parameters by means of an IARMA model for the time-domain realization of the frequency dependence using the independent program ARMAFIT.

Table 4.1 ATP built-in components.

Type of component	ATP option
Linear branches	• TYPE 0: Uncoupled lumped series RLC elements • TYPE 1, 2, 3: Mutually coupled RLC elements • CASCADED PI – Type 1, 2, 3 element (for steady state solution) • TYPE 51, 52, 53: Mutually coupled RL elements • PHASOR BRANCH [Y] – Type 51, 52, 53 element (for steady state solution) • TYPE-1, -2, -3: Distributed parameters elements • Constant-parameter line model (LINE CONSTANTS, CABLE CONSTANTS) • Frequency-dependent parameter line model (SEMLYEN, JMARTI, and NODA models) • Special double-circuit distributed line • SATURABLE TRANSFORMER COMPONENT • BCTRAN supporting routine • KIZILCAY F-DEPENDENT (High-order line admittance model)

Nonlinear branches	• TYPE 91
	  o Multiphase time-varying resistance
	  o TACS/MODELS-controlled resistance
	• TYPE 92
	  o Exponential ZnO surge arrester
	  o Multiphase, piecewise-linear resistance with flashover
	• TYPE 93: True-nonlinear inductance
	• TYPE 94: MODELS-controlled branch component
	• TYPE 96: Pseudo-nonlinear hysteretic inductor
	• TYPE 97: Staircase time-varying resistance
	• TYPE 98: Pseudo-nonlinear inductance
	• TYPE 99: Pseudo-nonlinear resistance
	• USER SUPPLIED FORTRAN NONLINEAR ELEMENT

Switches	• STAND ALONE SWITCHES
	  o Time-controlled switch
	  o Voltage-controlled switch
	  o Measuring switch
	• STATISTICAL SWITCHES
	  o STATISTIC switch
	  o SYSTEMATIC switch
	• CONTROLLED SWITCHES
	  o TYPE 11: Switch for modelling diodes and thyristors
	  o TYPE 12: Switch for modelling spark gaps and triacs
	  o TYPE 13: Simple TACS/MODELS-controlled switch

Sources	• EMPIRICAL SOURCES
	• ANALYTICAL SOURCES
	  o TYPE 10: Analytical source
	  o TYPE 11: Step function
	  o TYPE 12: Ramp function
	  o TYPE 13: Two-slope function
	  o TYPE 14: Cosine function/Trapped charge
	  o TYPE 15: Surge function
	  o TYPE 16: Simplified AC/DC converter model
	  o TYPE 18: Ideal transformer/Ungrounded voltage source
	• TACS/MODELS-CONTROLLED SOURCES
	  o TYPE 17: TACS/MODELS modulation
	  o TYPE 60: TACS/MODELS controlled source
	• ROTATING MACHINES
	  o TYPE 59: Three-phase synchronous machine (Prediction method)
	  o TYPE 58: Three-phase synchronous machine (Phase-domain solution)
	  o TYPE 19: Universal Machine module

Control systems	TACS (Transient Analysis of Control Systems) MODELS

Transformers An accurate representation of a transformer over a wide frequency range is very difficult, despite its relatively simple design. Several modelling approaches for use in transients programs have been developed:

- The representation of single- and three-phase *n*-winding transformers is made in the form of a branch impedance or admittance matrix [15]. This approach is used to derive models for low-frequency and slow-front transients, and cannot include nonlinear effects of iron cores, they are incorporated by connecting nonlinear inductances at winding terminals.
- Detailed models incorporating core nonlinearities, accurate enough for low-frequency and slow-front transients, can be derived by using the principle of duality from a topology-based magnetic model [16, 17].
- At high frequency, a transformer has a frequency-dependent behaviour. Accurate models for simulation of high-frequency electromagnetic transients can be divided into two groups: models with a detailed description of internal windings and terminal models (i.e. based on the fitting of the elements of a circuit that represent the transformer as seen from its terminals) [18].

ATP capabilities for representing transformers can be classified into two groups: built-in models and supporting routines. Built-in models can be used for analysing the behaviour of a transformer during low-frequency and slow-front transients [1]. A short description of these models and the supporting routines is presented below.

a. *Ideal transformer*. It is a branch component that can be used to represent a lossless ideal transformer characterized only by its turn ratio. This component has a very simple input format. One of its main advantages is that it can be used to obtain the representation of more complex transformers whose models are not available in the ATP.
b. *Saturable transformer component* (STC). This built-in model can be used to represent single- or three-phase transformers using data from the excitation and short-circuit tests at rated frequency. This model separates resistive and inductive parts internally, and ignores stray capacitances. It is valid from direct current (dc) up to 2 kHz. Nonlinear behaviour easily can be included, as well as the excitation losses. Single- and three-phase core-form transformers with low homopolar reluctance can be handled by this model. Three-phase three-legged transformers should be handled by the extended model taking into account the homopolar reluctance. However, this option is not recommended for modelling three-phase core-form transformers, as it cannot represent inductive coupling between phases.

Supporting routines (e.g. BCTRAN, XFORMER, SATURA, HYSDAT, HYSTERESIS HEVIA) can be used to build transformer models from manufacturers' data. The so-called *Hybrid Model* is a capability implemented in ATPDraw that takes advantage of TPBIG capabilities for creating more sophisticated low- and mid-frequency saturable transformer models [19]. Linear branch components available in TPBIG can be used to create transformer models adequate for high-frequency (fast and very fast front) transients.

Rotating machines Rotating machines are complex components whose functioning combines mechanical, electrical, and magnetic phenomena. In addition, controllers can be added in some machine models using either TACS (Transient Analysis of Control Systems) or MODELS. ATP built-in models include saturation effects and can represent rotating machines at low-frequency transients only. The representation of a rotating machine at high frequencies can also be complex, but it can be based on the basic components available in ATP. Although ATP models have been generally used to represent three-phase synchronous and asynchronous (induction) machines, ATP supports other conventional machine models adequate for low-frequency transients [1, 4].

Solution methods implemented for solving rotating machine transients can be classified into two groups.

- *Compensation-based method.* Each subnetwork connected to the machine is replaced by its Thevenin equivalent circuit seen from the machine terminals. These equivalents are calculated at each time step without knowledge of the machine variables. Since only the rotor angular speed is predicted, the iterations are confined to the machine equations, and the approach results in a scheme free of any error amplification. However, it requires the subnetwork connected to the armature side to be linear and the calculation of the equivalent circuits at every time step of the transient loop.
- *Prediction-based method.* It is based on the prediction of some armature-side variables at every time step from the knowledge of the variables calculated at previous time steps. Essentially, the machine is represented as voltage sources behind resistances.

The first model implemented in ATP was the synchronous machine (SM) model, the so-called Type 59 SM. This model was initially based on compensation [20] and later on prediction [21]. When using this model, a user can specify either manufacturer's data or electrical parameters. The SM is always automatically initialized based on the armature phase voltage and angle provided by the user. The control of this machine is carried out by specifying the per unit values of field voltage and mechanical power from the corresponding control systems. Internal machine variables can be passed to the control systems.

The Universal Machine (UM) module can represent up to 12 different machine designs (see Chapter 7) [22]. Unlike the SM module, whose mechanical system is embedded into the model, the mechanical system of the UM module is external and user-defined. The UMs can be initialized either automatically (by providing terminal phase voltage and angle for SMs or slip for induction machines) or manually (by providing internal winding currents). When automatic initialization is selected, ATP calculates the mechanical torque and field voltage; for this purpose, torque and field sources in the external mechanical system model have to be of AC-type with a very low frequency, and activated at $t < 0$. The solution method can be either compensation or prediction. The control of the UM is made by setting the absolute values of the mechanical torque and the field voltage. Internal machine variables can be passed into the control system.

Two phase-domain rotating machine models developed by TEPCO (Tokyo Electric Power Company) are available in ATP: the Type 58 three-phase SM and the Type 56 three-phase induction machine. These machines are, in general, more stable and they can be simulated with larger time steps.

Control systems: TACS and MODELS

ATP supports two options for representing control systems, TACS [23] and MODELS [6]. The main characteristics of both options are summarized below.

1. TACS is a simulation module for time-domain analysis of control systems based on a block diagram representation. A TACS section within an input data file consists of predefined blocks categorized into sources, transfer functions, devices, and FORTRAN statements; see Table 4.2. One time-step delay exists in the interface between a TACS-based model and the electrical network. An extra time delay can also occur due to nonlinearities in the control system; the number of this type of delay can be reduced by optimal sorting of the control blocks. A source in TACS can come from the electrical circuit or from other control systems based on MODELS language. Input variables can be node voltages, switch currents, internal

Table 4.2 TACS options.

Type of module	Module
Sources	Type 11: DC source
	Type 14: AC source
	Type 23: Pulse train
	Type 24: Ramp sawtooh
	Type 90: Node voltage
	Type 91: Switch current
	Type 92: Internal variable of special components (e.g. rotating machine angle)
	Type 93: Switch status
Transfer functions	General: With or without limits (either fixed or named)
	Limiter: Static, dynamic
Devices	Type 50: Frequency sensor
	Type 51: Relay switch
	Type 52: Level triggered switch
	Type 53: Transport delay
	Type 54: Pulse delay
	Type 55: Digitizer
	Type 56: User defined non-linearity
	Type 57: Multi-operation time-sequenced switch
	Type 58: Controlled integrator
	Type 59: Simple derivative
	Type 60: Input-IF component
	Type 61: Signal selector
	Type 62: Sample and track
	Type 63: Instantaneous min/max
	Type 64: Min/max tracking
	Type 65: Accumulator and counter
	Type 65: RMS meter
	Type 68: Hysteretic loss/residual mmf component
FORTRAN statements	• Algebraic operators: +, −, *, /, **
	• Logical operators: .OR., .AND., .NOT.
	• Relational operators: .EQ., .NE., .LT., .LE., .GE., .GT.
	• Functions (sin, cos, tan, cotan, sinh, cosh, tanh, asin, acos, atan)
	• Special functions (trunc, minus, invers, rad, deg, sign, not, seq6, ran)

machine variables and, switch statuses. Special internal sources can also be defined in TACS. FORTRAN statements can be used within TACS to define variables with a FORTRAN-like syntax; arguments in operations and functions can be other TACS functions. FORTRAN statements can also be used to define sources and relations between inputs and outputs. The sequence of the calculations follows the order of the blocks in the input file.

2. MODELS is a symbolic language that provides an option for numerical and logical manipulation of variables. Its syntax allows users to represent a system to either reflect its physical implementation, using a block approach, or describe its operation, using a description of its functional structure. The main MODELS features are:

   a. A distinction between model description and model use: A MODELS application requires the description of the model to be simulated and its interaction with the environment in which it is embedded; in addition, it is necessary to indicate the directives with which the simulation of the model is to be made.

b. The decomposition of a large model into sub-models with a hierarchical organization: each sub-model performs an easier task, in this way, reducing the complexity of a model and facilitating the modification of each sub-model without affecting the main model. A model can access FORTRAN subroutines that can be used to implement already-developed models/algorithms.

c. The self-documenting nature of a model description, which can be easily understood and can serve as a reference document.

When using MODELS, the following aspects must be taken into account:

- what elements are present in a model and how they interact with each other;
- how the model interacts with the environment;
- how the simulation is to be performed.

The elements of a model can be of two types: value-holding elements, represented by constants and variables of the model, and composite elements, represented as sub-models. The internal interaction of the elements is described by specifying the relations in which these elements are assembled. MODELS syntax permits a great flexibility by means of capabilities such as indexed repetition (FOR … DO … ENDFOR), conditional branching (IF statements), conditional repetition (WHILE … ENDWHILE), implicit repetition (DO … ENDO), user-defined functions.

The simulation of a model is described at the initialization phase, which establishes the history of the model prior to the instant of its first execution. The execution phase updates the value of the model variables and the interaction between a model and the electric system. A MODELS section occupies the same location as a TACS section, although both options can be used in the same case study. The MODELS section may include several types of declarations with the following general structure:

```
BEGIN NEW DATA CASE
Miscellaneous Data Cards
MODELS
STORAGE
INPUT
OUTPUT
MODEL ... ENDMODEL
MODEL ... ENDMODEL
USE ... ENDUSE
USE ... ENDUSE
ENDMODELS
ATP Cards
BEGIN NEW DATA CASE
```

With this language, the organization of a model is usually based on the following pattern:

```
MODEL <name>
INPUT   i1, i2[1..n]
DATA    d1, d2
OUTPUT o1[1..n], o2
VAR     o1[1..n], o2, v1, v2, v3[1..n]
DELAY CELLS -- specify past values for variables
```

```
HISTORY        -- assign variables for t < 0
INIT           -- assign variables for t = 0
ENDINIT
EXEC
               -- execute every local time step
ENDEXEC
ENDMODEL
```

As with TACS, a MODELS section brings electrical and mechanical variables of a system and performs the control of some variables of this system. The interface between the electrical network and a TACS/MODELS section is established by exchanging signals like node voltages, switch currents, switch statuses, time-varying resistance values, or voltage and current source values. A special interface between MODELS and the electrical circuit, the Type 94 component, addresses the time-delay problem and allows users modelling their own electrical components. See below.

Supporting routines

Several routines are available in ATP to support users when building models for lines, cables, transformers, and surge arresters. A summary of their main characteristics is provided below [1, 2].

- LINE CONSTANTS is a supporting routine to compute electrical parameters of overhead lines. It is linked to the supporting routines SEMLYEN SETUP, JMARTI SETUP and NODA SETUP to create overhead line models.
- CABLE CONSTANTS and CABLE PARAMETERS are supporting routines to compute electrical parameters of insulated cables and overhead lines. Both routines are linked to the supporting routines SEMLYEN SETUP, JMARTI SETUP and NODA SETUP to generate models of lines and cables.
- SEMLYEN SETUP is a supporting routine that generates frequency-dependent models for overhead lines and cables. Modal theory is used to represent unbalanced lines in time-domain. Modal propagation step response and surge admittance are approximated by second-order rational functions with real poles and zeros.
- JMARTI SETUP generates frequency-dependent models for overhead lines and cables. The fitting of modal propagation function and surge impedance is performed by asymptotic approximation of the magnitude by means of a rational function with real poles.
- BCTRAN is a supporting routine that can be used to derive a linear $[R]-[\omega L]$ or $[A]-[R]$ matrix representation for single- and three-phase transformers using data from excitation and short-circuit tests at rated frequency [1,15]. For three-phase transformers, both shell-type (low homopolar reluctance) and core-type (high homopolar reluctance) designs can be handled by the routine. Since BCTRAN can only generate linear models, nonlinear inductors must be included in the data deck to account for the nonlinear behaviour when this routine is used to derive a transformer model.
- XFORMER is used to derive a linear representation for single-phase, two- and three-winding transformers by means of an RL coupled branch model.
- SATURA is a conversion routine to derive flux–current saturation curve of a nonlinear inductor from either rms voltage-current characteristic or current-incremental inductance characteristic.
- HYSDAT is used to derive the characteristic of a Type 96 hysteretic inductor. This option can handle only one type of material (ARMCO Mn). The only information to be supplied by the user is the scaling (i.e. the point of the first quadrant where the hysteresis loop changes

from being multi-valued to being single-valued). HYSTERESIS HEVIA is a more recently implemented routine aimed at circumventing some of these limitations [24].

- ZNO FITTER can be used to derive a true non-linear representation for a zinc-oxide surge arrester, using manufacturer's data. This routine approximates the voltage-current characteristic by a series of exponential functions.
- DATA BASE MODULE allows users to combine sections of a data file in a single module, which may contain circuit and/or control system elements. Some data, such as node names or numerical data, may have fixed values inside the module, whereas other data can be treated as parameters that are passed to the module when it is inserted/included into the input data file.

Creation of custom-made models

Several TPBIG capabilities can be used to create custom-made models not available in the program (e.g. relay models) or to facilitate the usage of other components (e.g. a converter for representing a solid state switch). In addition to built-in components (see Table 4.1), the options to be considered for these types of tasks are the supporting routine DATA BASE MODULE and $PARAMETER. The development of new components can also be based on the Type 94 [25, 26].

The routine DATA BASE MODULE, presented above, can be used for developing integrated modules (e.g. power electronics converters) or creating components not available in TPBIG. In general, this routine is applied to facilitate the usage of large and complex components. With the $PARAMETER option, users can expand the application of DATA BASE MODULE by means of more powerful and flexible modules; for instance, some parameters of a given module can be internally calculated when $PARAMETER is combined with DATA BASE MODULE. Chapter 9 will illustrate how these routine and other ATP capabilities can be used for creating a library of modules aimed at analysing some specific case studies [27].

Nonlinear components (i.e. resistances and inductances) can be represented in ATP by a *true nonlinearity*, based on the compensation method, or a *pseudo nonlinearity*, represented by a piecewise linear approach. ATP also covers a third approach, the *controlled nonlinearity*: the Type 94 option is a MODELS-based component that connects directly into the circuit description and allows users to choose amongst several options [2]:

- *Thevenin.* The input is the Thevenin equivalent (resistance and voltage) of the circuit to which it is connected; the model returns the branch current. This option is for branches connected to ground and is treated as a true nonlinear component.
- *Norton.* The input is the node voltage of the circuit to which it is connected; the model returns a Norton equivalent (conductance and current). This option is for branches connected to ground and is treated as a pseudo-nonlinear component.
- *Norton transmission.* This option is similar to the previous one but it can be used when the nonlinear component has to be connected between two nodes, none of which is ground.
- *Iterated.* The input is the node voltage; the model returns the current and the *di/dv* conductance to be solved iteratively together with the circuit.

This nonlinearity can cause the conductance matrix to change frequently, requiring re-triangularization and time-consuming simulations.

4.2.2.3 Solution Techniques

Time-domain solution techniques

ATP uses the Dommel scheme to convert the differential equations of the network components into algebraic equations: a discrete companion model of a lumped parameter component is obtained from its differential equation using the trapezoidal integration; the discrete companion model of a component with distributed parameters (e.g. a lossless transmission line) is deduced from the application of the Bergeron method. See Chapter 3.

With the Dommel scheme, the discretized equations of network components are assembled using the nodal admittance matrix of the equivalent companion circuit [3, 4]:

$$[G][v(t)] = [i(t)] - [I(t)] \tag{4.1}$$

where $[G]$ is the nodal conductance matrix, $[v(t)]$ is the vector of node voltages, $[i(t)]$ is the vector of current sources, and $[I]$ is the vector of *history* terms. The nodal conductance matrix is symmetrical and remains unchanged since the integration is performed with a fixed time-step size. The solution of the transient process is obtained using triangular factorization. Once the matrices of the above expression are built, they are triangularized before the transient loop begins, and whenever a switch changes its status. This procedure can be applied to networks of arbitrary size in a very simple fashion. If the network contains voltage sources to ground, then the equation is split up into part A with unknown voltages and part B with known voltages

$$[G_{AA}][v_A(t)] = [i_A(t)] - [I_A] - [G_{AB}][v_B(t)] \tag{4.2}$$

Since this scheme can only be used to solve linear networks, some approaches have been implemented to cope with nonlinear and time-varying elements. They can be based on a piecewise-linear representation or the compensation method [28]. Using the first approach, the conductance matrix is changed and re-triangularized whenever the solution moves from one straight-line segment to another. Using compensation, nonlinear elements are represented as current injections which are superimposed to the solution of the linear network after this solution has been computed.

The trapezoidal rule is simple and numerically stable; however, it has some drawbacks: it uses a fixed time-step size and can originate numerical oscillations. In switching operations or transitions between segments in piecewise linear inductances, the trapezoidal rule acts as a differentiator, and introduces sustained numerical oscillations. Several techniques have been proposed to control or reduce these oscillations (see Chapter 3).

Frequency-domain solution techniques

The incorporation of frequency-domain solution techniques in a transients tool such as ATP can be very useful and crucial; they can be used to obtain the steady-state solution that exists prior to the disturbance, detect resonance conditions, or analyse harmonic propagation.

ATP can obtain a steady-state phasor solution, provided sinusoidal sources are present in the electrical network. The solution method implemented in ATP is based on nodal equations (complex phasor quantities for currents and node voltages) and assumes the electrical system is linear:

$$[Y][V] = [I] \tag{4.3}$$

As for the transient solution, this equation is partitioned when the network contains voltage sources to ground:

$$[Y_{AA}][V_A] = [I_A] - [Y_{AB}][V_B] \tag{4.4}$$

The implementation of a technique to obtain the steady-state solution can be very complex in the presence of nonlinear components and variable-topology circuits, which can produce steady-state harmonics. Although a significant effort has been made to develop more powerful initialization procedures, not many have yet been implemented; see for instance [29]. No solution method has yet been implemented in any EMTP-type tool to obtain the steady-state in systems with nonlinear components, switching devices, and control systems. However, some simple approaches can be used. The simplest one is known as the 'brute force' approach: the system is started from standstill and the simulation is carried out long enough to allow the transients to settle down to steady-state conditions. This approach can have a very slow convergence if the network has components with light damping. A more efficient method is to perform an approximate linear ac steady-state solution with nonlinear branches disconnected or represented by linearized models. ATP has a 'start again' feature: using a 'brute force' initialization, a snapshot is taken and saved once the system reaches the steady-state, later runs can be started at this point.

A different solution method is needed when initial operating conditions are specified as power constraints. The available power flow option, known as FIX SOURCE, can be used to iteratively adjust the magnitudes and angles of some node voltages until the specified constraints (e.g. active and reactive power, active power and voltage magnitude) are matched. This can be used to obtain the initial conditions for the subsequent transient simulation.

The FREQUENCY SCAN (FS) option performs repetitive steady-state phasor solutions, as the frequency of sinusoidal sources is incremented between a lower and upper frequency; a frequency-response of node voltages, branch currents or driving-point impedances/admittances is obtained. Typical applications of FS are the analysis and identification of resonant frequencies, and the frequency response of driving-point network impedances or admittances seen from a busbar (e.g. positive-sequence or zero-sequence impedance).

HARMONIC FREQUENCY SCAN (HFS) is a companion to FS [30]: FS solves the network for the specified sources, incrementing in each subsequent step, the frequency of the sources but not their amplitudes; HFS performs harmonic analysis by executing a string of phasor solutions determined by a list of sinusoidal sources entered by the user [1]. This capability is similar to the harmonic analysis implemented in some tools. The main advantage of HFS compared with the time-domain harmonic analysis is a reduction in runtime and avoidance of accuracy problems with Fourier analysis. ATP models developed for HFS analysis are the frequency-dependent RLC elements, the frequency-dependent load based on the CIGRE type C model, and some harmonic current/voltage sources.

Control systems dynamics

Control systems based on the TACS option are represented by means of block diagrams with interconnection between elements. Control elements can be transfer functions, FORTRAN algebraic functions, logical expressions and some special devices [1, 4, 23]. The solution method is also based on the trapezoidal rule. A control block in the s-domain can be described by the following relationship:

$$X(s) = G(s)U(s) \qquad \left(G(s) = K \frac{N_0 + N_1 s + \cdots + N_m s^m}{D_0 + D_1 s + \cdots + D_n s^n} \qquad m \le n \right) \tag{4.5}$$

where $U(s)$ and $X(s)$ are respectively the input and the output in the Laplace domain, and $G(s)$ is a rational transfer function.

Transfer functions are converted into algebraic equations in the time-domain

$$cx(t) = Kdu(t) + hist(t - \Delta t) \tag{4.6}$$

where K is the gain, while constants c and d are obtained from the coefficients of the rational transfer function [3].

A control system with many linear blocks results in a system of equations with the following general form:

$$[A_{xx}][x(t)] + [A_{xu}][u(t)] = hist(t - \Delta t) \tag{4.7}$$

Since the resulting matrices of this equation are by nature unsymmetrical, the electric network and the control system are solved separately. The network solution is first advanced, network variables are next passed to the control section; finally, control equations are solved. Finally, the network receives control commands. The whole procedure introduces a time-step delay.

When a nonlinear block is inside a closed-loop configuration, a true simultaneous solution is not possible. The procedure is simultaneous only for linear blocks, and sequential for nonlinear blocks. Therefore, the loop is broken and the system is solved by inserting a time delay.

Table 4.3 shows a summary of ATP-solution methods for linear networks.

4.2.3 The Graphical User Interface – ATPDraw

4.2.3.1 Overview

ATPDraw is an interactive Windows-based graphical preprocessor written in CodeGear Delphi 2007 (object-oriented Pascal), and developed to create and edit the data file of the system to be simulated [2, 5]. The user can build an electric circuit by selecting predefined components from the available menu; the program creates the corresponding data file in ATP/TPBIG code. Node naming is automatically administrated by the program, although the user may give name to selected nodes.

All package tools can be run from the ATPDraw menu. Users are also allowed to edit and add other batch jobs to expand the capabilities of the package. ATPDraw supports multiple documents that facilitate work on several circuits simultaneously and copy information between them. All kinds of standard circuit editing facilities (copy/paste, grouping, rotate, export/import, undo/redo) are available. In addition, ATPDraw supports the Windows clipboard and metafile export. A circuit is stored on disk in a single project file, which includes all the simulation objects and options needed to run the case. The project file is in zip-compressed format that makes file sharing with others very simple. Components have an individual icon in either bitmap- or vector-graphic style and an optional graphic background.

Most ATP standard components, including TACS, are supported; in addition, the user can create new objects based on MODELS or DATA BASE MODULE. Special modules for creating line/cable, transformer, and rotating machine models are also supported by ATPDraw, which also incorporates models not available in TPBIG; for instance, the hybrid transformer model and the WindSyn routine for rotating machines.

ATPDraw has been developed by H. Høidalen (SINTEF, Norway) since 1991. The latest versions have added capabilities for creating new models of transformers and machines, incorporating an optimization module, and has made available some ATP features, such as POCKET CALCULATOR VARIES PARAMETERS (PCVP). An update of the ATPDraw development was presented in [31].

4.2.3.2 Main Functionalities

The list of the main ATPDraw features includes, among others, the following items: built-in editor for creating and correcting data files; support of Windows clipboard for metafile/bitmap; output of Windows metafile format; copy/paste; rotate; import/export; group/ungroup; undo; print facilities; on-line help; icon editor for user-specified objects; multiple windows.

The user assembles the circuit by selecting components from the *Selection menu* (right-click in open space) or by copying them from any other circuit. Nodes are automatically connected if they overlap or if a *Connection* is drawn between nodes. Typically, the user stores the circuit in a project file with extension.acp. ATPDraw creates names of all nodes and writes the circuit

Table 4.3 ATP solution methods for linear systems.

Algorithm	Equations – functions	Notation – scheme
Transient solution of electrical networks	$[G][v(t)] = [i(t)] - [I(t)]$ $[G_{AA}][v_A(t)] =$ $[i_A(t)] - [I_A] - [G_{AB}][v_B(t)]$	$[G]$ the nodal conductance matrix $[v(t)]$ the vector of node voltages $[i(t)]$ the vector of current sources $[I]$ the vector of 'history' terms A part with unknown voltages B part with known voltages
Steady-state initialization of electrical networks	$[Y][V] = [I]$ $[Y_{AA}][V_A] = [I_A] - [Y_{AB}][V_B]$	$[Y]$ the nodal complex admittance matrix $[V]$ the vector of node voltages $[I]$ the vector of current sources A part without voltage sources to ground B part with voltage sources to ground
Transient analysis of control systems	$[A_{xx}][x(t)] + [A_{xu}][u(t)] = hist(t - \Delta t)$	the discretized control system matrix $[u]$ the input vector $[x]$ the output vector $[hist]$ the history vector
Interface electrical network-control system		

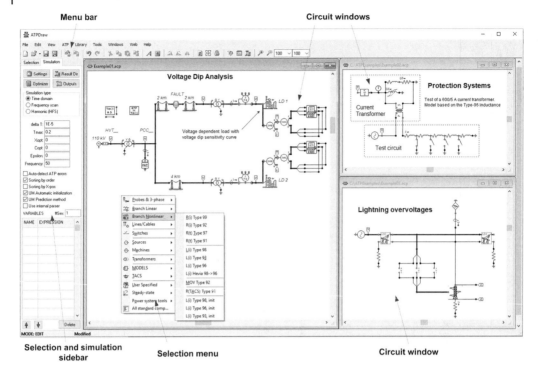

Figure 4.3 Main ATPDraw window.

specification to disk in ASCII code readable by ATP/TPBIG. ATPDraw gives warnings if a node has duplicate names or if the same name is given to non-connected nodes. Figure 4.3 shows the main windows of ATPDraw. The tool bar at the top of the main window is customizable and by default contains menu buttons for the most frequently used options; see Table 4.4. ATPDraw supports multiple undo-redo steps in strict sequence. Auto-saving and backup of the circuit can be performed at a user-selectable interval. This tool offers a large help file system in the compiled html help format. All default values of standard data of all components can be set by the user under the *Library* menu. This also applies to the icons. Icons can be in vector or bitmap graphic and have an accompanied bitmap (bmp, png) attached. Most of the components have an input dialogue similar to that shown in Figure 4.4 with data values given in the top-left grid and node names listed in the top-right grid.

ATPDraw manipulates different types of files: (i) circuit files in binary code, where the program stores the information about the equivalent graphical picture of circuits; (ii) input files in ASCII code, created as input for TPBIG simulations; (iii) support files for ATPDraw objects in binary code that specifies the data and nodes for an object with the icon and help information included; (iv) files created by DATA BASE MODULE and called by user-specified objects; (v) MODELS files in ASCII code, where the description of a model is contained.

ATPDraw allows users to create custom-made models by using the option *Compress*: a group of components can be replaced by a user-editable icon with connection nodes and data specification. The new icon can be treated as a conventional model to which a help menu can be attached. A nested hierarchical approach is also possible; that is, this option can also be used within a group (i.e. the user can compress a selected group of components within a group) or to combine two or more groups in a single group with an almost unlimited numbers of layers. The opposite operation is *Extract*: the group is extracted on the current project level once the group

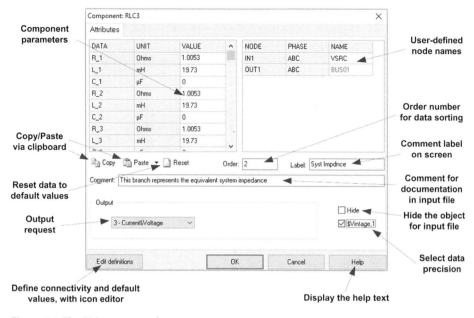

Figure 4.4 The *RLC* component box.

icon has been selected. Once the new icon has been created, the option *Edit Group* can be used to display the contents of a group: with this option, users can edit/modify the internal group, although deletion of reference components is not permitted. The option *Edit Circuit* displays the circuit where the current group belongs.

4.2.3.3 Supporting Modules for Power System Components

This subsection presents a short summary of ATPDraw capabilities for representing power components and creating a MODELS sections.

Line and cable modelling

The LCC module can be used to calculate parameters and create line/cable models: the user specifies geometry and physical properties of the line/cable; the electrical parameters are calculated and the required model is created. LCC can also plot the cross section (*View*) and verify the line/cable model (*Verify*). The *Verify* feature calculates the open- and short-circuit input impedance either at power frequency or as function of frequency (LINE MODEL FREQUENCY SCAN feature) and allows the user to compare the model to the exact PI model for positive-sequence or zero-sequence excitations. For long lines, this method gives inaccurate results, so a *LineCheck* feature is added externally to compensate for the receiving-end conditions [32]. ATPDraw users can also input, if available, line and cable models.

Transformer modelling

ATPDraw supports all transformer models available in TPBIG: STC, XFMR, and BCTRAN, and incorporates the Hybrid model as a built-in option [19]. When using the STC, the conversion from saturation curve to the flux-linkage/current peak characteristic is internally made by means of SATURA routine. The BCTRAN module supports some core configurations (Wye, Delta, and Auto) with all phase shifts for two and three windings. The XFMR module requires input similar to BCTRAN but also supports up to four windings and zigzag coupling.

Table 4.4 ATPDraw menus.

Menu	Item	Description	Shortcut
File	New	Open an empty circuit file	—
	Open	Load a circuit file into a new window	CTRL+O
	Save	Save the active circuit window to the current project file	CTRL+S
	Save As	Save the active circuit window to a new project file	—
	Close	Close the active circuit window	—
	Close All	Close all circuit windows	—
	Import	Insert a stored circuit into the current circuit	—
	Export	Export the selected circuit to an external project file	—
	Save Metafile	Save selected objects to a Windows metafile	—
	Print	Print the active circuit	—
Edit	Undo	Undo the previous operation	CTRL+Z
	Redo	Redo the previous operation	CTRL+Y
	Cut	Copy the current selected circuit to the clipboard and delete it	CTRL+X
	Copy	Copy the current selected circuit to the clipboard	CTRL+C
	Paste	Paste the ATPDraw content from the clipboard to the circuit	CTRL+V
	Duplicate	Copy + Paste	CTRL+D
	Delete	Delete the selected objects	Delete
	Select/All	Select the entire circuit	CTRL+A
	Edit Text	Enter Edit Text mode for adding and selecting text	CTRL+T
	Rotate-R	Rotate 90° clockwise	CTRL+R
	Rotate-L	Rotate 90° counter clockwise	CTRL+L
	Flip	Flip left-to-right (nodes change position; text is not flipped)	CTRL+F
	Copy Graphics	Copy the selected objects to the clipboard as a metafile object	CTRL+W
	Compress	Single icon replacement of a selected group	—
	Extract	The opposite operation to Compress	—
	Edit Group	Display the contents of a group	CTRL+G
	Edit Circuit	Display the circuit where the current group belongs	CTRL+H
View	Side Bar	Show or hide the sidebar at the right of the window	—
	Refresh	Redraw circuit	CTRL+Q
	Centre circuit	Draw circuit centred in the window	—
	Lock circuit	Turn on 'child safety'; only input, no edit	—
	Zoom In	Zoom in 20%	NUM+
	Zoom Out	Zoom out in 20%	NUM−
ATP	Settings	Specify/change simulation settings (Δt, tmax, Xopt, Copt, etc)	F3
	Run ATP	Make node names, create the input file and run ATP/TPBIG	F2
	Run Plot	Run the Plot Command	F8
	Sub-process	List intermediate steps: make node names, make file, run ATP	—
	Output Manager	List all output requests found in the project	F9
	Edit ATP File	Invoke text editor	F4
	View LIS File	Enable users to view the current LIS file.	F5
	Find Node	Search for a specific node name in the entire circuit	F6
	Find Next Node	Continue the search for a node name	F7
	Optimization	Evaluate a single cost function and adjust circuit variables	—
	Edit Commands	Enable users to specify executable files to run from ATP menu	—

In addition, the structure of the core (triplex, three-legged, five-legged, shell-form) can be taken into account if relative core dimensions are provided. The magnetic B/H relationship for the hybrid model is assumed to follow an expanded Frolich equation and is the same for both legs and yokes.

Rotating machine modelling
ATPDraw supports the Type 56 Induction Machine (IM), the Type 58 and 59 SMs, and some options of the UM module (Type 1: three-phase SM; Type 3: three-phase IM; Type 4: three-phase doubly fed IM; Type 6: single-phase IM; Type 8: DC machine). To facilitate the use of some UM options, the WindSyn program created by G. Furst can be used [33]; this program provides calculation of electrical parameters from manufacturers' data, adds options regarding rotor type and dampers, and offers start-up and control functionalities. This module also offers exciter and governor controls and plotting of torque-slip characteristics for induction machines.

MODELS module
ATPDraw facilitates the integration of a MODELS-based section into the input data file by the automatic identification of Input, Output and Data variable names in the declaration section, and the conversion of this interface into a component with proper nodes and data. Once the icon is constructed, the user must click on the input nodes to define the type of input (current, voltage, switch status, machine variable, TACS, Im(V), Im(I), Model, ATP). In order to use a model output as input to another model, the first one must also come first in the ATP file. This can be accomplished by the *Order* and *Sort by Order* options.

4.2.4   The Postprocessor – TOP

Several plotting programs are available to ATP users for displaying simulation results. A list of the postprocessors available in 2000 was analysed in reference [34]. This subsection details TOP (The Output Processor), a free postprocessing tool used in this book and developed by Electrotek Concepts [8].

TOP reads data from a variety of sources and generates high-quality graphics. Data formats supported include, amongst others ASCII Text, COMTRADE (IEEE C37.111-1999), PQDIF (Power Quality Data Interchange Format-IEEE Std 1159.3-2002), Dranetz 8010 and 8020 PQNode, Dranetz 65x Series, Electrotek SuperHarm, Electrotek FerroViewTM, EPRI/DCG EMTP for Windows, ATP, PSCAD, EPRI HARMFLO, Cooper Power Systems V-HARM, EPRI SDWorkstation, EPRI LPDW (CFlash, DFlash, TFlash), EPRI PQ Diagnostic System (Capacitor Switching and Lightning), Square D PowerLogic (using DADisp format), and Fluke 41.

The program is written in C++ to provide the necessary features for object-oriented programming. The code associated with manipulating a particular file format is compiled and linked into a Windows Dynamic-link Library (DLL), so new or updated data formats can be added to the program by simply providing a new DLL. This section summarizes TOP capabilities for data import and processing. For more details, see [8].

4.2.4.1   Data Management
The program uses a simple approach to handle data from various sources: first, the user selects the file that contains the data to be displayed; the program then provides an appropriate dialogue box (see Figure 4.5) that displays all of the quantities available in the current data source; finally, the user selects the items of interest that are then loaded into memory. The user may reactivate the dialogue box at any time and load more data objects onto the stack, or select another file and load data from that file onto the stack. The result is a stack of data objects of different types and origins that are available for display.

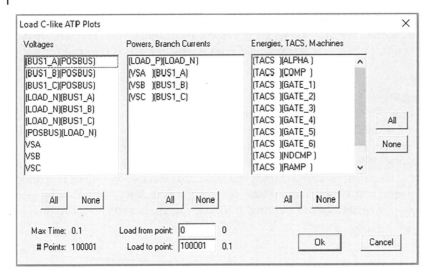

Figure 4.5 The stack-load box.

4.2.4.2 Data Display

TOP provides a variety of ways to visualize the data of interest in the form of tables and graphs; see Figure 4.6. Display options include: waveform and spectrum plots, frequency response plots, summary tables (including IEEE Std 519 application), summary bar/column charts, cumulative probability plots, probability density charts, 3D magnitude-duration histograms, background (CBEMA, ITIC, etc.) curves for magnitude-duration plots. It also allows users to view several different plots in multiple windows simultaneously. The program uses *windows* and *frames* to display selected data; that is, one or more quantities can be plotted on the same axis, and multiple sets of axes can be displayed in a single window. This option can display measured and simulated data in the same window, so it can be very useful for validating simulation results.

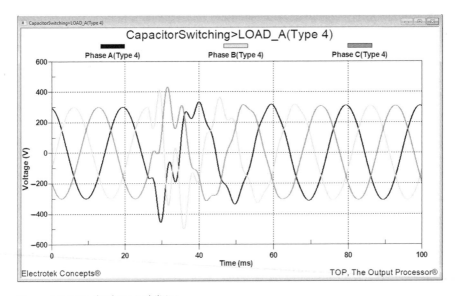

Figure 4.6 Data display capabilities.

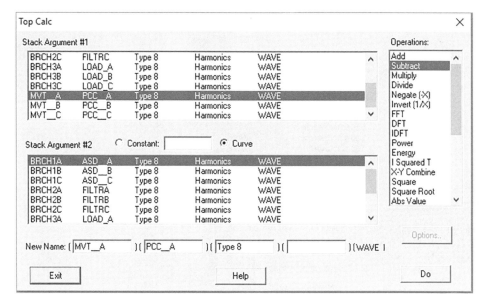

Figure 4.7 Data processing using TOPCalc.

4.2.4.3   Data Processing

The feature called 'TOPCalc' allows users to perform mathematical operations on the data objects supported by the program and provides a means for creating new objects. Operations include: addition; subtraction; multiplication; division; inversion; absolute value; Fast Fourier and Discrete Fourier Transforms (FFT, DFT); inverse Discrete Fourier Transform (IDFT); power, energy, and I^2t; integration; square; square root; X-Y combine (plots Y value of 1st argument versus Y value of 2nd argument); filter (high, low pass); time shift; transfer function; V, I, and power dB ratio; auto correlation; cross correlation; cumulative probability; probability distribution. Objects derived through the TOPCalc function can themselves be used as arguments for subsequent TOPCalc operations. TOPCalc requires one or two arguments, depending on the operation selected. Argument #1 is always a stack object; Argument #2, if required, may be either an object or a constant; see Figure 4.7.

4.2.4.4   Data Formatting

Although the program automatically assigns units to the X and Y variables when an object is created (loaded from input file), there are instances when the user will need to change the values. When the program loads a stack object it prepares information for a data block. The content of the data block varies depending on the object and input file type. TOP provides the capability to format data based on user preferences. Functions under user control include: base quantities (per unitizing), units (axis labels and multipliers), display colours, graph legend, axis scaling, grid lines, labels; see Figure 4.8. The set per-unit base quantities option allows the user to control scaling for each stack object individually. Measured and simulated data can be displayed on the same axis. The units command can be used when the user needs to assign a different set of units to the displayed object. The user can control the display colours, X and Y axis zooming, tick marks and grid lines, and axis labels.

4.2.4.5   Graphical Output

The data being visualized can be exported to a variety of formats, including Windows Metafile (.wmf), Windows Bitmap (.bmp), Joint Photographic Graphics (JPG), Portable Network Graphics (.PNG), and ASCII Tabbed Text (.TXT).

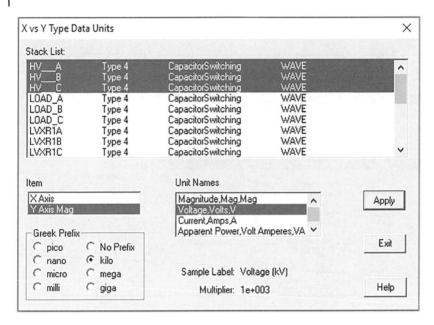

Figure 4.8 Data formatting functions – per-unit base quantities dialog.

4.3 Applications

The capabilities implemented in ATP can be used to perform a large number of transients studies in power systems. These capabilities can duplicate electromagnetic, electromechanical, and control system transients with a high accuracy. Table 4.5 shows a summary of the most important applications [35].

These applications can be classified as follows (see Chapter 1):

- *Time-domain simulations.* They are generally used for simulation of transients, such as switching or lightning overvoltages. They can also be used for analysing harmonic distortion created by power electronics devices.
- *Frequency-domain simulations.* ATP capabilities can be used to obtain the driving point impedance at a particular node versus frequency, detect resonance conditions, design filter banks, or analyse harmonic propagation.
- *Parametric studies.* They are usually performed to evaluate the relationship between variables and parameters; when one or more parameters cannot be accurately specified, this analysis will determine the range of values for which they are of concern.
- *Statistical studies.* In addition to statistical switches, users can combine some ATP capabilities to perform all types of Monte Carlo simulations, not covered by statistical switches.

ATP capabilities can also be used to expand the fields of applications; with this tool, a data case can be simulated several times before deactivating the program, parameters of the system under simulation can be changed according to a given law, some components can be either disconnected or activated, and some calculations can be carried out by external programs. In addition, it is possible, if required, to modify 'on line' the simulation time or the number of runs. Several concepts (e.g. multiple run option, data symbol replacement, data module) that can be of paramount importance for expanding the ATP applications were defined in Chapter 1. See also Chapter 11.

Table 4.5 ATP Applications.

Study	Requirements
Ferroresonance	• Basic power components
Parallel line resonance	• Steady-state initialization (SSI) • Basic power components
Harmonics	• Frequency scan/Harmonic power flow • Basic power components • Frequency-dependent models
Subsynchronous resonance	• Steady-state initialization • Basic power components • Three-phase synchronous machine
Transient stability	• SSI + Basic power components • Three-phase synchronous machine • Control system dynamics
Switching studies	• SSI + Basic power components • Statistical switches/tabulation • Frequency-dependent line model
Lightning studies	• Basic power components • Frequency-dependent line model
HVDC	• Basic power components • Semiconductor devices • Control system dynamics
FACTS devices	• Basic power components • Semiconductor devices • Control system dynamics
Adjustable speed drives	• Basic power components • Semiconductor devices + TACS • Rotating machine models
System protection	• SSI + Basic power components • Control system dynamics • Frequency-dependent line model
Secondary arc	• SSI + Basic power components • Control system dynamics • Frequency-dependent line model

4.4 Illustrative Case Studies

4.4.1 Introduction

The previous section has listed the main ATP applications and summarized some of the capabilities that can expand the list. Given the potential of the usual package tools and the possibilities that can be opened by adding other tools, the number of case studies in which the ATP can be successfully applied is practically unlimited. For instance, by using some simple rules and taking advantage of some capabilities (e.g. TO SUPPORTING PROGRAM feature to run supporting routines, DO loops to serialize power components, string replacement), it is possible to develop a data section aimed at creating the code of a component taking into account the transient process to be simulated and the information available.

In addition, ATP/TPBIG can be part of custom-made environments for both analysis and design applications; see Chapter 11. Since the subsequent chapters of this book will present some of the most common case studies to which ATP is currently applied (i.e. overvoltage calculation, impact of power electronics devices in transmission and distribution systems, subsynchronous resonance, power quality studies, protection of power systems), this section is used to show how the package can be successfully applied in other fields. The three case studies selected for this section will show how the ATP can be used to (i) allocate a capacitor bank for optimum reduction of power losses in a distribution system, (ii) detect resonance between parallel overhead transmission lines when shunt reactors are installed for compensating reactive power, (iii) select surge arrester rated voltage when protecting a substation transformer against lightning overvoltages. Note that the first two examples do not involve any transient calculation, although the first one uses a time-domain solution, while the second one takes advantage of the frequency-domain solution method available in ATP.

4.4.2 Case Study 1: Optimum Allocation of Capacitor Banks

The optimum location of a capacitor bank aimed at reducing losses in a distribution feeder has been the subject of several works [36]. In this document, the ATP will be used to estimate both the optimum size and the optimum location.

Figure 4.9 shows the single–phase diagram of a distribution feeder with a uniformly distributed load. To represent this load, the feeder has been divided into 20 sections of equal length, and one-twentieth of the load has been located at the end of each section. It is well known that the solution to this problem is based on the '2/3 rule': the optimum rating of the capacitor bank is two-thirds of total reactive load, and it must be located at two-thirds of the distance from the origin to the end of the feeder [36]. Since the total three-phase reactive load is 6 MVAr, the optimum rating of the capacitor bank should be about 4 MVAr. By moving such a capacitor bank along the feeder, the ATP provides the results presented in Figure 4.10. The study has been carried out by means of a single-phase model of the feeder, and the figure presents the single-phase losses. It is clear that the optimum location is about 2/3 of the feeder length, as predicted by the rule. A further study aimed at estimating both the location and the rated power of the capacitor bank that minimizes feeder losses has been carried out. Note that, in this case, two variables have to be changed. Figure 4.11 depicts the losses that originate as a function of the capacitor bank size and its location, proving again that the rule is acceptable. Although

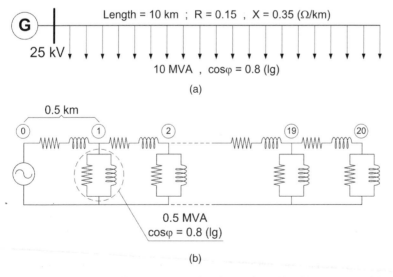

Figure 4.9 Case Study 1: feeder with a uniformly distributed load. (a) Diagram of the feeder. (b) Equivalent circuit.

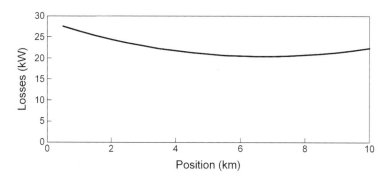

Figure 4.10 Case Study 1: feeder losses as a function of the capacitor location.

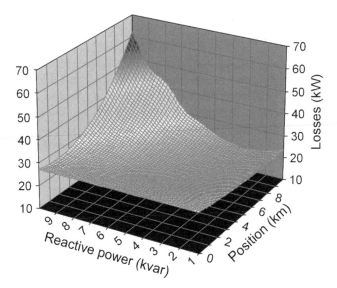

Figure 4.11 Case Study 1: feeder losses as a function of the capacitor rating and position.

loads were represented as constant impedances, the conclusion is similar to that derived from using voltage independent loads [36].

The problem was solved using a single input file and a two-step procedure in both studies. First, the model of the network was created by taking advantage of the DO loop feature to serialize the feeder sections; next, a parametric analysis was performed by changing the capacitor bank reactive power and its location, and using string replacement capabilities. PCVP is applied to move the capacitor bank model along the feeder nodes.

To obtain the feeder losses every time the capacitor bank is moved, the active power is measured at the feeder head and at every section load; active power losses are the difference that results after subtracting the sum of local active powers from the total active power measured at the feeder head.

Although this problem can easily be solved by using tools faster than the ATP, this package has an obvious advantage as it can be applied to any type of network (e.g. a line feeder with different conductor sections, with any type of load, including power electronics-based loads).

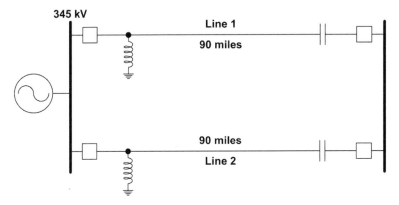

Figure 4.12 Case Study 2: diagram of the test system.

4.4.3 Case Study 2: Parallel Resonance Between Transmission Lines

Very high overvoltages can be originated in series-compensated parallel transmission lines when they are also compensated by means of shunt reactors. Figure 4.12 shows the diagram of the system configuration analysed in this study [37]. Consider the following scenario: (i) a single-phase-to-ground fault occurs in Line 1; (ii) circuit breakers that protect Line 2 open due to the fault, but the shunt reactors remain connected; (iii) the protection system of Line 1 opens the circuit breaker at the receiving end, but fails to open the circuit breaker at the sending end. The fault condition remains.

Resonant overvoltages could occur on the de-energized line due to coupling with the energized line. The value of these overvoltages will depend on the reactive power of the shunt reactors.

A parametric study can be carried out to identify for which range of values of the shunt reactor power a resonance condition could be produced.

The line geometry of the test system is shown in Figure 4.13. Since the resonance conditions can be analysed using steady-state calculations, the transmission lines are represented by means of 10-mile PI sections cascaded in series. The transmission lines are transposed every 30 miles.

This study is based on the combination of PCVP (the multiple run option) and $PARAMETER features.

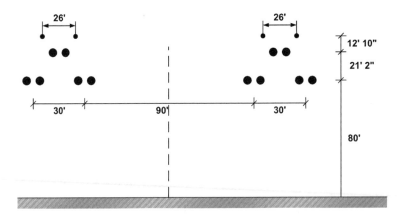

Figure 4.13 Case Study 2: shunt compensated parallel transmission lines – geometry.

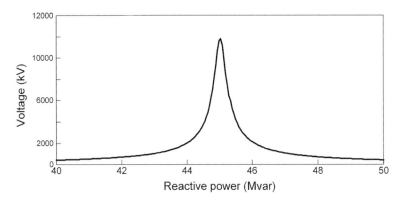

Figure 4.14 Case Study 2: parallel resonance – parametric study.

Figure 4.14 shows the voltage at the sending end of Line 2 as a function of the reactive power of these reactors. One can easily deduce for what range of values resonant overvoltages will be induced on the de-energized line.

There are several solutions to the problem of resonance between parallel lines, as it has been analysed in this example [37]: avoid reactor powers which could cause resonance, use switches to disconnect shunt reactors from de-energized lines, or install switched reactors in the sub-station rather than on the lines.

4.4.4   Case Study 3: Selection of Surge Arresters

To protect equipment against overvoltages, surge arresters are connected in parallel from each phase to ground, and located as closely as possible to the protected equipment. However, some distance between arrester and equipment cannot be always avoided. As a consequence, the voltage at the equipment will be higher than that at the arrester terminals due to oscillations on connecting leads. This is usually known as *separation effect* [38].

Figure 4.15 shows the scheme of a transformer-terminated distribution line with an arrester located at a certain distance from the transformer. A lightning stroke hits the overhead line at the span close to the single-transformer substation. The connecting lead between the arrester and the transformer can be either an overhead line or an insulated cable [39]. The study will be performed considering an idealized representation for all components: lines and cables are represented by means of non-dissipative single-phase travelling-wave models; a transformer is represented as an open circuit; an ideal current source is used to represent a lightning stroke, and only double-ramp waveforms are considered; an ideal gapless metal-oxide surge arrester will provide an infinite impedance for surge voltages lower than its discharge voltage. The length of lead wires that connect the arrester to the overhead line and ground are ignored, and it is assumed that there will not be flashover in the line, which is supposed to have an infinite length.

Figure 4.16 shows the equivalent circuit of this system as analysed in this chapter [40]. Figure 4.17 depicts some simulation results that illustrate the effect of the rate-of-rise of the stroke current; they correspond to a case with the following parameters: line and lead surge impedances $= 400\,\Omega$, arrester discharge voltage $= 100\,\text{kV}$. The lightning stroke hits the overhead line on the span close to the transformer; see Figures 4.15 and 4.16.

The procedure for selecting the surge arresters will be based on the simplified models described above. A $10\,\text{kA}/2\,\mu\text{s}$-to-crest stroke is used to select the arrester.

The following items are required for applying this procedure [41]: (i) a library of arrester modules: (ii) an algorithm to decide when the target has been achieved and how to select arresters when it has not been achieved; (iii) a look-up table to assign arrester data file.

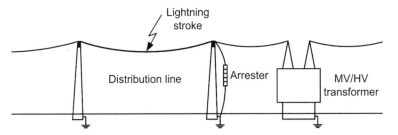

Figure 4.15 Case Study 3: scheme of a transformer-terminated line.

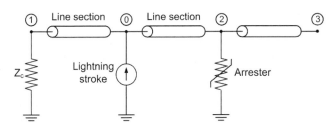

Figure 4.16 Case Study 3: equivalent circuit.

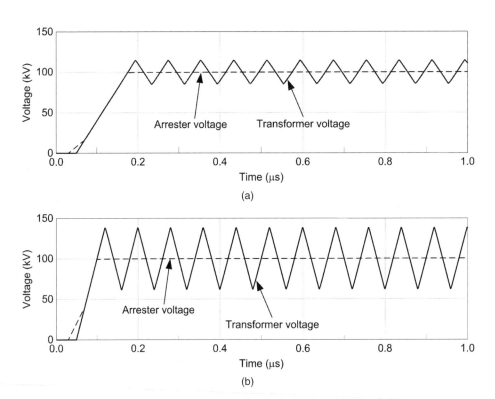

Figure 4.17 Case Study 3: arrester and transformer voltages. (a) Lead length: 6 m, stroke: 10 kA crest, 2 kA/μs. (b) Lead length: 6 m, stroke: 10 kA crest, 5 kA/μs.

The target in this work is the so-called *protective margin*, which is defined as follows [41]:

$$PM = \left(\frac{BIL}{MV} - 1 \right) \cdot 100 \qquad (4.8)$$

where *BIL* is the Basic Impulse Insulation Level of the transformer, and *MV* the maximum voltage originated at the transformer.

If the residual voltages of the arresters are rated from 30 to 130 kV, the selection could be based on the following algorithm [41]:

> IF (*PM* < 20) *THEN*
>> *Select a lower residual voltage*
>> *ELSIF Stop*
> *ENDIF* $\hspace{6cm}$ (4.9)

When using the models described above, the voltage originated at the transformer, V_{tmax}, will be given by [41]:

$$V_{tmax} = V_{res} + 2S_v\tau \qquad (4.10)$$

where V_{res} is the arrester residual voltage, S_v is the rate-of-rise of the lightning surge, and τ the propagation time between the arrester and the transformer. The MV that can be originated at the transformer is $2V_{res}$ [41].

Assume that the transformer BIL is 170 kV, and the distance between the arrester and the transformer is 12 m. A very simple mathematical calculation is needed to deduce that the residual voltage of the arrester should be 72 kV when a 20% protective margin is chosen. The selection procedure made, based on ATP simulations will produce the following results:

```
ARRESTER RESIDUAL VOLTAGE        PROTECTIVE MARGIN
          100.                        4.49234126
ARRESTER RESIDUAL VOLTAGE        PROTECTIVE MARGIN
           90.                        1.13206929
ARRESTER RESIDUAL VOLTAGE        PROTECTIVE MARGIN
           80.                        7.40722598
ARRESTER RESIDUAL VOLTAGE        PROTECTIVE MARGIN
           70.                       22.1110406
   SELECTED RESIDUAL VOLTAGE = 70.          KV
```

In practice, a much more rigorous analysis is needed, and the selection of the arresters could be based on the Mean Time Between Failures (MTBF). This value can be deduced from the selected arrester by performing a statistical analysis and a method similar to that presented in [42]; that is, a Monte Carlo simulation based on the probabilistic distribution of several variables (lightning surge current, lightning surge rate of rise, phase potential). A probabilistic distribution could also be considered for the transformer insulation.

It is obvious that more accurate models should be used for representing power components and lightning strokes, and several line spans from the arrester location should be included in the model. In addition, it could be important to consider failures caused by both direct and indirect strokes. However, the procedure could be as simple as that presented in this paper and the simulation time as short as that needed with the study detailed in this section.

4.5 Remarks

ATP capabilities that have been analysed in previous sections were shown in Figure 4.4. However, users can take advantage of some other capabilities for adding custom-made models, expanding applications, and speeding up simulations. In addition, there are some applications not listed in Table 4.5 that are opening new fields to ATP users. This section presents a very short description of these capabilities and applications.

- TACS 69, also known as 'Compiled TACS', is an option that can be used to implement TACS-based models by means of SUBROUTINE DEVT69. A template of the routine is available to develop FORTRAN-coded files, which can be compiled and linked to the executable TPBIG. In this manner, users can implement new models or assembly already-tested code to speed up simulations. This option has been used in some cases studies presented in Chapters 6 and 9.
- TACS POCKET CALCULATOR can be used as a pocket calculator for TACS supplemental variables. It can also be used for introducing conditional branching (i.e. IF-THEN-ELSE-ENDIF) within a TACS-coded section.
- Users can develop and implement a compiled model that can be inserted within a MODELS section as a *foreign* model [1] (see Figure 1.2). As with TACS 69, the *foreign* option can be used for implementing new models and algorithms, although it has recently been used for implementing real-time applications [43, 44].
- Several real-time applications for testing protection and control devices have been presented using ATP; see for instance [43–45]. Many of these applications take advantage of the MODELS-based foreign option and the advantages of C language for linking ATP to external platforms.
- It is possible to develop simulation environments in which ATP capabilities are combined with capabilities from other simulation tools and create powerful tools that can significantly expand ATP applications. Chapter 11 shows how parallel computing can be used to shorten the simulation time of parametric and statistical studies or applied to optimization problems.

References

1 Can/Am Users Group (2000). *ATP Rule Book*. Can/Am Users Group.

2 Kizilcay, M. and Hoidalen, H.K. (2015). EMTP-ATP. In: *Numerical Analysis of Power System Transients and Dynamics*, Chapter 2 (ed. A. Ametani). Stevenage, (United Kingdom): The Institution of Engineering and Technology.

3 Dommel, H.W. (1969). Digital computer solution of electromagnetic transients in single and multiphase networks. *IEEE Transactions on Power Apparatus and Systems* 88: 388–399.

4 Dommel, H.W. (1986). *ElectroMagnetic Transients Program. Reference Manual (EMTP Theory Book)*. Portland (OR, USA): Bonneville Power Administration.

5 Prikler, L. and Høidalen, H.K. (2009). ATPDRAW Version 5.6 for Windows 9x/NT/2000/XP/Vista. http://www.atpdraw.net (accessed 02 June 2019).

6 Dubé, L. and Bonfanti, I. (1992). MODELS: a new simulation tool in the EMTP. *European Transactions on Electrical Power Engineering* 2 (1): 45–50.

7 Ceraolo, M. (1998). PlotXY-a new plotting program for ATP. *EEUG News* 4 (1–2): 41–47.

8 Grebe, T. and Smith, S. (1999). Visualize system simulation and measurement data. *IEEE Computer Applications in Power* 12 (3): 46–51.

9 Celikag, D. and Kizilcay, M. (2002). *ATP Control Center*. EEUG Meeting, December 2002, Sopron, Hungary.

10 Ametani, A. (1980). A general formulation of impedance and admittance of cables. *IEEE Transactions on Power Apparatus Systems* 99 (3): 902–910.

11 Semlyen, A. and Dabuleanu, A. (1975). Fast and accurate switching transient calculations on transmission lines with ground return using recursive convolutions. *IEEE Transactions on Power Apparatus Systems* 94 (2): 561–571.

12 Marti, J.R. (1982). Accurate modeling of frequency-dependent transmission lines in electromagnetic transient simulations. *IEEE Transactions on Power Apparatus Systems* 101 (1): 147–155.

13 Noda, T., Nagaoka, N., and Ametani, A. (1996). Phase domain modeling of frequency dependent transmission lines by means of an ARMA model. *IEEE Transactions on Power Delivery* 11 (1): 401–411.

14 Noda, T., Nagaoka, N., and Ametani, A. (1997). Further improvements to a phase-domain ARMA line model in terms of convolution, steady-state initialization, and stability. *IEEE Transactions on Power Delivery* 12 (3): 1327–1334.

15 Brandwajn, V., Dommel, H.W., and Dommel, I.I. (1982). Matrix representation of three-phase n-winding transformers for steady-state and transient studies. *IEEE Transactions Power Apparatus Systems* 101 (6): 1369–1378.

16 Martinez, J.A. and Mork, B. (2005). Transformer modeling for low- and mid-frequency transients – a review. *IEEE Transactions on Power Delivery* 20 (2): 1625–1632.

17 Jazebi, S., Zirka, S.E., Lambert, M. et al. (2016). Duality derived transformer models for low-frequency electromagnetic transients – part I: topological models. *IEEE Transactions on Power Delivery* 31 (5): 2410–2419.

18 Martinez-Velasco, J.A. (2013). Basic methods for analysis of high frequency transients in power apparatus windings. In: *Electromagnetic Transients in Transformer and Rotating Machine Windings*, Chapter 2 (ed. C.Q. Su). IGI Global.

19 Mork, B.A., Gonzalez, F., Ishchenko, D. et al. (2007). Hybrid transformer model for transient simulation – part I: development and parameters. *IEEE Transactions on Power Delivery* 22 (1): 248–255.

20 Gross, G. and Hall, M.C. (1978). Synchronous machine and torsional dynamics simulation in the computation of electromagnetic transients. *IEEE Transactions on Power Apparatus Systems* 97 (4): 1074–1086.

21 Brandwajn, V. and Dommel, H.W. (1977). A new method for interfacing generator models with an electromagnetic transients program. IEEE PES PICA Conference, Toronto.

22 Lauw, H.K. and Meyer, W.S. (1982). Universal machine modeling for the representation of rotating machinery in an electromagnetic transients program. *IEEE Transactions on Power Apparatus Systems* 101 (6): 1342–1352.

23 Dube, L. and Dommel, H.W. (1972). Simulation of control systems in an electromagnetic transients program with TACS. IEEE PES PICA Conference, San Francisco.

24 Hevia, O.P. (2000). HYSTERESIS HEVIA: a new routine to generate input data for inductors with hysteresis. *EEUG News* 6 (1–2): 19–28.

25 Dube, L. (1995). MODELS: new type-94 Norton component. *EEUG News* 1 (3–4): 30–32.

26 Martinez-Velasco, J.A. (1996). A new method for EMTP implementation of non-linear components. 12th PSCC in Dresden, Germany (August 1996).

27 Martinez-Velasco, J.A. and Martin-Arnedo, J. (2001). EMTP modular library for power quality analysis. *EEUG News* 7 (4): 54–62.

28 Dommel, H.W. (1971). Nonlinear and time-varying elements in digital simulation of electromagnetic transients. *IEEE Transactions on Power Apparatus Systems* 90 (6): 2561–2567.

29 Dommel, H.W., Yan, A., and Wei, S. (1986). Harmonics from transformer saturation. *IEEE Transactions on Power Apparatus Systems* 1 (2): 209–215.

30 Furst, G. (1998). Harmonic analysis using the HARMONIC FREQUENCY SCAN in ATP. *EEUG News* 4 (1–2): 18–25.

31 Høidalen, H.Kr. (2017). ATPDraw – Fundamental source code changes in version 7. EEUG Meeting in Kiele, Germany (September 2017).

32 Høidalen, H.Kr. (2003). ATPDraw Line Check module. EEUG Meeting in Graz, Austria (December 2003).

33 Furst, G. and Hoidalen, H.Kr. (2008). WindSyn for ATPDraw. EEUG Meeting in Izmir, Turkey (September 2008).

34 Prikler, L. (2000). Main characteristics of plotting programs for ATP. *EEUG News* 6 (3–4): 28–33.

35 Gole, A. Martinez-Velasco, J.A. and Keri, A. (eds.) (1998). *Modeling and Analysis of Power System Transients Using Digital Programs*. IEEE Special Publication TP-133-0, IEEE Catalog No. 99TP133-0.

36 Martinez-Velasco, J.A. (1996). Teaching surge arrester protection using the ATP. EEUG Meeting in Budapest, Hungary (November 1996).

37 Gönen, T. (2008). *Electric Power Distribution System Engineering*, 2e. Boca Raton (FL, USA): CRC Press.

38 EPRI Report EL-4202. (1985). Electromagnetic Transients Program (EMTP) Primer.

39 IEEE C62.22-1991, (1992). IEEE Guide for the Application of Metal Oxide Surge Arresters for AC Systems.

40 Clayton, R.E., Grant, I.S., Hedman, D.E., and Wilson, D.D. (1983). Surge arrester protection and very fast surges. *IEEE Transactions on Power Delivery* 102 (8): 2400–2412.

41 Martinez, J.A. and Martin-Arnedo, J. (2003). Expanding capabilities of EMTP-like tools: from analysis to design. *IEEE Transactions on Power Delivery* 18 (4): 1569–1571.

42 Furst, G. (1996). Monte Carlo lightning backflash model for EHV lines. A MODELS-based application example. EEUG Meeting in Budapest, Hungary (November 1996).

43 Fabián Espinoza, R.G., Molina, Y., and Tavares, M. (2018). PC implementation of a real-time simulator using ATP foreign models and a sound card. *Energies* 11: 2140.

44 Wehrend, H. and Hartwig, C. (2018). Real-time hybrid simulator using ATP-EMTP. EEUG Meeting in Arnhem, The Netherlands.

45 Perez, E. and de la Ree, J. (2016). Development of a real time simulator based on ATP-EMTP and sampled values of IEC61850-9-2. *International Journal of Electrical Power & Energy Systems* 83: 594–600.

To Probe Further

ATP data files corresponding to the three case studies presented in this chapter are provided in the website of this book. Although readers might not be familiar with all the ATP features used in those case studies, the files can be used to explore those applications, or create new case studies.

5

Introduction to the Simulation of Electromagnetic Transients Using ATP

Juan A. Martinez-Velasco and Francisco González-Molin

5.1 Introduction

A systematic procedure for the study of transients in power systems might be outlined as follows (see Chapters 1 and 4, and references [1, 2]):

1. Define the goal of the study and select the results that are of concern.
2. Decide the system zone to be represented in the model and the model to be used for representing each component. This step must be based on the events and phenomena to be studied, since they can serve to estimate the frequency range of phenomena that can occur in the simulation, and consequently, the most adequate representation of each component.
3. Select the time-step size and the length of the simulation from the information collected in the previous step. Remember that solution techniques implemented in Alternative Transients Program (ATP) work with a fixed time-step size selected by the user.
4. Collect data and draw the test system circuit using ATPDraw capabilities. This step is straightforward for simple test systems; however, if the test system is very large and some component models very sophisticated, some previous preparation (for which it can be crucial knowing ATP data formats) will be required. For many case studies, mainly those that deal with large system models, this is the most time-consuming step [3].
5. Estimate the expected results. For large system models this is not an easy task.
6. Run the input data file. Take into account that in many studies, the initial steady-state solution is required. Therefore, an alternative path may consist of running ATP to obtain only the steady-state solution, which can be useful to check the connectivity of the system model and verify some input data [1].
7. Compare simulation and expected results. Based on this comparison, it is easy to understand that this step can be based on an iterative approach in which the input data file (or the circuit diagram implemented in ATPDraw) is progressively debugged.
8. Run all cases that could be required for a clear understanding of the transient phenomena under study. To achieve this goal, some parameter variation can be required.

The studies analysed in this chapter are intended to help ATP users with the process of learning transient analysis of power systems by beginning with very simple cases. The cases have a common organization: each study begins with a short description (in some cases supported by mathematical analysis) of the phenomena of interest and, in some cases, an analysis of the expected results; continues with the preparation of input data and a picture of the implemented circuit in ATPDraw; the study ends with a discussion of the selected simulation results, which, in some cases, are compared to the expected results. Since these case studies are mostly aimed at introducing the way in which the ATP package can be used to analyse (electromagnetic)

Transient Analysis of Power Systems: A Practical Approach, First Edition. Edited by Juan A. Martinez-Velasco.
© 2020 John Wiley & Sons Ltd. Published 2020 by John Wiley & Sons Ltd.
Companion Website: www.wiley.com/go/martinez/power_systems

transients in power systems, several cases are not connected to actual system transients, so the comparison between simulation results and expected results should be seen as a validation of the results rather than a validation of the modelling approaches.

The cases analysed in this chapter cover test systems with linear and nonlinear, lumped- and distributed-parameter component models. The system models do not include rotating machines, power electronics converters or control systems; case studies including any or all of these components will be analysed in subsequent chapters.

The chapter has been organized as follows. Sections 5.2 and 5.3 are respectively dedicated to introduce the input data formats used in ATP and discuss some important issues to be accounted for when editing input data files and analyse simulation results [1, 2]. Section 5.4 presents a detailed analysis of both the series and parallel *RLC* circuits; the study of many practical transients in power systems can be reduced to any of these two simple configurations, and they provide an excellent background to introduce the application of a simulation package like the ATP [4–6]. Sections 5.5 and 5.6 analyse transients caused when switching currents predominately capacitive and/or inductive; for this second type, reactances can either be linear or nonlinear. Section 5.7 analyses transients in circuits with distributed-parameter models considering that other components may have either linear or nonlinear behaviour [4–7].

5.2 Input Data File Using ATP Formats

The most popular option for creating a new ATP input file is the path through ATPDraw, a friendly graphical user interface (GUI) that allows ATP users to take advantage of most ATP capabilities. There will, however, be many cases for which the knowledge of the traditional text code-based formats can be crucial; see, for instance, Chapter 9. This section describes the structure of an ATP input data with the formats used to model basic source, branch, and switch components that can be required to study switching transients.

The basic structure of an ATP input data file without control system data is as follows [8]:

```
BEGIN NEW DATA CASE
    Fixed point Miscellaneous Data Line
    Integer Miscellaneous Data Line
    Branch, Transformer and Line/Cable Data
BLANK CARD ENDING BRANCH CARDS
    Switch Data
BLANK CARD ENDING SWITCH CARDS
    Source Data
BLANK CARD ENDING SOURCE CARDS
    Output Request Data
BLANK CARD ENDING OUTPUT CARDS
BLANK
BEGIN NEW DATA CASE
BLANK CARD ENDING ALL CASES
```

Note that most data sections are separated by the so-called blank cards; alternatively, the user can enter a blank line as 'BLANK' line starting in column 1. Comment lines can be used to document the input data file by entering a 'Cblank' in Columns 1 and 2; ATP prints these lines in the record of input data, but ignores them during the simulation loop.

A summary of information to be specified with the basic components (i.e. branches, switches, sources) is presented below [8].

- BEGIN NEW DATA CASE: This line always precedes all input data cases.
- *Fixed-point miscellaneous data line.* This line can contain up to seven real number parameters to be entered in fields eight columns wide. In general, users specify the first four:
  o DELTAT is the time step used in the simulation. It must always be greater than zero.
  o TMAX is the length of time to be simulated. It can be equal to or less than zero, in which case, the ATP performs a calculation of the steady-state solution only.
  o XOPT is the power frequency for purposes of inductance specification. If it is zero or blank, all inductances are entered in millihenries; if it is larger than zero (e.g. 50 or 60 Hz), all inductances are entered as ohms at the specified frequency.
  o COPT is the power frequency for purposes of capacitance specification. If it is zero or blank, all capacitances are entered in microfarads; if it is larger than zero, all capacitances are entered as microohms at the specified frequency.
- *Integer miscellaneous data line.* This line can contain up to 10 integer parameters, although users will generally input the following seven parameters:
  o IOUT specifies the rate at which output variables are printed during the simulation. If IOUT is zero or one, each time step is printed. If IOUT = k, then every kth time step is printed.
  o IPLOT specifies the rate at which output variables are saved for plotting in the same way that IOUT controls printed output.
  o IDOUBL causes a network topology listing to be printed out when it is equal to 1. It is useful in checking branch and switch connections when setting up a case.
  o KSSOUT causes a complete steady-state voltage and current solution to be printed for each branch in the network when it is equal to 1, only switch and source steady-state solutions when it is equal to 2, and switch, source, and requested output variable steady-state solutions when it is equal to 3.
  o MAXOUT causes to print the maximum values attained by each output variable during the transient simulation when it is to 1.
  o ICAT causes all generated plot data to be saved for future plotting by a separate program when it is equal to 1.
  o NENERG causes a probabilistic switching simulation to be performed when it is greater than 0.
- *Branch data. RLC* branch parameters (resistance, inductance, and [series] capacitance) are entered in fields six columns wide. At least one of these parameters must be non-zero. Resistances are specified in ohms; inductances in ohms or millihenries, and capacitances in microohms or microfarads. Six-characters are used to name nodes. If one node name is left blank, the branch is assumed to be connected from a node to ground. A request to output branch current and branch voltage variables is specified by a nonzero entry in column 80: use a 1 to request branch current output, 2 to have the branch differential voltage output, 3 to have both branch current and voltage output, and 4 to have branch power and energy output.
- *Switch data.* The format is similar to that of branches, except that switch closing and opening times must be specified. ATP switch models can perform only one close-open operation. The switch is connected between two nodes, one of which may be left blank to indicate a switch to ground. The values of T-CLOSE and T-OPEN must be specified in seconds. A negative T-CLOSE indicates the switch is closed before the simulation begins. If T-OPEN is left blank or is assigned a value greater than TMAX, the switch will never open during the simulation. Switch current and voltage outputs may be requested in the same way as for branches.
- *Source data.* A variety of source types are available in the ATP. The sources can be either node voltages to ground or currents injected at the node. The user has to specify the type

(e.g. 11 = step function, 14 = cosine function), the node name where the source is connected (sources are always grounded), whether it is a voltage or a current source, and the amplitude (i.e. the current or voltage level of the step function, or the line-to-ground peak amplitude of the cosine wave). For cosine function waveforms, the frequency in Hertz and the phase angle in degrees are also specified. Both fields are left blank for a step function. A source becomes inactive when $t <$ TSTART or when $t >$ TSTOP. Sinusoidal (Type 14) sources with negative TSTART are active in the pre-transient steady state.

- *Node voltage output request data.* Some data lines can be used to specify the names of nodes where the voltage to ground is desired. More than one line may be used if the voltage of more than 13 nodes is required. A single data line can be used with a '1' in Column 2 to request all node voltages to be output.

5.3 Some Important Issues

There are several important issues related to the simulation of a case study that must be considered before and after simulating a test case [2].

5.3.1 Before Simulating the Test Case

The user has to decide about the part of the test system that must be included in the model, how to represent each system component and how to configure the system model (to avoid connectivity problems and numerical oscillations), what set of units is the most appropriate, how to select the time step and simulation time, or what outputs should be requested.

5.3.1.1 Setting Up a System Model

One of the initial steps concerns how much of the power system has to be modelled. Once a basic system model has been implemented and tested, the user can expand the model to observe their effect on the results. It is worth emphasizing the importance of starting out with simple models.

5.3.1.2 Topology Requirements

The test system model must be configured to avoid numerical oscillations (see Chapter 3) and floating subnetworks. The following rules can be applied:

- Do not connect switches in series with an inductance unless current can flow through the inductance, regardless of the status of the switches, and do not place voltage sources across a capacitance.
- When modelling inductance losses, put part of the losses in a series resistance to ensure that dc currents trapped in the inductance will eventually decay. Similarly, resistors can be connected in parallel with capacitors so that voltages trapped on the capacitor will also decay.
- When modelling semiconductors as ideal switches, include current limiting reactors and parallel snubber circuits; these components limit excessive di/dt and dv/dt, and enhance numerical stability.
- Floating subnetworks occur when switches open and disconnect part of the system model from any path to ground (e.g. unloaded delta tertiary windings); the voltages of these networks are undefined, and the ATP sets the voltage at one node in the floating subnetwork equal to zero. This may be acceptable for delta tertiary windings but, in general, it is best to include (very small) stray capacitances to ground at each terminal of an unloaded delta winding, or at one node in each subnetwork which could become floating.

5.3.1.3 Selection of the Time-Step Size and the Simulation Time

The following rules are suggested for estimating an acceptable time step.

1. Determine the shortest travel time among the modelled lines and cables. To obtain the maximum time step, divide this travel by 10 if the line is important to the study, or by 4 if the line forms part of the outlying system.
2. Estimate the period of oscillation for each LC loop according to $T = 1/f = 2\pi\sqrt{LC}$. The time step should be no more than one-twentieth of the oscillation period if the shortest T comes from a loop that plays an important role; otherwise, the time step could be as high as 1/4th of T.
3. Calculate the (RC and L/R) time constants for the lumped-parameter elements. The time step should not exceed the shortest of these time constants.
4. When simulating harmonics, the time step should be equal to, or shorter than, one degree of a power frequency cycle.

In general, smaller time steps than those derived from these rules are preferable. In each study, the user can compare cases with different time steps to ensure that using a smaller time step has no significant effect on the results of interest.

To limit interpolation errors during the simulation, it is generally preferable to choose the time step so that each transmission line's travel time consists of an integer number of time steps; a time step which is not a 'round' number can be selected in order to satisfy this condition.

There is always a trade-off between computational effort and the accuracy that must be achieved. The length of the simulation time, TMAX, can be selected according to the following aspects:

- Several cycles of pre-transient conditions must be simulated to ensure that the correct initial conditions are reached. This is particularly important when the steady-state solution contains harmonics.
- At least 10 cycles of the dominant switching transient should be simulated to observe damping rates and ensure that resonances do not occur.
- When multiple switching operations are to be simulated, sufficient time between switching operations must be allowed for the transients to decay, unless the scheduled switching times were decided by a control strategy.
- It is not necessary to duplicate the physical time delays, which may be several seconds or minutes in real time.

Proper values of TMAX can often be determined from natural frequencies and time constants. Field test results and some previous experimentation with the ATP are valuable means for determining both Δt and TMAX. As guidance in selecting the time step, TMAX and the most adequate representations, the user can consult widely accepted tables that provide the frequency ranges associated with the most common transients in power systems [3].

5.3.1.4 Units

Entering the input data in physical units is the recommend option; however, very often data are in per-unit or percent. Per-unit impedances can be used if all values are on the same MVA base and transformers are properly represented. Synchronous machine data can be a problem when input data are in per-unit: the machine parameters are on the machine's own MVA and kV base; to use per-unit impedances in the rest of the system, a change of base on the machine data is required. Output voltages and currents will also be provided in physical values rather than the more convenient per-unit values; the postprocessor can be used to rescale outputs and obtain per-unit values.

5.3.1.5 Output Selection

The phasor steady-state solution and the network connectivity table should be generally selected as outputs. A printed time-loop solution is of limited use in debugging a case; plotted variables will always be much more useful for this purpose. The variable minima and maxima and the times of switch operation are also useful information. Even though only a few variables may be of significance, it is convenient to plot a large number of variables to assist in verifying the implemented system model. When plotting voltages and currents, all three phases should be plotted.

5.3.2 After Simulating the Test Case

Once the simulation results are available it is crucial to verify that they are as expected. In addition, debugging the input data file can be important, even if results are correct, to explore the system model performance and to better understand the transient phenomena under study.

5.3.2.1 Verifying the Results

The following rules of thumb that can be useful to check results:

- A basic knowledge of the phenomena to be simulated is crucial, so it always advisable to be aware about the (basics of the) phenomenon before simulating it.
- It is important to check the input parameter interpretation and the network connectivity table provided in the .LIS file.
- The steady-state phasor solution should be checked for bus voltage magnitudes, injected source currents and power, and the MW/Mvar loads. For example, line and switch currents should be (nearly) balanced if this is the expected result. If there is a message warning about a nonlinear inductance operating outside the linear flux region, make adjustments. Pre-switching steady-state waveforms should be examined for characteristic harmonic content if they should reach a stable undistorted condition before a transient is initiated.
- Transient frequencies can be verified according to $f = 1/(2\pi\sqrt{LC})$ for lumped-parameter circuits, $f = 1/4\tau$ for distributed-parameter models for which τ is the propagation time. The surge impedance, $Z = \sqrt{L/C}$, is useful for relating transient voltage and current peak magnitudes.
- Travelling wave reflections can be used to check the response of systems with distributed-parameter models. Remember that inductive terminations initially appear as open circuits, while capacitive terminations initially appear as short circuits. Damping provided by lines, cables, transformers, and loads are generally low, especially at high frequencies; some exceptions are frequency-dependent overhead lines and resistive shunts to ground.

5.3.2.2 Debugging Suggestions

If there is an error and the cause is not obvious, or if the case runs but results seem to be erroneous, the following steps can be taken:

1. Check the interpretation provided in the .LIS file.
2. Check the network connectivity table provided in the .LIS file against each element in the system model diagram, and each element against the network connectivity table.
3. Check the steady-state solution for consistency.
4. If any subnetwork is floating, or is weakly tied to ground, strengthen its path to ground.
5. Repeat the case with a smaller time step.

5.4   Introductory Cases. Linear Circuits

5.4.1   The Series and Parallel RLC Circuits

Both the series and parallel *RLC* circuits are used in this chapter to introduce the simulation of electromagnetic transients with the ATP package. The two linear schemes are interesting because networks involved in many practical transients can be reduced to such simple configurations for the purpose of analysis [4]. The series *RLC* circuit can be seen as a very simplified single-phase model of a system to which a capacitor bank is connected: parameters *R* and *L* represent the equivalent impedance measured from the node to which the capacitor bank, represented by the parameter *C*, is connected; the goal of the study could be to estimate the transient current that will flow through the breaker and the voltage that can occur at the capacitor terminals. The parallel *RLC* circuit could be seen as the single-phase model of an unloaded transformer: the *L* branch represents the unsaturated core inductance, the *R* branch represents core losses, while the *C* branch is the winding capacitance; the usual goal is to estimate the transient recovery voltage (TRV) that can occur between the terminals of the breaker that opens the circuit.

The study of the two circuits is carried out without a connection to any real test system: the series circuit is analysed assuming the transient is caused when the circuit is connected to a dc voltage source; the parallel circuit is analysed when the transient is caused by the disconnection of the circuit from an ac voltage source. Some theoretical results are provided for both of them; they are further supported by the simulation results. The two circuits are also used to illustrate how to implement a case study in ATPDraw. Some information about the formats to be used when editing the data input files is also provided.

5.4.2   The Series RLC Circuit: Energization Transient

5.4.2.1   Theoretical Analysis
A simple transient can be caused by the act of closing a switch. Figure 5.1 shows the scheme of the series *RLC* circuit supplied from an ideal dc voltage source.

In this study, the source has a constant voltage and negligible impedance compared with the *RLC* circuit. When the switch is closed, the voltage equation is:

$$e(t) = Ri(t) + L\frac{di(t)}{dt} + \frac{1}{C}\int_{-\infty}^{t} i(t)dt \qquad (e(t) = V) \qquad (5.1)$$

Upon the application of the Laplace transform, the following forms are derived if the initial values of the circuit variables are assumed zero:

$$E(s) = RI(s) + sLI(s) + \frac{I(s)}{sC} \qquad (5.2a)$$

$$I(s) = \frac{s}{L}\frac{E(s)}{s^2 + \alpha s + \omega_0^2} = \frac{V}{L}\frac{1}{s^2 + \alpha s + \omega_0^2} \qquad \left(E(s) = \frac{V}{s}\right) \qquad (5.2b)$$

Figure 5.1 Series *RLC* circuit – Energization of the circuit from a dc voltage source.

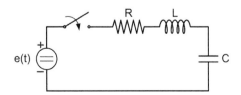

where

$$\alpha = \frac{R}{L} = \frac{1}{\tau_S} \qquad\qquad \omega_0 = \frac{1}{\sqrt{LC}} \quad \to \quad f_0 = \frac{1}{2\pi\sqrt{LC}} \tag{5.3}$$

τ_S is the time constant of the series RLC circuit and f_0 is the natural frequency.

For a thorough analysis of this circuit, readers are referred to reference [4]. The general solution of this circuit is as follows:

$$i(t) = \frac{V}{L}\frac{2\tau_S}{(4\lambda^2 - 1)^{1/2}}e^{-t/2\tau_S}\sin\left(\frac{(4\lambda^2 - 1)^{1/2}}{2\tau_S}t\right) \tag{5.4}$$

where

$$\tau_S = \frac{L}{R} \qquad\qquad \lambda = \frac{1}{R}\sqrt{\frac{L}{C}} = \frac{Z}{R} \tag{5.5}$$

and Z is the surge impedance.

The solution (5.4) for the circuit current depends on the value of parameter λ. The above form remains as shown when $\lambda > \frac{1}{2}$; however, it must be changed for other values of λ [4]:

- When $\lambda = \frac{1}{2}$, the circuit is said to be *critically damped*, and Eq. (5.4) becomes

$$i(t) = \frac{V}{L}te^{-t/2\tau_S} \tag{5.6}$$

- When $\lambda < \frac{1}{2}$, the circuit is said to be *critically overdamped*, and Eq. (5.4) becomes

$$i(t) = \frac{V}{L}\frac{2\tau_S}{(1 - 4\lambda^2)^{1/2}}e^{-t/2\tau_S}\sinh\left(\frac{(1 - 4\lambda^2)^{1/2}}{2\tau_S}t\right) \tag{5.7}$$

Note that in all solutions, there is a term that decreases with time and a term that either decreases or oscillate with time. From a quick look at the test circuit, it is obvious that, after a long period, the current flowing through the switch and the circuit will become zero and the voltage at the capacitor will be that of the source. In other words, the circuit will reach a steady state after a transient period whose characteristics will depend of the RLC parameters.

If the solution is to be obtained by means of the numerical technique implemented in ATP, and described in Chapter 3, then the discretized equivalent of the test circuit must first be derived. Figure 5.2 depicts the companion circuit once the switch is closed.

The current equations of this circuit can be expressed as follows:

$$\frac{v_0(t) - v_1(t)}{R} = \frac{v_1(t) - v_2(t)}{2L/\Delta t} + I_L(t) = \frac{v_2(t)}{\Delta t/2C} + I_C(t) \qquad (v_0(t) = e(t)) \tag{5.8}$$

where

$$I_L(t) = \left[\frac{\Delta t}{2L}v_L(t - \Delta t) + i_L(t - \Delta t)\right] \qquad (v_L(t) = v_{12}(t) = v_1(t) - v_2(t)) \tag{5.9a}$$

$$I_C(t) = -\left[\frac{2C}{\Delta t}v_C(t - \Delta t) + i_C(t - \Delta t)\right] \qquad (v_C(t) = v_2(t)) \tag{5.9b}$$

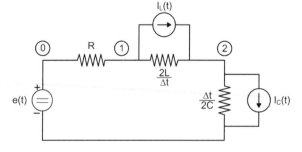

Figure 5.2 Series RLC circuit – Discretized equivalent.

where $v_L(t-\Delta t)$ and $v_C(t-\Delta t)$ are respectively the voltages across the inductor and the capacitor at the instant $(t-\Delta t)$, while $i_L(t-\Delta t)$ and $i_C(t-\Delta t)$ are respectively the currents through the inductor and the capacitor at the instant $(t-\Delta t)$.

The following matrix form results from the solution of the Eq. (5.8):

$$
\begin{bmatrix}
\dfrac{1}{R} + \dfrac{\Delta t}{2L} & -\dfrac{\Delta t}{2L} \\[2ex]
-\dfrac{\Delta t}{2L} & \dfrac{\Delta t}{2L} + \dfrac{2C}{\Delta t}
\end{bmatrix}
\begin{bmatrix}
v_1(t) \\[1ex]
v_2(t)
\end{bmatrix}
=
\begin{bmatrix}
\dfrac{v_0(t)}{R} - I_L(t) \\[1ex]
-I_C(t)
\end{bmatrix}
\tag{5.10}
$$

5.4.2.2 ATP Implementation

This subsection provides an introduction on how to create the input file for the test system analysed in this section using text code-based traditional formats and the more friendly ATPDraw approach respectively. Assume the parameters of the test circuit shown in Figure 5.1 are:

- Source: DC source, $V = 1\,\mathrm{V}$
- *RLC* Branch: $R = 1\,\Omega$, $L = 1\,\mathrm{mH}$, $C = $ Variable
- Switch: *Tclose* $= 0$, *Topen* $= 1\,\mathrm{s}$

ATP input data file
The input data file for this case could be the following one:

```
BEGIN NEW DATA CASE
C
C Test system: Energization of a series RLC circuit
C
C         1         2         3         4         5         6         7         8
C 34567890123456789012345678901234567890123456789012345678901234567890123456789 0
C Miscellaneous Data Cards
C   dT  >< Tmax >< Xopt >< Copt >
   1.E-5    .015
        50        1         1         1         1         0         0         1         0
C ----- Branch cards
C < n1 >< n2 ><ref1><ref2>< R   >< L   >< C   >
   NODE_1                    1.    1.  100.                                       0
BLANK card ending BRANCH cards
C ----- Switch cards
C < n1 >< n2 >< Tclose ><Top/Tde ><   Ie   ><Vf/CLOP ><  Type  >
   SRCE_1NODE_1                 1.                                                1
C ----- Source cards
BLANK card ending SWITCH cards
C < n 1><>< Ampl. >< Freq. ><Phase/T0><  A1   ><   T1   >< TSTART >< TSTOP  >
11SRCE_1 0        1.                                          -1.     1.E3
BLANK card ending SOURCE cards
C ----- Node voltage output request
   SRCE_1
BLANK OUTPUT
BLANK PLOT
BEGIN NEW DATA CASE
BLANK
```

Note that the simulation is carried out for 15 ms with a time step of 10 μs; the switch closes at the moment the simulation begins ($t = 0$), and opens at 1 s, a period long enough to have the switch closed during the simulation. The source is active before the simulation begins

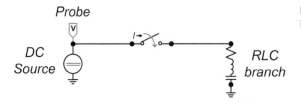

Figure 5.3 Series *RLC* circuit – ATPDraw implementation.

(TSTART = −1.0), although this is irrelevant as the switch that connects the source to the *RLC* branch is closed at the time the simulation begins.

ATPDraw implementation

To create a new circuit, users have to select the *New* command in the *File Menu* or pressing the 'page' symbol in the *Component Toolbar*. From this moment, every new component is added to the circuit by selecting its icon from the *Component Selection Menu*, which appears with a right mouse click on the open area of the circuit window [9].

Figure 5.3 displays the circuit implemented in ATPDraw for simulating this case study. Note that for building this circuit, up to four objects have to be captured and linked: the dc source, the switch, the *RLC* branch, and the voltage probe. The procedure is the same for all objects: after calling to the *Component Selection Menu*, the user has to go to the specific group (i.e. Sources, Switches, Branch Linear, and Probes) and select the right component by clicking on it with the left mouse button, holding down, and dragging it to the desired position of the circuit window. Once the icon of one component has been inserted in the circuit window, the menu window of that component has to be opened by clicking on the component icon with the right mouse button. After opening the window, the specification of parameters is straightforward. The user can take advantage of the *Help* available for each component menu. Figure 5.4 shows the menu windows for all the components needed in this test circuit.

5.4.2.3 Simulation Results

ATP creates a .LIS file during the simulation. This file can be very useful and it is important to understand the information it provides. The following aspects can be noticed in this file: (i) while reading the file, ATP shows the interpretation of each line/card sorted by class (BRANCH, SWITCH, SOURCE, etc.); (ii) after finishing the reading, the program provides the requested information (network connectivity, the simulated values for the requested variables during the time-domain loop, extreme of output variables, etc.); (iii) the file finishes with memory storage and timing figures.

In this simple case, the simulation begins after closing the switch at $t = 0$; in other scenarios the simulation can begin with the steady-state solution.

Data cards do not have to be ordered as shown above. The sorting of input data by class, built around the character '/' in column 1 can be used to edit an input file with any card order: whenever a data class changes, it is possible to mark the discontinuity with a '/' card (e.g. /TACS, /BRANCH) which indicates what type of data follows until a new '/' is found. Actually, this is the option implemented in ATPDraw (i.e. cards in the data file are sorted by class).

Figure 5.5 shows results derived from different capacitance values: 100, 1000, 4000, and 8000 μF, remaining constant the values of R (= 1 Ω) and L (= 1 mH). These results can be checked by taking into account the equations that correspond to the transient caused when closing the switch; that is, Eqs. (5.4, 5.6, and 5.7).

Since the values of R and L are constant, the time constant τ_S also remains constant and equal to 1 ms; see Eq. (5.5). The following values result from the capacitance values used in this case:

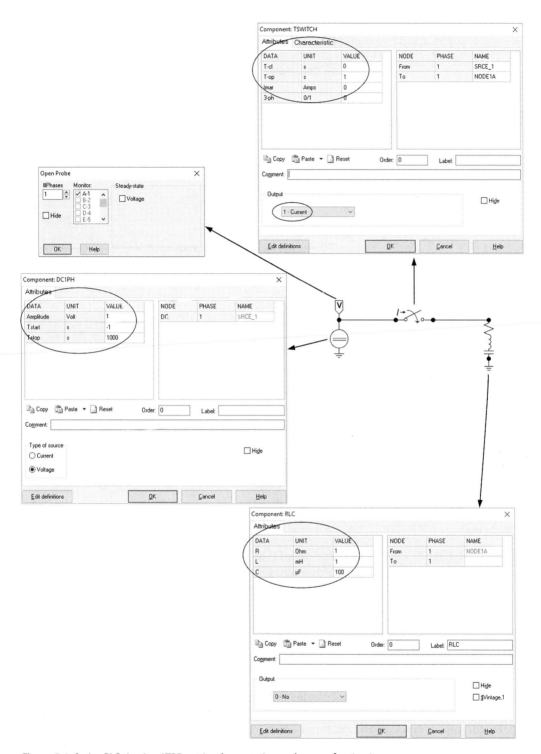

Figure 5.4 Series *RLC* circuit – ATPDraw implementation and menus for circuit components.

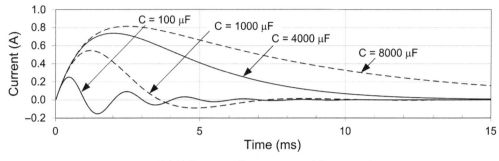

(a) Voltages at the source and the capacitor

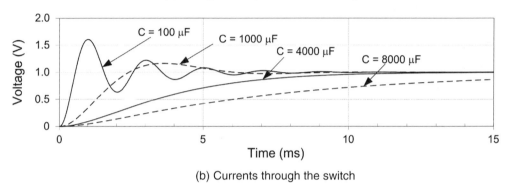

(b) Currents through the switch

Figure 5.5 Series *RLC* circuit – Simulation results with different values of the capacitance. (a) Voltages at the source and the capacitor. (b) Currents through the switch.

- $C = 100\,\mu\text{F}$ $f_0 \approx 503\,\text{Hz}$ $Z = 3.162\,\Omega$ $\lambda = 3.162$
- $C = 1000\,\mu\text{F}$ $f_0 \approx 159\,\text{Hz}$ $Z = 1\,\Omega$ $\lambda = 1$
- $C = 4000\,\mu\text{F}$ $f_0 \approx 80\,\text{Hz}$ $Z = 0.5\,\Omega$ $\lambda = 0.5$
- $C = 8000\,\mu\text{F}$ $f_0 \approx 56\,\text{Hz}$ $Z = 0.353\,\Omega$ $\lambda = 0.353$

Note that

- when $C < 4000\,\mu\text{F}$, $\lambda > 0.5$ and the transient is oscillatory;
- when $C = 4000\,\mu\text{F}$, $\lambda = 0.5$ and the transient is critically damped;
- when $C > 4000\,\mu\text{F}$, $\lambda < 0.5$ and the transient is critically overdamped.

This behaviour can easily be confirmed by checking the results depicted in Figure 5.5.

5.4.3 The Parallel RLC Circuit: De-energization Transient

5.4.3.1 Theoretical Analysis

Figure 5.6 shows the switch-off of a parallel *RLC* circuit. The aim of this example is to analyse the TRV developed between the switch terminals during the opening operation. Note that the equivalent impedance of the power supply, seen from the load side, has been neglected again. The TRV will then be the difference between voltages at switch contacts. Since the voltage at the source side is known and is not altered by the switch operation, only the transient caused at the load side has to be analysed.

Figure 5.6 Parallel *RLC* circuit – Switch-off of the circuit supplied from an ac voltage source.

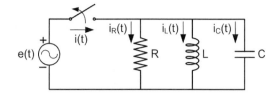

The equations of the *RLC* circuit once the switch is open can be written as follows:

$$v_R(t) = v_L(t) = v_C(t) = v(t) \tag{5.11a}$$

$$v_R(t) = Ri_R(t) \; ; \; v_L(t) = L\frac{di_L(t)}{dt} \; ; \; v_C(t) = \frac{1}{C}\int_{-\infty}^{t} i_C(t)dt \tag{5.11b}$$

$$i_R(t) + i_L(t) + i_C(t) = 0 \tag{5.11c}$$

The solution upon the application of the Laplace transform is as follows [4]:

$$RI_R(s) = sLI_L(t) = \frac{I_C(s)}{sC} + \frac{V_C(0)}{s} = V(s) \tag{5.12a}$$

$$\left(\frac{1}{R} + \frac{1}{sL} + sC\right)V(s) - CV_C(0) = 0 \tag{5.12b}$$

$$V(s) = \frac{sV_C(0)}{s^2 + as + \omega_0^2} \tag{5.12c}$$

$$v(t) = V_C(0) \cdot e^{-t/2\tau_P}\left(\cos\left(\frac{(4\eta^2 - 1)^{1/2}}{2\tau_P}t\right) - \frac{1}{(4\eta^2 - 1)^{1/2}}\sin\left(\frac{(4\eta^2 - 1)^{1/2}}{2\tau_P}t\right)\right) \tag{5.12d}$$

where $V_C(0)$ is the voltage trapped across the capacitor when the switch opens and

$$\tau_P = RC = \frac{1}{a} \tag{5.13a}$$

$$\omega_0 = \frac{1}{\sqrt{LC}} \;\; \rightarrow \;\; f_0 = \frac{1}{2\pi\sqrt{LC}} \tag{5.13b}$$

$$\eta = R\sqrt{\frac{C}{L}} = \frac{R}{Z} \tag{5.13c}$$

τ_P is the time constant, f_0 is the natural frequency, and Z is the surge impedance of the parallel *RLC* circuit.

As for the series circuit, the solution for the parallel circuit voltage shown in (5.12d) depends on the value of parameter η. Note that this parameter is the inverse of parameter λ defined in the series *RLC* circuit. The above form remains as shown when $\eta > \frac{1}{2}$; however, it must be changed for other values of η [4]:

- When $\eta = \frac{1}{2}$, the circuit is said to be *critically damped*, and Eq. (5.12d) becomes

$$v(t) = V_C(0) \cdot e^{-t/2\tau_P}\left(1 - \frac{t}{2\tau_P}\right) \tag{5.14}$$

- When $\eta < \frac{1}{2}$, the circuit is said to be *critically overdamped*, and Eq. (5.12d) becomes

$$v(t) = V_C(0) \cdot e^{-t/2\tau_P}\left(\cosh\left(\frac{(1 - 4\eta^2)^{1/2}}{2\tau_P}t\right) - \frac{1}{(1 - 4\eta^2)^{1/2}}\sinh\left(\frac{(1 - 4\eta^2)^{1/2}}{2\tau_P}t\right)\right) \tag{5.15}$$

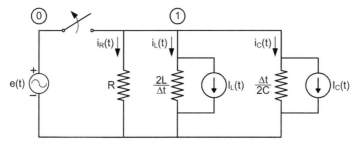

Figure 5.7 Parallel *RLC* circuit – Discretized equivalent.

Since the energy stored in the right side of the circuit, once the breaker has opened, will be dissipated in the resistor, currents and voltages will be zero after a long enough period (that will depend on the circuit parameters). The TRV is given by

$$TRV = e(t) - v(t) \tag{5.16}$$

The discretized equivalent of the parallel *RLC* circuit can be that displayed in Figure 5.7. The current equation of this circuit after the switch opens is as follows:

$$\frac{v(t)}{R} + \frac{v(t)}{2L/\Delta t} + I_L(t) + \frac{v(t)}{\Delta t/2C} + I_C(t) = 0 \tag{5.17}$$

where

$$I_L(t) = \left[\frac{\Delta t}{2L}v(t - \Delta t) + i_L(t - \Delta t)\right] \tag{5.18a}$$

$$I_C(t) = -\left[\frac{2C}{\Delta t}v(t - \Delta t) + i_C(t - \Delta t)\right] \tag{5.18b}$$

Upon solving Eq. (5.17), the following form results:

$$\left[\frac{1}{R} + \frac{\Delta t}{2L} + \frac{2C}{\Delta t}\right][v(t)] = -[I_L(t) + I_C(t)] \tag{5.19}$$

5.4.3.2 ATP Implementation

Assume the parameters of the test circuit are:

- AC source: $V_{peak} = 1$ V, Phase $= -90°$, Frequency $= 50$ Hz
- *RLC* branches: $R =$ Variable, $L = 2$ H, $C = 50$ pF
- Switch: $T_{close} = -1$, $T_{open} = 1$ ms

According to (5.13), the time constant of this circuit is $\tau_p = 1$ μs, and the natural frequency is $f_0 = \omega_0/2\pi \approx 15.92$ kHz.

ATP input data file

The simulation will be carried out for 10 ms with a time step of 1 μs. Since the switch is now closed prior to the moment the simulation begins, and opens at 1 ms, the simulation begins from the steady-state solution of the circuit. The input data file for this new test case when using the formats discussed above might be as follows:

```
BEGIN NEW DATA CASE
C
C Test system: De-energization of a parallel RLC circuit
C
```

```
C          1         2         3         4         5         6         7         8
C 345678901234567890123456789012345678901234567890123456789012345678901234567890
C Miscellaneous Data Cards
   1.E-6      .01
      500        1         1         1         1         0         0         1         0
C          1         2         3         4         5         6         7         8
C 345678901234567890123456789012345678901234567890123456789012345678901234567890
C ----- Branch cards
C < n1 >< n2 ><ref1><ref2>< R    >< L    >< C   >
   NODE_1                                   5.E-5                                   0
   NODE_1                          2.E3                                             1
   NODE_1                2.E4                                                       0
BLANK card ending BRANCH cards
C ----- Switch cards
C < n 1>< n 2>< Tclose ><Top/Tde ><    Ie    ><Vf/CLOP >< type  >
   SOURCENODE_1         -1.        .001                                            3
BLANK card ending SWITCH cards
C < n 1><>< Ampl.  >< Freq.  ><Phase/T0><   A1   ><    T1   >< TSTART >< TSTOP  >
14SRCE_1            1.        50.       -90.                       -1.       100.
BLANK card ending SOURCE cards
   SOURCENODE_1
BLANK card ending SOURCE cards
BLANK OUTPUT
BLANK PLOT
BEGIN NEW DATA CASE
BLANK
```

ATPDraw implementation

The ATPDraw-based implementation for the parallel test circuit displayed in Figure 5.6 might be that shown in Figure 5.8. For building this circuit, up to seven objects have to be captured and linked: the ac source, the switch, the R, L, and C branches, and the two voltage probes. The procedure for building the circuit is similar to that applied with the series test circuit: after calling to the *Component Selection Menu*, the user has to go to the specific group (i.e. Sources, Switches, Branch Linear, and Probes) and select the right component by clicking on it with the left mouse button, holding down and dragging it to the desired position of the circuit window. Once the icon of one component has been inserted in the circuit window, the menu window of that component has to be opened by clicking on the component icon with the right mouse button. After opening the window, the specification of parameters is straightforward. The user can take advantage of the *Help* available for each component menu [9]. Although the menu windows for the components of this test circuit have some differences with respect to those needed in the previous test circuit, the procedure for each component is straightforward.

5.4.3.3 Simulation Results

Figure 5.9 shows results derived with constant values of L ($= 2\,H$) and C ($= 50\,pF$), and three different resistance values: 20, 100, and 800 kΩ. All the results correspond to the TRV developed between the switch terminals. They can be checked by taking into account the equations that correspond to the transient caused when opening the switch; see Eqs. (5.12)–(5.15).

Figure 5.8 Parallel *RLC* circuit – ATPDraw implementation.

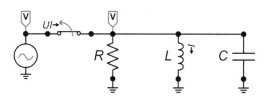

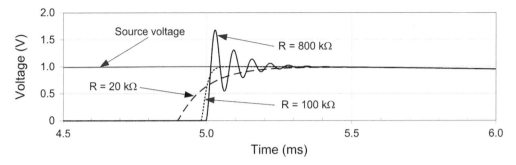

Figure 5.9 Parallel *RLC* circuit – Simulation results with different resistance values.

The following values for Z and η result from the parameter values used in this case:

- $R = 20\,k\Omega$ $L = 2\,H$ $C = 50\,pF$ $f_0 = 15.92\,kHz$ $Z = 200\,k\Omega$ $\eta = 0.10$
- $R = 100\,k\Omega$ $L = 2\,H$ $C = 50\,pF$ $f_0 = 15.92\,kHz$ $Z = 200\,k\Omega$ $\eta = 0.50$
- $R = 800\,k\Omega$ $L = 2\,H$ $C = 50\,pF$ $f_0 = 15.92\,kHz$ $Z = 200\,k\Omega$ $\eta = 1.25$

Since the value of R is not the same for the three runs, the value of the time constant τ_p varies; in turn, since the values of L and C remain the same, the natural frequency will be the same for the three cases (i.e. 15.92 kHz). Note that from the selected parameters the values of η are respectively below, equal, and above 0.5. Consequently, the simulation results will correspond respectively to overdamped, damped, and oscillatory scenarios, which can be confirmed from Figure 5.9. A simple calculation would conclude that the inductance of the test circuit is dominant; that is, the current supplied by the source is basically the current through the inductor, and it is in quadrature and lagging with respect to the source voltage. Therefore, at the time the switch opens the current through the inductor is zero, and the voltage across the capacitor is at the peak, being its magnitude the peak value of the supply voltage. The sign in this value can be either positive or negative since the current across the switch can be cancelled when it is decreasing (from positive) or increasing (from negative) to a zero value.

The transient phenomenon that occurs after the interruption of the circuit current can be summarized as follows:

- The capacitor is fully charged and begins its discharge through the inductor and the resistor. The electric energy stored in this capacitor begins to decrease as the voltage decreases, while the magnetic energy stored in the inductor (which is zero at the instant the switch opens) begins to increase as its current increases. When the capacitor is completely discharged, part of the energy is stored in the inductor, and part has been dissipated in the resistor. A reversal of the above process begins, but due to the energy dissipated by the resistor, the peak value of the new oscillation will be lower than that of the previous one.
- Unless the resistance value is infinite, the transient is damped, and so the oscillations will be; that is, they will fade away because of losses.
- The frequency of the oscillations originated in the parallel *LC* circuit is the natural frequency of this circuit, which is given by the expression (5.13b). For the case under study ($L = 2\,H$, $C = 50\,pF$), the natural frequency f_0 is 15.92 kHz.
- The TRV (i.e. the voltage across the breaker/switch contacts) is the difference between the source voltage and the voltage developed across the parallel circuit. This voltage has a component at power frequency, 50 Hz, and a component at the natural frequency, f_0, whose value can be higher than that of the power source in case of an oscillatory performance. Depending

on the damping, the TRV can reach a peak value close to twice the peak value of the source voltage and exhibit a steep front. Both aspects, a high peak and a steep wave front, can be the cause of a dielectric failure of a real circuit breaker.

5.5 Switching of Capacitive Currents

5.5.1 Introduction

Capacitor banks are installed at almost all voltage levels to improve the power factor of industrial plants or regulate voltage in power systems. Usually, they are switched in and out during different periods of the load. The switching of a capacitor bank can cause hazardous conditions for the equipment in the vicinity of the bank since voltage and current stresses can occur. Current stresses can be caused when a fault occurs in the vicinity of the bank and when the bank is energized; voltage stresses can be of concern not only when the bank is energized but also when it is de-energized in case of breaker failure during the opening operation, since a restrike in the circuit breaker can create overvoltages much higher than the ones encountered when the bank is energized.

Transient studies are made to identify high-amplitude and high-frequency currents that can affect the specification of other equipment connected in the vicinity, and to check the necessity of current limiting reactors. When performing a transient study involving the switching of a shunt capacitor bank or a fault in its vicinity, it is necessary to decide upon how the system components have to be modelled. In general, there is no need for a very large representation of the power system, and modelling the main components in the vicinity of the capacitor bank will generally suffice.

The examples presented in this section introduce the study of transient phenomena caused during the energization and de-energization of capacitor banks. Only single-phase lumped-parameter models are used to represent the test systems. The section has been organized into two parts dedicated to analyse very simple capacitive circuits supplied by either a dc or an ac source.

5.5.2 Switching Transients in Simple Capacitive Circuits – DC Supply

5.5.2.1 Energization of a Capacitor Bank

Figure 5.10 shows a simplified circuit where a capacitor bank is energized from a dc voltage source. The parameters R and L represent the equivalent impedance of the system seen from the point of common coupling. The basic equation for this circuit is the same that was used for the series RLC circuit; see Eq. (5.1), and its solution is given by expression (5.4). Two particular cases, when the parameter λ is equal or larger than 0.5 were also obtained; see (5.6, 5.7). Remember that the solutions are obtained by assuming that there is no initial current in the circuit and the capacitor is initially discharged.

Figure 5.10 Energization of a capacitor bank – DC supply.

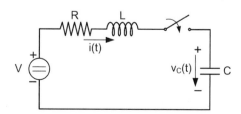

Consider a lossless circuit; that is, a particular case of the series RLC circuit with $R = 0$. If initial conditions are again zero, the equation of the circuit and its Laplace transform can be written as follows:

$$V = L\frac{di(t)}{dt} + \frac{1}{C}\int_{-\infty}^{t} i(t)dt \quad \Rightarrow \quad \frac{V}{s} = LsI(s) + \frac{I(s)}{sC} \tag{5.20}$$

The following form for the current is obtained from the second equation:

$$I(s) = \frac{V}{L}\frac{1}{s^2 + \omega_0^2} \tag{5.21}$$

The time-domain solution of this equation can be written as follows:

$$i(t) = \frac{V}{Z}\sin\omega_0 t \tag{5.22}$$

where Z and ω_0 are respectively the surge impedance and the natural frequency of the circuit.

The equation for the voltage across the capacitor and its Laplace transform can be written as follows:

$$V = L\frac{di(t)}{dt} + v_C(t) \quad \Rightarrow \quad \frac{V}{s} = LsI(s) + V_C(s) \tag{5.23}$$

Upon substitution of the expression of $I(s)$, the capacitor voltage equation becomes:

$$V_C(s) = V\left(\frac{1}{s} - \frac{s}{s^2 + \omega_0^2}\right) \tag{5.24}$$

Finally, the following expression is obtained:

$$v_C(t) = V[1 - \cos\omega_0 t] \tag{5.25}$$

According to this expression, the voltage across the capacitor is oscillatory and can reach twice the source voltage V. Assume the peak value of the source voltage is 1 V, and the circuit parameters are $L = 1\,\text{mH}$, $C = 100\,\mu\text{F}$. The values of the surge impedance and the natural frequency are $Z = 3.162\,\Omega$ and $f_0 = \omega_0/2\pi \approx 503\,\text{Hz}$. Figure 5.11 shows the simulation results obtained with ATP using a time-step size of 10 μs. Note that, although the supply voltage is constant, both the circuit current and the capacitor voltage are oscillatory. The frequency of the oscillations corresponds to the natural frequency obtained above, the peak of the capacitor voltage is 2 V, and the peak of the current is that of Eq. (5.22); that is, $1/Z = 0.316\,\text{A}$.

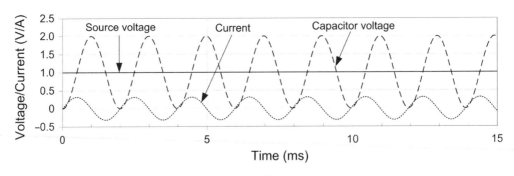

Figure 5.11 Energization of a capacitor bank – DC supply. Simulation results.

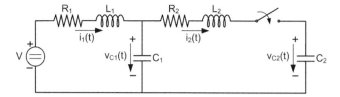

Figure 5.12 Switching of a back-to-back capacitor bank – DC supply.

5.5.2.2 Energization of a Back-to-Back Capacitor Bank

Figure 5.12 shows the electrical circuit that illustrates this situation: a second capacitor bank is energized from the node where a first already-energized capacitor bank is present. The inrush transient, in this case, mainly consists of the interchange of currents between the two banks; very often, the current from the supply system can be ignored, although this will not be the case when the supply system impedance is comparable to the impedance between the banks being switched back to back. When the second capacitor bank is switched on, the inrush current in both banks can reach a peak larger than the peak of the first connection, with a frequency (of current and voltage) higher than the frequency of oscillations when the first bank is switched on.

Assume $R_1 = R_2 = 0$. At the moment the switch is closed, the current through inductor L_1 can be non-zero, but the current through inductor L_2 is zero. The voltages across capacitor C_1 and C_2 can be written as follows:

$$v_{C_1}(t) = V - L_1 \frac{di_1(t)}{dt} = \frac{1}{C_1} \int_{-\infty}^{t} (i_1(t) - i_2(t)) dt \tag{5.26a}$$

$$v_{C_2}(t) = V - L_1 \frac{di_1(t)}{dt} - L_2 \frac{di_2(t)}{dt} = \frac{1}{C_2} \int_{-\infty}^{t} i_2(t) dt \tag{5.26b}$$

Upon application of the Laplace transform the above equations become:

$$V_{C_1}(s) = \frac{V}{s} - sL_1 I_1(s) + L_1 I_1(0) = \frac{V_{C_1}(0)}{s} + \frac{1}{sC_1}[I_1(s) - I_2(s)] \tag{5.27a}$$

$$V_{C_2}(s) = \frac{V}{s} - sL_1 I_1(s) + L_1 I_1(0) - sL_2 I_2(s) = \frac{I_2(s)}{sC_2} \tag{5.27b}$$

The two capacitor voltages can be obtained from the following system of equations:

$$\begin{bmatrix} sL_1 + \dfrac{1}{sC_1} & -\dfrac{1}{sC_1} \\ sL_1 & sL_2 + \dfrac{1}{sC_2} \end{bmatrix} \begin{bmatrix} I_1(s) \\ I_2(s) \end{bmatrix} = \begin{bmatrix} \dfrac{V}{s} + L_1 I_1(0) - \dfrac{V_{C_1}(0)}{s} \\ \dfrac{V}{s} + L_1 I_1(0) \end{bmatrix} \tag{5.28a}$$

$$V_{C_1}(s) = \frac{1}{sC_1}[I_1(s) - I_2(s)] + \frac{V_{C_1}(0)}{s} \tag{5.28b}$$

$$V_{C_2}(s) = \frac{I_2(s)}{sC_2} \tag{5.28c}$$

First, a solution of the two currents is obtained from Eqs. (5.28a); next, the solution for the two capacitor voltages is derived. Obviously, a time-domain solution for the capacitor voltages becomes very complicated, although similar case studies have been analysed in the literature; see for instance [4].

Figure 5.13 shows the circuit implemented in ATPDraw. Note that the inductor models may include a series resistance; this can be used to explore the differences between simulation results with or without damping. To simulate this case, a couple of aspects have to be accounted for:

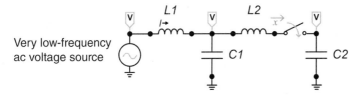

Figure 5.13 Back-to-back capacitor switching – DC supply. ATPDraw implementation.

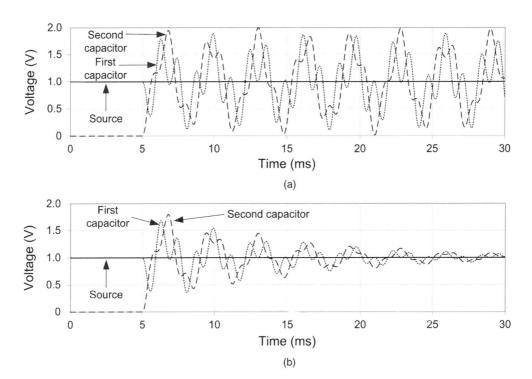

Figure 5.14 Energization of a back-to-back capacitor bank with dc supply – Simulation results: Source voltage and voltage across capacitors C_1 and C_2. (a) $R_1 = R_2 = 0\,\Omega$, $L_1 = L_2 = 1$ mH, $C_1 = C_2 = 100\,\mu$F, (b) $R_1 = R_2 = 1\,\Omega$, $L_1 = L_2 = 1$ mH, $C_1 = C_2 = 100\,\mu$F.

- The transient solution of this case begins with the circuit formed by the dc voltage source, the inductor L_1 and the capacitor C_1 operating under steady state. However, there is no solution technique implemented in ATP to obtain the steady-state solution in a circuit supplied by dc sources. A solution to this limitation is to use an ac voltage source of a very low frequency; see Figure 5.13.
- To check that the steady-state solution prior to the closing of the switch is correct, the switch will be closed 5 ms after beginning the simulation. This short period can be used to show that the solution suggested above work properly.

Assume the following parameters: $V = 1$ V, $L_1 = L_2 = 1$ mH, $C_1 = C_2 = 100\,\mu$F. Figure 5.14 shows some simulation results corresponding to cases with and without damping, using a time-step size of $10\,\mu$s. Note that voltage at the second capacitor node reaches peaks higher than those at the first capacitor node.

5.5.3 Switching Transients in Simple Capacitive Circuits – AC Supply

5.5.3.1 Energization of a Capacitor Bank

Figure 5.15 shows a simplified circuit where a capacitor bank is energized from an ac voltage source. As with the previous case, the series impedance RL could represent the equivalent impedance of the power system seen from the point of common coupling. This case study is aimed at estimating the transient voltage that will develop between capacitor terminals.

The basic equation for the circuit is again that of the series RLC circuit:

$$v(t) = Ri(t) + L\frac{di(t)}{dt} + \frac{1}{C}\int_{-\infty}^{t} i(t)dt \tag{5.29}$$

If there is no initial current in the circuit and the capacitor is initially discharged, the application of the Laplace transform gives the same forms that were obtained in subsection 5.4.2.1. The circuit current equation can be written as follows:

$$I(s) = \frac{1}{L}\frac{sV(s)}{s^2 + \alpha s + \omega_0^2} \tag{5.30}$$

where Z and ω_0 are again the surge impedance and the natural frequency of the circuit.

Assuming a general expression for the source voltage at the time the switch closes, $v(t) = V\cdot\cos(\omega t + \varphi)$, the circuit current equation can be written as follows:

$$I(s) = \frac{V}{L}\left(\frac{s\cos\varphi}{s^2 + \omega^2} - \frac{\omega\sin\varphi}{s^2 + \omega^2}\right)\frac{s}{s^2 + \alpha s + \omega_0^2} \qquad \left(V(s) = V\left(\frac{s\cos\varphi}{s^2 + \omega^2} - \frac{\omega\sin\varphi}{s^2 + \omega^2}\right)\right) \tag{5.31}$$

For the particular case with $R = 0$ (and $\alpha = 0$), the above equation becomes:

$$I(s) = \frac{V}{L}\left(\frac{\cos\varphi s}{s^2 + \omega^2}\frac{s}{s^2 + \omega_0^2} - \frac{\sin\varphi\omega}{s^2 + \omega^2}\frac{s}{s^2 + \omega_0^2}\right) \tag{5.32}$$

Given the relationship between the capacitor voltage and current

$$V_C(s) = \frac{I(s)}{sC} \tag{5.33}$$

the capacitor voltage equation can be written as

$$V_C(s) = \omega_0^2 V\left(\frac{\cos\varphi s}{s^2 + \omega^2}\frac{1}{s^2 + \omega_0^2} - \frac{\omega\sin\varphi}{s^2 + \omega^2}\frac{1}{s^2 + \omega_0^2}\right) \tag{5.34}$$

from which the following form is obtained:

$$v_C(t) = V\left(\cos\varphi\frac{\cos\omega t - \cos\omega_0 t}{1 - \left(\omega/\omega_0\right)^2} - \text{sen}\varphi\frac{\text{sen}\omega t - \left(\omega/\omega_0\right)\text{sen}\omega_0 t}{1 - \left(\omega/\omega_0\right)^2}\right) \tag{5.35}$$

Figure 5.15 Energization of a capacitor bank – AC supply.

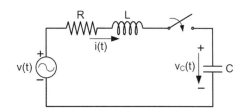

If $\omega_0 \gg \omega$, the above expression can be simplified to

$$v_C(t) \approx V(\cos(\omega t + \varphi) - \cos \varphi \cdot \cos \omega_0 t) \tag{5.36}$$

The time-domain expression for the circuit current would be that shown below:

$$i(t) = C\frac{dv_C(t)}{dt} \approx \frac{V}{Z}\left(\cos \varphi \cdot \sin \omega_0 t - \frac{\omega}{\omega_0}\sin(\omega t + \varphi)\right) \tag{5.37}$$

Note that the voltage between the capacitor terminals depends on the circuit parameters and the instant at which the switch is closed (i.e. the value of the phase angle φ).

- If $\varphi = 0$ then

$$v_C(t) \approx V(\cos \omega t - \cos \omega_0 t) \tag{5.38}$$

That is, the voltage at the capacitor is oscillatory and can reach twice the voltage source.

- If $\varphi = \pm\pi/2$ then

$$v_C(t) \approx \mp V \sin \omega t \tag{5.39}$$

As the supply voltage at the time the switch closes is zero, the peak voltage between capacitor terminals will basically be that of the supply source.

Assume the source that supplies the circuit is a 50 Hz voltage source with a peak voltage of 1 V. The value of R varies while the values of L and C remain constant and equal to 0.1 mH and 500 µF, respectively. Figure 5.16 shows some simulation results obtained with a time-step size of 10 µs and two different closing times: (i) when the supply voltage is at a peak; (ii) when the supply voltage crosses the zero value. One can observe from these results that:

- both the circuit current and the capacitor voltage will exhibit oscillations superimposed to the power frequency waveform,
- the worst scenario (i.e. largest capacitor voltage and source current) is obtained when the switch closes at the moment the source voltage is at its peak (note the differences between voltages and currents when the switch closes at a peak and at a zero),
- without damping ($R = 0$), the peak of the capacitor voltage can reach 2 V in the worst case (i.e. switch closes at 5 ms), and the current peak is that of Eq. (5.37),
- these oscillations have a frequency that corresponds to the natural frequency; for $L = 0.1$ mH and $C = 500$ µF, $f_0 \approx 712$ Hz.

5.5.3.2 Energization of a Back-to-Back Capacitor Bank

The electrical circuit that illustrates this situation is that shown in Figure 5.17. The procedure to obtain the expressions of currents and voltages is the same that was presented in subsection 5.5.2.2; the circuit is now supplied from an ac voltage source.

Assume the following parameters:

- Source: AC source, $V_{peak} = 1$ V, Phase $= -90°$, Frequency $= 50$ Hz
- Circuit 1: $L_1 = 0.5$ mH, $C_1 = 500$ µF
- Circuit 2: $L_2 = 0.5$ mH, $C_2 = 500$ µF

As for the case with a dc source, the value of the resistances will be varied. The transient solution of this case begins with the circuit formed by the ac voltage source, the inductor L_1 and the capacitor C_1 operating under steady state; the switch is closed at the moment the voltage

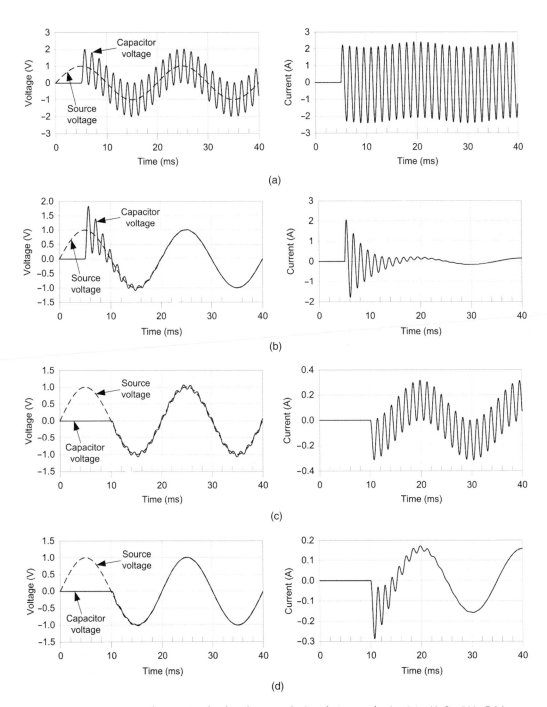

Figure 5.16 Energization of a capacitor bank with ac supply. Simulation results: $L = 0.1$ mH, $C = 500$ µF. (a) Switch closes at 5 ms – $R = 0\,\Omega$, (b) Switch closes at 5 ms – $R = 0.05\,\Omega$, (c) Switch closes at 10 ms – $R = 0\,\Omega$, (d) Switch closes at 10 ms – $R = 0.05\,\Omega$.

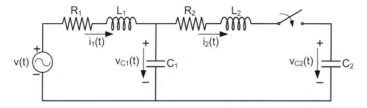

Figure 5.17 Switching of a back-to-back capacitor bank – AC supply.

at capacitor C_1 is at the peak. Figure 5.18 shows some simulation results with three different values of the resistances.

The main conclusions from these results may be summarized as follows:

- Oscillations of voltages and currents remain undamped and reach the highest values without losses ($R_1 = R_2 = 0$). Without damping, peak voltages can be twice the source voltage peak.
- The damping effect of the resistance is very obvious: the higher the resistances, the faster the damping and the lower the peak values of voltages and currents.

5.5.3.3 Reclosing into Trapped Charge

The tendency of circuit breakers to open nearly zero currents virtually ensures that the instant of breaker opening in a capacitive circuit will coincide with a peak voltage condition across the capacitor. Thus, a breaker opening will normally leave a capacitor charged to its peak voltage. Since the reclosing of a breaker can occur at any random instant, depending on the exact instant of breaker closing, overvoltages greater than 2 pu can develop. The mechanism by which these overvoltages can develop is illustrated with the simple circuit shown in Figure 5.19 [2].

Assume the source voltage is $v(t) = V \cdot \cos\omega t$. While the breaker is closed, the circuit current and the voltage across the capacitor can be expressed as follows:

$$i(t) = \frac{\omega C}{1 - \omega^2 LC} V \cos \omega t = \frac{\omega C}{1 - \left(\omega/\omega_0\right)^2} V \cos \omega t \tag{5.40a}$$

$$v(t) = \frac{1}{1 - \omega^2 LC} V \cos \omega t = \frac{1}{1 - \left(\omega/\omega_0\right)^2} V \cos \omega t \tag{5.40b}$$

where ω_0 is again the natural frequency of the circuit; see formula (5.3).

Regardless of the exact time of contact parting, effective opening will not occur until the first zero-current crossing. This will only happen at $t = k \cdot \pi/\omega$, with k integer. At any of these times, the trapped voltage is [2]:

$$V_{trapped} = \frac{V}{1 - \left(\omega/\omega_0\right)^2} \tag{5.41}$$

If $\omega_0 \gg \omega$, then $i(t) \approx \omega CV \cos\omega t$, $v(t) \approx V \cos\omega t$, and $V_{trapped} \approx V$.

To show the overvoltages that can occur when reclosing into a trapped charge, the circuit depicted in Figure 5.20 has been implemented in ATPDraw. Remember the entire process requires that the switch that represents the breaker is initially closed; it opens leaving some trapped voltage in the capacitor, and finally it recloses. To obtain such sequence, two parallel switches, as shown in Figure 5.20, are required: the first switch is initially closed and opens at 1 ms; the reclosing operation is modelled with the second switch that closes into a capacitor with a trapped charge. The closing instant of this second switch is delayed until the moment the source voltage is a peak of opposite sign to that of the trapped voltage. Note that the source inductance is represented with a parallel resistance that is required to avoid

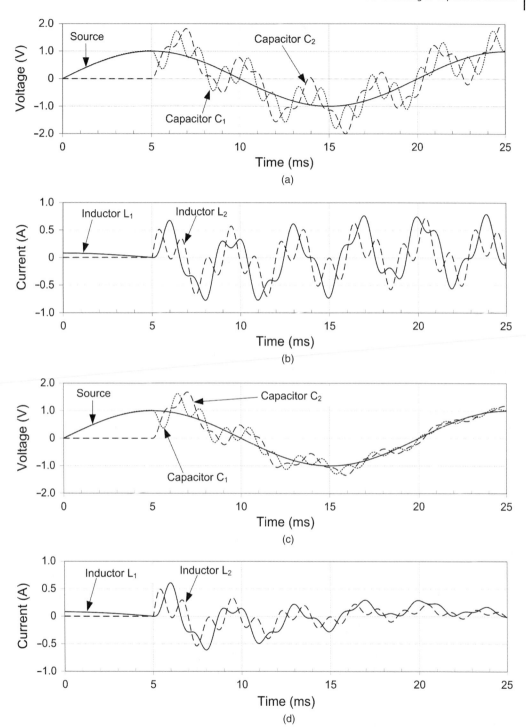

Figure 5.18 Energization of a back-to-back capacitor bank – AC supply – Switch closes at 5 ms. Simulation results ($L_1 = L_2 = 0.5$ mH, $C_1 = C_2 = 250$ μF). (a) $R = 0\,\Omega$, – Source, C_1 and C_2 voltages, (b) $R = 0\,\Omega - L_1$ and L_2 currents, (c) $R = 0.1\,\Omega$ – Source, C_1 and C_2 voltages, (d) $R = 0.1\,\Omega$, – L_1 and L_2 currents.

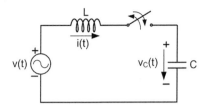

Figure 5.19 Reclosing into trapped charge – Diagram of the test system.

Two switches are used to represent the reclosing operation. The first is initially closed and used to model the opening; the second switch is used to model the reclosing

Figure 5.20 Reclosing into trapped charge – ATPDraw diagram of the test system.

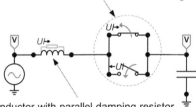

Inductor with parallel damping resistor to avoid numerical oscillations during the opening of the first switch

voltage oscillations after opening the switch. This option is available in ATPDraw; to obtain the parallel resistance value, the user has to specify the value of factor K_p (see Chapter 3). In this study $K_p = 5$.

Assume the parameters of this circuit are:

- Source: AC source, $V_{peak} = 1$ V, Phase $= 0°$, Frequency $= 50$ Hz
- Inductor: $L = 0.5$ H
- Capacitor: $C = 500$ μF
- Switch 1: $T_{close} = -1$, $T_{open} = 1$ ms
- Switch 2: $T_{close} = 20$ ms, $T_{open} = 1$ s

Figure 5.21 shows some simulation results. The conclusions derived from these results can be summarized as follows:

- The first switch opens at the moment the circuit current is zero (i.e. at $t = 10$ ms). The trapped voltage in the capacitor is the peak voltage of the source (i.e. $V_{trapped} = 1$ V).
- Since the second switch closes at the moment the source voltage is at its peak value (i.e. at $t = 20$ ms), and this voltage is equal and opposite to the trapped voltage, the resulting voltage at the capacitor reach a peak value of 3.
- The voltage at the capacitor oscillates with a frequency equal to the natural frequency of the test circuit (i.e. $f_0 = \omega_0/2\pi \approx 318$ Hz). Since the test circuit has no damping (except for the parallel damping resistance), the oscillation is undamped.

5.5.4 Discharge of a Capacitor Bank

The simple circuit shown in Figure 5.22 will be used to analyse the discharge of a capacitor. Assume the capacitor is initially charged at a voltage $V_C(0)$, and at a certain moment the switch closes.

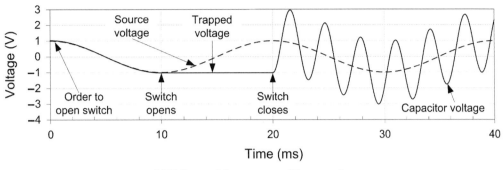

(a) Voltages at the source and the capacitor

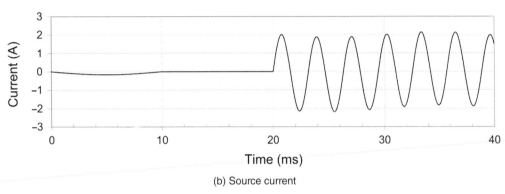

(b) Source current

Figure 5.21 Reclosing into trapped charge – Simulation results. (a) Voltages at the source and the capacitor. (b) Source current.

Figure 5.22 Discharge of a capacitor bank.

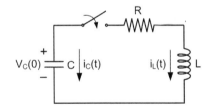

After the switch closes, the equations of the circuit can be written as follows:

$$v_R(t) + v_L(t) = v_C(t) \tag{5.42a}$$

$$i_L(t) + i_C(t) = 0 \tag{5.42b}$$

From the relations between voltages and currents in all elements, Eq. (5.42a) becomes

$$Ri_L(t) + L\frac{di_L(t)}{dt} = \frac{1}{C}\int_{-\infty}^{t} i_C(t)dt \tag{5.43}$$

Upon application of the Laplace transform, the following equation results:

$$RI_L(s) + sLI_L(s) = \frac{I_C(s)}{sC} + \frac{V_C(0)}{s} \tag{5.44}$$

Taking into account (5.42b) and the relationships

$$i(t) = i_L(t) = -i_C(t) \quad \Rightarrow \quad I(s) = I_L(s) = -I_C(s) \tag{5.45}$$

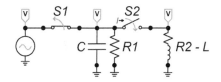

Figure 5.23 Discharge of a capacitor bank – ATPDraw implementation.

Equation (5.44) becomes:

$$RI(s) + sLI(s) + \frac{I(s)}{sC} = \frac{V_C(0)}{s} \tag{5.46}$$

from where the following expression for the circuit current is obtained:

$$I(s) = \frac{V_C(0)}{L} \frac{1}{s^2 + \alpha s + \omega_0^2} \tag{5.47}$$

The parameters α and ω_0 are those obtained in the study of the series RLC circuit.

This result is very similar to that obtained from the study of the series RLC circuit, and so it is its time-domain Laplace antitransform.

The circuit implemented in ATPDraw is displayed in Figure 5.23: the capacitor is initially connected to an ac voltage source, so the opening of switch S_1 will leave a trapped voltage in the capacitor.

This configuration will allow simulation of the discharge of the capacitor considering different combinations of values for R_1, R_2, and L:

- if R_1 is very large or infinite, the closing of S_2 will cause the discharge of the capacitor into the circuit R_2-L;
- if R_1 is very large and R_2 is zero, the circuit will have no losses and the discharge of the capacitor will cause a permanent oscillatory voltage;
- if the branch R_2-L is not connected, the capacitor will discharge into the parallel resistor R_1.

Figure 5.24 shows some simulation results that are summarized below. In all cases, the source voltage, which is depicted in all plots, is given by $v(t) = \sin(2\pi 50 \cdot t)$ and *Topen* for switch S_1 is 1 ms. Since the source current crosses the zero value at 5 ms, it is at this moment when the ideal switch actually opens.

a. Switch S_2 closes at 6 ms, resistor R_1 is very large (i.e. it is assumed infinite), the value of R_2 is variable, while $L = 1\,\text{mH}$. The cases included in this plot correspond to three different scenarios for which $\omega_0 > \alpha$, $\omega_0 = \alpha$, $\omega_0 > \alpha$. Note that the discharge of the capacitor is, as expected, oscillatory, critically damped and damped, respectively.

b. Switch S_2 closes at 6 ms, resistor R_1 is very large (i.e. it is assumed infinite), $R_2 = 0$, and the value of L is varied. All the cases included in this plot show the same behaviour: the capacitor voltage exhibits undamped oscillations whose frequency is given by $f_0 = \omega_0/2\pi$.

c. Switch S_2 remains open, and the value of resistance R_1 is varied. The three cases included in this plot show how the discharge of the capacitor depends on the time constant whose value is the product RC; since the capacitance C remains constant (i.e. $100\,\mu\text{F}$), the larger the value of R_1, the longer the discharge period of the capacitor.

d. Switch S_2 closes at 6 ms, the value of inductance L is varied, while the values of resistances R_1 and R_2 remain constant (i.e. $R_1 = 100$ and $R_2 = 1\,\Omega$). In all cases included in this plot, switch S_2 closes after the capacitor begins its discharge, and this discharge exhibits a damped and oscillatory behaviour whose frequency depends of the value of ω_0.

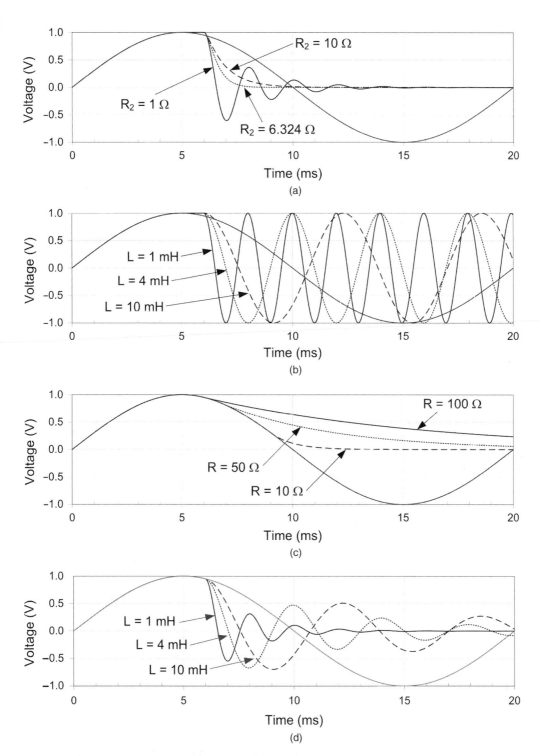

Figure 5.24 Discharge of a capacitor bank ($C = 100\,\mu$F) – Simulation results. (a) $R_1 = 1$ MΩ, R_2 is variable, $L = 1$ mH, (b) $R_1 = 1$ MΩ, $R_2 = 0$, L is variable, (c) $R_1 =$ variable, (d) $R_1 = 100\,\Omega$, $R_2 = 1\,\Omega$, L is variable.

5.6 Switching of Inductive Currents

5.6.1 Introduction

Energization and de-energization of circuits in which an inductive current is predominant are common operations in power systems; the current can be that of a reactor or an unloaded transformer. High overvoltages can occur when breaking inductive currents if restrikes or reignitions occur during the interruption process. An ideal interruption occurs when the current passes through zero. If the current is interrupted by the circuit breaker before it reaches the zero crossing, the phenomenon is known as *current chopping*. The magnitude of the chopped current depends on the type of breaker and the circuit where the breaker is connected. The behaviour of the arc in the circuit breaker has a fundamental influence on the transient phenomenon.

The switching phenomenon involves the interchange of energy stored in the inductances and the capacitances of the electrical circuit; this interchange can usually cause high-frequency oscillations at the reactance side and hazardous recovery voltages across the breaker that interrupts the inductive current. A voltage escalation across the breaker terminals can show up if a dielectric failure occurs several times during the competition between the recovery voltage and the withstand voltage across the breaker.

Of particular interest is the switching of inductive currents in nonlinear reactors. The inrush current caused during the energization of a nonlinear transformer core can reach high values with a significant harmonic content. When a line and a transformer are energized together, resonance overvoltages can occur: the inrush currents interact with the power system, whose resonant frequencies are a function of the series inductance (associated with the short circuit strength of the system) and the shunt capacitances of lines and cables. This may result in long-duration resonant temporary overvoltages.

As with the switching of capacitive currents, when performing a transient study involving the switching of inductive currents, it is necessary to decide upon how the system components have to be modelled. The examples analysed in this section are aimed at introducing the study of transient phenomena caused during the energization and de-energization of inductive currents using simplified single-phase lumped-parameter system models. The section has been organized into two parts dedicated to analysing very simple circuits in which the reactances are respectively linear and nonlinear.

5.6.2 Switching of Inductive Currents in Linear Circuits

5.6.2.1 Interruption of Inductive Currents

Consider the circuit shown in Figure 5.25: a voltage source with predominantly inductive equivalent impedance supplies an inductive load. The series $R_l L_l$ circuit may represent different types of loads depending on the values of both parameters. The parallel resistance R_p represents losses, usually associated to saturable reactances, while the parallel capacitance C may represent the interwinding parasitic capacitance of the reactor. The impedances of both the parallel resistance and the parallel capacitance (at the source frequency) are larger than the load impedance, so the total current is inductive and it lags the supplying voltage by about 90°.

The transient caused when the switch opens can be analysed following a procedure similar to that applied with the parallel RLC circuit in subsection 5.4.3; although the circuit configuration is not the same, the effect of those parameters that were not present in the circuit depicted in Figure 5.6 may not be too important and rather predictable; for instance, the effect of the series $R_{eq}L_{eq}$ is unimportant if the breaker does not reignite during the opening operation and it is assumed that the breaker opens when the source current crosses the zero value. On the other hand, the effect of the resistance R_l can easily be predicted if its value is lower than the

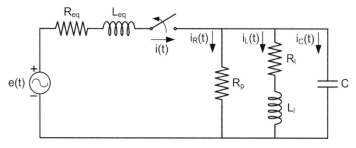

Figure 5.25 Interruption of an inductive current.

impedance of the load reactance (represented by L_l): it will limit the load current and add some damping during the transient that will follow the opening of the breaker.

Without reignition, the current flowing through the capacitor and the inductor branches at the time the breaker opens is zero, while it is very small in the parallel resistor; however, some charge will be stored in capacitance. This charge is mostly discharged through the inductance load and will force the voltage at the load terminals of the breaker to oscillate at the natural frequency of the circuit C-L_l; see expression (5.13b).

Given that the source-side terminal of the breaker will follow the source voltage, if the natural frequency of the load is higher than the source frequency, the voltage across the breaker can reach a value close to twice the source voltage peak. Consider the following parameters:

- AC source: $V_{peak} = 1$ V, Phase $= -90°$, Frequency $= 50$ Hz
- Source impedance: $R_{eq} = 0.001\,\Omega$, $L_{eq} = 0.01$ mH
- Parallel capacitance: $C = 5\,\mu F$
- Load impedance: $L_l = 10$ mH (with a series resistance $R_l = 0.02\,\Omega$)
- Switch: $T_{close} = -1$, $T_{open} = 0.1$ ms

Note that the parameter values of the source impedance are rather low; this is unimportant if the switch can successfully open after the zero crossing of the load current.

Figure 5.26 shows the configuration of the circuit implemented in ATPDraw to simulate this case study. Two different combinations of the series and parallel resistances have been simulated to compare two cases with different levels of damping: (i) $R_p = 500\,\Omega$; (i) $R_p = 5000\,\Omega$. Figure 5.27 shows some simulation results. The first plot confirms that in both scenarios the source current is interrupted when the voltage value is at its positive peak. On the other hand, the voltage at the load side starts oscillating at the natural frequency (i.e. $f_0 \approx 712$ Hz), and the damping depends on the resistance values. As expected, and given that f_0 is larger than the source frequency, the TRV across breaker terminals in both cases reaches a value close to twice the source voltage peak.

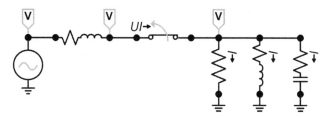

Figure 5.26 Interruption of an inductive current. ATPDraw implementation.

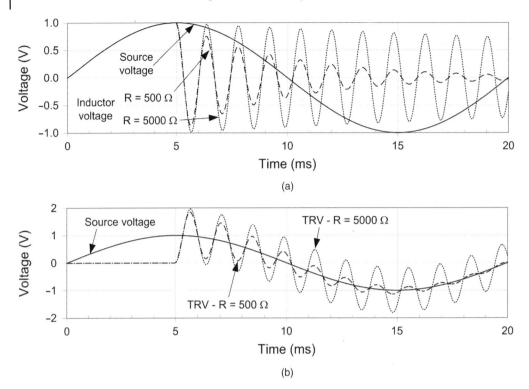

(a)

(b)

Figure 5.27 Interruption of a small inductive current. Simulation results. (a) Source and load voltage, (b) Source voltage and transient recovery voltage.

5.6.2.2 Voltage Escalation During the Interruption of Inductive Currents

When the current approaches its zero crossing in a gas-filled circuit breaker, the arc is resistive, so arc voltage and current reach the zero crossing simultaneously. The problem of current interruption during this phase is governed by a competition between the electric power input due to the recovery voltage and the thermal losses from the electric arc; this thermal recovery phase has a duration of a few microseconds. If the energy input is such that the gas molecules dissociate into free electrons and positive ions, the plasma state is created again and current interruption fails; this is known as *thermal breakdown*. The thermal recovery phase is characterized by rate of rise of recovery voltage (dV/dt) versus a rate of current decay characteristic (di/dt); the point of intersection gives the limit of the *thermal current*.

If the current interruption is successful, the gas channel cools down and the post-arc current disappears. However, a later failure can still occur when the dielectric strength of the gap between the breaker contacts is not sufficient to withstand the TRV. This dielectric recovery phase can be characterized by a boundary that separates fail and clear conditions on a maximum restrike voltage (V_{max}) versus rate of current decay (di/dt) diagram. The point of intersection gives the limit of the *dielectric current* [10].

A failure during the thermal recovery phase must be based on a dynamic arc representation, while a failure during the dielectric recovery phase can be based on precomputed voltage withstand characteristic [10]. The second option is analysed in this case as detailed below.

Due to the high frequency with which the load voltage oscillates, the TRV between circuit breaker terminals can rise very quickly. A race between the withstand voltage (that depends on the mechanical separation between breaker terminals) and the voltage stress (caused by

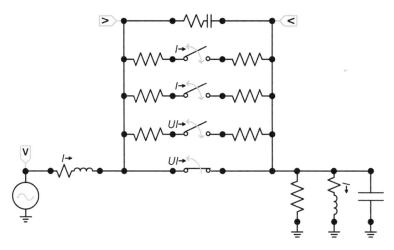

Figure 5.28 Voltage escalation during the interruption of an inductive current. ATPDraw implementation.

the steeped front of the recovery voltage) can finish with a dielectric failure between circuit breaker contacts if the breaker cannot develop enough withstand voltage capability. The failure does not have to occur at the peak value of the TRV if the voltage front is quick enough. On the other hand, the failure can occur several times if the TRV front is steeped enough or the mechanical separation is very slow. The transient that follows a dielectric failure can occur with very high-frequency oscillations that will provoke frequent zero crossing of the source current within a short period. If the breaker failure occurs several times, the voltage peak across the load can escalate to very high values.

Figure 5.28 shows the circuit implemented in ATPDraw to illustrate the phenomenon known as *voltage escalation*. To cause consecutive circuit breaker failures, the breaker has been represented by means of several parallel switches that consecutively close and open. The closing times can be accurately obtained by comparing the withstand voltage (modelled as a function of the mechanical speed with which the breaker contacts separate) and the TRV that the circuit parameters can cause during the opening transient. To illustrate the danger that this phenomenon can represent, this case is simulated by assuming the worst scenario: consecutive failures occur when the voltage between breaker contacts after a failure is at its peak value. In this study, all failures occur during positive voltage. To obtain these values and simulate all failures with a single input data case, the test circuit has been simulated several times to estimate the time at which the TRV developed after a failure reaches its peak value.

An important feature of the circuit implemented in ATPDraw is the parallel snubber circuit need to avoid the numerical oscillations that can occur in the TRV across the circuit breaker.

The circuit parameters are as follows (see Figure 5.25):

- AC source: $V_{peak} = 1\,\mathrm{V}$, Phase $= -90°$, Frequency $= 50\,\mathrm{Hz}$
- Source impedance: $R_{eq} = 0.05\,\Omega$, $L_{eq} = 0.5\,\mathrm{mH}$
- Load impedance: $R_l = 0.3\,\Omega$, $L_l = 10\,\mathrm{mH}$
- Parallel capacitance: $C = 5\,\mu\mathrm{F}$
- Parallel resistance: $R = 500\,\Omega$
- Snubber circuit: $R_{snub} = 500\,\Omega$, $C_{snub} = 0.1\,\mu\mathrm{F}$

Figure 5.29 shows some simulation results that are discussed below.

- The TRV developed between circuit breaker contacts exhibits the escalation phenomenon: the voltage starts oscillating until the moment at which it reaches a peak, and a failure occurs. Note that, in this case, the breaker fails up to three times and the TRV peak exceeds five times the source voltage peak.
- The voltage developed at the load side of the breaker after the first breaker opening is a mirror of the TRV. The physical process in this part of the circuit has to be separated into the oscillations caused after a breaker opening and the oscillations caused after a breaker closing (i.e. breaker failure). The oscillation after a breaker opening can be explained by means of the discussion presented with the previous case study; their frequency is the natural frequency of the parallel $L_l - C$ circuit (i.e. 712 Hz). However, the oscillations that occur after a breaker closing have a much higher frequency because they are now governed by the natural frequency of the series $L_{eq} - C$ (i.e. 3.18 kHz). Note the steeped front that follows each breaker failure.
- Without breaker failures, the source current would be zero after the breaker opening. However, after the first zero crossing, the current exhibits spikes that occur every time the breaker fails. Two important aspects of these spikes are the (very high) peak values they reach and their (very short) duration. Both values are governed by the circuit parameters. Due to these short durations, a breaker opening follows almost immediately follows a breaker failure since the zero crossing of the source current caused by a breaker closing oscillates with a very high frequency.

5.6.2.3 Current Chopping

The cases studied above illustrate the phenomena often associated to the interruption of the magnetizing current of a reactor or a transformer. If the current is suppressed when it passes through zero (as in the above cases), there is no energy accumulated in the inductance but only in the capacitance. In practice, a premature interruption is possible as the current approaches zero. This phenomenon is called *current chopping* and its occurrence depends on the type of circuit breaker. The circuit parameters used for the new case study are as follows:

- AC source: $V_{peak} = 1\,\text{V}$, Phase $= -90°$, Frequency $= 50\,\text{Hz}$
- Source impedance: $R_{eq} = 0.05\,\Omega$, $L_{eq} = 0.5\,\text{mH}$
- Parallel *RLC* load: $R = 450\,\Omega$, $L = 10\,\text{mH}$, $C = 5\,\mu\text{F}$
- Snubber circuit: $R_{snub} = 5000\,\Omega$, $C_{snub} = 0.1\,\mu\text{F}$

Figure 5.30 shows the diagram of the circuit implemented in ATPDraw for analysing the effect of current chopping. The new circuit is similar to the circuits shown in Figures 5.26 and 5.28, although the series resistance of the load has been neglected, and the parallel snubber circuit has been added to avoid numerical oscillations when the circuit breaker current chops.

A parameter that has to be changed in this circuit with respect to that of Figure 5.26 is the current margin (i.e. parameter *Imargin*) with which the ideal switch that represents the breaker can open. The behaviour of the test circuit in case of current chopping is shown in Figure 5.31. Actually, the figure compares the circuit response with and without current chopping; that is, the simulation results from the two cases correspond to values of *Imargin* equal to 0 and 0.05 A, respectively (see the first plot in which the currents from both cases are interrupted with different values). The effect of current chopping is evident from the comparison of the

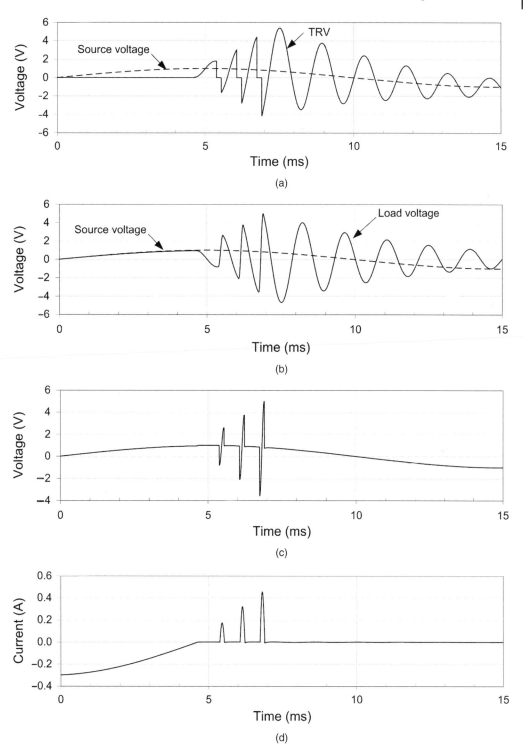

Figure 5.29 Voltage escalation during the interruption of an inductive current. Simulation results. (a) Source voltage and transient recovery voltage, (b) Source and load voltage, (c) Voltage at the source-side breaker terminal, (d) Source current.

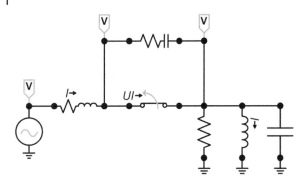

Figure 5.30 Current chopping during the interruption of an inductive current. ATPDraw implementation.

TRVs that result with and without this phenomenon (see Figure 5.31b): if the current is interrupted before zero crossing, the transient voltage developed across the circuit breaker will be higher than that developed after a zero crossing.

Current chopping can cause high-frequency reignitions and overvoltages. When the breaker chops the current, the voltage increases almost instantaneously, if this overvoltage exceeds the specified dielectric strength of the circuit breaker, reignition takes place. When this process is repeated several times, a rapid escalation of the TRV can occur, as illustrated by the previous case study. The high-frequency oscillations are governed by the electrical parameters of the concerned circuit, circuit configuration, and interrupter design.

The importance of current chopping can easily be understood by neglecting the influence of losses at the load side. After interruption at current zero, the energy stored in the load side is the energy stored at the capacitance, whose voltage is at the maximum. This energy is transferred

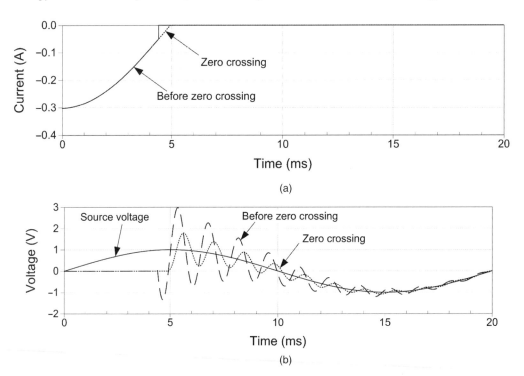

Figure 5.31 Current chopping during the interruption of an inductive current. Simulation results. (a) Source current, (b) Source voltage and transient recovery voltage.

to the inductance and vice versa; the following expression holds:

$$\frac{1}{2}CV_{max}^2 = \frac{1}{2}LI_{max}^2 \tag{5.48}$$

where V_{max} and I_{max} are the maximum values of the voltage across the capacitance and of the current through the inductance, respectively.

At the time the current is interrupted, V_{max} is equal to the voltage at the source side, and can be assumed equal to the source voltage peak. The frequency of the oscillations for both voltage and current is the natural frequency of the test circuit (i.e. the loop LC at the load side).

When premature current interruption occurs, the energy stored at the load side is stored in both the inductance and the capacitance, and it is equal to

$$\frac{1}{2}LI_0^2 + \frac{1}{2}CV_0^2 \tag{5.49}$$

where V_0 and I_0 are the values of the voltage across the capacitance and the current through the inductance at the moment the current is interrupted.

The frequency of the oscillations is again the natural frequency of the circuit, but the maximum voltage at the capacitance is obtained from the following equation:

$$\frac{1}{2}CV_{max}^2 = \frac{1}{2}LI_0^2 + \frac{1}{2}CV_0^2 \tag{5.50}$$

from where

$$V_{max} = \sqrt{\frac{L}{C}I_0^2 + V_0^2} = \sqrt{(ZI_0)^2 + V_0^2} \tag{5.51}$$

where Z is the surge impedance.

This equation can be normalized with respect to the source voltage as follows:

$$V_{max(pu)} \approx \sqrt{1 + \left(\frac{ZI_0}{V_0}\right)^2} \tag{5.52}$$

Although the value of I_0 will usually be small, the value of Z is in the range $10–100\,k\Omega$, so the voltage change caused by current chopping can be very high. The voltage magnitude given by (5.52) might not be a problem for the equipment insulation, but the TRV in the circuit breaker increases and the chance of voltage escalation is higher. As shown above, successive breaker failures can cause severe overvoltages that could damage equipment insulation.

In the preceding analysis, it was considered that all energy stored in the inductor is recoverable and released to the capacitor. In practice, this is seldom the case due to the effect of unavoidable losses (e.g. in the case of a transformer this energy is partially lost as hysteresis losses). Typical value for the transferable energy is in the range 40–60%.

5.6.2.4 Making of Inductive Currents

Consider the circuit shown in Figure 5.32. The configuration is similar to that used for analysing the interruption of inductive currents; however, the transient is now that caused when the parallel RLC circuit is energized.

The equations for this circuit can be written as follows:

$$e(t) = \left(R_{eq}i(t) + L_{eq}\frac{di(t)}{dt}\right) + v(t) \tag{5.53a}$$

$$v(t) = Ri_R(t) = L\frac{di_L(t)}{dt} = \frac{1}{C}\int_{-\infty}^{t} i_C(t)dt \tag{5.53b}$$

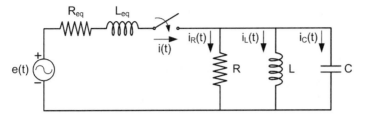

Figure 5.32 Making of an inductive current.

$$i(t) = i_C(t) + i_R(t) + i_L(t) \tag{5.53c}$$

Upon Laplace transformation and further manipulation of the resulting equations, the following form is derived for the voltage between reactor terminals:

$$V(s) = \left(1 + \frac{(s + \alpha_1)(s + s\alpha_0 + \omega_0^2)}{s\omega_1^2}\right)^{-1} E(s) \tag{5.54}$$

where

$$\alpha_0 = \frac{1}{RC} \qquad \omega_0 = \frac{1}{\sqrt{LC}} \qquad \alpha_1 = \frac{R_{eq}}{L_{eq}} \qquad \omega_1 = \frac{1}{\sqrt{L_{eq}C}} \tag{5.55}$$

The antitransform of this result is too complex and will not be analysed here. There are, however, a couple of issues that must be taken into account to understand the simulation results.

- As the source current is the result of adding the currents of the three parallel branches, up to three different solutions for currents can be considered. The expression for each current and the total source can be derived from Eqs. (5.53b), (5.53c) and (5.54). Since the source resistance is common, there is damping in the three paths.
- The waveforms of the source current and the voltage across the inductance can also exhibit some oscillations. The frequency of these oscillations can be higher than the natural frequency of the parallel LC circuit ω_0, and much higher than the source frequency. Consequently, overvoltages between the inductance terminals can occur.

The circuit parameters for this case are as follows:

- AC source: $\qquad\qquad V_{peak} = 1\,\text{V}$, Phase $= -90°$, Frequency $= 50\,\text{Hz}$
- Source impedance: $\qquad R_{eq} = 0.05\,\Omega$, $L_{eq} = 0.5\,\text{mH}$
- Parallel RLC load: $\qquad R = 100\,\Omega$, $L = 10\,\text{mH}$, $C = 5\,\mu\text{F}$

Figure 5.33 shows the diagram of the circuit implemented in ATPDraw. Figure 5.34 shows some simulation results that confirm the above discussion: the source current and the voltage at the reactance node oscillate with a frequency that depends on the source reactance and the load capacitance while the voltages reach peak values close to twice the source voltage peak. The frequency of the oscillations in the source current is about 3.18 kHz.

5.6.3 Switching of Inductive Currents in Nonlinear Circuits

Nonlinear components in ATP can be classified into the following groups (see Chapters 3 and 4):

- *True nonlinear component.* This option assumes the rest of the network is linear and represented by means of a Thevenin equivalent. The network equations are solved using compensation. This approach avoids time-step delays but requires the system to be solved twice; see

Figure 5.33 Making of an inductive current. ATPDraw implementation.

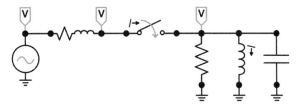

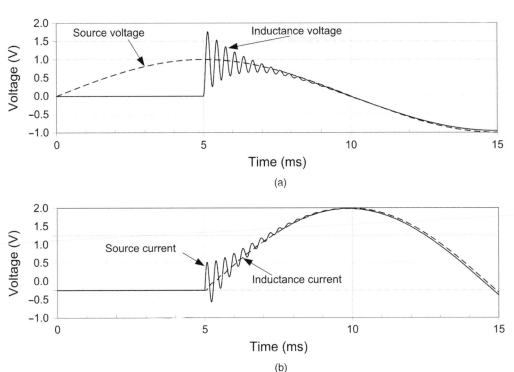

(a)

(b)

Figure 5.34 Voltage escalation during the interruption of an inductive current. Simulation results. (a) Source and inductance voltages, (b) Source and inductance currents.

Chapter 3 [11]. In addition, it puts some topological constraints on the nonlinearities, since this model cannot be connected in series or delta, and even parallel connections can be a problem.

- *Pseudo nonlinear component.* This nonlinear approach is supposed to operate on a linear segment modelled by a constant conductance and a history source until the segment border is violated. Then the conductance and current source are changed. This will result in a time-step delay between each segment shift and possible inaccuracies especially for resistors. In general, this approach is robust enough and allows parallel connection of several nonlinearities.
- *Controlled nonlinear component.* This nonlinearity involves values (of resistance and current source) specified by the control system. This nonlinearity can cause the conductance matrix to change frequently, requiring re-triangularization and time-consuming simulations. The inherent time-step delay between measured voltage/current and the response can also cause some numerical problems and inaccuracies. Otherwise, this approach is very flexible when it comes to nonlinear modelling. The Type 94 component combines the control and nonlinearity to avoid the time delay.

ATP supports the following options for representing nonlinear inductances [8]: Type 93 true nonlinear inductance; Type 96 pseudo-nonlinear hysteretic inductance; Type 98 pseudo-nonlinear inductance.

The specification of a nonlinear inductance requires a characteristic curve that relates flux linkage and current [8]. A supporting routine called SATURA is available in ATP to derive a relation between peak values of flux linkage and current from either a relationship between rms values of voltage and current, or from a current–incremental inductance curve. The typical approach is to use the rms v-i curve obtained from a standard test [12, 13]. If such a test curve is available, it is important to take into account that the exciting current could include core loss and magnetizing components. In addition, it can be important to remember that winding capacitances can affect low-current data, and hysteresis biases saturation curve.

The following procedure is suggested to obtain the excitation curve:

1) Extract loss component from excitation current for each current point:

$$I_m \approx \sqrt{I_{exc}^2 - (P_{exc}/V_{exc})^2} \qquad (5.56)$$

2) Convert the voltage-current rms curve to an instantaneous flux-current relationship. This conversion is carried out by means of the SATURA supporting routine [11].
3) Compensate for effect of winding capacitance. Winding capacitance can dominate magnetizing reactance causing 'cobra' flux-current curves and cancel much of the magnetizing current [14]. These capacitances are seldom given by the manufacturers.

In case of modelling a hysteretic core, one of the routines available in ATP (i.e. HYSDAT and HYSTERESIS HEVIA) should be used.

This subsection is dedicated to analysing transients caused when switching nonlinear inductors. The case studies will be based on the circuits analysed in previous sections but replacing the reactor model, which will now be nonlinear. Only anhysteretic characteristics are analysed in this chapter. Hysteresis can be of paramount importance in some transients involving nonlinear reactances.

5.6.4 Transients in Nonlinear Reactances

Consider the circuit shown in Figure 5.35. The aim is to obtain the steady-state and transient solutions that occur when the nonlinear reactor is either disconnected or connected. The capacitor represents the parasitic capacitance of the reactor winding, while the resistor is used to include reactor losses, although a constant resistance is an idealization as losses depend on the frequency and the voltage across the nonlinear reactor. As illustrated below, the effect of both parameters can be crucial. The saturation characteristic to be specified is that provided in Table 5.1.

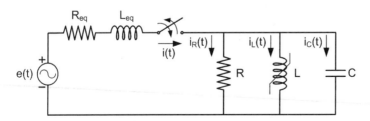

Figure 5.35 Switching of an inductive current. Nonlinear inductance.

Table 5.1 Switching of a nonlinear inductance – excitation test characteristic.

RMS voltage (kV)	RMS current (A)
19.515	0.182
24.120	0.266
27.282	0.378
30.357	0.562
34.520	1.204

The data input file for the routine SATURA can be that shown below: the values to be specified are the frequency, the single-phase base rms voltage (in kV), the single-phase base power (in Mvar), a parameter for controlling the output, and the $I_{rms}-V_{rms}$ break points in per unit.

```
SATURATION
$ERASE
C 3456789 123456789 123456789 123456789 123456789 123456789 123456789 12345678
C FREQ >< VBASE>< SBASE>          <KTHIRD>
    50.0   14.434   0.3333                 0
C    IRMS(PU)    ><    VRMS(PU)    >
       0.00788          0.7806
       0.01152          0.9648
       0.01637          1.0913
       0.02434          1.2143
       0.05213          1.3808
              9999
$PUNCH
BLANK card ending all  "SATURATION"   data cases
```

The output created by this routine is the following one

```
C  <+++++++>   Cards punched by support routine on   02-Jan-18   13:28:32   <+++++++>
C SATURATION
C $ERASE
C C 3456789 123456789 123456789 123456789 123456789 123456789 123456789 12345678
C C FREQ >< VBASE>< SBASE>          <KTHIRD>
C    50.0   14.434   0.3333                 0
C C    IRMS(PU)    ><    VRMS(PU)    >
C       0.00788          0.7806
C       0.01152          0.9648
C       0.01637          1.0913
C       0.02434          1.2143
C       0.05213          1.3808
C             9999
   2.57329649E-01   5.07201318E+01
   4.27160040E-01   6.26886794E+01
   6.67049174E-01   7.09081217E+01
   1.04951163E+00   7.89001486E+01
   2.48029368E+00   8.97186240E+01
             9999
```

Assume the parameters of the supply source are:

- AC source: $V_{peak} = 20.41$ kV, Phase $= 0°$, Frequency $= 50$ Hz
- Source impedance: $R_{eq} = 0.1\,\Omega$, $L_{eq} = 2$ mH

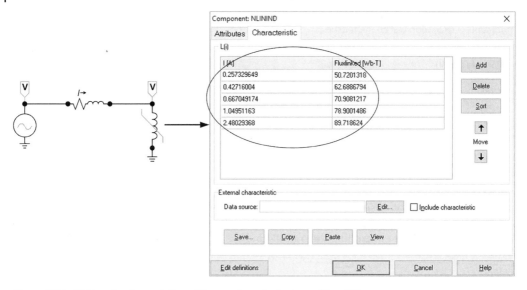

Figure 5.36 ATPDraw implementation of the nonlinear inductance (Type 98).

Figure 5.36 depicts the menus that correspond to the nonlinear inductance when it is modelled with the Type 98 component (pseudo-nonlinear inductance). Note that the user has to specify the current, in (A), and the flux, in (Wb-turn), at steady state, and the current-flux characteristic. ATPDraw allows users to visualize this characteristic, as shown at the bottom of the figure.

5.6.4.1 Interruption of an Inductive Current

The interruption of a small inductive current can cause high overvoltages, as illustrated above. If the current is interrupted after zero crossing, the physical phenomenon is not usually dangerous, the charge trapped in the capacitance discharges into the resistance and the nonlinear inductance; the circuit starts oscillating, and depending on the damping provided by the resistance, the stored energy and oscillations fade away rather quickly. However, a premature current interruption can occur, and this can be a particularly dangerous phenomenon. This new study is carried out to include the influence of the parallel resistance and capacitance.

Figure 5.37 displays the circuit implemented in ATPDraw for testing the transients that can occur during the interruption of the current. The following case studies have been simulated:

- *Imargin* = 0 (A) $R = \infty/5\,\mathrm{M\Omega}, C = 200\,\mathrm{pF}$
- *Imargin* = 0.05 (A) $R = \infty/5\,\mathrm{M\Omega}, C = 200\,\mathrm{pF}$

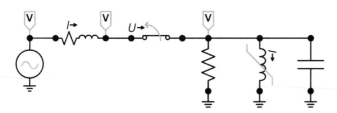

Figure 5.37 Case Study: Interruption of a small inductive current – ATP implementation.

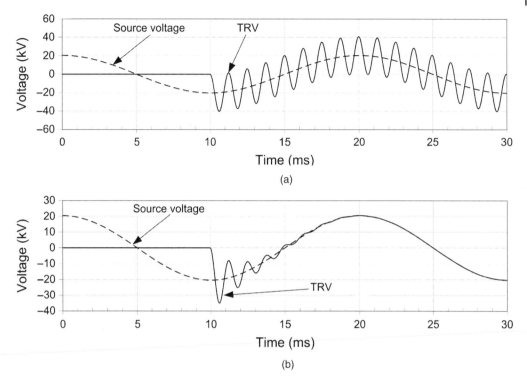

Figure 5.38 Interruption of small inductive current after zero crossing – Simulation results: Source voltage and TRV across the breaker (*Imargin* = 0 A). (a) $R = \infty$, (b) $R = 5$ MΩ.

Figures 5.38 and 5.39 provide some simulation results. The conclusions are rather evident, when the current is interrupted after the zero crossing: (i) the highest stresses occur in a lossless circuit ($R = 0$): due to the undamped oscillations that occur at the reactor side, the peak value of the TRV across the breaker will double the peak value of the source voltage; (ii) in a more realistic scenario, with $R \neq 0$, the oscillations will quickly be damped and the TRV will always be below 2 pu.

As in the previous case study, if the current is interrupted before the zero crossing, then the transient phenomenon can be dangerous, depending on the chopped current. However, one must be careful with these results because the parallel resistance affects not only the damping of the oscillations but also the peak values of both the voltage developed across the nonlinear reactor and the TRV across the breaker. A realistic modelling of the reactor could provide stresses lower than those obtained from this study.

5.6.4.2 Energization of a Nonlinear Reactance

The transients caused during the energization of a nonlinear reactor may lead to very high peaks of the inrush currents. Consider the energization of a lossless nonlinear reactor from an infinite bus. To analyse the inrush current, assume that the magnetizing characteristic is that given in Figure 5.40.

If the applied voltage is $v = V\cos\omega t$, the established flux linkage will be a function of the applied voltage and the flux linkage offset λo [15]. Using the notation $\theta = \omega t_0$, the flux linkage offset is made up of the remanent flux linkage λ_r and the component $-(V/\omega)\sin\theta$ due to the

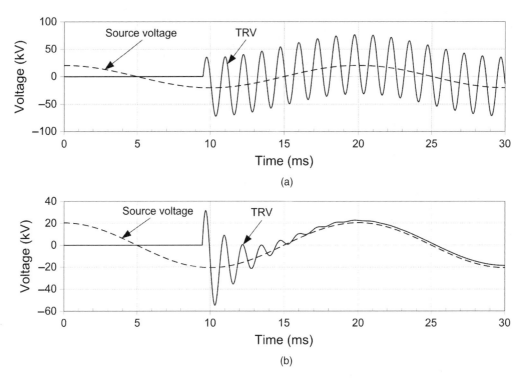

Figure 5.39 Interruption of small inductive current before zero crossing – Simulation results: Source voltage and TRV across the breaker (*Imargin* = 0.05 A). (a) $R = \infty$, (b) $R = 5$ MΩ.

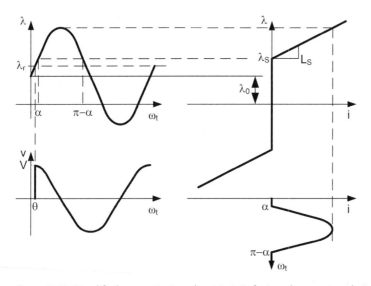

Figure 5.40 Simplified magnetization characteristic for inrush current analysis.

linkage mismatch condition at energization:

$$\lambda = \frac{V}{\omega} \sin \omega t + \lambda_o \qquad \rightarrow \qquad \lambda_o = \lambda_r - \frac{V}{\omega} \sin \theta \qquad (5.57)$$

Under normal system conditions, the reactor core would be driven temporarily into saturation asymmetrically when the flux linkage exceeds the saturation level λ_S, being the inrush current:

$$i = \frac{1}{L_S} (\lambda - \lambda_S) \qquad (5.58)$$

The circuit implemented in ATPDraw is that shown in Figure 5.37, although this time the switch connects the source to the reactor. Figure 5.41 displays some simulation results.

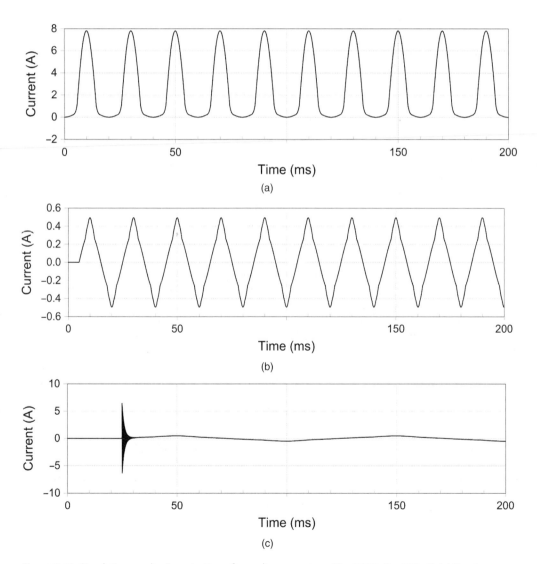

Figure 5.41 Simulation results. Energization of a nonlinear reactor – ($R = 5$ MΩ, $C = 200$ pF). (a) Reactance current – $t_{close} = 0$, (b) Reactance current – $t_{close} = 5$ ms, (c) Source current – $t_{close} = 5$ ms.

The main conclusions can be summarized as follows:

1) The waveform and peak value of the transient current through the nonlinear reactance depends on the closing time, or more specifically, the phase angle of the source voltage at the time the breaker closes. The highest peak value is reached when the breaker closes at the time the source voltage crosses zero, and the lowest one when the breaker closes at the time the source voltage is at its peak.

2) The source current is the sum of the currents through the three parallel branches, and it can exhibit a high-frequency component due to the transient current through the capacitance branch. The parallel resistance has a negligible impact on the waveform and peak value of the source current, irrespective of the time at which the switch closes. However, both can be significantly affected by the parallel capacitance when the switch does not close at the time the source voltage crosses zero. The frequency of this current depends on the values of L_{eq} and C; in this case study it is about 252 kHz. Associated to this current there is a voltage oscillation that can duplicate the initial voltage at the reactor side. As with the current through the reactance, the transient current and voltage caused by the parallel capacitance branch depends on the closing time.

5.6.5 Ferroresonance

Ferroresonance is a term applied to a wide variety of interactions between capacitors and iron-core inductors that can result in high overvoltages and cause failures in transformers, cables, and arresters [16]. Ferroresonance generally occurs during a system imbalance and/or a switching operation that places a capacitance in series with a transformer magnetizing impedance. Although ferroresonance can be caused by a parallel association of a capacitor and a nonlinear inductor, in most situations, it is a resonance that results from a series connection of a nonlinear inductance and a capacitance. Ferroresonance in power systems can involve large substation transformers, distribution transformers, or instrument transformers.

The capacitance can be in the form of capacitance of underground cables, transmission lines, capacitor banks, coupling capacitances between double circuit lines, or voltage grading capacitors in high voltage (HV) circuit breakers. In fact, ferroresonance may also arise due solely to transformer winding capacitance. System events that may initiate ferroresonance include single-phase switching or fusing, or loss of system grounding. Due to high-peak currents and high core fluxes, ferroresonance can lead to heating of the transformer; high temperatures inside the transformer may weaken the insulation and cause a failure under electrical stresses.

Consider the series LC circuit with a nonlinear inductance shown in Figure 5.42. Assuming a linear circuit and using phasor quantities, the following form is derived for the inductance voltage:

$$E = V_C + V_L = jX_C I + V_L \quad \Rightarrow \quad V_L = E - jX_C I \tag{5.59}$$

This equation can be used to obtain the voltage-current diagram shown in Figure 5.43, which provides a justification of how ferroresonance can be originated. The ac source is represented by its rms voltage, E, while V_L and V_C are the rms voltages across the inductance and the

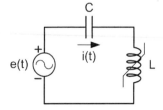

Figure 5.42 Nonlinear series *LC* circuit.

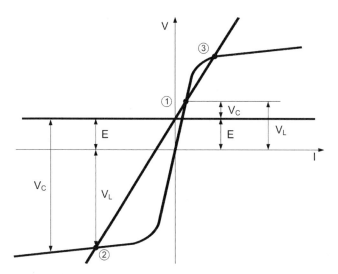

Figure 5.43 Voltage-current diagram of a nonlinear series *LC* circuit.

capacitance, respectively. The nonlinear inductance is represented by its rms $V-I$ characteristic. The solutions of the circuit correspond to the intersection of V_L and V_C. Depending on the operating point, the voltages across both the inductance and the capacitance can be much larger that the source voltage.

When analysing this phenomenon, it is important to account for the following aspects: (i) Ferroresonance is not strictly a resonance phenomenon since the source frequency is constant; a natural circuit resonance cannot be defined as for a linear circuit, and superposition, as used to obtain the diagram of Figure 5.43, cannot be applied. (ii) The current through a saturated reactance is distorted and contains harmonics, so this diagram is not rigorous; as it was derived by assuming that only source frequency voltages and currents do exist in the circuit (otherwise ((5.59)) is not correct). (iii) The solution depicted in Figure 5.43 gives three possible operating points. A different number of intersections can be obtained with a different capacitance slope. In addition, only operating points 1 and 2 represent stable steady-state solutions.

Figure 5.44 shows the diagram of the circuit implemented in ATPDraw for this case study. Note that the circuit includes the equivalent impedance of the supply source and a parallel resistance, which in this case, will be used to represent not only reactance losses but also any parallel damping provided by additional loads. The parameters of the source, the equivalent impedance and the nonlinear inductance are the same that were used in previous case studies. The values of the series capacitance and the parallel resistance will be varied to analyse its impact on the ferroresonant behaviour. Simulation results obtained with several capacitance and resistance values are shown in Figure 5.45. Note that, according to these results, high voltages can develop across the inductance. The two values of the parallel resistance allows understanding of the impact that parallel damping can have on ferroresonance occurrence: with little or no damping, ferroresonance can occur even with small capacitance values; if the

Figure 5.44 Series ferroresonance. ATPDraw implementation.

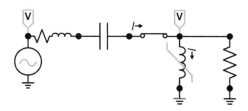

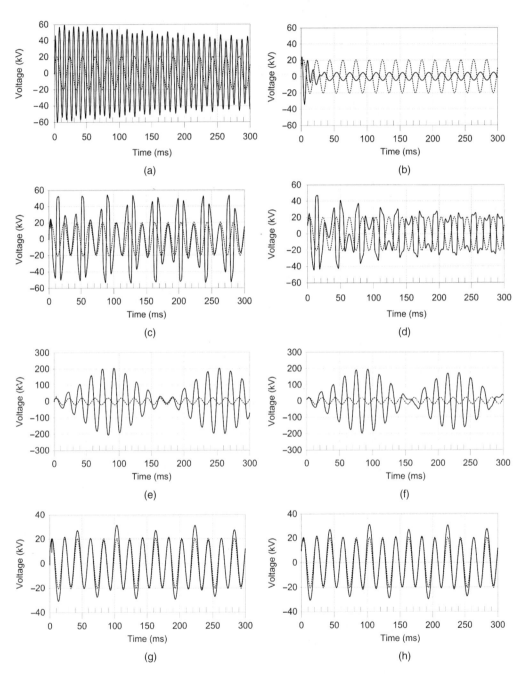

Figure 5.45 Series ferroresonance – Simulation results. (a) $C = 0.01\,\mu F - R = 50\,M\Omega$, (b) $C = 0.01\,\mu F - R = 0.5\,M\Omega$, (c) $C = 0.09\,\mu F - R = 50\,M\Omega$, (d) $C = 0.09\,\mu F - R = 0.5\,M\Omega$, (e) $C = 0.81\,\mu F - R = 50\,M\Omega$, (f) $C = 0.81\,\mu F - R = 0.5\,M\Omega$, (g) $C = 2.43\,\mu F - R = 50\,M\Omega$, (h) $C = 2.43\,\mu F - R = 0.5\,M\Omega$.

damping increases, a ferroresonant behaviour will occur within a certain range of capacitance values. With lower resistance values (e.g. equal to or lower than $10\,k\Omega$), ferroresonance does not occur, except with high capacitance values, with which some distorted voltages can appear.

Although the general requirements for ferroresonance are a voltage source, a saturable inductance, some capacitance and little damping, there are many situations in which this phenomenon can occur. System conditions that can lead to ferroresonance are: ungrounded neutral, small power demand or no load at all, one or two switch poles are open, power supply made through a cable that provides the capacitance.

5.7 Transient Analysis of Circuits with Distributed Parameters

5.7.1 Introduction

The representation of a power component in transients studies must be based on the range of frequencies associated to the transient process under study. One of the aspects to be considered is the distributed nature of parameters: for frequencies above 1 kHz, it is usually recommended to represent some components (e.g. lines and cables) with a distributed-parameter model, and for higher frequencies to include the dependence of parameters with frequency.

This section provides an introduction to transients in power systems with components represented by means of distributed-parameter models considering very simple modelling approaches. Several concepts, such as wave propagation or propagation velocity, are used when studying transients in power systems with distributed-parameter component models. This section begins with the study of electromagnetic transients in linear systems and continues with the study of systems whose models also include also nonlinear components. The next subsection introduces the study of transients in systems with components based on distributed-parameter models by simulating very simple systems; the subsection is divided into two parts: the first part deals with the energization or lines and cables; the second one with the analysis of the TRV that can occur between breaker terminals when a permanent fault is cleared. The last subsection provides an introduction to the protection of power equipment against lightning overvoltages by means of surge arresters. It is emphasized that, as in previous sections, the case studies presented here are very simple and all components are based on idealized models; the main goal is to provide an introduction to studies that will be covered in more detail in subsequent chapters.

5.7.2 Transients in Linear Circuits with Distributed-Parameter Components

5.7.2.1 Energization of Lines and Cables

Figure 5.46 shows the configuration of the system under study: an unloaded line is energized from a source whose representation includes the series equivalent impedance. Although the source can be of any type, a dc source with a series resistance will facilitate the description

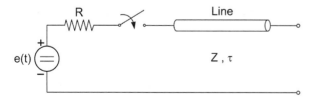

Figure 5.46 Energization of an unloaded ideal transmission line.

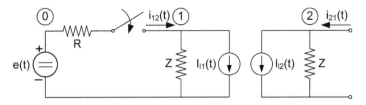

Figure 5.47 Energization of an unloaded ideal transmission line. Companion circuit.

of the physical phenomena that can occur when the line is represented by an ideal (lossless) distributed-parameter model (see Chapter 3). Assume the line is characterized by its surge impedance, Z, and its propagation time, τ. The companion circuit for this system (in which the source impedance is reduced to the equivalent resistance) could be that shown in Figure 5.47. Remember that the surge impedance of an ideal line is a pure resistance.

The nodal equations of this circuit once the switch is closed are as follows:

$$
\begin{bmatrix} \dfrac{1}{R} + \dfrac{1}{Z} & 0 \\ 0 & \dfrac{1}{Z} \end{bmatrix}
\begin{bmatrix} v_1(t) \\ v_2(t) \end{bmatrix} =
\begin{bmatrix} \dfrac{e(t)}{R} - I_{l1}(t) \\ -I_{l2}(t) \end{bmatrix}
\tag{5.60}
$$

where the history terms are

$$
I_{l1}(t) = -\left[\frac{v_2(t-\tau)}{Z} + i_{21}(t-\tau) \right]
\tag{5.61a}
$$

$$
I_{l2}(t) = -\left[\frac{v_1(t-\tau)}{Z} + i_{12}(t-\tau) \right]
\tag{5.61b}
$$

Solving for the voltages, the equations become:

$$
\begin{bmatrix} v_1(t) \\ v_2(t) \end{bmatrix} =
\begin{bmatrix} \dfrac{RZ}{R+Z} & 0 \\ 0 & Z \end{bmatrix}
\begin{bmatrix} \dfrac{e(t)}{R} - I_{l1}(t) \\ -I_{l2}(t) \end{bmatrix}
\tag{5.62}
$$

The following expressions result for voltages and currents:

$$
v_1(t) = \frac{RZ}{R+Z}\left[\frac{e(t)}{R} - I_{l1}(t) \right] \qquad\qquad i_{12}(t) = \frac{e(t)}{R+Z} + \frac{Z}{R+Z}I_{l1}(t)
\tag{5.63a}
$$

$$
v_2(t) = -ZI_{l2}(t) \qquad\qquad i_{21}(t) = 0
\tag{5.63b}
$$

Assume the line is initially dead; immediately after the switch closing, the current and voltage at the open end of the line are:

$$
i_{12} = \frac{e}{R+Z} \qquad v_1 = e \cdot \frac{Z}{R+Z}
\tag{5.64}
$$

A period of time τ after, the value of the controlled current source and the terminal voltage at the right side are (see (5.61b and 5.63b)):

$$
I_{l2} = -2\frac{e}{R+Z} \qquad v_2 = 2e\frac{Z}{R+Z} = 2v_1
\tag{5.65}
$$

This will be the result for every surge arriving to the open line end; that is, the incident voltage coming from the sending line end will be doubled. The controlled current source at the sending end of the line will change from zero after a roundtrip of the travelling surges (i.e. at $t = 2\tau$ and every new roundtrip). As for the first surge reflected from the open end, since the current at the receiving end is permanently zero, the current will be:

$$I_{l1} = -\frac{2v_1}{Z} \tag{5.66}$$

Consequently, the voltage variation at the sending end will be:

$$\frac{2v_1}{Z} \cdot \frac{RZ}{R+Z} = v_1 \frac{2R}{R+Z} = v_1 \left(1 + \frac{R-Z}{R+Z}\right) \tag{5.67}$$

The physical phenomena that these equations suggest is well known: as soon as the switch closes, a voltage surge v_1, given by ((5.64)), appears at the sending end, Node 1. This surge propagates without distortion and damping to the open-line terminal where it will arrive after a period equal to the propagation time, τ, and a new voltage surge (known as *reflected wave* and equal to v_1) will start its propagation to Node 1 to where it will arrive after a new period equal to τ. That is, after a round trip of duration 2τ, the voltage at the sending end of the line will notice a variation due to incident surge coming from the unloaded terminal. As soon as this incident wave arrives to Node 1, a new (reflected) surge is caused and starts its propagation back to Node 2, where it will cause a new reflected wave after a period equal to the propagation time. The voltage variation at the sending end, Node 1, is the result of adding the incident wave, coming from Node 2, and the reflected wave. Taking into account (5.67), one can conclude that the reflection coefficient at Node 1 is given by the following relationship:

$$\frac{R-Z}{R+Z} \tag{5.68}$$

The propagation of waves reflected at both line terminals will continue. Since the line is unloaded and some energy loss is caused by the source resistance, at the end the voltage at both line terminals will reach a steady state which in the case of a dc source will be that of the source.

Figure 5.48 shows the ATPDraw implementation of an ideal transmission line energized from both a dc and an ac source, respectively. In addition to the line length, to be specified in m, and

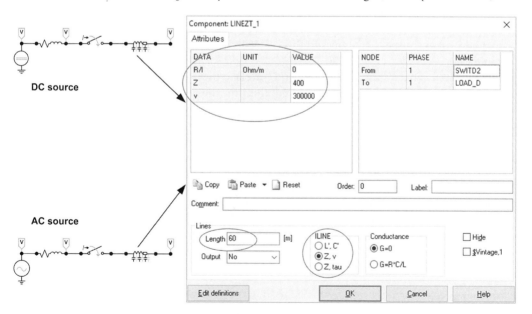

Figure 5.48 Energization of an unloaded ideal transmission line. ATPDraw implementation.

the resistance per unit length, to be specified in ohm/m, the user has to introduce two additional parameters, which can be specified in three different ways [8, 9]:

1. Inductance (in mH/m if Xopt = 0, in ohm/m if Xopt = power frequency), and capacitance (in μF/m if Copt = 0 or in μMho/m if Copt = power frequency).
2. Surge impedance, in ohm; and propagation velocity in, m/s.
3. Surge impedance, in ohm, and propagation time, in s.

The relationships between these parameters for an ideal line/cable are as follows:

$$Z = \sqrt{\frac{L}{C}} \qquad \upsilon = \frac{1}{\sqrt{LC}} \qquad \tau = \frac{\upsilon}{\ell} \qquad (5.69)$$

where L and C are the inductance and capacitance per unit length respectively, Z is the surge impedance, υ is the propagation velocity, τ is the propagation time, and ℓ is the line/cable length.

Note that if two of these parameters (i.e. L and C, Z and υ, Z and τ) are known, the other two can be easily obtained.

Consider that the test line has a surge impedance of 400 Ω and a propagation velocity of 300 000 km/s (if the length is in km; see Figure 5.48). Assume the parameters of the voltage sources are:

- DC source: Voltage = 1 V
- AC source: Frequency = 50 Hz, Peak voltage = 1 V, Phase angle = 0°.

Figure 5.49 shows simulation results that correspond to two different lengths of the line when it is energized from the two sources. In both cases, the source resistance has 50 Ω. The following results and conclusions derive from the above discussion:

- The value of the voltage at the sending end of the line as soon as the switch closes is given by (5.64); that is, $v_1 = 0.8889$ V.
- The voltage at the receiving (open) end, as soon as this surge reaches this terminal, is twice this value; that is, $v_2 = 1.7778$ V. Since the reflection coefficient at the open end is unity, any incident voltage surge that arrives to this end will cause a new reflected voltage surge of the same value.
- The reflected voltage surge at the open end travels to the sending end; as soon as it arrives to this end, a new voltage surge is reflected but this time the new voltage surge is of the opposite sign as the reflection coefficient at the receiving end is −0.1111; see (5.68). Therefore, as soon as a voltage surge reflected from the open end arrives to the receiving end, a newly reflected voltage surge of opposite sign is created and starts travelling to the open end.
- The new voltage at the sending end will be the result of adding the voltage from the source (i.e. 0.8889 V), the voltage of the incident surge coming from the other line end (i.e. 0.8889 V), and the voltage of the new reflected surge (i.e. −0.6914 V). For this test case, voltage surges arriving to the open end are alternatively positive and negative; this is the reason why the voltage at this end, irrespective of the type of source, oscillates around the source voltage value (see the two plots of Figure 5.49).
- The frequency of the oscillations that the voltage developed at the open exhibits depends on the propagation time and is given by the following form:

$$f = \frac{1}{4\tau} \qquad (5.70)$$

This can easily be checked from the oscillations shown in Figure 5.49.

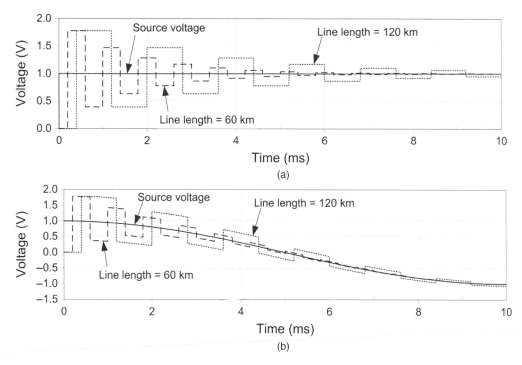

Figure 5.49 Energization of an unloaded ideal transmission line – Simulation results. (a) Voltage at the unloaded line terminal – DC source, (b) Voltage at the unloaded line terminal – AC source.

- Given that there is some damping caused by the source resistance, after some oscillations the voltage at the open end will reach a value close to that of the source.

5.7.2.2 Transient Recovery Voltage During Fault Clearing

The transient caused by the interruption of a current is analysed here when the system model includes distributer-parameter component models. The case studied below deals with the current interruption that follows a permanent fault condition.

The simulation of this type of scenario can be carried out by simply opening the breaker/switch that clears the fault, as in all case studies analysed in previous sections. There is, however, an alternative that, in simple cases, can be useful to better understand the transient process. Transients caused by switching operations in linear systems can be analysed by using the *superposition principle*: the switching process caused by an opening operation is obtained by adding the steady-state solution, which exists prior to the opening operation, and the transient response of the system that results from short-circuiting voltage sources and open-circuiting current sources to a current injected through the switch contacts, see Figure 5.50. Since the current through the switch terminals after the opening operation will be zero, the injected current must be equal to the current that was flowing between switch terminals prior to the opening operation. This approach is also known as the *current injection method*.

To obtain the TRV waveform, analysis of the transient response may suffice, since this voltage is zero during steady state; i.e. prior to the opening operation, see Figure 5.50. However, it is important to keep in mind that the recovery voltage will consist of two components: a transient component and a steady-state component, which is the voltage that remains after the transient dies out. The actual waveform of the voltage oscillation is determined by the parameters of the power system. Its rate of rise and amplitude are of vital importance for a successful operation of

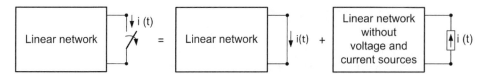

Figure 5.50 Application of the current injection method.

the interrupting device. If the rate of recovery of the contact gap at the instant of current zero is faster than the rate of *rise of the recovery voltage* (RRRV), the interruption is successful in the thermal region (i.e. first 4–8 μs of the recovery phase). If, however, the RRRV is faster than the recovery of the gap, then there will be a failure.

The example included in this subsection is simulated using the two approaches: (i) once the system model has been implemented in ATPDraw the simulation is carried out by simply opening the switch that clears the fault; (ii) prior to the ATP simulation the model implemented in ATPDraw is run under fault condition, the calculated fault current is used to represent a current source that replaces the switch (as indicated in Figure 5.50) and the model is simulated to obtain the TRV, which is the voltage developed between the current source terminals.

Consider the system depicted in Figure 5.51, where several transmission lines emanate from a bus, which is fed from a transformer. A three-phase grounded fault occurs on the bus, as shown in the figure. Figure 5.52 shows the one-line equivalent circuit used to obtain the TRV across the circuit breaker by means of the current injection method. Transmission lines are represented as one-phase ideal lines with constant and distributed parameters, and characterized by means of their surge impedance and travel time. The system equivalent upstream, the substation bus, has been represented by the corresponding inductance. This scheme assumes that system parameters are referred to a single voltage level.

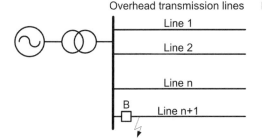

Figure 5.51 TRV analysis. System configuration.

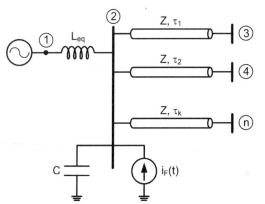

Figure 5.52 TRV analysis. One-line equivalent circuit for TRV calculation.

Figure 5.53 TRV analysis. Simplified circuit for calculation of the initial TRV.

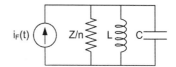

To estimate initial TRV and RRRV, line models are replaced by resistances whose values are equal to their surge impedance. A simplified circuit, shown in Figure 5.53, can be applied before the first reflection from any remote line end arrives to the substation. In this circuit, the resistance is Z/n, where Z is the surge impedance of all lines and n is the number of parallel lines connected to the substation node that will be included in the model. A further simplification ignores the capacitance, which is justified on the basis that the surge impedance of the lines provides enough damping. The resulting equivalent circuit consists of a parallel combination of only inductance (L) and resistance ($= Z/n$).

The solution of the circuit is approximated by assuming that the injected current is represented as a ramp whose slope is equal to the initial slope of the fault current that would have continued to flow at the instant of current zero; that is, when the current interruption takes place. Assume the time-domain of the fault current is:

$$i_F(t) = \sqrt{2}I \sin(\omega t) \tag{5.71}$$

where I is the rms value of the phase current, the approximation used in this analysis is

$$i_F(t) \approx \omega\sqrt{2}It \tag{5.72}$$

whose Laplace transform has the following form

$$I_F(s) \approx \omega\sqrt{2}\frac{I}{s^2} \tag{5.73}$$

The operational impedance, for the circuit without capacitance, is given by the following expression:

$$Z(s) = \frac{1}{\frac{1}{sL} + \frac{1}{R}} = \frac{sR}{s + \frac{R}{L}} \qquad (R = Z/n) \tag{5.74}$$

The resulting voltage, which happens to be the TRV of the circuit, is then given by:

$$V_{RV}(s) \approx \omega\sqrt{2}I\frac{R}{s\left(s + \frac{R}{L}\right)} \tag{5.75}$$

The time-domain solution of this equation is as follows [1]:

$$v(t) \approx \sqrt{2}I \cdot \omega L \cdot (1 - e^{-(Z/nL)t}) \tag{5.76}$$

This result is only valid for the initial TRV developed across the circuit breaker, since it ignores the effect of the reflected surges from the receiving ends of transmission lines. According to this result, the RRRV is directly proportional to the fault current and the frequency of the source, and inversely proportional to the number of transmission lines that remain connected to the bus.

Assume that four transmission lines are connected to the substation bus, and the fault occurs at the beginning of the fourth line. Relevant data for the test system is as follows:

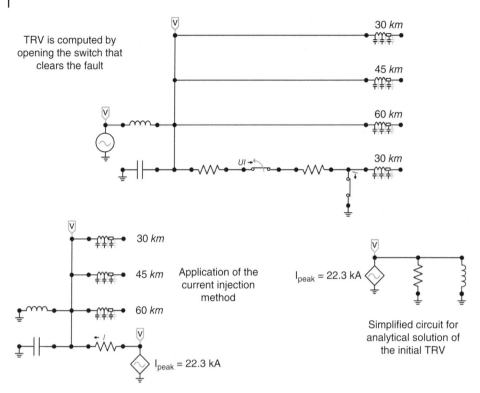

TRV is computed by opening the switch that clears the fault

30 km
45 km
60 km
30 km

$UI \rightarrow$

Application of the current injection method

$I_{peak} = 22.3$ kA

Simplified circuit for analytical solution of the initial TRV

30 km
45 km
60 km

$I_{peak} = 22.3$ kA

Figure 5.54 Case Study for TRV and RRRV estimation. ATPDraw implementation.

- Rated frequency 50 Hz
- Substation transformer voltages 400/220 kV
- Short-circuit capacity at 220 kV substation terminals 6 000 MVA
- Substation capacitance measured at 220 kV 4 nF
- Lines Surge impedance: 390 Ω

  Propagation velocity: 300 000 km/s

From these values, the equivalent impedance of the system measured at the 220 kV side of the substation is 25.67 mH.

Figure 5.54 shows the diagrams of two circuits implemented in ATPDraw to obtain the TRV across the breaker by means of both a direct calculation of the transient voltage and the application of current injection method, plus a third circuit to estimate the RRRV following the method detailed above. Figure 5.55 compares the TRVs that result from opening the switch, from the application of the current injection method and from the simplified scheme. Remember that this latter approach is only valid for estimating the initial rate-of-rise of the TRV.

Take into account that the current injection in the second circuit has been delayed until the moment at which the TRV in the first circuit (i.e. TRV results from opening the switch that clears the fault) begins. One can observe that:

- There is a perfect match between the TRV obtained with both methods.
- The initial rise obtained with the simplified circuit follows an exponential function and matches the results obtained with the two previous approaches until the moment the

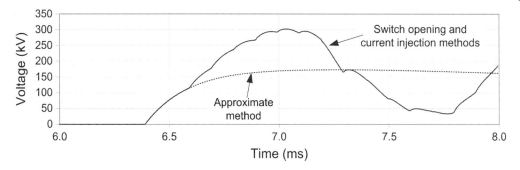

Figure 5.55 Transient recovery voltage – Simulation results.

first reflection from the open end (i.e. node 3) arrives at the substation; this occurs after a round-trip travel in the shortest line (i.e. that with 30 km) after current interruption. In this case, it is about 200 μs. Since the new wave is reflected with a positive coefficient, it increases the TRV. This trend remains until a new reflected wave from the end of the second line arrives at the substation, which occurs 300 μs after the TRV started to be developed. A third reflected wave from the end of the third line arrives 400 μs after.

- Obviously, since the upper line is much shorter, more than one reflected wave from its terminal node will arrive to the substation bus during this interval. The maximum value of the recovery voltage is approximately 302 kV, and it is reached about 0.7 ms after the current was interrupted.

200 μs after current interruption, the value of the recovery voltage obtained from expression (5.76) is about 114.5 kV, which reasonably agrees with the value shown in Figure 5.55.

The initial rate-of-rise is an important parameter for the selection of the circuit breaker. It can be estimated from (5.76) by taking the derivative and setting $t = 0$. It results:

$$RRRV \approx \sqrt{2}I \cdot \omega \frac{Z}{n} \tag{5.77}$$

The value for this example is 0.91 kV/μs.

5.7.3 Transients in Nonlinear Circuits with Distributed-Parameter Components

5.7.3.1 Surge Arrester Protection

The metal oxide surge arrester (MOSA) is a very efficient device for lightning protection of power equipment. An arrester provides a low-resistance path and its goal is to limit overvoltages below the corresponding insulation level of equipment: the surge arrester should act like an open circuit during normal operation of the system, limit transient voltages to a safe level, and bring the system back to its normal operational mode as soon as the transient voltages are suppressed [17]. Therefore, a surge arrester must have a nonlinear voltage-current characteristic: an extremely high resistance during normal system operation and a relatively low resistance during overvoltages.

Figure 5.56 illustrates the way in which an ideal gapless MOSA behaves: if the maximum value of the wave that arrives to an arrester is larger than the discharge voltage, the maximum value of the wave that propagates further will be limited to the discharge voltage, see Figure 5.56a. This effect can be analysed by assuming that the wave is not distorted, but an additional wave, known as the *relief wave*, is generated and the effect of both waves matches the effect of the actual wave, see Figure 5.56b [17, 18]. Once the arrester voltage has equalled its discharge value, this voltage remains unchanged unless its value decreases below the discharge voltage.

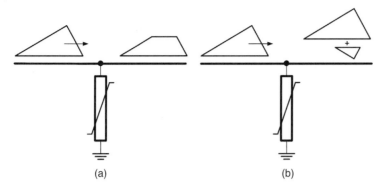

Figure 5.56 Ideal gapless arrester behaviour. (a) Limitation of lightning overvoltages, (b) Generation of the relief wave.

This subsection presents one case study of surge arrester protection against overvoltages caused by lightning strokes. The study is based on a very simplified representation of the test system and power components; the main goals are to show how ATP can be applied to simulate systems with nonlinear and distributed-parameter based component models, and illustrate surge arrester behaviour during transients caused by lightning strokes. Remember that for carrying out these types of case studies, equipment models must be adequate for fast-front transients, which justify the use of distributed-parameter models.

5.7.3.2 Protection Against Lightning Overvoltages Using Surge Arresters

A surge arrester is installed in parallel and as close as possible to the equipment to be protected. However, some distance between the surge arrester and the protected equipment cannot always be avoided. As a consequence, the voltage produced across equipment terminals will be higher than the residual voltage of the surge arrester. This is known as *separation distance effect*. Figure 5.57 shows the diagram of a substation that is protected at its Medium Voltage (MV) side by surge arresters [19]. These arresters have been installed at a certain distance from the MV side of the transformer. This example assumes that only one feeder is connected to the transformer; more complex situations (i.e. multi-feeder substations) can be found in actual systems.

The impact of a lightning stroke on the distribution line, at the span close to the transformer, will cause two travelling waves that will propagate along the line in opposite directions from the point of impact. The goal is to analyse the protection provided by the surge arrester to the MV

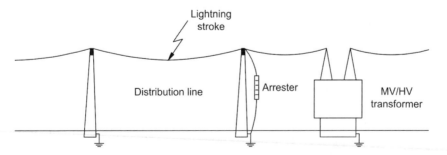

Figure 5.57 Diagram of a substation protected by surge arresters. Separation distance effect.

Figure 5.58 Lightning stroke current waveform.

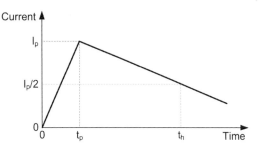

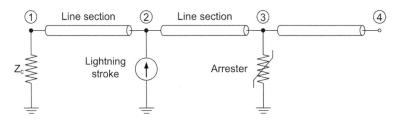

Figure 5.59 Scheme of the test system. Separation distance effect.

side of the transformer, taking into account that there will be a separation distance between the surge arrester and the transformer. See Case Study 3 in Chapter 4.

The study will be performed using a single-phase equivalent scheme and assuming an idealized behaviour of all components of the system:

1. Line sections are represented as single-phase lossless constant distributed-parameter lines.
2. The transformer is represented as an open circuit.
3. The lightning stroke is represented as an ideal current source, with a double-ramp waveform, see Figure 5.58.
4. The surge arrester behaves as an open circuit when the voltage across its terminals is below the residual voltage, and its voltage remains constant when this value is reached and the discharge current is large enough.

Figure 5.59 shows the scheme that will be used to analyse the separation distance effect:

- the lightning stroke impacts the distribution line at the midspan, node 2;
- the line span, connected to nodes 1 and 3, has been split into two sections and matched at node 1 (i.e. the resistance installed at this point is equal to the surge impedance of the line) to prevent the effect of reflections at this node;
- the surge arrester has been installed at node 3;
- the transformer, located at node 4, is represented as an open circuit;
- the lead section between the surge arrester and the transformer has the same characteristic parameters as the distribution line.
- the effect of arrester leads (i.e. connections from arrester terminals to ground and line conductors) is neglected.
- the power frequency voltages are also neglected, so the study will be carried out by assuming that the system is not energized when the lightning stroke impacts the line.

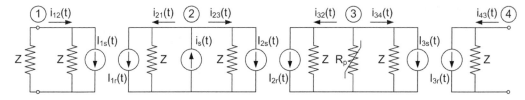

Figure 5.60 Separation distance effect. Companion circuit.

To illustrate the protection provided by the surge arrester, it is assumed that there will not be line flashover. In reality, flashovers can occur between phase conductors or across insulators.

The peak magnitude of any wave that meets the arrester is limited to the residual value. As soon as the voltage magnitude exceeds this value, a relief wave is generated. The addition of this wave and the incident wave must equal the wave that propagates beyond the surge arrester, so the analysis can be performed by superposing the effects of both the incident and the relief waves.

Figure 5.60 shows the resistive companion network that is obtained from substituting circuit elements of Figure 5.59 by their companion equivalents. Note that Z is the surge impedance of the distribution line and the surge arrester has been represented as a nonlinear resistance R_p.

The nodal equations corresponding to the circuit shown in Figure 5.60 are:

$$
\begin{bmatrix}
\dfrac{2}{Z} & 0 & 0 & 0 \\
0 & \dfrac{2}{Z} & 0 & 0 \\
0 & 0 & \dfrac{2}{Z}+\dfrac{1}{R_p} & 0 \\
0 & 0 & 0 & \dfrac{1}{Z}
\end{bmatrix}
\begin{bmatrix}
v_1(t) \\ v_2(t) \\ v_3(t) \\ v_4(t)
\end{bmatrix}
=
\begin{bmatrix}
-I_{1s}(t) \\
i_s(t) - I_{1r}(t) - I_{2s}(t) \\
-I_{2r}(t) - I_{3s}(t) \\
-I_{3r}(t)
\end{bmatrix}
\tag{5.78}
$$

where

$$
I_{1s}(t) = -\left[\frac{v_2(t-\tau_l)}{Z} + i_{21}(t-\tau_l)\right] \qquad I_{1r}(t) = -\left[\frac{v_1(t-\tau_l)}{Z} + i_{12}(t-\tau_l)\right] \tag{5.79a}
$$

$$
I_{2s}(t) = -\left[\frac{v_3(t-\tau_l)}{Z} + i_{32}(t-\tau_l)\right] \qquad I_{2r}(t) = -\left[\frac{v_2(t-\tau_l)}{Z} + i_{23}(t-\tau_l)\right] \tag{5.79b}
$$

$$
I_{3s}(t) = -\left[\frac{v_4(t-\tau_d)}{Z} + i_{43}(t-\tau_d)\right] \qquad I_{3r}(t) = -\left[\frac{v_3(t-\tau_d)}{Z} + i_{34}(t-\tau_d)\right] \tag{5.79c}
$$

where τ_l and τ_d are the travel times of the sections that represent, respectively, a half-span of the distribution line and the separation distance between the arrester and the transformer.

The following relationships are derived from the circuit shown in Figure 5.60:

$$
i_{12}(t) = -\frac{v_1(t)}{Z} \qquad i_{43}(t) = 0 \tag{5.80}
$$

The substitution of these expressions into the history terms yields:

$$
I_{1r}(t) = 0 \qquad I_{3s}(t) = -\frac{v_4(t-\tau_d)}{Z} \tag{5.81}
$$

The history term from the matched end is zero, so the value of the travel time of this line section is unimportant as the voltages and currents originated at node 1 will not affect the rest of the network.

The behaviour of the surge arrester can also be reproduced by representing it as an open switch when the voltage across its terminals is below the residual voltage, and as a fixed voltage source with $v = V_{res}$, when the voltage reaches the residual value. Considering the two possibilities discussed above, the set of equations for each situation will be as follows:

- Before the surge arrester voltage reaches the residual value ($R_p \approx \infty$)

$$\begin{bmatrix} v_1(t) \\ v_2(t) \\ v_3(t) \\ v_4(t) \end{bmatrix} = Z \begin{bmatrix} 1/2 & 0 & 0 & 0 \\ 0 & 1/2 & 0 & 0 \\ 0 & 0 & 1/2 & 0 \\ 0 & 0 & 0 & 1 \end{bmatrix} \begin{bmatrix} -I_{1s}(t) \\ i_s(t) - I_{1r}(t) - I_{2s}(t) \\ -I_{2r}(t) - I_{3s}(t) \\ -I_{3r}(t) \end{bmatrix} \tag{5.82}$$

- After the surge arrester voltage reaches the residual value ($v_3(t) = V_{res}$)

$$\begin{bmatrix} v_1(t) \\ v_2(t) \\ v_4(t) \end{bmatrix} = Z \begin{bmatrix} 1/2 & 0 & 0 \\ 0 & 1/2 & 0 \\ 0 & 0 & 1 \end{bmatrix} \begin{bmatrix} -I_{1s}(t) \\ i_s(t) - I_{1r}(t) - I_{2s}(t) \\ -I_{3r}(t) \end{bmatrix} \tag{5.83}$$

Note that in the second set, the residual voltage $v_3(t) = V_{res}$ must be included in the history terms $I_{2s}(t)$ and $I_{3r}(t)$. Since power frequency voltage is neglected, all history terms are initially zero; that is, all components are not energized before the lightning strokes impact the line.

The voltage stresses at the transformer location can be different with different stroke parameters and a different separation distance between arrester and transformer.

Consider the following parameters:

- Lightning stroke: Double-ramp current source (I_p) = 10 kA
- Overhead line: Surge impedance (Z_c) = 390 Ω

Propagation velocity (v) = 300 m/µs
- Surge arrester: Residual voltage (V_{res}) = 100 kV

Two stroke waveforms will be assumed: 8/50 and 1/20 µs. In both cases, the first and the second value respectively indicate the time to crest (t_f) and the tail time (t_h), see Figure 5.58.

Figure 5.61 shows the ATPDraw implementation of the circuit depicted in Figure 5.59. The figure shows the menus corresponding to the arrester model represented by means of the Type-81 nonlinear resistor. Note the nonlinear characteristic of the resistance: the discharged current through the arrester is negligible with voltages below the residual voltage, and as soon as the voltage reaches the residual value, the current starts flowing through the arrester and its value depends on the system configuration and parameters.

Figure 5.62 shows simulation results obtained with different values of these parameters. The most important conclusions derived from these results can be summarized as follows:

- the longer the separation distance between arrester and transformer, the higher the voltage caused at the transformer;
- the steeper the wavefront of the stroke current, I_{max}/t_{max}, the higher the voltage caused at the transformer;
- as expected, the maximum overvoltage that can occur at the transformer is twice the residual voltage of the surge arrester.

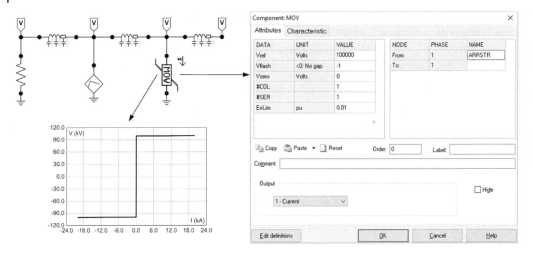

Figure 5.61 Separation distance effect. ATPDraw implementation and surge arrester menus.

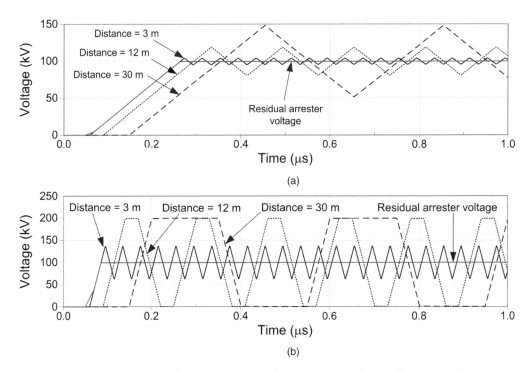

Figure 5.62 Separation distance effect – Simulation results. (a) Stroke waveform: 10 kA, 8/50 μs, (b) Stroke waveform: 10 kA, 1/20 μs.

A critical distance can be defined as the separation distance beyond which the voltage at the transformer will reach a value twice the arrester residual voltage. This distance is given by the following expression [19]:

$$d_{cri} = \frac{V_{res}}{ZS_i} v \tag{5.84}$$

where S_i $(= I_{max}/t_f)$ is the initial slope of the stroke current, see Figure 5.58, and v is the propagation velocity of the lead between the arrester and the transformer.

Upon application of (5.84), the following critical distances are obtained for each stroke waveform: (i) with a waveform of 10 kA, 8/50 μs, d_{cri} is 61.5 m; (ii) with a waveform of 10 kA, 1/20 μs, d_{cri} is 7.7 m. Figure 5.62 confirms these results: the voltage at the transformer location will not reach twice the value of the arrester residual voltage in any of the three scenarios with a 8/50-μs waveform, but it doubles the arrester residual voltage with a 1/20-μs waveform when the separation distance is longer than 7.7 m.

References

1 EPRI Report EL-4202 (1985). *Electromagnetic Transients Program (EMTP) Primer*. EPRI.

2 EPRI Report EL-4650 (1986). *Electromagnetic Transients Program (EMTP) Application Guide*. EPRI.

3 Martinez-Velasco, J.A. (2010). Parameter determination for electromagnetic transient analysis in power systems, Chapter 1. In: *Power System Transients. Parameter Determination* (ed. J.A. Martinez-Velasco), 17–135. Boca Raton, FL, USA: CRC Press.

4 Greenwood, A. (1991). *Electrical Transients in Power Systems*, 2e. New York, NY, USA: Wiley.

5 Shenkman, A.L. (2005). *Transient Analysis of Electric Power Circuits Handbook*. Dordrecht (The Netherlands): Springer.

6 Das, J.C. (2010). *Transients in Electrical Systems. Analysis, Recognition, and Mitigation*. New York, NY, USA: McGraw-Hill.

7 Bewley, L.V. (1951). *Traveling Waves on Transmission Systems*. New York (NY, USA): Wiley.

8 Can/Am Users Group (2000). *ATP Rule Book*. Can/Am Users Group.

9 Prikler, L. and Høidalen, H.K. (2009). ATPDRAW Version 5.6 for Windows 9x/NT/2000/XP/Vista. www.atpdraw.net (accessed 28 June 2019).

10 Martinez-Velasco, J.A. and Popov, M. (2009). Circuit breakers, Chapter 7. In: *Power System Transients. Parameter Determination* (ed. J.A. Martinez-Velasco), 447–555. CRC Press.

11 Dommel, H.W. (1986). *Electromagnetic Transients Program. Reference Manual (EMTP Theory Book)*. Portland, OR, USA: Bonneville Power Administration.

12 IEEE Std. C57.12.91-2011. (2011). IEEE Standard Test Code for Dry-Type Distribution and Power Transformers.

13 IEC Standard 60076-1. (2011). *Power Transformers – Part 1: General*. IEC.

14 Walling, R.A., Barker, K.D., Compton, T.M., and Zimmerman, I.E. (1993). Ferroresonant overvoltages in grounded padmount transformers with low-loss silicon-steel cores. *IEEE Transactions on Power Delivery* 8 (3): 1647–1660.

15 Sybille, G., Gavrilovic, M.M., Belanger, J., and Do, V.Q. (1985). Transformer saturation effects on EHV system overvoltages. *IEEE Transactions on Power Apparatus and Systems* 104 (3): 671–680.

16 Iravani, M.R., Chaudhary, A.K.S., Giewbrecht, W.J. et al. (2000). Modeling and analysis guidelines for slow transients: part III: the study of ferroresonance. *IEEE Transactions on Power Delivery* 15 (1): 255–265.

17 Martinez, J.A. and Gonzalez-Molina, F. (2000). Surge protection of underground distribution cables. *IEEE Transactions on Power Delivery* 15 (2): 756–763.

18 Martinez-Velasco, J.A. and Castro-Aranda, F. (2009). Surge arresters, Chapter 6. In: *Power System Transients. Parameter Determination* (ed. J.A. Martinez-Velasco). CRC Press.

19 Martinez-Velasco, J.A. and Marti, J.R. (2018). Electromagnetic transients analysis, Chapter 12. In: *Electric Energy Systems. Analysis and Operation*, 2e (eds. A. Gómez-Expósito, A.J. Conejo and C. Cañizares), 509–582. CRC Press.

Acknowledgement

Section 5.3 of this chapter is based on material published in an EPRI report; see Reference [2].

To Probe Further

ATP data files for the case studies presented in this chapter are provided in the companion website. For readers who are not familiar with ATP they could be used as an introduction to the simulation package. Those who are already familiar with ATP could use them to explore the transient performance of the case studies analysed in this chapter.

6

Calculation of Power System Overvoltages

Juan A. Martinez-Velasco and Ferley Castro-Aranda

6.1   Introduction

An overvoltage is a voltage between one phase and ground, between two phases, or across a longitudinal insulation with a crest value exceeding the corresponding crest of the maximum system voltage. Standards distinguish several classes and shapes of overvoltages [1–4]:

1) *Temporary overvoltages*: They are undamped or weakly damped overvoltages of relatively long duration (seconds, even minutes). Temporary overvoltages (TOVs) are originated by faults, load rejection, resonance, and ferroresonance conditions, or by a combination of these factors. The representative TOV is characterized by a standard short duration (one minute) power-frequency waveshape.
2) *Slow-front overvoltages*: They are unidirectional or oscillatory highly-damped overvoltages, with a slow front and a short duration. These overvoltages are caused by switching operations, fault initiation, or remote lightning strokes. They exhibit time to crest of about 2÷5 ms and time to half value of up to 20 ms. The representative slow-front overvoltage is characterized by a standard switching impulse, and a peak voltage or a probability distribution of overvoltage amplitudes.
3) *Fast-front overvoltages*: Fast-front shaped overvoltages are caused primarily by lightning strokes, although they can also be caused by some switching operations or fault initiation. Their time to peak value can vary between 0.1 and 20 μs.
4) *Very fast-front overvoltages*: Very fast-front transients belong to the highest frequency range of transients in power systems (from 100 kHz up to 50 MHz). Their shape is usually unidirectional with time to peak below 0.1 μs, total duration below 3 ms, and with superimposed oscillations at frequencies of up to 50 MHz. In general, they are the result of switching operations or faults. Very fast-front overvoltages (VFFO) are usually associated with high-voltage disconnect switch operation in gas insulated substations (GIS), and with cable-connected motors.

Standards also include continuous power-frequency voltages, which originate with the system under normal operating conditions [1]. For systems whose maximum voltage exceeds that given in the standards, the actual maximum system operating voltage should be used.

This chapter presents a short description of the main causes of overvoltages and a summary of the modelling guidelines to be used when calculating overvoltages with a transients tool like ATP. For more details on these topics, readers are referred to the specialized literature on overvoltage calculation and insulation coordination studies [5–11]. The main core of this chapter is dedicated to illustrating how to apply ATP to the calculation of different types of overvoltages and to the estimation of the probability distribution of some random-nature overvoltages. The

Transient Analysis of Power Systems: A Practical Approach, First Edition. Edited by Juan A. Martinez-Velasco.
© 2020 John Wiley & Sons Ltd. Published 2020 by John Wiley & Sons Ltd.
Companion Website: www.wiley.com/go/martinez/power_systems

description of each case study includes a summary of modelling guidelines, some details about the ATP implementation, and a selection of simulation results.

6.2 Power System Overvoltages: Causes and Characterization

This section presents a short introduction to the main causes of overvoltages and their description. This will later be used to justify the approaches recommended for representing power system components when calculating overvoltages or their statistical distribution; see [5–11].

Temporary Overvoltages: A summary of the most frequent causes that lead to TOVs and their consequences is presented below.

- *Fault overvoltages*: Phase-to-ground faults can produce power-frequency phase-to-ground overvoltages on the unfaulted phases. The overvoltage magnitude depends on the system grounding and on the fault location. In effectively grounded systems, the overvoltage is about 1.3 pu and the duration (including fault clearing) is generally less than one second. In resonant grounded systems, the TOVs is about 1.73 pu or even greater, and the duration is generally less than 10 seconds.
- *Load-rejection overvoltages*: They are a function of the rejected load, the system topology after disconnection, and the characteristics of the sources (e.g. speed and voltage regulators of generators). The longitudinal TOVs depend on whether phase opposition is possible; such phase opposition can occur when the voltages on each side of the open switching device are not synchronized. A distinction should be made between system configurations when large loads are rejected: (i) a system with relatively short lines and high short-circuit power at terminal stations will have low overvoltages; (ii) a system with long lines and low short-circuit power at generating sites will have high overvoltages.
- *Resonance and ferroresonance overvoltages*: TOVs may arise from the interaction of capacitive elements (lines, cables, series capacitors) and inductive elements (transformers, shunt reactors). The resonant overvoltage is initiated by a sudden change in the system configuration (e.g. load rejection, single-phase switching of a transformer terminated line, isolation of a bus potential transformer through breaker capacitance). These overvoltages can reach magnitudes higher than 2 pu and last until the condition is cleared.
- *Transformer energization*: Resonance overvoltages can occur when a line and an unloaded or lightly loaded transformer are energized together. The transformer can cause high-magnitude highly-distorted inrush currents due to the nonlinear behaviour of its core; these currents can interact with the power system, whose frequency response may exhibit a resonance at a frequency included in the inrush currents. The consequence may be a long-duration resonant TOV [12].
- *Longitudinal overvoltages* may occur during *synchronization* due to phase opposition at both sides of the switch. The representative longitudinal TOVs are derived from the expected overvoltage, which has amplitude equal to twice the phase-to-ground operating voltage and a duration that can vary from several seconds to some minutes. When synchronization is frequent, the probability of occurrence of a ground fault and consequent overvoltage shall be considered; in such cases, the representative overvoltage amplitudes are the sum of the assumed maximum ground-fault overvoltage on one terminal and the continuous operating voltage in phase opposition on the other [1, 2].

TOVs are used to select surge arresters; that is, arresters are selected to withstand these overvoltages. Resonant and ferroresonant overvoltages are an exception and they should not be used for arrester selection; instead, they should be limited by detuning the system from the resonant frequency, changing the system configuration, or installing damping resistors. The combination of TOVs of different origin may lead to higher arrester stresses and perhaps to higher insulation levels.

Slow-Front Overvoltages: They are generally caused by switching operations. The most frequent causes of slow-front overvoltages are discussed below.

- *Line/cable energization and reclosing overvoltages*: Energization and reclosing of lines/cables can originate slow-front overvoltages on all phases. Their calculation has to consider trapped charges left on the phases in case of high-speed reclosing. The longitudinal insulation between non-synchronous systems can be subject to overvoltages of one polarity at one terminal and the crest of the operating voltage of the other polarity at the other terminal; consequently, the longitudinal insulation can be exposed to voltages higher than the phase-to-ground insulation. In synchronized systems, the highest switching overvoltage and the operating voltage have the same polarity, and the longitudinal insulation is exposed to voltages lower than the phase-to-ground insulation.
- *Fault overvoltages*: Slow-front overvoltages can be produced during phase-to-ground fault initiation and clearing. If the switching overvoltages for energizing and reclosing are controlled to below 2 pu, fault and fault clearing may produce higher overvoltages. A conservative estimate may assume that the maximum overvoltage during fault clearing is about 2 pu, and the maximum value caused by a fault initiation is about $(2k - 1)$ pu, where k is the ground fault factor in per unit of the peak line-to-ground system voltage.
- *Load-rejection overvoltages*: Load rejection can cause longitudinal voltage stresses across switching devices and phase-to-ground insulator stresses. If arresters are used to limit switching overvoltages to below 2 pu, the energy discharged by the arresters should be carefully estimated, especially when generators, transformers, long transmission lines, or series capacitors are present.
- *Inductive and capacitive current switching overvoltages*: Capacitor bank energization produces overvoltages at the capacitor location, line terminations, transformers, remote capacitor banks, and cables. The energizing transient at the switched capacitor location is usually less than 2 pu phase-to-ground. The highest phase-to-phase overvoltages are most commonly associated with energizing ungrounded capacitor banks. Restrikes or reignitions during the interruption of capacitive currents can produce extremely high overvoltages, which can also occur as a consequence of the chopping of inductive currents due to the transformation of magnetic energy to capacitive energy.

Fast-Front Overvoltages: They are generally produced by lightning discharges, although switching of nearby equipment may also produce fast-front waveshapes.

- *Fast-front lightning overvoltages* can be caused by strokes to phase conductors (shielding failure), strokes to line shield wires, or by nearby strokes to ground. Induced voltages by nearby strokes are generally below 400 kV and are important only for lower (distribution) voltage systems. Lightning-caused overvoltages on an overhead line generate surge voltages that can reach a substation, being those surges caused by a backflash more severe than those caused by a shielding failure.
- *Fast-front switching overvoltages*: The connection and disconnection of nearby equipment can produce oscillatory short-duration fast-rising surges with similar waveshapes to lightning. However, as their magnitudes are usually smaller than those caused by lightning, their importance is restricted to special cases. Their maximum value is approximately 3 pu with restrike and 2 pu without restrike.

Very Fast-Front Overvoltages: Causes that can originate VFFOs are disconnector operations and faults within GISs, switching of motors and transformers with short connections to the switchgear, and certain lightning conditions.

6.3 Modelling for Simulation of Power System Overvoltages

6.3.1 Introduction

The simulation of transient phenomena requires a representation of network components valid for a frequency range that varies from direct current to several MHz. Modelling of power components taking into account the frequency-dependence of parameters can currently be achieved through mathematical models which are accurate enough for a specific range of frequencies. Each range of frequencies usually corresponds to some particular transient phenomena. One of the most accepted classifications is that proposed by the IEC and CIGRE, in which frequency ranges are classified into four groups, as shown in Table 6.1.

If a representation is already available for each frequency range, the selection of the model may suppose an iterative procedure: the model must be selected based on the frequency range of the transients to be simulated; however, the frequency ranges of the test case are not usually known before performing the simulation. This task can be alleviated by using widely accepted classification tables. Table 6.2 shows a list of common transient phenomena.

An important effort has been dedicated to clarify the main aspects to be considered when representing power components in transient simulations. Users of electromagnetic transients tools can nowadays obtain information on this field from several sources; see [13–15].

The simulation of a transient phenomenon implies not only the selection of models but the selection of the system area that must be represented. The rules to be followed when selecting

Table 6.1 Classification of frequency ranges.

Group	Frequency range	Shape designation	Representation mainly for
I	0.1 Hz–3 kHz	Low-frequency oscillations	Temporary overvoltages
II	50 Hz–20 kHz	Slow-front overvoltages	Switching overvoltages
III	10 kHz–3 MHz	Fast-front overvoltages	Lightning overvoltages
IV	100 kHz–50 MHz	Very fast-front overvoltages	Restrike overvoltages

Table 6.2 Origin and frequency ranges of transients in power systems.

Origin	Frequency Range
Ferroresonance	0.1 Hz–1 kHz
Load rejection	0.1 Hz–3 kHz
Fault clearing	50 Hz–3 kHz
Line switching	50 Hz–20 kHz
Transient recovery voltages	50 Hz–100 kHz
Lightning overvoltages	10 kHz–3 MHz
Disconnector switching in GIS	100 kHz–50 MHz

the models and the system area for a transient study have been discussed in previous chapters. Although there could be some differences, the steps when simulating overvoltages could be listed as follows [9]: (i) select the system zone taking into account the frequency range of the transients; (ii) minimize the part of the system to be represented; (iii) implement an adequate representation of losses; (iv) consider an idealized representation of some components if the system to be simulated is too complex; (v) perform a sensitivity study if one or several parameters cannot be accurately determined.

This section summarizes the recommended modelling guidelines for representing power system components when simulating the causes that lead to any type of overvoltage.

6.3.2 Modelling Guidelines for Temporary Overvoltages

TOVs arise with frequencies close to the power frequency, usually below 1 kHz, so the models required for their analysis are power-frequency models for which the frequency dependence of parameters is not usually required. A summary of the guidelines proposed in the literature is presented below [11–17].

- The power supply model will depend on the case study. It can be represented as an ideal voltage source in series with a three-phase impedance (specified by its positive- and zero-sequence impedances), as a synchronous generator, or as a network equivalent whose impedance has been fitted in a frequency range typically below 2 kHz. If a synchronous generator model is required, then it has to include saturation, control units and the mechanical part.
- Lines and cables will be represented by a PI equivalent with parameters calculated at power frequency, although in some cases, zero-sequence parameters must be fitted in a frequency range of up 1 kHz. The number of PI sections required for representing a line/cable will depend on the length and the frequency range of the transients to be analysed. Line transpositions and cable cross-bonding will also affect the number of sections. Corona effect is required only when the overvoltage can exceed the ionization threshold. Line towers and insulator models are not required. Modelling the grounding impedances of a line may be required in some fault calculations; in such cases, a low-frequency low-current model will suffice.
- Models for transformers, shunt reactors and capacitor banks will be usually required. Transformer models should be implemented with caution, mainly in ferroresonant studies, for which it is important to properly model the transformer core and its saturation characteristics. A saturable reactance can be a source of harmonics which may cause resonance problems.
- Substation busbar models are not required as it can be assumed that the voltage is the same in the whole substation. However, the models of some substation equipment and the substation ground grid may be required. For instance, the model of a voltage transformer can be critical in some ferroresonance studies.
- Models for loads and power electronic converters can be required. As a rule of thumb, no load condition will usually represent the most conservative scenario, since load adds damping. However, in some cases, a load model may be required to limit the conditions under which overvoltages can arise. Different approaches for representing loads were suggested in [18]. Models of power electronic converters are required when the converter can be the source of harmonics that may cause resonance overvoltages; in such cases, including filter models is mandatory.

Guidelines for ferroresonance studies are discussed in the case study included in this chapter.

6.3.3 Modelling Guidelines for Slow-Front Overvoltages

A discussion of the extent of the system to be modelled and details about equipment models typically used for the simulation of switching transients are presented below; see [11, 18].

6.3.3.1 Lines and Cables

The most accurate representations for lines and cables are based on distributed-parameter models. The frequency dependence of the line parameters may be an important consideration, particularly when the ground return mode is involved (e.g. during a line-to-ground fault). In these cases, a frequency-dependent distributed-parameter line model gives a very accurate representation for a wide range of frequencies. Lumped-parameter models (PI circuits) are less accurate and computationally more expensive, since a number of cascaded short-sections are needed to approximate the distributed nature of the physical line/cable. The use of nominal PI circuits is usually restricted to the case of very short lines when the travelling time is smaller than the time-step size, although cascaded PI sections can be used without excessive loss of accuracy for some studies; e.g. line energization. The number of PI circuits will usually depend on the desired accuracy. In cable studies, in which the frequencies span a large bandwidth and the cable parameters significantly vary within this range, a frequency-dependent parameter model must be used.

6.3.3.2 Transformers

A lumped-parameter coupled-winding model with a sufficient number of RLC elements that fit the impedance characteristics at the terminal within the frequency range of interest will suffice. The nonlinear characteristic of the core should usually be included, although the frequency characteristic of the core is often ignored. This may be an oversimplification as the eddy current effect prevents the flux from entering the core steel at high frequencies, thereby making the transformer appear to be air-cored. This effect begins to be significant even at frequencies in the order of 3–5 kHz. For switching surge studies, several approaches may be used to model transformers [19]; they can be based on transformer nameplate, a model synthesized from the measured impedance versus frequency response of the transformer, or a detailed model obtained from the transformer geometry and material characteristics.

6.3.3.3 Switchgear

In switching transient studies, the switch is often modelled as an ideal conductor (zero impedance) when closed, and an open circuit (infinite impedance) when open. A discussion about the modelling approaches to be used in closing and opening operations, as well as for representing faults, is presented below [20]. See also Chapter 2.

Opening: Studies are mostly based on an ideal switch model that opens at a current zero. The dynamic characteristic of the arc is not usually modelled, although it can be useful in some cases [21]. In certain instances (e.g. when small inductive currents are being interrupted), the current in the switch can extinguish prior to its natural zero crossing; severe voltage oscillations that can stress the circuit breaker can result due to this current chopping. For a detailed description of this phenomenon, see [22].

Closing: Circuit breakers can close at any time (angle) on the power frequency wave. Transient voltage and current magnitudes depend on the instant of the voltage waveform at which the circuit breaker contacts close electrically. The withstand strength between circuit breaker contacts decreases as the contacts come closer. If the field stress across the contacts exceeds this withstand strength, prestrike occurs. If this is taken into account, the distribution of closing angles is confined to the rising and peak portions of the voltage waveforms. Some modern devices can control the closing angle of the poles to close at or near the voltage zero between the contacts; they can reduce overvoltages and inrush currents [23]. A statistical switching method can be applied to the breaker poles over the timespan around the voltage zero, within the tolerance of the closing time. A statistical switching case typically consists of several hundred separate simulations, each using a different set of circuit breaker closing times. For a single-phase circuit, the set of closing times can be represented as a uniform distribution from 0 to 360° with reference to the power frequency period. A three-phase (pole) circuit breaker can be modelled as three single-phase circuit breakers, each with an independent uniform distribution covering 360°. However, an alternative (dependent) model can be used if the three poles are mechanically linked and adjusted so that each pole attempts to close at the same instant. In reality, there will be a finite time or pole span between the closing instants of the three poles. The pole span can be modelled with an additional statistical parameter, typically a normal (Gaussian) distribution. Statistical cases with pre-insertion resistors or reactors require a second set of three-phase switches. The first set is modelled as described above; the closing times of the second set (which shorts the resistors or reactors) depend upon the first set plus a fixed time delay; typically one-half to one cycle for pre-insertion resistors used with circuit breakers, and 7–12 cycles (depending on application voltage class) for pre-insertion reactors used with circuit-switchers closing in air through high-speed disconnect blades.

Faults: Their simulation may include both closing and opening operations. They are usually modelled as ideal switches in series with other elements, if necessary. The switch is closed during the steady-state solution at a specific time or voltage. Several runs with variations in the closing instant should be carried out since the point on wave of switching can affect the transient. Faults may also be modelled with flashover-controlled switches to represent a gap; the switch is operated when the gap voltage exceeds a fixed value. More sophisticated models include a voltage-time characteristic. Faults generally involve arcs, which can be modelled as an ideal switch, a constant or variable voltage, a constant or variable resistance, or a combination of voltage and resistance. The ideal switch is the most common option because the arc voltage is usually small compared with voltage drops elsewhere (i.e. along the transmission line). Arc modelling can be important when studying the causes of secondary arc phenomena, such as single-pole reclosing.

6.3.3.4 Capacitors and Reactors

Capacitor banks are usually modelled as lumped elements. However, some studies require the modelling of secondary parameters such as series inductance and loss resistance. The inductance of the buswork is sometimes important when studying the back-to-back switching of capacitor banks, or in the study of faults on the capacitance bus. The damping resistance of this inductance should be estimated for the natural frequency of oscillations. Reactors are usually modelled as lumped inductors with a series resistance. A parallel resistance may be added for realistic high-frequency damping. Core saturation characteristic may also have to be modelled. A parallel capacitance across the reactor should be included for reactor opening studies (chopping of small currents). The total capacitance includes the bushing capacitance and the equivalent winding-to-ground capacitance. For series reactors, there is a capacitance from the terminal to ground and from terminal to terminal. More sophisticated models may be developed for determining internal stresses [24].

6.3.3.5 Surge Arresters

A gapless metal oxide surge arrester (MOSA) can be modelled as a nonlinear resistance. The preferred representation is a true nonlinear element which iterates at each time step to a convergent solution and is thus numerically robust [25–27]. The V—I characteristic should be modelled with 5–10 (preferably exponential) segments. Waveform-dependent characteristics are not usually required for slow-front transient simulations. The surge arrester lead lengths and separation effects can be ignored.

6.3.3.6 Loads

In general, the power system load is represented using an equivalent circuit with parallel-connected resistive and inductive elements. The power factor of the load determines the relative impedance of both elements. Shunt capacitance is represented with the resistive and inductive elements of the load if power-factor correction capacitors are used. Whenever loads are lumped at a load bus, the effects of lines, cables, and any transformers downstream from the load bus need to be considered [15]. Certain types of load may require specific representation of some components (e.g. induction motors, adjustable-speed drives, fluorescent lighting loads, etc.). The need for such detailed representation is determined by the phenomenon being investigated. A load model will be included in the study only when it can add crucial information; otherwise, the model is not considered and the most conservative results are derived.

6.3.3.7 Power Supply

The power supply model depends on the phenomenon being investigated. In general, a generator can be modelled as a voltage behind the subtransient impedance. If a power system supplies the zone under study, the supply system can be modelled as an ideal sine-wave source in series with its equivalent impedance. Often a network equivalent is used in order to simplify the representation of the portion of the power network not under study. Several network equivalents can be considered [18]: (i) the first type represents the Thevenin equivalent of the connected system, with a X/R ratio selected to adequately represent the damping; (ii) the second type represents the surge impedance of connected lines. The second equivalent may be used to reduce the connected lines to a simple equivalent surge impedance when the lines are long enough so that reflections are not of concern in the system under study. If the connected system consists of a known Thevenin equivalent and additional transmission lines, the two impedances may be combined in parallel. More complex equivalents which properly represent the frequency response characteristic may be required. For an update of the work performed on network equivalents, see [28].

6.3.4 Modelling Guidelines for Fast-Front Overvoltages

This section describes the models of power system components to be used in lightning studies. Since lightning-related surge voltages and currents cannot be easily measured or verified, the models presented should be treated as the recommended approach in representing the behaviour of the power system components within the specified frequency range [29, 30].

6.3.4.1 Overhead Transmission Lines

The model of an overhead line in lightning studies must include the representation of phase conductors and wires, towers, insulators, and grounding impedances.

Phase Conductors and Shield Wires: The overhead line is represented by means of multi-phase untransposed distributed- and frequency-dependent parameter line sections for each span. Bundled-phase conductors can be represented by one equivalent conductor. The line parameters are calculated by means of the LINE CONSTANTS routine, using the tower structure geometry and conductor data as input. The parameters can be calculated at 400/500 kHz, including skin effect.

Line Length and Termination: In lightning overvoltage calculations, the peak voltage at the struck tower may be influenced by reflections from the adjacent towers, so a sufficient number of adjacent towers at both sides of the struck tower should be modelled to accurately determine overvoltages. This can be achieved by selecting the number of line spans modelled such that the travel time between the struck tower and the farthest tower is more than one-half of the lightning surge front time. The number of line spans modelled must be increased when the effects due to the tail of the lightning surge are considered, especially when evaluating the insulator flashovers with the leader propagation method or the energy discharged by arresters. In substation design studies, a similar approach can be followed. Furthermore, it may be desirable to determine the first reflections of overvoltages accurately at any point inside the substation. This criterion may require detailed modelling of additional towers further away from the substation depending upon the distance from the struck tower and substation layout. In both transmission line and substation design studies, the line extended beyond the last tower can be represented with a matrix of self and mutual resistances equal to the corresponding line surge impedances. A simpler option is to add a line section long enough to avoid that the reflections from the open point could reach the last tower included in the model.

Towers: The representation of a tower is usually made in circuit terms. The simplest model represents a tower as a single-phase distributed-parameter lossless line, whose surge impedance depends on the structure details [31]. Typical values range from 100 to 300 Ω, and the velocity of propagation can be assumed to be equal to the speed of light. Since the surge impedance of the tower varies as the wave travels from top to ground, more complex models have been developed that represent a tower by means of several line sections and circuit elements that are assembled taking into account its structure. These models are based on non-uniform transmission lines, or on a combination of lumped- and distributed-parameter circuit elements [31]. The latter approach is also motivated by the fact that, in many cases, it is important to obtain the lightning overvoltages across insulators located at different heights above ground; this is particularly important when two or more transmission lines with different voltage levels share the same tower.

Grounding Impedance: The peak overvoltage on the tower depends on the grounding impedance, whose influence on the tower top voltage is determined by its response time and current dependence. The response time is usually only important in cases where counterpoises with distances greater than 30 m from the tower base are installed. In that case, a frequency-dependent distributed-parameter model should be considered [7, 31]. Within 30 m of the tower base, the time response can generally be neglected and the tower grounding impedance is determined by using the current dependence of the grounding resistance. In general, it is recommended to consider the waveform dependence of tower foundation and counterpoise grounding. A lumped resistive model may not be adequate as compared to more detailed models of counterpoise grounding. The counterpoise can be represented either as a nonlinear resistance or as a distributed-parameter line with dispersed conductive connections to earth [31].

Insulators: The insulators may be represented by voltage-dependent flashover switches in parallel with capacitors connected between the respective phases and the tower. The capacitors simulate the coupling effects of conductors to the tower structure. For a simplified analysis, an idealized representation can also be adequate: the flashover mechanism of the insulators is represented by voltage-time curves, whose characteristics are a function of the insulator length and applicable only within the range of parameters covered experimentally [29, 30, 32–35]; the insulator is represented as a switch that closes when the voltage exceeds the flashover voltage calculated from voltage-time curve. The start-up time for the voltage-time characteristics must be synchronized to the instant at which lightning stroke hits the shielding wire or the tower top. However, the behaviour of insulation under the stress of the standard impulse cannot accurately predict its performance when exposed to any non-standard lightning impulse. Furthermore, it is inaccurate to assume that flashover will occur when a voltage wave just exceeds the voltage-time curve at any time; the experimental characteristic is only adequate for relating the peak of the standard impulse voltage to the time of flashover. To obtain more accurate results, a representation of air gap (insulator strings and spark gaps) breakdown subject to standard and non-standard lightning impulses is necessary. Analytical procedures to predict the performance of insulation as a function of the impulse voltage waveform, the time to flashover, the gap configuration and others, have been developed and validated by tests performed in the high-voltage laboratories. The most widely used procedures are the integration method and the leader progression model [31, 33, 35].

Corona: Corona has a significant effect on overvoltage surges associated with lightning strokes to overhead lines [36–40]: (i) for high-magnitude surges, the corona effect is independent of the conductor size and geometry; (ii) for low voltages, the effect differs due to different corona inception voltages; (iii) weather conditions have no significant impact on corona distortion; (iv) the coupling factor between phases increases with increasing surge voltage; (v) the tail of the surge is not influenced by corona. Corona introduces a time delay to the front of the impulse corresponding to the loss of energy necessary to form the corona space charge around the conductor. This time delay takes effect above the corona-inception voltage and varies with surge magnitude. This variation with voltage can be expressed as a voltage-dependent capacitance which is added to the geometrical capacitance of the transmission line. The corona-inception voltage, V_i, for a single conductor above earth can be estimated by means of the following form [32]:

$$V_i = 23 \cdot \left(1 + \frac{1.22}{r^{0.37}}\right) \cdot r \cdot \ln \frac{2h}{r} \tag{6.1}$$

where r is the conductor radius, in cm, and h is the conductor height, in cm.

The modelling details of corona can be expressed by curves of charge (q) versus impulse voltage (V). Several corona models have been proposed to represent the dynamic capacitance region of the q/V curve in a piecewise linear fashion. The approach presented in [37] can be used to estimate the variation of the steepness of lightning overvoltages impacting on substations with travel length. This approach relies upon the observation that the time delay as a function of travel distance becomes linear for voltages substantially higher than the corona-inception level; that is, in this region, the steepness of the overvoltage is independent of the voltage value. This yields the following relationship [2, 4]:

$$S = \frac{1}{\frac{1}{S_o} + A \cdot d} \tag{6.2}$$

where S_o is the original steepness of the overvoltage, S is the new steepness after the waveform travels for a distance d, and A is a constant. The constant A is a function of the line geometry only and is also dependent on the surge polarity. Typical values are given in [2, 4, 37]. Although corona effects may reduce the peak of lightning-related overvoltages by more than 20%, in many studies corona is neglected to obtain conservative results.

6.3.4.2 Substations

The overall substation model can be derived from the substation layout, and must include buswork, insulators and other substation equipment [30].

Buswork and Conductors: The buswork and conductors between the discontinuity points inside the substation, and connections between the substation equipment are explicitly represented by line sections. These line sections are modelled by untransposed distributed-parameter sections if they are longer than 3 m; otherwise, a lumped-parameter inductance of 1.0 μH/m can be used. The line parameters can be calculated using the LINE CONSTANTS routine [20]. Remember that the minimum conductor length with distributed-parameter representation dictates the simulation time step.

Substation Equipment: The substation equipment, such as circuit breakers, substation transformers, and instrument transformers, are generally represented by their stray capacitances to ground. The open/close status of circuit breakers/switches should be considered. In lightning studies conducted to design transformer protection against fast-front overvoltages, a conservative approach is to represent the transformer as an open circuit. To increase the level of accuracy, it is recommended to account for the capacitance of the winding. A resistance equal to the surge impedance of the winding can be placed across the capacitance. The model can be enhanced by adding the inductive transformer model and relevant capacitances between windings, windings to core and windings to ground, as well as bushing capacitances. Such a model can also be used to calculate the surge transfer from winding-to-winding as in the case of generator (or motor) protection studies. Capacitances can be determined from the geometry of the coil and core structure, or from manufacturers and tables. Values of the winding capacitance, together with capacitance values for outdoor bushings, are provided in [6]. Some transformer models can accurately determine how a voltage applied to one set of terminals is transferred to another set of terminals, see [19, 41, 42]. These models duplicate the transformer behaviour over a wide range of frequencies, and act like a filter suppressing some frequencies and passing others. The use of a frequency-dependent transformer model is important when determining the surge that appears on a generator bus when a steep-fronted surge impinges on the high-voltage terminals of the step-up transformer. Some models allow the simulation of any type of multi-phase, multi-winding transformer as long as its frequency characteristics are known, either from measurements or from calculations based on the physical layout of the transformer.

Insulators and Bus Support Structures: The bus support structures are represented by a distributed-parameter model with surge impedances calculated from the structure geometry, and with velocity of propagation equal to the speed of light [29, 30]. The representative grounding resistance inside the substations is usually between 0.1 and 1 Ω. Comparative simulations have shown that the support structures do not have much impact on the simulation results, and can be neglected. The capacitance to ground of all insulators should be represented, since the substation capacitance is one of the critical parameters that modify lightning surge waveforms.

6.3.4.3 Surge Arresters

The voltage–current characteristics of MOSAs are a function of the incoming surge steepness. Protective characteristics for surge arresters are available from manufacturers. Since arrester terminal voltage and current do not reach their peak values at the same time, the frequency-dependent characteristics of arresters may be of significant importance when excited by fast-front transient surges [26, 27]. The arrester must be modelled as a nonlinear resistor. Several frequency-dependent surge arrester models have been developed [27]. These models can reproduce arrester characteristics over a wide range of frequencies. The nonlinear arrester characteristics need to be modelled up to at least 20–40 kA, since high current surges initiated by close backflashovers can result in arrester discharge currents above 10 kA. The arrester lead lengths must be considered to account for the effects of additional voltage rise across the lead inductance. A lumped-element representation with an inductance of 1.0 μH/m will suffice. The energy discharged by the arrester must be monitored to verify that the maximum allowed energy dissipation is not exceeded.

6.3.4.4 Sources

Two types of inputs must be considered in lightning studies, the instantaneous phase voltage at the time the stroke hits the line and the lightning stroke current. The insulator stress is a combination of the voltage due to the lightning current and the power-frequency voltage.

Initial Conditions: The instant of lightning stroke with respect to the instantaneous steady-state ac voltage must be coordinated to maximize its impact for worst-case conditions. This can be achieved by properly selecting the magnitude and phasing of the three-phase sinusoidal voltage sources at the terminating point of the transmission line. In transmission line design studies, one of the objectives is to determine the highest line outage rate, which is generally maximized by finding the minimum critical lightning current that causes insulator backflashovers. If the lightning hits the tower when the contribution of ac power-frequency voltage to the insulation stress is maximum, the backflashover can occur with a smaller lightning current.

Lightning Stroke: The lightning stroke is represented as a current source. Its parameters (crest, front time, maximum current steepness, duration, and polarity) are determined by a statistical approach, since they are all statistical in nature, generally characterized by log-normal distributions [33, 43]. The peak current can be statistically correlated to the steepness and the time to crest of the current wave form. Both the steepness and the front time increase as the peak current increases. The detailed calculation procedure for these parameters is shown in the CIGRE Guide [33]; see also [7]. A rigorous approach requires the front of the lightning current source to be upwardly concave, although for practical purposes, a linearly rising front at the selected maximum current steepness can be sufficient. In this case, a negative triangular wave shape for the lightning current source can be selected. The double exponential impulse model should be used with caution since this model does not accurately reflect the concave wave shape of the wave front [29, 30].

6.3.5 Modelling Guidelines for Very Fast-Front Overvoltages in Gas Insulated Substations

The models of GIS components for VFFOs are based on electrical equivalent circuits that combine lumped elements and distributed-parameter lines. Although at very high frequencies the skin losses can produce a noticeable attenuation, these losses are usually neglected due to the geometrical structure of GIS and the enclosure material. This gives conservative results. Only the dielectric losses in some components (e.g. capacitive-graded bushing) need be taken into account.

Modelling guidelines for representing GIS equipment in computation of internal transients are discussed below [13–15, 44–47]. The models are based on single-phase representations; however, depending on the substation layout and the study to be performed, three-phase models for inner conductors should be used [48, 49]. For the calculation of internal transients, the distributed-parameter line model takes into account the internal mode (conductor-enclosure) only, assuming that the external enclosure is perfectly grounded. If transient enclosure voltages are of concern, then a second mode (enclosure-ground) is to be considered.

A short description of most important GIS component models follows [45–49].

1) *Bus ducts*: For a range of frequencies below 100 MHz, a bus duct can be represented as a lossless transmission line. A different approach should be used for vertical bus sections. As for the propagation velocity, empirical corrections are usually required to adjust its value.

Experimental results show that the propagation velocity in GIS ducts is close to 0.95–0.96 of the speed of light [50]. The error committed by ignoring skin effect losses is usually negligible. Other devices, such as elbows, can also be modelled as lossless lines.

2) *Surge arresters*: Experimental results have shown that switching operations in GIS do not usually produce voltages high enough to cause surge arresters to conduct, so the arrester can be modelled as a capacitance-to-ground. However, when the arrester conducts, the model should take into account the steep front-wave effect, since the voltage developed across the arrester for a given discharge current, increases as the time to crest of the current increases, but reaches crest prior to the crest of the current. A detailed model must represent each internal shield and block individually, and include the travel times along shield sections, as well as capacitances between these sections, capacitances between blocks and shields, and the blocks themselves.

3) *Circuit breakers*: A closed breaker can be represented as a lossless transmission line, whose electrical length is equal to the physical length, being the propagation velocity reduced to 0.95–0.96 of the speed of light. The representation of an open circuit breaker is more complicated due to internal irregularities. In addition, circuit breakers with several chambers contain grading capacitors, which are not arranged symmetrically. The electrical length must be increased above the physical length due to the effect of a longer path through the grading capacitors, while the speed of progression must be decreased due to the effects of the higher dielectric constant of these capacitors.

4) *Gas-to-air bushings*: A bushing gradually changes the surge impedance from that of the GIS to that of the line. A detailed model of the bushing must consider the coupling between the conductor and shielding electrodes, and include the representation of the grounding system connected to the bushing. A simplified model consists of several transmission lines in series with a lumped resistor representing losses. The surge impedance of each line section increases as the location goes up the bushing. If the bushing is distant from the point of interest, the resistor can be neglected and a single-line section can be used.

5) *Transformers*: At very high frequencies, a winding of a transformer behaves like a capacitive network consisting of series capacitances between turns and coils, and shunt capacitances between turns and coils to the grounded core and transformer tank. The terminal capacitance to ground, which mostly comes from the capacitance of the terminal bushing to ground, must be added to obtain the total capacitance of the winding. A common practice is to model a transformer as a capacitor representing the capacitance of the winding to ground. At very high frequencies, the saturation of the magnetic core and leakage impedances can be neglected. When voltage transfer has to be calculated, interwinding capacitances and secondary capacitance-to-ground must also be represented. If voltage transfer is not of concern, an accurate representation can be obtained by developing a circuit that matches the frequency response of the transformer at its terminals. The modelling of transformers for the analysis of very fast-front transients (VFFTs) has been the subject of several works; see [41, 42].

6) *Current transformers*: Insulating gaps are usually installed in the vicinity of current transformers. During high-voltage switching operations, these gaps flashover, establishing a continuous path. Travelling waves propagate with little distortion. Current transformer models can often be neglected. Several approaches have been proposed to represent these transformers; see for instance [44].

7) *Spark dynamics*: The behaviour of the spark in disconnector operations can be represented by a dynamically variable resistance, with a controllable collapse time. In general, this representation does not affect the magnitude of the maximum VFFO, but it can introduce a significant damping on internal transients [51].

6.4   ATP Capabilities for Power System Overvoltage Studies

The calculation of power system overvoltages, regardless of their causes, must usually be based on a time-domain simulation, an adequate modelling of the system components, and a large enough model of the system zone to be analysed. A simulation package to be applied for such purposes must also provide capabilities that could allow users to estimate the performance of a system or a component by means of some index (e.g. the lightning flashover rate of an overhead transmission line). Consequently, the simulation tool used to carry out these calculations and studies should be able to provide models and capabilities that could fulfil the following requirements.

- Built-in models for representing the causes of overvoltages and the various system components that must be modelled in a given study. From the modelling guidelines proposed in Section 6.3, it is evident that, depending on the frequency range associated to an overvoltage study, more than one representation is required for each power system component (i.e. lines, cables, transformers, machines): the response of a system or component depends on the frequency range associated to the transient under study. The capabilities of the simulation package should be able to provide frequency-dependent representations of all power system components.
- Capabilities for developing custom-made tools and models. If built-in capabilities cannot cover the models and applications that are required in overvoltage studies, the simulation tool must have capabilities that will allow users to develop and implement custom-made modules that can facilitate some simulation tasks or represent components not directly available in the tool. The continuous improvements introduced in ATPDraw facilitate these tasks to users: the current ATPDraw version allows users to create custom-made models by taking advantage of the routine DATA BASE MODULE and MODELS language, as well as to compress several components into a single group which can be nested as part of a higher-level group. All these options allow users to customize edition and simulation tasks. Details about the ATP capabilities that can be used for these purposes are discussed in Chapter 9.
- Capabilities for carrying out multiple runs while one or more test system parameters are varied. Parametric and statistical studies are very often useful options to understand the effect that some parameters have on the transient performance of a system or a component. For the topics covered in this chapter, a multiple run option is crucial for carrying out a statistical study aimed at estimating the performance of an overhead transmission line in front of lightning discharges or a parametric study aimed at determining a resonance condition.

Table 6.3 presents a summary of ATP capabilities that can be used in overvoltage studies.

6.5   Case Studies

6.5.1   Introduction

This section presents several case studies that analyse different causes of overvoltages in power systems. For each case study, a description of the test system and the modelling guidelines followed when implementing the system in ATP are provided. Simulation results will illustrate the importance that overvoltage calculations can have in power system design.

6.5.2   Low-Frequency Overvoltages

The two case studies selected for illustrating low-frequency overvoltages are related to two typical causes of TOVs in power systems, namely resonance and ferroresonance. Actually,

Table 6.3 ATP capabilities for overvoltage studies.

ATP capability	Remarks
Built-in components	Linear and nonlinear branches, frequency-dependent line and cable models, switches, and source models, are the most common ATP options in power system overvoltage studies. Built-in models for representing transformers and rotating machine can only be used in low-frequency overvoltages (e.g. load rejection, resonance, and ferrresonance).
Supporting routines	Several ATP routines can be used to create models of lines, cables, transformers, and arresters.
Development of custom-made modules	ATPDraw capabilities has several options that can be applied to create custom-made models: Compress-Group, user-specified model based on the DATA BASE MODULE routine, and user-specified model based on MODELS code. TACS provides a powerful option (TACS 69) for implementing new models and capabilities. MODELS language can be used for complementary tasks, such as representing insulator strings in lightning overvoltage studies. ATP does not have a built-in model for representing transformers in high-frequency transients, but users can create such models using other capabilities.
Multiple runs	POCKET CALCULATOR VARIES PARAMETERS (PCVP) is the ATP option for multiple runs; it can be used for either parametric or statistical studies. Built-in switch models for statistical studies are also available.

the study of the first case may not involve the simulation of any transient phenomenon (i.e. calculations can be carried out under steady-state conditions). The second case analyses a ferroresonance case in a distribution system. Ferroresonance is a very nonlinear and a very common phenomenon, mainly at distribution levels, for which some care is required when modelling the saturable core of the involved transformer. For an introduction to the origin and the mitigation of these overvoltages, see [52].

6.5.2.1 Case Study 1: Resonance Between Parallel Lines

Very high overvoltages can originate in series-compensated parallel transmission lines when they are also compensated by means of shunt reactors due to a resonance condition. Figure 6.1 shows the diagram of the test analysed here [53]. This system was presented in Chapter 4, and is analysed again in this chapter with more details about the implementation of the test system in ATPDraw.

The geometry of the test lines is shown in Figure 6.2. The two transmission lines are transposed every 30 miles. Since the resonance conditions can be analysed using steady-state

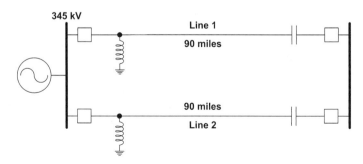

Figure 6.1 Case Study 1: Diagram of the test system.

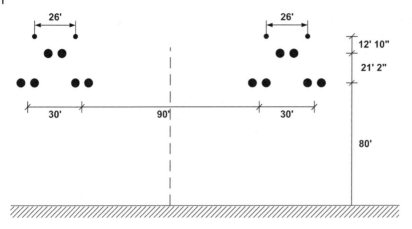

Figure 6.2 Case Study 1: Line geometry.

calculations, the transmission lines are represented by means of 10-mile constant-parameter PI sections cascaded in series.

Assume a single-line-to-ground fault occurs in Line 1, and consider the following sequence of events: (i) the single-phase-to-ground fault occurs; (ii) the circuit breakers that protect Line 2 open due to the fault, but the shunt reactors remain connected; (iii) the protection system of Line 1 opens the circuit breaker at the receiving end, but fails to open the circuit breaker at the sending end. The fault condition remains.

Resonant overvoltages could occur on the de-energized line due to the coupling with the energized line. The value of these overvoltages will depend on the reactive power of the shunt reactors.

A parametric study is used to find out for which range of values of the shunt reactor power a resonance condition can occur. This study is based on the combination of several ATP options: PCVP (i.e. the multiple run option), $PARAMETER, and a MODELS-based section for printing/reporting the results.

Figure 6.3a depicts the circuit implemented in ATPDraw to estimate the shunt reactor power values with which the scenario described above can lead to resonance overvoltages in Line 2. The ATP features required to carry out the parametric analysis are shown in Figure 6.3b: a $PARAMETER section is used to calculate the per phase reactance value of the shunt reactors whose reactive power is varied between two limits (in this example between 40 and 60 Mvar); the reactance value is passed to the reactor models included in the test system (see Figure 6.3b), while the reactive power is passed to a resistance branch for output purposes; the voltage developed across this resistance is passed to a MODELS section for further printing.

Figure 6.4 shows the application of the LCC option in ATPDraw to obtain the model of a line section (i.e. 10-mile section). Note that the PI model has been selected to represent the test line.

Figure 6.5 shows the voltage at the sending end of Line 2 as a function of the reactive power of the reactor (see Figure 6.3). The figure provides the results corresponding to three different distances between the parallel lines (see Figure 6.2): 90, 100, and 120 ft. One can easily deduce for what range of values resonant overvoltages will be induced on the de-energized line. As expected, the highest voltages are developed when the distance between lines is the shortest one; however, the differences between peak values are not too large with the three distances analysed in this case study.

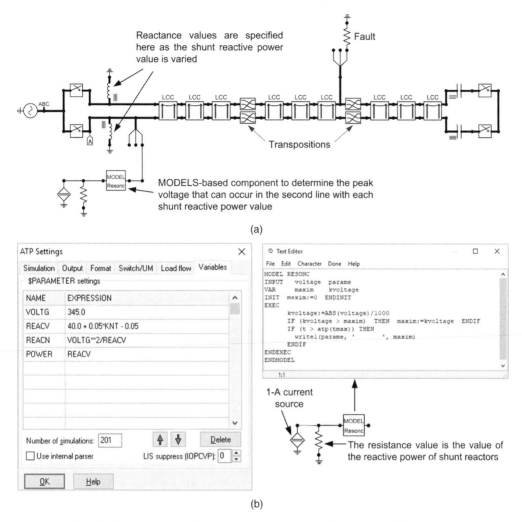

Figure 6.3 Case Study 1: ATPDraw implementation. (a) ATPDraw implementation of the test system, (b) Selected options for parametric analysis.

6.5.2.2   Case Study 2: Ferroresonance in a Distribution System

Introduction

The general requirements for ferroresonance are a source voltage, a saturable reactance, a capacitance, and little damping [16, 17]. Ferroresonance can involve large substation transformers, distribution transformers, or instrument transformers. System events that may initiate ferroresonance include single-phase switching or fusing, or loss of system grounding. Damping added to the circuit will attenuate the ferroresonant voltage and current. Although some damping is always present in the form of resistive source impedance, transformer losses, and even corona losses in high-voltage systems, most damping is due to the load applied to the secondary of the transformer.

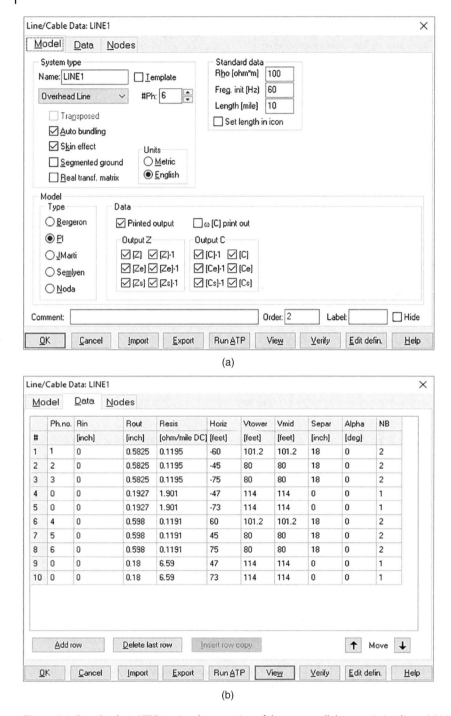

Figure 6.4 Case Study 1: ATPDraw implementation of the test parallel transmission lines. (a) Line model selection, (b) Line data input.

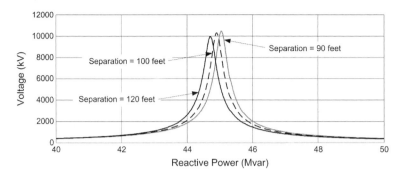

Figure 6.5 Case Study 1: Parallel resonance – Parametric study.

Ferroresonance is rarely seen provided all three source phases are energized, but may occur when one or two of the source phases are lost while the transformer is unloaded or lightly loaded. The loss of one or two phases can happen due to clearing of single-phase fusing, operation of single-phase reclosers or sectionalizers, or when energizing/de-energizing using single-phase switching procedures. If one or two poles of the switch are open and either the capacitor bank or the transformer have grounded neutrals, a series path through capacitance(s) and magnetizing reactance(s) exists and ferroresonance is possible. If both neutrals are simultaneously grounded or ungrounded, then no series path exists and there is no clear possibility of ferroresonance [16]. In all of these cases, the voltage source is the applied system voltage.

Ferroresonance is possible for any transformer core configuration. Three-phase core types provide direct magnetic coupling between phases, so voltages can be induced in the open phase(s) of the transformer. Ferroresonance may occur depending on the interrupting device, the type of transformer, the load on the secondary of the transformer, and the length and type of line/cable. The use of single-phase interruption and switching practices in systems containing multi-legged core transformers is a common condition for initiating ferroresonance. Replacement of all single-phase switching and interrupting devices with three-phase devices can eliminate this problem. An alternate solution is to replace multi-legged core transformers with single-phase banks or triplex designs wherever there is a small load factor.

Modelling for ferroresonance

The transformer model is the most critical part of any ferroresonance study. In addition, simulation results have a great sensitivity to the model used and errors in nonlinear model parameters. Although much effort has been made in refining equivalent circuit models for transformers, determining nonlinear parameters is still a difficult task: a different model and a different approach for determining model parameters are required for each type of core. Another critical part is the system zone that must be represented in the model. Both aspects are discussed in the following paragraphs.

The Study Zone: Parts of the system that must be simulated are the source impedance, the line(s)/cable(s), the transformer, and any other capacitance not already mentioned. Source representation is not generally critical, unless the source contains nonlinearities; the steady-state Thevenin impedance and open-circuit voltage will suffice. Lines and cables may be represented as coupled PI equivalents, cascaded for long lengths. Shunt or series capacitors may be represented as a standard capacitance, paralleled with the appropriate resistance. Stray capacitance may also be incorporated either at the corners of open-circuited delta transformer winding or midway along each winding. Other capacitance sources are transformer bushings, interwinding capacitances, and busbar capacitances.

Single-Phase Transformers: They are typically modelled by means of the so-called T-model [19], which is topologically correct only when the primary and secondary windings are not concentrically wound. Errors in leakage representation are not significant unless the core saturates; however, obtaining the linear parameters may not be easy, and a judgement must be made as to how it is divided between the primary and secondary windings. Model performance depends mainly on the representation of the nonlinear elements. The magnetizing inductance is typically represented as a piecewise linear characteristic or as a hysteretic inductance [19, 54, 55]. The linear value of this inductance (below the knee of the curve) does not much affect the simulation results, although great sensitivities are seen for the shape of the knee and the final slope in saturation. Factory test data provided by the transformer manufacturer may be insufficient to obtain the core parameters. The SATURA supporting routine is used to convert the rms $V-I$ open circuit characteristic to the instantaneous λ-i characteristic of the magnetizing inductance. However, it is important that open circuit tests be performed for voltages as high as the conditions being simulated; that is, open circuit tests should be therefore made for 1.3 or higher pu voltage values. Modern low-loss transformers have comparatively large interwinding capacitances which can affect the shape of the excitation curve and cause significant errors when calculating core parameters. In these cases, factory tests must be performed to get the $V-I$ curve before the coils are placed on the core.

Three-Phase Transformer: A complete representation of the windings can be obtained by using a coupled inductance matrix [56], to which the core equivalent is attached. The inductance matrix is obtained from standard short-circuit tests. Problems can arise for rms short-circuit data involving windings on different phases, since the current may be non-sinusoidal. The hybrid model presented in [57, 58] is based on this approach. A method of obtaining topologically correct models is based on the duality between magnetic and electrical circuits [59, 60]. This approach results in models that include saturation in each individual leg of the core, interphase magnetic coupling, and leakage effects. Several models based on the principle of duality have been presented in the literature [19, 54–56]. However, since tests suggested in the literature cannot always be performed and no standard tests have been developed for determining the parameters specified in some models, the use of these models is presently limited.

Distribution transformer ferroresonance

Figure 6.6 shows the diagram of the test system. The opening of the three phases of the circuit breaker CB is not simultaneous, so one or two poles remain closed for a certain period during the operation. The objective of the study is to estimate the cable length that can initiate ferroresonance when not all poles of the circuit breaker are open and the load at the LV side of the distribution transformer is very low. This system configuration exhibits the prerequisites for ferroresonance: capacitance provided by the insulated cable, saturable inductance provided by the distribution transformer, and little damping (i.e. lightly loaded transformer). The main parameters of the components that are of concern for a ferroresonance study are detailed below:

1) HV network: 110 kV, 1500 MVA, 50 Hz, $X/R = 10$, $X_0/X_1 = 1.1$.
2) Substation transformer: Triplex core, 110/25 kV, 10 MVA, 9%, Yd11, grounded through a zig-zag reactance with 75 Ω per phase.
   - No load test (positive sequence, MV side): $V_0 = 100\%$; $I_0 = 0.296\%$; $W_0 = 18.112$ kW.
   - Short-circuit test (positive sequence, HV side): $V_{sh} = 12\%$; $I_{sh} = 83.34\%$; $W_{sh} = 125.6$ kW.

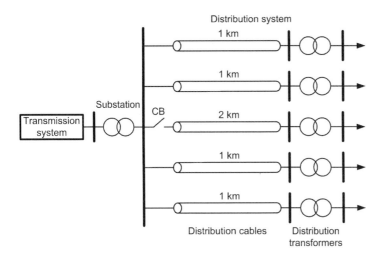

Figure 6.6 Case Study 2: Diagram of the test system.

Figure 6.7 Case Study 2: Configuration of the distribution cable system.

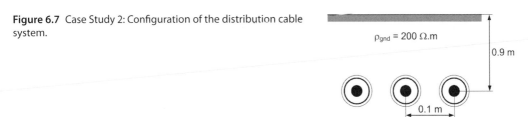

3) Cable: Al RHV, $3 \times (1 \times 240\,\text{mm}^2)$, 18/30 kV (see Figure 6.7).
4) Distribution transformer: Three-legged stacked core, 25/0.4 kV, 1 MVA, 6%, Dyn11.
   - Short-circuit test (positive sequence, MV side): Vsh = 6%; Ish = 100%; Wsh = 12 kW.
   - No load test (homopolar sequence, LV side): Vh = 100%; Ih = 0.5%; Wh = 1.8 kW.
   - Saturation curves are shown in Figure 6.8.

The models selected have the following features:

a) The HV transmission network is represented as an ideal balanced and constant three-phase voltage source in series with a three-phase impedance specified by its symmetrical impedances Z_1 and Z_0.

Figure 6.8 Case Study 2: Saturation curves of the distribution transformer core.

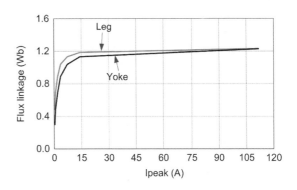

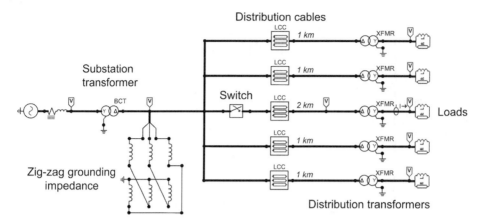

Figure 6.9 Case Study 2: ATPDraw implementation of the test system.

b) The cable is represented by its PI equivalent, whose parameters are obtained at power frequency.

c) Since the model and parameters of the substation transformer are not critical, this transformer is represented without including nonlinearities.

d) The switch needed to open the phases that originate ferroresonance has an ideal behaviour.

The circuit implemented in ATPDraw for simulating this test system is shown in Figure 6.9. This implementation has been based on the following ATP capabilities:

- The substation transformer is represented using the BCTRAN model.
- The distribution cables are represented by a constant distributed-parameter model using the ATPDraw LCC option (i.e. the CABLE CONSTANTS routine).
- The distribution transformers are represented by the Hybrid Transformer model [57, 58].
- Loads are modelled as parallel *RL* branches by means of a custom-made model. The procedure to implement this type of model is detailed in Chapter 9.

Figure 6.10 shows the application of the BCTRAN option to obtain the substation transformer. Remember that the saturable core of a transformer model based on this option is not included, so the model is linear. Figure 6.11 shows the application of the LCC option to obtain the models of distribution cables; note that the Bergeron model has been selected for representing the cables.

Figure 6.12 shows the main menu of the Hybrid Transformer model used to represent the distribution transformers. In this case, the characteristics of the saturable cores have been specified.

All the scenarios analysed in this case study consider a lightly loaded transformer, assuming only active power load. The simulation results presented in Figure 6.13 correspond to different cable lengths and a very low-load level of the affected transformer. The plots include the three voltages that occur at the secondary terminals of the transformer. The effect of the load level of any other transformer is not important for this study. According to these results, the voltages depend on the length of the cable and the transformer load level. As expected, a very light load favours ferroresonance, which can be avoided by increasing the transformer load. Ferroresonance cannot occur when the load level of the distribution transformer is equal or above 1% of the rated power. When only one pole is open, ferroresonance cannot occur if the transformer load increases above 0.5% of the rated power. The effect of transformer capacitances was not considered. Although these capacitances could have some influence on the conditions that can

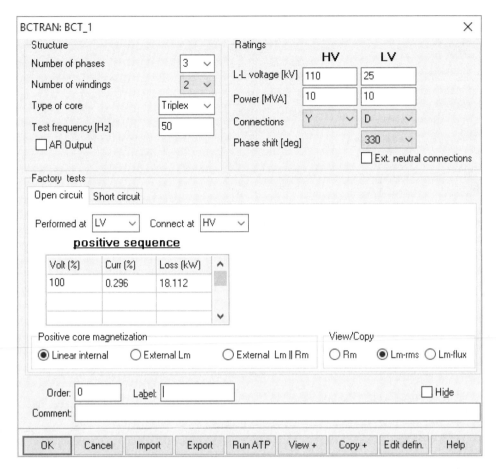

Figure 6.10 Case Study 2: Data input for the substation transformer (BCTRAN model).

originate ferroresonance, one should not expect large differences with respect to the results presented here since the cable capacitances are much larger than the transformer capacitances.

An important aspect that is worth mentioning is the representation of saturable cores. In this study, such representation is based on the anhysteretic characteristic. As shown in [61], this can be an inaccurate approach, since only a hysteresis-based model can provide accurate results; therefore, the results presented here should be analysed with some care.

6.5.3 Slow-Front Overvoltages

Switching transients in power systems are caused by the operation of breakers and switches. The switching operations can be classified into two categories: energization and de-energization. The former category includes energization of lines, cables, transformers, reactors, or capacitor banks; the latter category includes current interruption under faulted or unfaulted conditions.

The results from the study of switching transients are useful to: (i) determine voltage stresses on equipment; (ii) select arrester characteristics; (iic) calculate the transient recovery voltage across circuit breakers; (iv) analyse the effectiveness of mitigating devices (e.g. pre-insertion resistors or inductors).

The level of detail required in the model varies with the study. For example, a line may be represented by a PI-section equivalent in some line energization studies; in other situations,

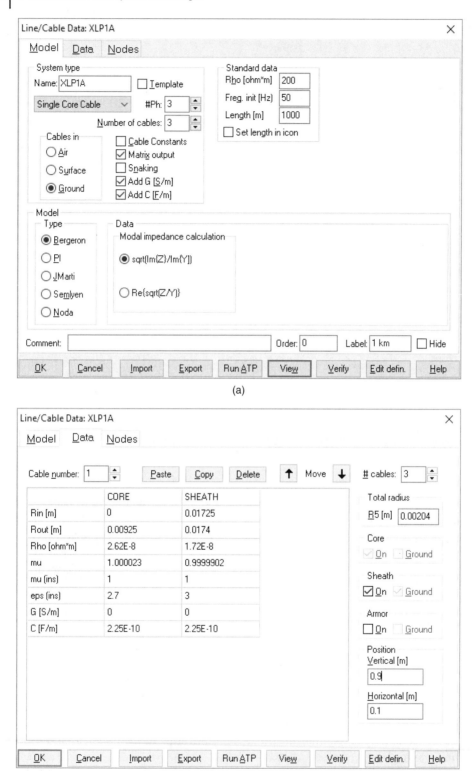

Figure 6.11 Case Study 2: Distribution cable data input. (a) Cable model selection, (b) Cable data input.

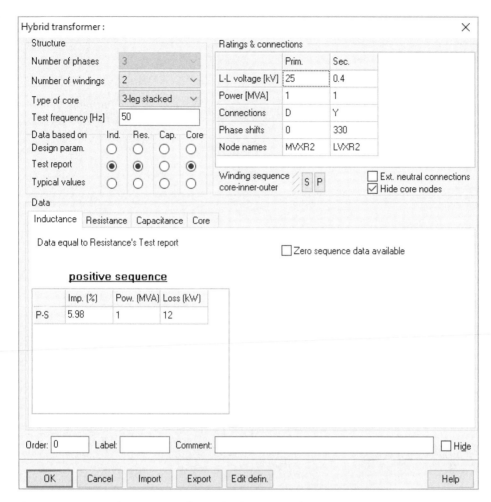

Figure 6.12 Case Study 2: Data input for the distribution transformers (Hybrid model).

a distributed-parameter model with frequency dependence may be necessary. In addition, the results are highly sensitive to the value of certain parameters; for example, the maximum overvoltage for a line energization depends on the exact point on the wave at which the switch contacts close. Thus, a number of runs for the same system have to be made with the time of energization being different in each run either in a predictable manner (i.e. for determining the peak overvoltage) or statistically (for obtaining an overvoltage probability distribution).

This section presents two case studies of slow-front overvoltages caused by switching operations. The first case study analyses the overvoltages that can be caused during the energization of an overhead transmission line; the study is carried out taking into account the random nature of the switching overvoltages. The second case study deals with capacitor bank switching, and is aimed at illustrating the overvoltages that can occur during capacitor bank energization and those caused by restrikes during capacitor bank de-energization.

6.5.3.1   Case Study 3: Transmission Line Energization

The energization of lines and cables by closing the circuit breaker may cause significant transient overvoltages. It is important to distinguish between energization and reclosing. In the

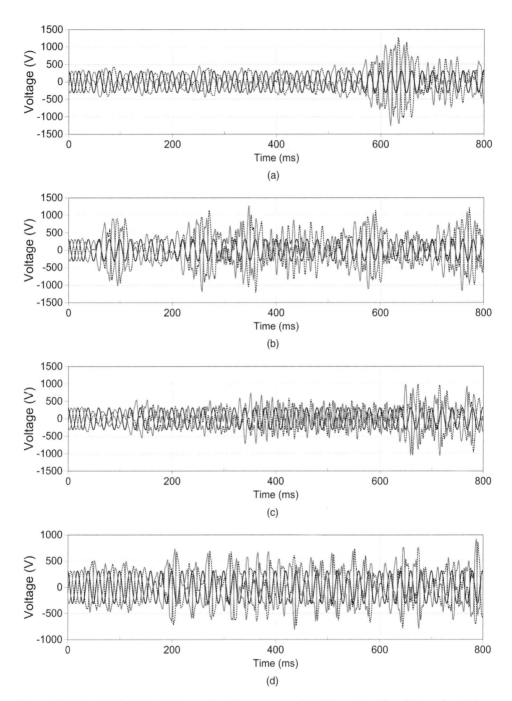

Figure 6.13 Case Study 2: Simulation results with two open poles. (a) Length = 2 km, (b) Length = 1.5 km, (c) Length = 1 km, (d) Length = 0.5 km.

former case, there is no trapped charge. In case of reclosing, the line/cable may have been left with a trapped charge after the initial breaker opening. Under such circumstances, the transient overvoltages can reach values of up to 4.0 pu. The aim of these studies is to determine the overvoltage stresses and choose the insulation strength in order to achieve an outage rate criterion [7, 62]. The representation of all components included in the system zone model (i.e. source, transformers, overhead lines, insulated cables, circuit breakers) must be adequate for slow-front transient simulations. A variety of line and cable models can be used in these studies, including PI-circuit and distributed-parameter models, as explained in Chapter 2 and Section 6.3.

Figure 6.14 shows the tower design of the test line, a 50 Hz, 300 km, 400 kV transmission line. Characteristics of phase conductors and shield wires are provided in Table 6.4.

The transmission line is represented by means of a distributed-parameter model. Although the most rigorous model should be based on a frequency-dependent parameter model, this case study compares results from both constant and frequency-dependent parameter models. The study discusses simulation results from both transposed and untransposed line models.

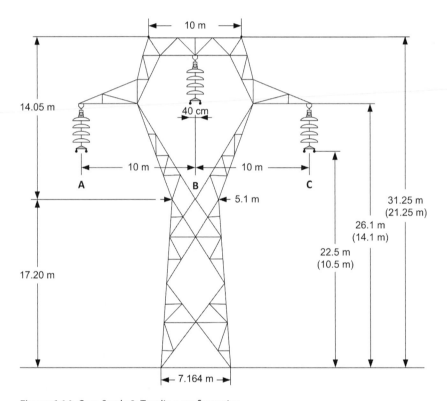

Figure 6.14 Case Study 3: Test line configuration.

Table 6.4 Case Study 3: Characteristics of wires and conductors.

	Conductor type	Diameter (mm)	DC Resistance (Ω/km)
Phase conductors	CURLEW	31.63	0.05501
Shield wires	94S	12.60	0.64200

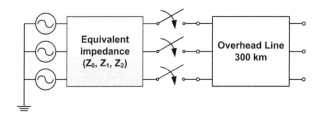

Figure 6.15 Case Study 3: Diagram of the test system.

To perform the calculations, the source side will be represented by a network equivalent with a short-circuit capacity of about 10 000 MVA (see Figure 6.15), and considering the following parameters of the network equivalent:

$R_1 = 1.3287\,\Omega$ $\qquad\qquad$ $X_1 = 15.9447\,\Omega$ $\qquad\qquad$ $L_1 = 50.7537\,\text{mH}$

$R_0 = 1.5331\,\Omega$ $\qquad\qquad$ $X_0 = 12.2652\,\Omega$ $\qquad\qquad$ $L_0 = 39.0413\,\text{mH}.$

The full study will consist of the following steps:

1) The various line models to be used in this case study will be implemented in ATPDraw.
2) The transient response of the line models will be analysed taking into account that it is crucial to evaluate the overvoltages that can occur during both closing and reclosing operations.
3) A statistical study will be carried out to obtain the distribution of overvoltages that can occur with closing and reclosing operations. A third option, reclosing with limiting resistors, will be also analysed. This part of the study will also determine the conditions under which the energization of the test line will cause the maximum overvoltages, estimate the maximum overvoltages, as well as the number of runs required to obtain an accurate enough distribution of overvoltages, by using both systematic and statistical switching.

ATP implementation
The model of the test line considering the various approaches mentioned above will be implemented in several steps. Firstly, the test-line configuration is implemented in ATPDraw using the LCC option. Then, each line model is created and used to develop the entire test system. Remember that the goal is to obtain the statistical distribution of switching overvoltages; in this part of the study only the implementation of the line models is covered. The transient analysis covered in the next part will require system models that could detail the line behaviour during closing and reclosing operations. The ATP capabilities for statistical studies will be analysed afterwards.

Figure 6.16 shows the menus of the line model implemented in ATPDraw using the LCC option. From the icon of the transmission line (i.e. LCC icon), the user can open the first window in which the following information has to be specified: name of the line model to be implemented (in the figure LINA1); the type of component to be modelled: line or cable; standard data (i.e. ground resistivity, initial frequency, length); units (i.e. English or metric); data for line model (i.e. phase transpositions, conductor bundle, ground wires, skin effect, real transformation matrix); type of model (i.e. Bergeron, PI model, JMarti, Semlyen, Noda); data for fitting (i.e. number of decades, points per decade, frequency matrix, frequency for steady-state calculations); see Figure 6.16a. In the second menu (i.e. Line Data Menu), the user has to specify geometrical and material data for overhead line conductors; see Figure 6.16b. Other menu windows, not shown in the figure, can be used to display, zoom, and copy to the clipboard the cross section of the specified line, name nodes, assign conductor number to nodes, establish a LINE MODEL FREQUENCY SCAN or a POWER FREQUENCY CALCULATION data case by means of the Verify option.

Figure 6.16 Case Study 3: ATPDraw implementation of the transmission line model. (a) Line model selection, (b) Line data input.

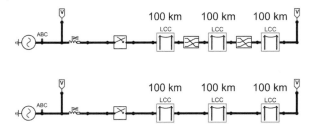

Figure 6.17 Case Study 3: ATPDraw implementation of the test line models – Closing operation.

Transient response during line energization

The analysis of the transient response of the test line can be useful to understand the subsequent statistical study. Several files have been created to simulate the two-line models implemented in ATPDraw. Since one of the goals is to compare the transient response of the transposed and untransposed line models, two test systems have been created with each line model.

Figure 6.17 shows the diagram of the two test systems for closing operations. Figure 6.18 displays some simulation results obtained when energizing the test line. The simulation was carried out by assuming that the three breaker poles close simultaneously. Note that the peak voltages at the open terminal can exceed 2 pu and the transient response is different for each line phase. This latter result can be explained by remembering the line geometry, which can cause some differences between voltages at the central and outer phases, and the different phase angle in the voltage of the three phases at the time the breaker closes. From the comparison of these results, it is evident that the highest overvoltages at the open end of the line are obtained with the transposed line, and although the peak values obtained with the two models are very similar,

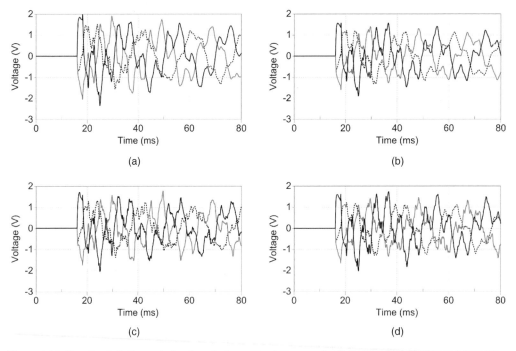

Figure 6.18 Case Study 3: Transmission line closing – Simulation results: Open terminal voltages. (a) JMarti line model: Transposed line, (b) JMarti line model: Untransposed line, (c) Bergeron line model: Transposed line, (d) Bergeron line model: Untransposed line.

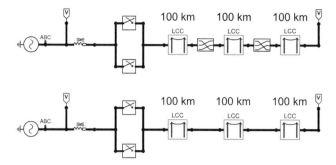

Figure 6.19 Case Study 3: ATPDraw implementation of the test line models – Reclosing operation.

the highest values correspond to the frequency-dependent model, which also gives a smoother transient response.

A reclosing operation is the operation that takes place after the line has been disconnected from the system. After the breaker opening, a trapped charge remains in all phases; reclosing into this trapped charge can cause overvoltages higher than those obtained with a closing operation such as that detailed above. Figure 6.19 shows the diagram of the two test systems for reclosing operations. Note that there are two breakers at the source side: the first one is needed to disconnect the line from the system; the second one recloses the line with trapped charge. Remember that the diagram is the same for both the JMarti and the Bergeron models. Figure 6.20 shows some simulation results obtained when reclosing the test line. The simulation was again carried out by assuming that the three breaker poles close simultaneously. Note that now the peak voltages at the open terminal can exceed 3 pu.

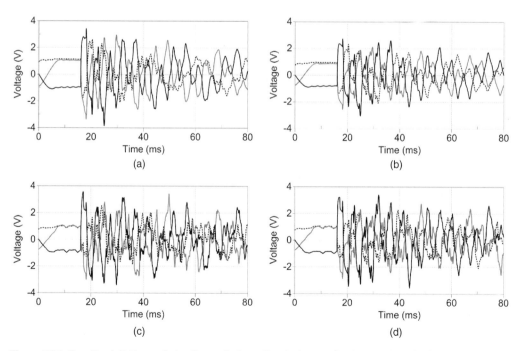

Figure 6.20 Case Study 3: Transmission line reclosing – Simulation results: Open terminal voltages. (a) JMarti line model: Transposed line, (b) JMarti line model: Untransposed line, (c) Bergeron line model: Transposed line, (d) Bergeron line model: Untransposed line.

From the analysis and comparison of these results, the main conclusions are similar to those obtained with a closing operation: the highest overvoltages at the open end are obtained with the transposed line; the peak values are very similar although the highest values are obtained with the frequency-dependent model, which causes again a smoother transient response.

One should be careful with the conclusions derived from the above study. Although it seems that the highest overvoltages can occur with a transposed line represented by the frequency-dependent parameter mode, the fact is that many more runs should be carried out (e.g. by changing the closing time of the breaker poles) in order to be sure that these conclusions are correct. On the other hand, it is important to keep in mind that the three poles of a breaker do not close simultaneously, as assumed in the statistical study presented below.

Reclosing overvoltages can be reduced by pre-inserting resistors. First, the auxiliary contacts of the pre-insertion resistors close; after a time interval of about one half-cycle of the power frequency the main contacts close and pre-insertion resistors are short-circuited. The three scenarios (energizing, reclosing, pre-insertion of resistors) are analysed below.

Statistical study

The goal now is to obtain a statistical distribution of overvoltages at the open terminal of the line using a transposed line represented by the frequency-dependent parameter model. The first step is aimed at estimating the conditions with which the highest overvoltage is obtained.

Maximum overvoltage during energization: In this study, it is assumed that the maximum overvoltage is produced when energizing an unloaded line and only phase-to-ground overvoltages are of concern. Since a three-phase closing operation produces three phase-to-ground overvoltages. Two methods can be used to obtain data from each operation [2]:

1) *phase-case method*, the highest peak value of the overvoltage of each phase-to-ground is taken into account;
2) *case-peak method*, only the highest peak value of the three overvoltages is collected.

Using the first method, each operation contributes three values, while only one value is collected with the second approach. This case study is based on the second method.

Figure 6.21 depicts the ATPDraw implementation of the circuit used to obtain the peak value of voltages obtained at each phase of the terminal node. The case study uses a multiple run option, with 20 runs that are specified in the Settings, and a MODELS-based section that measures the three peak voltages and determines the maximum one. The phase-to-ground overvoltage magnitudes at various closing times are calculated over the duration of one 50 Hz cycle, assuming a cosine wave for phase A of the voltage source and the corresponding displaced voltage waves for phases B and C.

The resulting overvoltage distribution is summarized in Figure 6.22. Since the model is linear, the simulations have been carried out with a peak value of the source voltage of 1 V, so the results can be assumed per unit. It is also evident that the maximum overvoltage occurs when the breaker is closed before the voltage wave of phase *C* is at the peak. It can also be seen that the maximum overvoltages (of value 2.34 pu) are produced when the poles close at 23 and 33 ms. Based on this result, the circuit breaker aiming time is chosen as 33 ms for further analysis. The study was repeated to estimate the aiming point with reclosing operations. The new aiming point occurs at 35 ms and the peak voltage reaches a value of 3.84 pu; see Figure 6.22.

Systematic switching: In order to find out the maximum overvoltage by using a systematic switch, the energizations were performed over the entire range of a cycle, from 0° to 360°. With $\Delta t = 0.1$ ms, 20 steps in phase A, 10 steps in phase B and 5 steps in phase C are chosen. This requires a total energization of 1000 runs. The t_{mid} is chosen as 33 ms for all the phases.

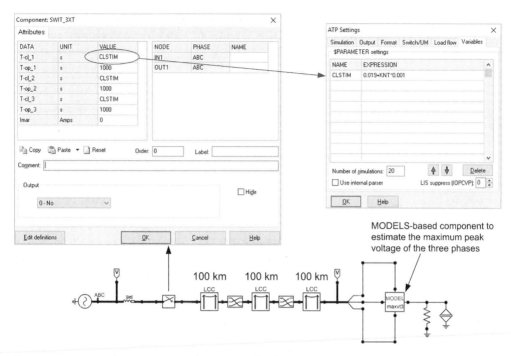

Figure 6.21 Case Study 3: ATPDraw implementation for obtaining the maximum overvoltage at the open end of the line caused by a closing operation.

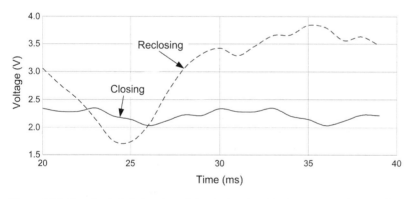

Figure 6.22 Case Study 3: Estimation of the aiming time Maximum overvoltage at the open line end.

The circuit implemented in ATPDraw is shown in Figure 6.23. Note that the three poles are assumed independent from each other. Although the number of time increments (NSTEP) is different for the three phases (i.e. 20, 5, and 5, respectively), the selected size of time increment (INCT), specified in seconds, is the same for the three phases.

The distribution of maximum overvoltages, the cumulative frequency and the cumulative density function of overvoltages (i.e. the probability that the peak voltage in any phase at the open end of the line can exceed a given value) are listed in the output file (i.e. the .LIS file). Figure 6.24 displays the cumulative density function. From the figure, one can observe that all peak voltages exceed 1.40 pu of the rated voltage, and the maximum peaks can reach values as high as 2.70 pu. Other useful information are the mean value and the standard deviation of the overvoltages. From this study, these values are respectively 2.11 and 0.211 pu. The figure compares results obtained with 1000 ($= 20 \times 10 \times 5$) and 500 ($= 20 \times 5 \times 5$). A summary of these results is provided below:

	Grouped data (500 runs)	Ungrouped data (500 runs)	Grouped data (1000 runs)	Ungrouped data (1000 runs)
Mean	2.1789	2.1787	2.1060	2.1059
Variance	0.0267	0.0270	0.0445	0.0446
Standard deviation	0.1635	0.1644	0.2109	0.2112

Statistic switching: The energizations were again performed over the entire range of a cycle, assuming that the three poles are independent. The parameters required to perform the simulations are the point of circuit breaker closing (aiming point of 33 ms) and the standard deviation, σ. To cover the entire range of a 50 Hz cycle, the standard deviation for a uniformly distributed switch must be $20/(2\sqrt{3}) = 5.77$ ms. If a normal distribution is truncated at -4σ and $+4\sigma$, more than 99.99% of the switching times will range within a cycle centred at the aiming instant; therefore, for a 50 Hz cycle, the selected standard deviation is $20/8 = 2.5$ ms.

Figure 6.25 shows the statistical distribution of overvoltages obtained with the two distributions when the total number of energizations is 1000 in both cases. It can be seen that although there is a higher probability of capturing the highest possible overvoltage with a normal distribution and 1000 runs, the differences are not too large.

In order to estimate the number of runs necessary to evaluate the maximum overvoltage, the study was repeated with 200 runs. Interestingly, the new results provide an accurate enough estimation using either distribution.

A further study was aimed at analysing the effect that the standard deviation value can have on a normally distributed switch. The values used in the study for σ ranged from 1 to 2.5 ms. From the results, one can deduce that a larger standard deviations helps to obtain a broader spectrum of overvoltage values and therefore the highest overvoltages.

The main conclusions of this study can be summarized as follows. If the circuit breaker poles are assumed to be independently distributed, the maximum overvoltages can be reached by selecting either a uniform or a normal distribution of the closing times; however, it is recommended a standard deviation is chosen that guarantees that the entire range of closing times within a cycle will be covered. Differences will not be significant with any number of runs above 200.

It is interesting to observe that both systematic and statistic switches, as used in this example, have captured the same maximum overvoltage when using the same number of runs.

Probability distribution of switching overvoltages

It is well known that the overvoltages that can occur when energizing an overhead transmission line may be greater than 2 pu. However, the most onerous scenario corresponds to a reclosing operation; that is, a line energization with trapped charge. The presence of the trapped charge can be due to a phase-to-ground fault; as a consequence of the fault, the breakers at both line ends open to clear the fault; and a charge of about 1 pu is trapped on the unfaulted phases. After reclosing the breakers at one end, the magnitude of the resulting voltages at the open end may be above 3 pu. Reclosing overvoltages can be reduced by preinserting resistors: the auxiliary contacts of the pre-insertion resistors initially close; after a time interval of about one half-cycle of the power frequency, the main contacts close.

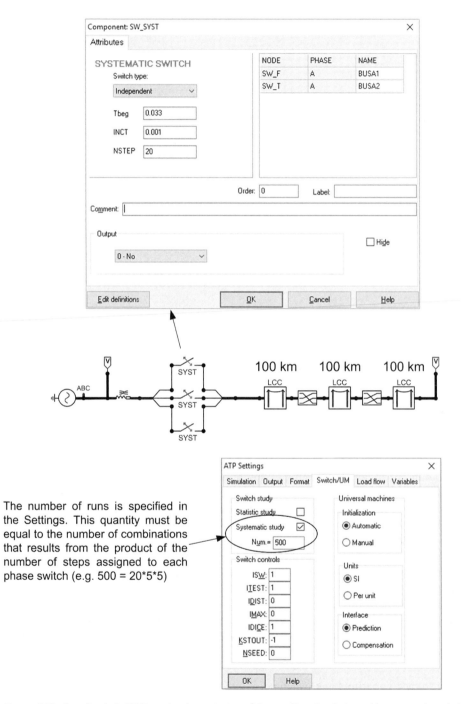

Figure 6.23 Case Study 3: ATPDraw implementation of the test line simulation with systematic switches.

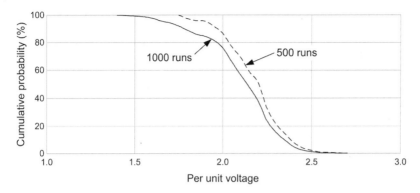

Figure 6.24 Case Study 3: Peak voltages with a systematic switch.

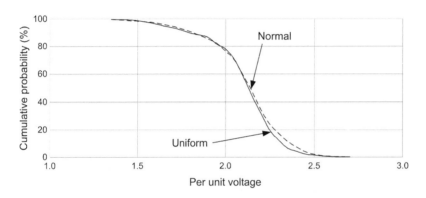

Figure 6.25 Case Study 3: Peak voltages with uniform and normal distributions – 1000 runs.

The three scenarios (energizing, reclosing, pre-insertion of resistors) are analysed with the following common features: (i) it is assumed that only phase-to-ground overvoltages are of concern, and only the highest peak value of the three overvoltages is collected from each run; (ii) the energizations are performed over the entire range of a cycle and assume that the three poles are independent; (iii) the closing time of each pole is randomly varied according to a normal probability distribution, with a standard deviation of 2.5 ms; (iv) the aiming time is chosen following the method discussed above for closing and reclosing operations.

Figure 6.26 shows the diagram of the circuits implemented in ATPDraw for simulating the three scenarios. Reclosing is analysed by assuming that a 1 pu voltage is trapped on each phase, and two different values of pre-insertion resistances, 200 and 400 Ω, are used. Table 6.5 summarizes the main results of each scenario. As expected, the lowest overvoltages are obtained when the line energization is carried out with pre-insertion resistors and the largest resistance value used in simulations. This can be better seen from the cumulative probabilities shown in Figure 6.27.

6.5.3.2 Case Study 4: Capacitor Bank Switching

Figure 6.28 shows the diagram of the test system that will be used to illustrate some of the overvoltages that can be caused by capacitor switching. The configuration of this test system is the same that was used to analyse ferroresonance, although some component parameters have been changed. The main parameters are now as follows:

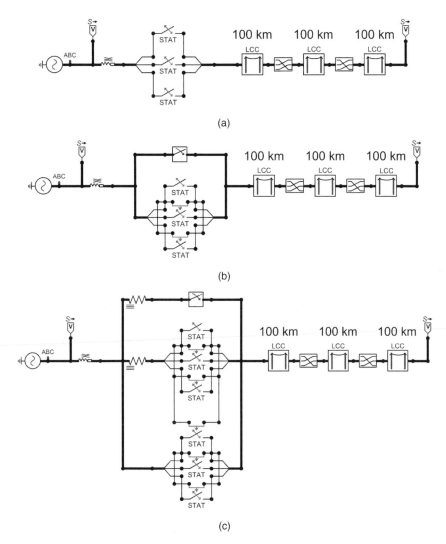

Figure 6.26 Case Study 3: ATPDraw implementation for statistical simulations. (a) Closing operation. (b) Reclosing operation. (c) Reclosing operation with presinsertion resistors.

Table 6.5 Case Study 3: Statistical distribution of phase-to-ground voltages.

Case	Mean value (pu)	Standard deviation (pu)
Energizing	2.045	0.204
Reclosing	2.966	0.637
Pre-insertion resistors – 200 Ω	1.712	0.327
Pre-insertion resistors – 400 Ω	1.568	0.244

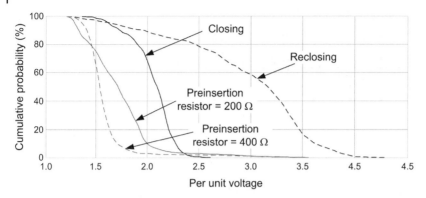

Figure 6.27 Case Study 3: Transmission line energization - Peak voltage distribution.

1) HV network: 110 kV, 1500 MVA 50 Hz, $X/R = 10$, $X_0/X_1 = 1.1$.
2) Substation transformer: Triplex core, 110/25 kV, 12 MVA, 9%, Yd11, grounded through a zig-zag reactance with 75 Ω per phase.
   - No load test (positive sequence, MV side): $V_0 = 100\%$; $I_0 = 0.296\%$; $W_0 = 18.112$ kW.
   - Short-circuit test (positive sequence, HV side): $V_{sh} = 9\%$; $I_{sh} = 83.34\%$; $W_{sh} = 125.6$ kW.
3) Cable: Al RHV, $3 \times (1 \times 240\,\text{mm}^2)$, 18/30 kV (see Figure 6.7).
4) Distribution transformer: Three-legged stacked core, 25/0.4 kV, 2 MVA, 6%, Dyn11.
   - Short-circuit test (positive sequence, MV side): $V_{sh} = 6\%$; $I_{sh} = 100\%$; $W_{sh} = 12$ kW.
   - No load test (homopolar sequence, LV side): $V_h = 100\%$; $I_h = 0.5\%$; $W_h = 1.8$ kW.
5) Capacitor banks: They are installed at the MV-side of the substation transformer and the terminals of Load 3.

Saturation curves of distribution transformers are similar to those shown in Figure 6.8. Since the rated power of the transformers is now 2 MVA, the only change to be introduced in that figure affects the scale of currents, which have been doubled. The load values are those provided in Table 6.6. Note that only a capacitor bank has been placed at LV nodes, namely in parallel with the Node 3 load.

The goal of this case study is to estimate the impact that the connection and disconnection of the capacitor bank located at the MV side of the substation transformer can have on the MV and LV nodes of the test system; mainly at the LV node at which a capacitor bank is already installed.

Table 6.6 Case Study 4: Rated values of the test system loads.

Load Node	Rated voltage (V)	Rated power (MVA)	Power factor
1	400	1.4	0.60
2	400	1.6	0.80
3	400	1.2	0.60
4	400	1.5	0.90
5	400	1.8	0.85

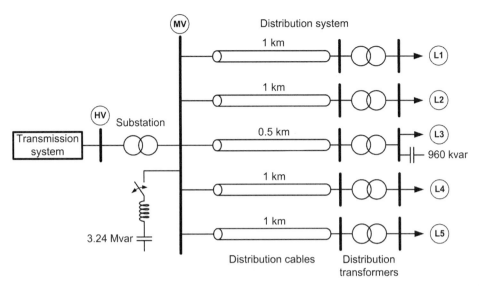

Figure 6.28 Case Study 4: Diagram of the test system.

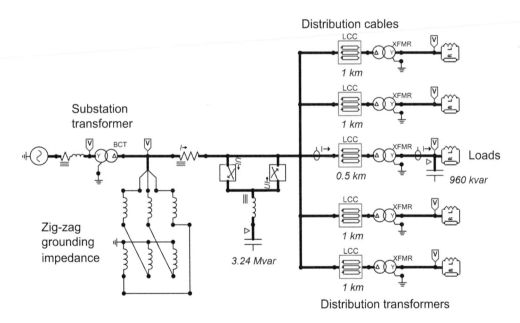

Figure 6.29 Case Study 4: ATPDraw implementation.

The circuit implemented in ATPDraw for simulating this test system is shown in Figure 6.29. This implementation has been based on the following ATP capabilities: (i) all system transformers (i.e. the substation transformer and distribution transformers) are represented with the Hybrid Transformer model [57]; (ii) The distribution cables are represented by a constant distributed-parameter model using the LCC option (i.e. the CABLE CONSTANTS routine); (iii) loads are again modelled as parallel *RL* branches by means of a custom-made model.

The first case analysed with this example is the impact that the energization of the capacitor bank to be installed at the MV side of the substation transformer can have on the test system

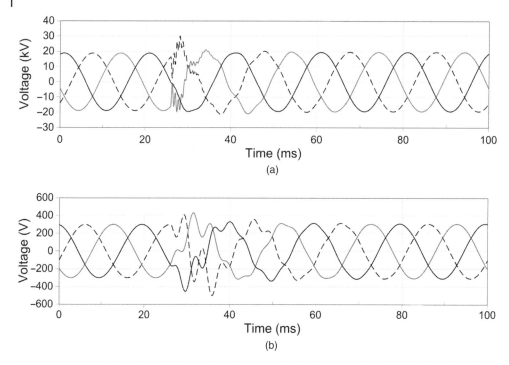

Figure 6.30 Case Study 4: Simulation results during capacitor bank energization. (a) MV-side substation transformer voltages, (b) Voltages at load L3 terminals.

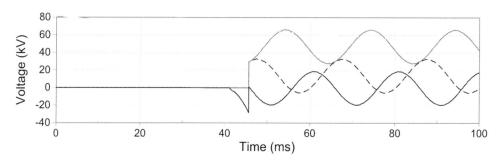

Figure 6.31 Case Study 4: Transient recovery voltages during capacitor bank de-energization.

voltages. A voltage magnification can occur at the lowest voltage node where a capacitor bank is already installed in parallel with the load. Figure 6.30 shows some simulation results, namely the voltages caused at nodes MV, L1, and L3 after switching the 25 kV/3.24 Mvar shunt capacitor bank. It is assumed that the three switch poles close simultaneously at the moment when the phase A voltage is at a peak; that is at $t = 40$ ms. The oscillograms show that the peak overvoltage is comparatively higher at the LV node L3, where a compensating capacitor has been installed.

More dangerous overvoltages can be caused if the same capacitor bank is de-energized and some restrike did occur during the opening operation. Consider that the capacitor bank is connected to the system and it is de-energized. The voltages developed across the terminals of each switch pole would be those shown in Figure 6.31. One can observe that the voltage across one switch pole can be more than 3 times the nominal voltage. This occurs in the phase *C* pole.

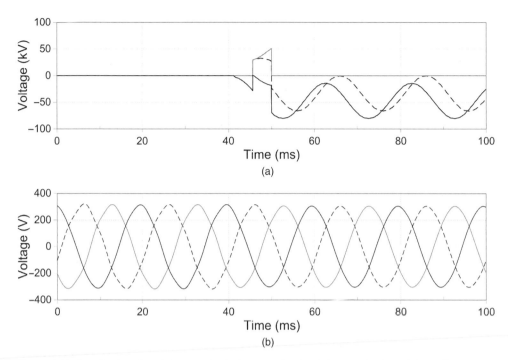

Figure 6.32 Case Study 4: Simulation results during capacitor bank de-energization with restrike in one switch pole. (a) TRVs between switch terminals, (b) Voltages at load L3 terminals.

Assume that a restrike occurs in phase C when the recovery voltage reached exceeds a value of 50 kV. In the test under study, this occurs at about 50 ms. Figure 6.32 shows the test system response after the phase-C pole restrikes. It is evident from these results that the restrike of this single pole will not have any impact on the system variables. The variations that can be observed in the plots are basically due to the fact that the capacitor bank has been separated from the system and this can affect the substation transformer currents (i.e. the voltage and the current at the secondary MV side are no longer in phase, and the current is larger than when the bank was connected to the system) and to a lesser extent to voltages, even those measured at LV levels.

However, immediately after the restrike of the phase C pole, the voltage across the phase A pole jumps to above 50 kV. Since, from a mechanical point of view, the withstand voltage across this phase has hardly increased, it can be assumed that this switch pole will also restrike. Figure 6.33 shows the new results: the system response after the restrike of the second pole exhibits higher peak voltages that in some cases (e.g. at the LV node to which a capacitor bank is installed) exceeds 2.5 times the nominal voltage. Although the peak voltages at other nodes can exceed twice the nominal voltage (e.g. at the MV terminals of the substation transformer and other LV node loads), the highest peak voltage develops at the LV node to which a capacitor bank (i.e. node L3 in Figure 6.28). This is known as voltage magnification [63].

Actually, the transients caused by the two restrikes could have been very different if the instants at which the switch poles did restrike were different.

6.5.4 Fast-Front Overvoltages

The main cause of fast-front transients in power systems is lightning. This section presents the application of ATP to lightning overvoltage calculations. The case study uses the modelling

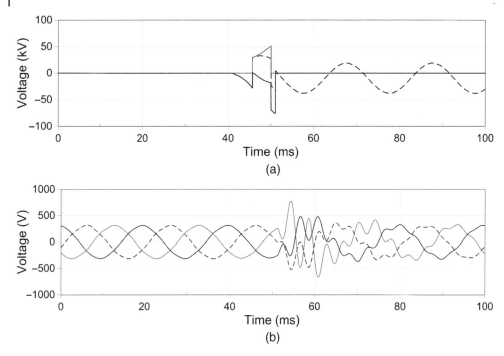

Figure 6.33 Case Study 4: Simulation results during capacitor bank de-energization with restrike in two switch poles. (a) TRVs between switch terminals, (b) Voltages at load L3 terminals.

guidelines recommended in Section 6.3 to implement the model of an overhead transmission line; the goal is to calculate lightning overvoltages and estimate the lightning performance of the test line taking into account the random nature of lightning. This section includes a parametric study of lightning-caused overvoltages.

6.5.4.1 Case Study 5: Lightning Performance of an Overhead Transmission Line

Figure 6.34 shows the tower design of the test transmission line with two conductors per phase and one shield wire, whose characteristics are presented in Table 6.7.

Data relevant for this example follows:

- Power system:
  Frequency = 50 Hz
  Rated voltage = 400 kV
  Grounding = Low impedance system, Earth fault factor = 1.4
  Duration of temporary overvoltage = one second
- Line:
  Span length = 390 m
  Strike distance of insulator strings = 3.066 m
  Conductor configuration = 2 conductor per phase
  Low-current low-frequency grounding resistance = 50 Ω
  Soil resistivity = 200 Ω.m.

The lightning performance depends on the line design (e.g. shielding, grounding) and the atmospheric activity. An unacceptable lightning flashover rate can be due either to a very high backflashover rate (BFOR), a very high shielding failure flashover rate (SFFOR), or both. One reason for a very high BFOR is a poor grounding, while the main reason for a very high SSFOR

Table 6.7 Case Study 5: Conductors characteristics.

	Conductor type	Diameter (mm)	DC Resistance (Ω/km)
Phase conductors	Cardinal	30.35	0.0586
Shield wire	7 N8	9.78	1.4625

is a poor shielding. An additional reason that can affect both rates is too short a striking distance of insulator strings. The grounding resistance specified above is not small, but it is not as high as to significantly increase the lightning flashover rate of the line. Soil ionization may decrease the effective grounding resistance value, so if the line model includes a high-current grounding model, the lightning performance of the test line is not so affected by the average grounding resistance value assumed for this line. However, the strike distance of insulator strings is rather

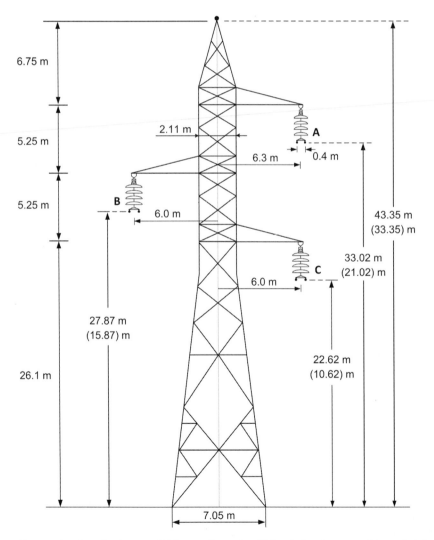

Figure 6.34 Case Study 5 : 400 kV line configuration (Values in brackets are midspan heights).

short for the rated voltage of this line, so it certainly affects the line behaviour. Finally, it is advisable to analyse the shielding provided to this line by the single sky wire.

This study has been organized as follows. Initially, the steps to implement a line model adequate for calculating lightning overvoltages are presented; this part of the study includes a summary of the modelling guidelines to be used in this type of study, the application of an incident model (i.e. the electrogeometric model) that will provide some information about the lightning performance of the test line, and the implementation in ATP of the line model. The second part presents a parametric study aimed at assessing the lightning overvoltages that can result as a function of some line parameters (e.g. peak current magnitude of the lightning stroke, grounding resistance). This will be complemented by the statistical study covered in the third part, whose main goal is to estimate the lightning performance of the test line. The last part is dedicated to analysing the performance of the line when protected by surge arresters; this part includes the selection of surge arresters, the ATP implementation of the arrester model, and an analysis of the energy discharged by arresters.

Calculation of lightning overvoltages

Modelling guidelines: Models used to represent the different parts of a transmission line are detailed in the following paragraphs. See Section 6.3 and Chapter 2 for more details on the guidelines.

1) The line (shield wires and phase conductors) is modelled by means of several spans at each side of the point of impact. Each span is represented as a multi-phase untransposed frequency-dependent and distributed-parameter line section.
2) The line termination at each side of the above model, needed to avoid reflections that could affect the simulated overvoltages around the point of impact, is represented by means of a long enough section, whose parameters are calculated as for the line spans.
3) A tower is represented as an ideal single-conductor distributed-parameter line. The surge impedance of the tower is calculated according to CIGRE recommendations [33].
4) The representation of insulator strings (when used) relies on the application of the leader progression model [33, 35, 64]. The leader propagation is deduced from the following equation:

$$\frac{d\ell}{dt} = kv(t)\left(\frac{v(t)}{g-\ell} - E_0\right) \tag{6.3}$$

where $v(t)$ is the voltage across the gap, g is the gap length, ℓ is the leader length, E_0 is the critical leader inception gradient, and k is a leader coefficient.

5) The grounding impedance is represented as a non-linear resistance whose value is approximated by the following expression [2, 3]:

$$R_T = \frac{R_0}{\sqrt{1+I/I_g}} \qquad \left(I_g = \frac{E_0\rho}{2\pi R_0^2}\right) \tag{6.4}$$

where R_0 is the grounding resistance at low current and low frequency, I is the stroke current through the resistance, and I_g is the limiting current to initiate sufficient soil ionization, being ρ the soil resistivity (Ω.m) and E_0 the soil ionization gradient (about $400\,kV/m$ [65]).

6) A lightning stroke is represented as an ideal current source with a concave waveform. In this work, return stroke currents are represented by means of the Heidler model [66]. A return stroke waveform is defined by the peak current magnitude, I_{100}, the rise time, t_f ($= 1.67$ ($t_{90} - t_{30}$)), and the tail time, t_h, that is the time interval between the start of the wave and the 50% of peak current on tail.

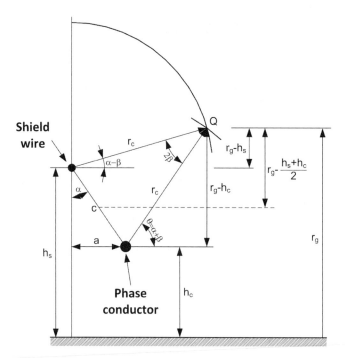

Figure 6.35 Application of the electrogeometric model.

Line shielding: The incidence model used in this example is the electrogeometric model proposed by Brown and Whitehead [67]. According to this model, the striking distances are calculated as follows:

$$r_c = 7.1I^{0.75} \qquad\qquad r_g = 6.4I^{0.75} \qquad\qquad (6.5)$$

where r_c, is the striking distance to both phase conductors and shield wires, r_g is the striking distance to earth, and I is the peak current magnitude of the lighting return stroke current.

Figure 6.35 shows the situation used to define the maximum shielding failure current, I_m; that is, the situation where all striking distances (from ground, from the shield wire, from the phase conductor for phase A) coincide at a single point Q.

The most exposed phase conductor in the line is the upper one (phase A) in Figure 6.34. Therefore, the values from the geometry of the test line to be used in these equations are $a = 6.3$ m, $h_s = 36.68$ m and $h_c = 26.35$ m. The resulting value for I_m is 26.3 kA. Although it is not a low value, this peak current is not so far from the values usually encountered for transmission lines whose rated voltage is similar to that of the test line. However, it will be important to consider this maximum value at the time of selecting the arrester energy withstand capability.

ATP implementation: Figure 6.36 displays the diagram implemented in ATPDraw to simulate the test line. The figure shows only the details at both sides of the point of impact. The line is represented by means of five spans at each side of the point of impact when the lightning stroke impacts on a shield wire (Figure 6.36a), and by four and a half spans at each side of the point of impact when the lightning stroke hits a phase conductor (Figure 6.36b). Note that the insulators are represented as open switches to record the voltage developed across each phase. The option used to monitor the voltages is the branch voltage probe available in ATDraw.

As mentioned above, the lightning stroke current has been represented by means of the Heidler model available in ATPDraw; each span is represented by a frequency-dependent distributed-parameter line section whose details are shown in Figure 6.37; the length of the line sections at both terminals of the line model is 30 km, long enough to avoid that waves reflected

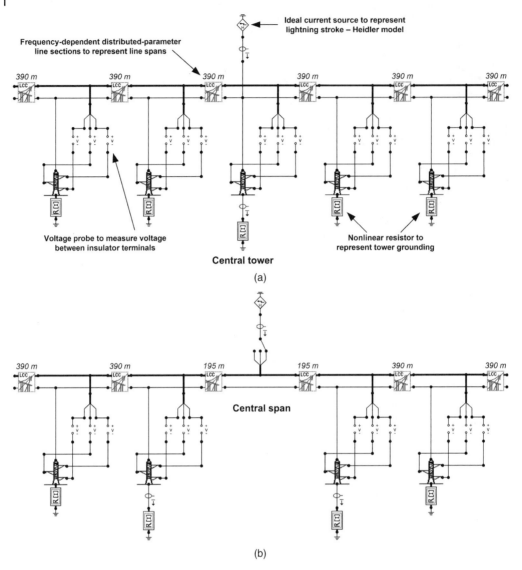

Figure 6.36 Case Study 5: ATPDraw implementation of the test line for lightning overvoltage calculation. (a) Lightning stroke impacts on a shield wire, (b) Lightning stroke impacts on a phase conductor.

at the open terminals could affect the voltages developed in the tower of concern. Note also that the lightning stroke current is represented by means of an ideal current source (i.e. with infinite parallel resistance). The model incorporates custom-made models to represent the tower and the grounding resistor. The tower model was implemented considering the model proposed by CIGRE [33]. As for the grounding, it is assumed nonlinear and represented by a nonlinear resistor whose resistance value is a function of the current; see Eq. (6.4).

Figures 6.38 and 6.39 show some simulation results. The parameters of the lightning stroke currents are as follows:

- Lightning stroke to a shield wire: $I_p = 90$ kA, $t_f = 1.2$ μs, $\tau = 50$ μs, $n = 5$
- Lightning stroke to a phase conductor: $I_p = 25$ kA, $t_f = 1.2$ μs, $\tau = 50$ μs, $n = 5$.

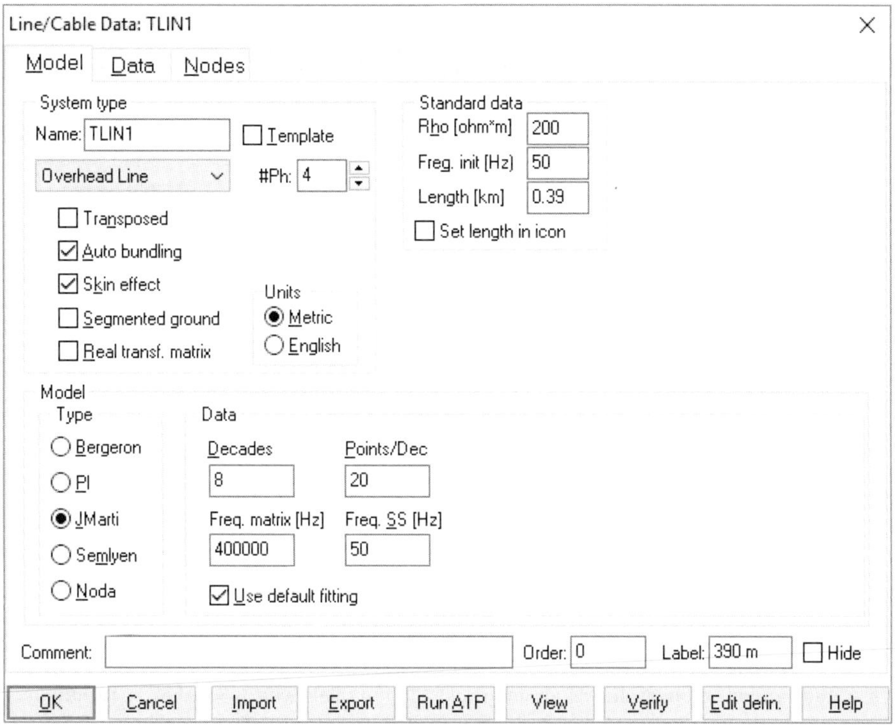

Line/Cable Data: TLIN1 — (a)

Model · Data · Nodes

System type
Name: TLIN1 · ☐ Template
Overhead Line ∨ · #Ph: 4
☐ Transposed
☑ Auto bundling
☑ Skin effect
☐ Segmented ground
☐ Real transf. matrix

Units
◉ Metric
○ English

Standard data
Rho [ohm*m]: 200
Freq. init [Hz]: 50
Length [km]: 0.39
☐ Set length in icon

Model
Type:
○ Bergeron
○ PI
◉ JMarti
○ Semlyen
○ Noda

Data:
Decades: 8 · Points/Dec: 20
Freq. matrix [Hz]: 400000 · Freq. SS [Hz]: 50
☑ Use default fitting

Comment: _____ · Order: 0 · Label: 390 m · ☐ Hide

OK · Cancel · Import · Export · Run ATP · View · Verify · Edit defin. · Help

(a)

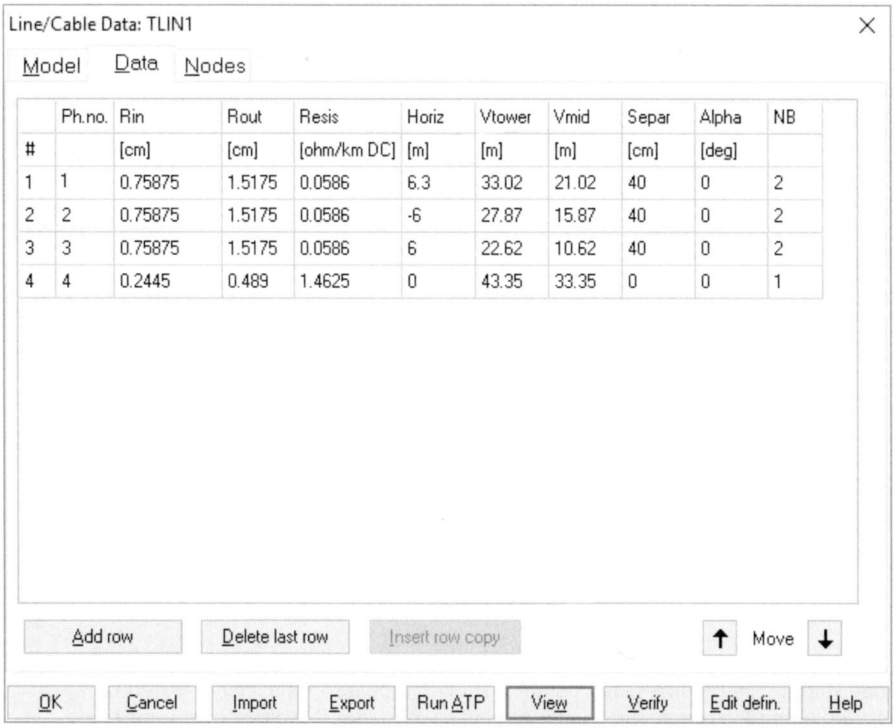

Line/Cable Data: TLIN1 — (b)

Model · Data · Nodes

#	Ph.no.	Rin [cm]	Rout [cm]	Resis [ohm/km DC]	Horiz [m]	Vtower [m]	Vmid [m]	Separ [cm]	Alpha [deg]	NB
1	1	0.75875	1.5175	0.0586	6.3	33.02	21.02	40	0	2
2	2	0.75875	1.5175	0.0586	-6	27.87	15.87	40	0	2
3	3	0.75875	1.5175	0.0586	6	22.62	10.62	40	0	2
4	4	0.2445	0.489	1.4625	0	43.35	33.35	0	0	1

Add row · Delete last row · Insert row copy · ↑ Move ↓

OK · Cancel · Import · Export · Run ATP · View · Verify · Edit defin. · Help

(b)

Figure 6.37 Case Study 5: LCC menus for the test line. (a) Selection of line model and some line specifications, (b) Conductor geometry and physical characteristics.

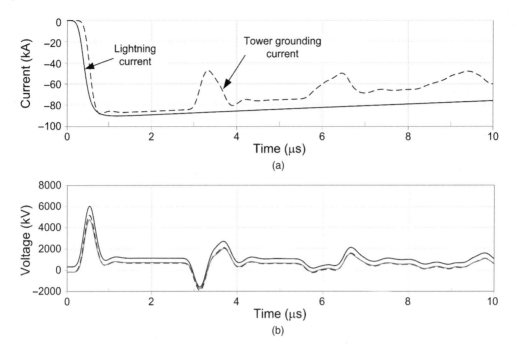

Figure 6.38 Case Study 5: Simulation results – Lightning stroke hits the shield wire (I_p = 90 kA). (a) Lightning stroke current and current through the grounding resistance, (b) Voltages across the line insulators.

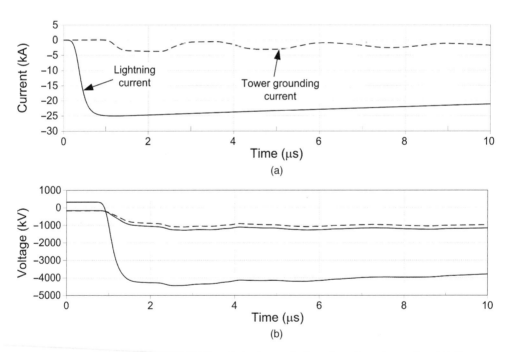

Figure 6.39 Case Study 5: Simulation results – Lightning stroke hits phase A conductor (I_p = 25 kA). (a) Lightning stroke current and current through the grounding resistance, (b) Voltages across the line insulators.

The stroke duration τ is the time at which the current amplitude has fallen to 37% of its peak value. In both cases, the angle of phase A of the voltage source that feeds the line is 0°.

The plots display the voltages obtained between terminals of insulators either at the tower to which the lightning stroke impacts (i.e. impact on a shield wire) or to the closer tower in case of impact on a phase conductor (i.e. at the midspan).

These results show that when the stroke impacts on the shield wire (at the top of a tower), most of the stroke current is derived to ground, as a consequence of the surge impedance values of the tower and the shield wires. As for the voltages developed between the terminals of the insulators installed in the impacted tower (and measured from the line conductors to the tower; see Figure 6.38), they begin with non-zero values due to the initial conditions in the line. For the operating conditions assumed in the simulated case, the highest peak value occurs between the terminals of phase insulators. In case of shielding failure (see Figure 6.39), the effect of the grounding model is negligible because the current that will flow through the grounding of the impacted tower is very small; however, the differences between the voltages developed between terminals of insulators is much larger, and the highest voltage is developed in the insulator string of the impacted phase (phase A in the simulated case).

Parametric study of lightning overvoltages

The goal is to analyse the relationship between the voltages developed across insulators and some parameters of the test line and lightning stroke. The study presented here aims to obtain the overvoltages as a function of the stroke current peak value. Any other line parameter remains constant. The study has been carried out following a procedure similar to that used with Case Study 1 (see Subsection 6.5.2.1). Figure 6.40 shows the results. According to these results, there is a linear relationship between the voltages developed across insulators and the peak current magnitude of the lightning stroke, although the grounding is assumed to have a nonlinear behaviour.

Lightning performance of the test line

Monte Carlo procedure: The goal is to estimate the lightning flashover rate of the test line. The implemented procedure is based on the Monte Carlo method, whose main features are summarized below [68]:

a) The calculation of random values includes the parameters of the lightning stroke (peak current, rise time, tail time, and location of the vertical leader channel), phase conductor voltages, the grounding resistance and the insulator strength.
b) The last step of a return stroke is determined by means of the electrogeometric model. The model used in this study is that proposed by Brown and Whitehead [67]. See above.
c) Overvoltage calculations are performed once the point of impact has been determined.
d) If a flashover occurs in an insulator string, the run is stopped and the flashover rate updated.
e) The convergence of the Monte Carlo method is checked by comparing the probability density function of all random variables to their theoretical functions; the procedure is stopped when they match within the specified error.

The following probability distributions are used:

- Lightning flashes are assumed to be of negative polarity, with a single stroke and with independently distributed parameters, being their statistical behaviour approximated by a log-normal distribution, whose probability density function is [7, 43]:

$$p(x) = \frac{1}{\sqrt{2\pi}x\sigma_{\ln x}} \exp\left[-\frac{1}{2}\left(\frac{\ln x - \ln x_m}{\sigma_{\ln x}}\right)^2\right] \tag{6.6}$$

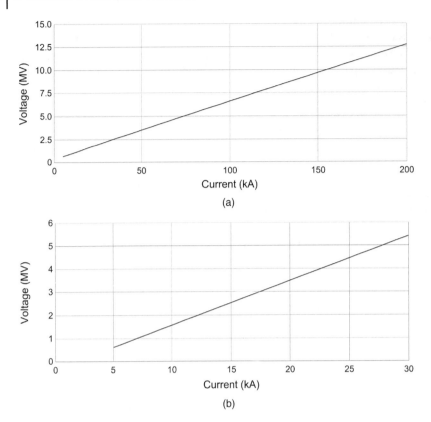

Figure 6.40 Case Study 5: Parametric study – Overvoltages versus peak current magnitude of the lightning stroke. (a) Overvoltages caused by a lightning stroke to the tower, (b) Overvoltages caused by a lightning stroke to a phase conductor.

Table 6.8 Case Study 5: Statistical parameters of negative polarity return strokes [43].

Parameter	Units	x	σ_{lnx}
I_{100}, kA	kA	34.0	0.740
t_f,	μs	2.0	0.494
t_h,	μs	77.5	0.577

where σ_{lnx} is the standard deviation of lnx, and x_m is the median value of x. Table 6.8 shows the values selected for each parameter.

- The power-frequency reference angle of phase conductors is uniformly distributed between 0 and 360°.
- The insulator models are based on the leader progression model (see Chapter 2 and Subsection 6.3.4). The parameters are determined according to a Weibull distribution. The mean value and the standard deviation of E_0 are 570 kV/m and 5% respectively. The value of the leader coefficient is $k = 1.3E\text{-}6\ m^2/(V^2\ s)$ [33].
- The grounding nonlinear resistance has a normal distribution with a mean value of 50 Ω and a standard deviation of 5 Ω. The value of the soil resistivity is 200 Ω.m.

- Before the application of the electrogeometric model, the stroke location is estimated by assuming a vertical path and a uniform ground distribution of the leader. Only flashovers across insulator strings are assumed.

ATP implementation: Figure 6.41 shows the flowchart of the procedure implemented in ATP [69]. The ATP capabilities that were used in the development and application of this procedure are listed below.

1) The multiple-run option is used to perform all the runs required by the Monte Carlo method. The procedure can be stopped when either the convergence is achieved or the maximum number of runs is reached.
2) A compiled routine was developed and embedded into the source code to obtain the values of the random variables/parameters that must be generated at every run according to the probability distribution function assumed for each one. The generation of random numbers is based on a linear congruential multiplicative generator with a loop of sequences greater than 2^{144} [70].
3) Voltages across insulators are continuously monitored; when the voltage stress in a single phase exceeds the strength, the flashover counter is increased and the simulation is stopped.
4) The Monte Carlo procedure is stopped when the maximum number of runs, specified by the user, is reached or when the probability density functions of all the random variables match their theoretical functions within the specified error. In this work, the resulting and theoretical distributions are compared at 10%, 30%, 50%, 70%, and 90% of the cumulative distribution functions. More than 10 000 runs were needed to match them within an error margin of 10%. For an error margin of 5% no less than 30 000 runs were needed.
5) A report showing the main input and output variables, as well as the progress of the flashover rate, can be printed with a frequency chosen by the user.

The compiled routines and modules developed for application of these routines can be classified into two groups: those developed for application of the Monte Carlo method (generation of random numbers, estimation of different random variables and parameters, application of the electrogeometric model, control of the convergence and information output) and those developed for representing the parts of a transmission line (three-phase source, insulator, and grounding resistance models) that are not available as built-in models.

Simulation results: After 20 000 runs, the flashover rates due to backflashovers and to shielding failures are respectively 1.740 and 0.085 per 100 km-year. Therefore, the total flashover rate is 1.825 per 100 km-year. These values were obtained with a ground flash density of $N_g = 1$ fl/km²-year. That is, if $N_g = 3$ fl/km²-year, the total rate will be about 5.475 flashovers per 100 km-year. These flashover rate values are high for a transmission line.

The procedure generates several files. One of them provides a short summary of the lightning performance with the following information:

```
Total number of runs                        = 20000.
Total number of strokes to ground           = 14575.
Total number of strokes to shield wire      = 5406.
Total number of strokes to conductors       = 19.
Total number of backflashovers              = 348.
Total number of shielding failure flashovers = 17.
Backflashover rate                          = 1.740/100 km/year
Shielding failure flashover rate            = 0.085/100 km/year
Flashover rate                              = 1.825/100 km/year
```

Figure 6.42 shows some simulation results that will help to understand the lightning performance of this line. A backflashover with a peak current magnitude of 80 kA or below is not

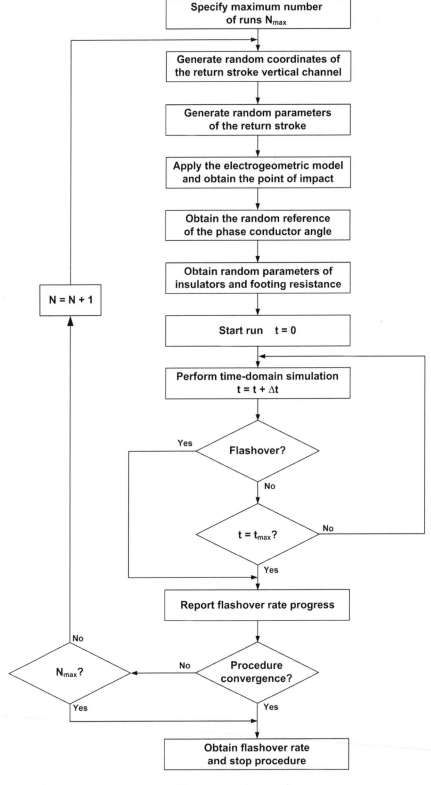

Figure 6.41 Case Study 5: Diagram of the Monte Carlo procedure.

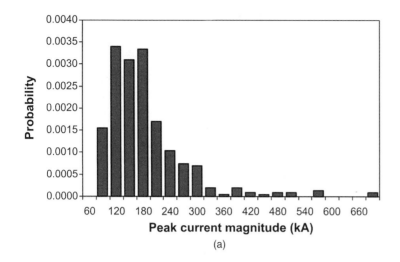

(a)

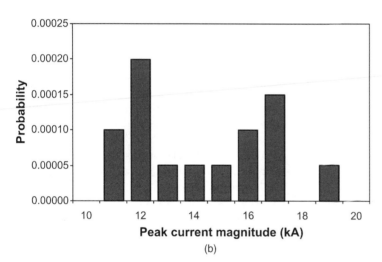

(b)

Figure 6.42 Case Study 5: Distribution of stroke currents that caused flashovers. (a) Strokes to shield wires, (b) Strokes to phase conductors.

unusual, but it can justify the backflashover rate of 1.740, which is a rather high value for a transmission line. This is mainly but not solely due to the short strike distance of the insulator strings. Figure 6.42a shows that a not-so-small percentage of backflashovers were caused by strokes with a peak current magnitude above 300 kA. These high values result from the theoretical statistical distribution of stroke parameters, but they will not frequently occur in reality, although peak current magnitudes higher than 800 kA have been measured [71].

One can observe that the highest peak current magnitude that causes shielding failure flashover is below the value deduced above (i.e. 26.3 kA). This is an indication that more runs should be performed to obtain a better accuracy with the Monte Carlo method. Take into account that only 19 strokes out of 20 000 impacted a phase conductor, and 17 of them caused flashover.

The grounding resistance specified above is not small, but it is not as high as to significantly increase the lightning flashover rate of the line. Soil ionization may significantly decrease the effective grounding resistance value, so if the line model includes a high-current grounding model, the lightning performance of the test line is not so affected by the average grounding resistance value assumed for this line.

Arrester protection

The installation of surge arresters can be justified in overhead lines with a poor lightning performance. By default, one can assume that the LFOR of any transmission line will become negligible if arresters are installed in parallel to all insulator strings. However, this scenario will be expensive. An alternative could be the installation of arresters in parallel to only one or two phases and, for long or very long transmission lines, only in those areas with a high keraunic level. There is an important aspect that has to be carefully analysed when arresters are installed to protect overhead lines: their energy withstand capability must be selected taking into account the actual energy stresses. This section presents the selection of surge arrester for the test line, the development and implementation of the arrester model and the estimation of the energy requirements of the selected arrester when it is installed in parallel to all insulator strings.

Arrester selection: The corresponding standard maximum system voltage specified for 400 kV in IEC is 420 kV. Since the earth-fault factor is 1.4, the following values are obtained, assuming that the duration of the TOV is one second:

- $MCOV - COV$ $COV = \frac{420}{\sqrt{3}} = 242.49$ (kV)
- TOV_C $TOV_C = 1.4 \cdot \frac{420}{\sqrt{3}} = 339.48$ (kV)
- TOV_{10} $TOV_{10} = 1.4 \cdot \frac{420}{\sqrt{3}} \cdot \left(\frac{1}{10}\right)^{0.02} = 309.61$ (kV)

Given the length of the test line, the switching surge energy discharged by arresters is much lower than the usual switching impulse energy ratings of station-class arresters. All the above values are rms values.

Table 6.9 presents ratings and protective data of surge arresters for this voltage level guaranteed by one manufacturer. The nominal discharge current is 20 kA according to IEC and 15 kA according to ANSI/IEEE. The arrester housing is made of silicone.

The only arrester that provides a safety margin a 20% in both MCOV and TOV capabilities is the last one, which is the selected arrester in this case study. Therefore, the ratings of the arrester are as follows:

- Rated voltage (rms): $U_r = 360$ kV
- Maximum continuous operating voltage (rms): $U_c = 267$ kV (291 kV according to IEEE)
- Temporary overvoltage capability at 10 seconds: $TOV_{10} = 396$ kV
- Nominal discharge current (peak): $I = 20$ kA (15 kA according to IEEE)
- Line discharge class: Class 4.

From the manufacturer data sheets, the height and the creepage distance of the selected arrester are respectively 3.216 and 10.875 m.

Table 6.9 Case Study 5: Surge arrester protective data.

Max. system voltage U_m kV$_{rms}$	Rated voltage U_r kV$_{rms}$	Maximum continuous operating voltage		TOV capability 10 s kV$_{rms}$	Maximum residual voltage with current wave	
		IEC U_c kV$_{rms}$	IEEE MCOV kV$_{rms}$		30/60 µs 3 kA kV$_{peak}$	8/20 µs 10 kA kV$_{peak}$
420	330	264	267	363	684	751
	336	267	272	369	696	765
	342	267	277	376	709	779
	360	267	291	396	746	819

Implementation of the arrester model: The arrester model is that proposed by D.W. Durbak [72] and adopted by the IEEE [73, 74]. Figure 6.43 shows the scheme: it incorporates two time independent non-linear resistors (A_0 and A_1), a pair of linear inductors (L_0 and L_1) paralleled by a pair of linear resistors (R_0 and R_1), and a capacitor C. L_1 and R_1 form a low-pass filter that sees a decaying voltage across it. For low-frequency surges, the impedance of the filter R_1 and L_1 is very low, and A_0 and A_1 are practically in parallel. In high frequency transients, the impedance of the filter becomes very high and the discharge current is distributed between the two nonlinear branches. Figure 6.44 shows V—I characteristics of A_0 and A_1, see also Table 6.10, where voltage values are in per unit of V_{10} [26].

Figure 6.43 IEEE surge arrester model for fast-front transient analysis.

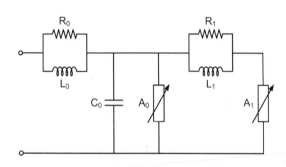

Figure 6.44 V—I characteristics for nonlinear resistors.

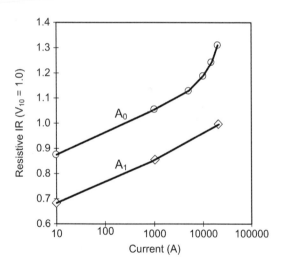

Table 6.10 Values for A_0 and A_1 in Figure 6.44.

Current (kA)	Voltage (per unit of V_{10})	
	A_0	A_1
0.01	0.875	0.681
1	1.056	0.856
5	1.131	—
10	1.188	—
15	1.244	—
20	1.313	1.000

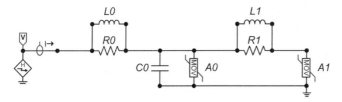

Figure 6.45 Case Study 5: ATPDraw implementation of the surge arrester model.

Formulas to calculate parameters of the circuit shown in Figure 6.43 were initially suggested in [72]. They are based on the estimated height of the arrester, the number of columns of MO disks, and the curves shown in Figure 6.44. The following information is required to determine the parameters of the fast front transient model:

- d = height of the arrester, in meters;
- n = number of parallel columns of MO disks;
- V_{10} = discharge voltage for a 10 kA, 8/20 μs current, in kV;
- V_{ss} = switching surge discharge voltage for an associated switching surge current, in kV.

Linear parameters are derived from the following equations [73]:

$$L_0 = 0.2\frac{d}{n} \ (\mu H) \qquad R_0 = 100\frac{d}{n} \ (\Omega) \qquad (6.7a)$$

$$L_1 = 15\frac{d}{n} \ (\mu H) \qquad R_1 = 65\frac{d}{n} \ (\Omega) \qquad (6.7b)$$

$$C = 100\frac{n}{d} \ (pF) \qquad (6.7c)$$

These formulas do not always give the best parameters, but provide a good starting point.

Figure 6.45 shows the general configuration of the surge arrester model in ATPDraw and the circuit that will be used to test its behaviour in fast-front transients.

Note: The current source in this test circuit is represented by means of the Heidler model; see Figure 6.45. The parameters that have to be specified in this model do not exactly correspond to the parameters specified in standards, and a conversion procedure should be made (to adapt the values) prior to the application of this source model. In this study, the parameters of the Heidler source current are assumed equal to those needed to define a standard waveform. Consequently, results should be used with care.

The procedure proposed by the IEEE WG to determine the parameters is as follows [73]:

1) Determine linear parameters (L_0, R_0, L_1, R_1, C) from the previously given formulas, and derive the nonlinear characteristics of A_0 and A_1.
2) Adjust A_0 and A_1 to match the switching surge discharge voltage.
3) Adjust the value of L_1 to match the V_{10} voltages.

It is applied to a one column arrester with the following characteristics: (i) overall height of 3.216 m, (ii) V_{10} = 819 kV, (iii) V_{ss} = 746 kV for a 3 kA, 30/60 μs current waveform.

- *Initial values*: The initial parameters that result from using Eqs. (6.7) are:

$$L_0 = 0.2 \cdot 3.216 = 0.6432 \ (\mu H) \qquad R_0 = 100 \cdot 3.216 = 321.6 \ (\Omega)$$

$$L_1 = 15 \cdot 3.216 = 48.24 \, (\mu H) \qquad R_1 = 65 \cdot 3.216 = 209.04 \, (\Omega)$$

$$C = 100 \cdot \frac{1}{3.216} = 31.095 \, (pF)$$

- *Adjustment of A_0 and A_1 to match switching surge discharge voltage*: The arrester model was tested to adjust the nonlinear resistances A_0 and A_1. A 3 kA, 30/60 μs double-ramp current was injected into the initial model. The result was a 742 kV voltage peak that matches the manufacturer's value within an error of less than 1%.
- *Adjustment of L_1 to match lightning surge discharge voltage*: To match the discharge voltages for an 8/20 μs current L_1 is modified until a good agreement between the simulation result and the manufacturer's value is achieved. A summary is shown in Table 6.11.

Figure 6.46 displays the implementation of the two nonlinear resistors in ATPDraw.

Arrester energy analysis: Figure 6.47 shows the diagram of the circuit implemented in ATP-Draw for analysing the effect of surge arresters on the lightning performance of the test line; the scenario shown in the figure corresponds to the impact of a lightning stroke on the shield wire. A custom-made group has been created to represent the surge arrester.

The calculation of overvoltages caused by lightning is irrelevant with arresters connected in parallel to all insulator strings, since all voltages between insulator terminals will be limited by the residual voltages of arresters and will be below the flashover level; that is, there will not be flashovers. In such a scenario, the calculations of concern are related to the energy that surge arresters will discharge. Obviously, a different scenario will result from the installation of arresters in parallel to only one or two insulators per tower. Observe that phase C conductors are somehow shielded by phase A conductors, so shielding failures should only occur in conductors of phases A and B, although any of these scenarios will not prevent that a backflashover could occur in phase C insulators if this phase is not protected by surge arresters.

In other words, the estimation of the lightning flashover rate is not necessary when arresters are connected in parallel to all insulator strings, but it is of concern if the arresters are connected in parallel to insulators of only one or two phases.

Some differences in the modelling guidelines must be accounted for when the energy requirement of surge arresters is the main concern:

- Line spans must be represented as multi-phase untransposed frequency-dependent distributed-parameter line sections, since the calculations with a constant-parameter model will not be accurate enough.
- For arrester energy evaluation, implementation of a line with no less than 6 spans at both sides of the point of impact is recommended. This is mainly due to the fact that the discharged energy is shared by arresters located at adjacent towers. Since corona effect is not included in the calculations, damping, and energy losses caused during propagation are not as high as when corona is included.
- The tail time of the return stroke current has a strong influence, being the effect of the rise time very small, or even negligible for low-peak current values. Therefore, the simulation time must always be longer than the tail time of the lightning stroke current.

Table 6.11 Case Study 5: Adjustment of surge arrester model parameters.

Run	L_1 (mH)	Simulated V_{10} (kV)	Difference (%)	Next value of L_1
1	48.24	857.8	4.74	24.12
2	24.12	818.5	0.06	—

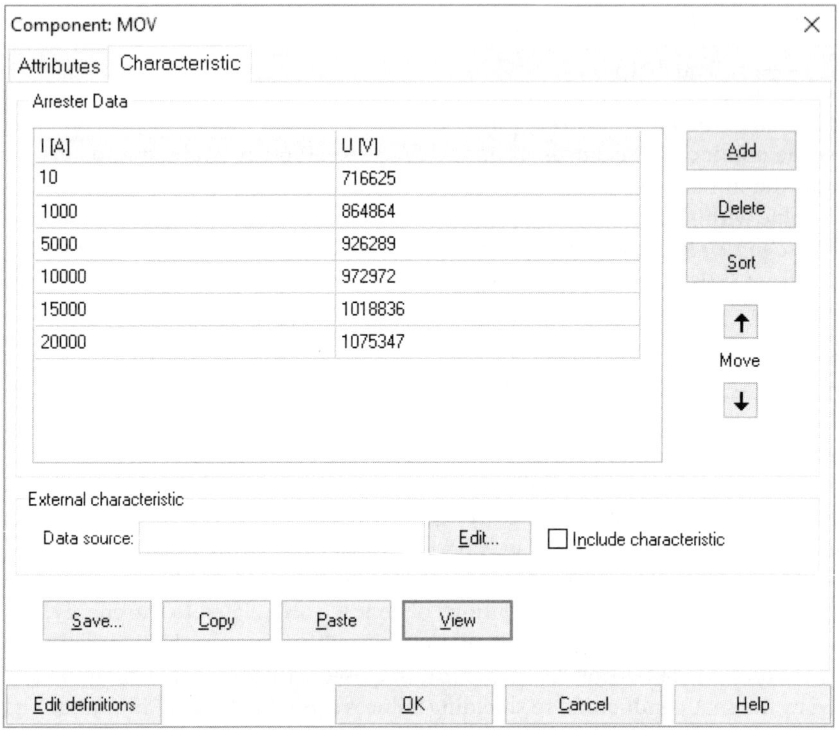

(a)

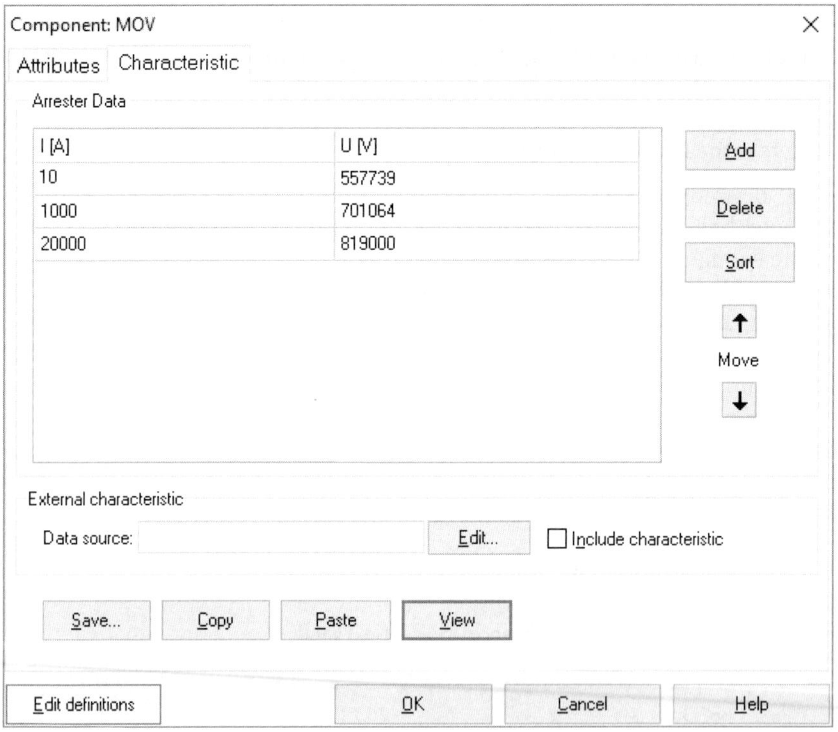

(b)

Figure 6.46 Case Study 5: Nonlinear characteristics of the surge arrester model. (a) Nonlinear characteristic – Resistor A_0, (b) Nonlinear characteristic – Resistor A_1.

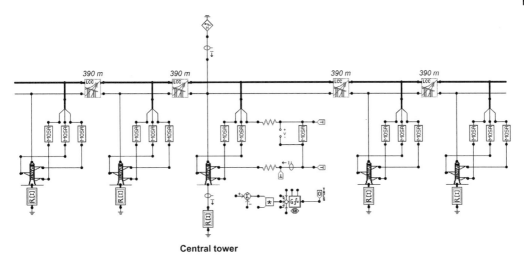

Figure 6.47 Case Study 5: ATPDraw implementation of the test line for analysing the effect of surge arresters – Impact of a lightning stroke to a shield wire.

The effect of phase conductor voltages is important for arrester energy evaluation [27]. Finally, arrester lead models can usually be neglected since it is assumed that the lead lengths are very short and they will not cause significant difference in both voltages and discharged energies.

Figure 6.48 shows the result of a parametric study whose goal was to estimate the energy discharged by arresters when the lightning stroke impacts on either a shield wire or a phase conductor. Although the peak current values of return strokes that can hit a tower can reach values larger than 200 kA, one can conclude from the first plot that the impact of a stroke on a tower will hardly cause arrester failure. The results obtained when the stroke hits a phase conductor show a higher energy requirement although the peak current values are much lower. In any case, the maximum energy for a peak current magnitude lower than 30 kA is below the maximum energy withstand capability of the selected arresters (i.e. $2U_r$ kJ/kV). Since the probability of a stroke with a peak current magnitude above 200 kA is very low, the conclusion is that the probability of arrester failure caused by a high-energy stress is virtually negligible.

Another parameter whose effect on the energy discharged by surge arresters must be analysed is the grounding resistance. However, since its influence when the lightning discharge hits a phase conductor is not significant, much larger energy values with higher grounding resistance values should not be expected. The energy values were obtained using positive-polarity lightning stroke currents and with a specific reference phase angle of the three-phase ac source that supplied the line (i.e. 180°); different results should be expected with negative-polarity currents and other phase angle values.

6.5.5 Very Fast-Front Overvoltages

Very fast-front transients (VFFTs) can be caused by disconnector operations and faults within GISs, switching of motors and transformers with short connections to the switchgear, or certain lightning conditions [2]. VFFT in GIS arise any time there is an instantaneous change in voltage. This change can be caused by disconnector/breaker operations, the closing of a grounding switch, or the occurrence of a fault. These transients have a rise time in the range of 4–100 ns, and are normally followed by oscillations with frequencies in the range of 1–50 MHz, and magnitudes in the range of 1.5–2.0 pu of the line-to-neutral voltage crest. These values are generally

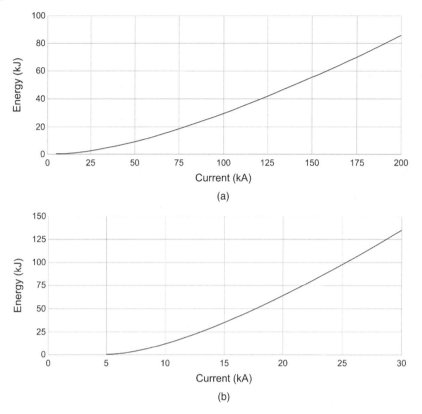

(a)

(b)

Figure 6.48 Case Study 5: Energy discharged by surge arresters. (Stroke waveform: Heidler waveform, $t_f = 1.2\,\mu s$, $t_h = 50\,\mu s$; Grounding resistance: $R_0 = 50\,\Omega$, $\rho = 200\,\Omega.m$). (a) Strokes to a tower, (b) Strokes to a phase conductor.

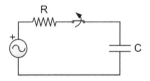

Figure 6.49 Case Study 6: Scheme of the circuit for analysing the variation of voltage during disconnector operation.

below the insulation level of the GIS and connected equipment of lower voltage classes. VFFO in GIS are of greater concern at the highest voltages, for which the ratio of the insulation level to the system voltage is lower. The generation and propagation of VFFT from their original location throughout a GIS can produce internal and external overvoltages. This section presents some case studies that illustrate the origin and propagation of VFFT as well as the generation of VFFO in GIS.

6.5.5.1 Case Study 6: Origin of Very Fast-Front Transients in GIS

VFFTs within a GIS are usually generated by disconnect switch operations, although there can be other causes, such as the closing of a grounding switch or a fault. A large number of pre- or restrikes can occur during a disconnector operation due to the relatively slow speed of the moving contact [75]. Figure 6.49 shows a very simple configuration used to explain the general switching behaviour and the pattern of voltages on opening and closing of a disconnector at a capacitive load [76]. During an opening operation, sparking occurs as soon as the voltage between the source and the load exceeds the dielectric strength across contacts. After a restrike, a high-frequency current will flow through the spark and equalize the capacitive load voltage to the source voltage. The potential difference across the contacts will fall and the spark will extinguish. The subsequent restrike occurs when the voltage between contacts reaches the new dielectric strength level that is determined by the speed of the moving contact and other disconnector characteristics. The behaviour during a closing operation is very similar, and the load side voltage will follow the supply voltage until the contact-make. For a discussion of the physics involved in the restrikes and prestrikes of a disconnect switch operation see [76].

Figure 6.50 shows the diagram of the circuit implemented in ATPDraw that will be used to simulate the behaviour of a disconnector: the disconnector is represented by means of an ideal switch controlled from a MODELS-based section. The figure shows the menu window and the MODELS code controlling the disconnector during opening.

According to this model, the strength between disconnector terminals varies linearly and it can be defined by a maximum value, a speed factor, a polarity factor (that establishes a relationship between the strength with positive and negative polarity voltages), and a current margin needed to decide when the switch/disconnector interrupts the current.

Figure 6.51 depicts the voltage developed across the capacitor during opening and closing operations. The simulation results were obtained with the following values:

- Voltage source: Frequency = 50 Hz, Voltage peak = 1 V, Phase angle = $-90°$
- Disconnector: Maximum strength = 1.2 V, Speed factor = $4\,s^{-1}$

  Polarity factor = 1, Current margin = 0.001 A.

The oscillograms obtained for both opening and closing operations duplicate the behaviour summarized above. Note that after an opening operation a trapped voltage/charge remains at the load side of the disconnector. A reclosing into this trapped charge may cause higher voltages, as illustrated in the example analysed in Chapter 5.

6.5.5.2 Case Study 7: Propagation of Very Fast-Front Transients in GIS

The scheme shown in Figure 6.52 illustrates the generation of VFFT due to a disconnector operation. The breakdown of a disconnector when it is closing originates two surges V_L and V_S which travel outward in the bus duct and back into the source side respectively. The magnitude of both travelling surges is given by:

$$V_L = \frac{Z_L}{Z_S + Z_L}(V_1 - V_2) \qquad V_S = -V_L \tag{6.8}$$

where Z_S and Z_L are the surge impedances on the source and load side, respectively. V_1 is the source-side voltage, while V_2 is the trapped voltage at the load side.

Figure 6.53 displays the circuit implemented in ATPDraw that will be used to analyse the generation of travelling surges according to (6.8) and how these surges can cause overvoltages within the GIS. The system is energized from a dc source SRC1, while source SRC2 is used to leave trapped charge at the load side of switch SW1. The polarity of the trapped charge is opposite of that of source SRC1. To leave trapped charge on the load side, an ac source (SRC2) of very low frequency is activated before the simulation begins and disconnected at the moment the simulation begins. The peak value of the two sources is 1 V.

All GIS sections are represented as lossless lines with a surge impedance of 45 Ω and a propagation velocity equal to that of light. The lengths of all sections are indicated in the figure.

Figure 6.54 shows some simulations' results, namely the voltage developed at nodes SR1, SR2, LD1, and LD2 (see Figure 6.53), when switch SW1 is closed at $t = 100$ ns. The analysis of these

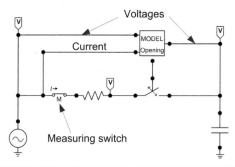

(a)

```
MODEL opening
DATA    speed           -- Speed factor
        dsmax           -- Maximum strength of the
                           disconnector/breaker
        factor          -- Polarity factor
        imargin         -- Current margin
VAR     vvoltage        -- Voltage across the switch
        vswitch         -- Absolute value of the voltage across
                           the switch
```

Figure 6.50 Case Study 6: Disconnector opening – ATPDraw implementation. (a) Menu window, (b) MODELS code.

```
        absc            -- Absolute value of the source current
        switch          -- Switch status
        strength        -- Variable strength of the
                           breaker/disconnector
        fire1           -- Control signal
 INPUT vsource          -- Source voltage
        vcapacitor      -- Capacitor voltage
        current         -- Source current
 OUTPUT fire1           -- Control signal
 INIT      switch:=closed        fire1:=1      ENDINIT
 EXEC
        vvoltage:=vsource - vcapacitor
        IF (vvoltage >= 0) THEN
            strength:=dsmax*speed*t
        ELSE
            strength:=dsmax*speed*t*factor
        ENDIF
        IF (strength > dsmax) THEN strength:=dsmax ENDIF
            vswitch:=abs(vvoltage)    absc:=abs(current)
switch:=open
        IF (vswitch > strength) THEN
            switch:=closed fire1:=1
            ELSE
             IF (absc < imargin) THEN fire1:=-1 ENDIF
        ENDIF
 ENDEXEC
 ENDMODEL
```

(b)

Figure 6.50 (Continued)

voltages confirms the expression (6.8) and proves that the propagation and reflection of surges can cause voltages higher than the source voltage.

- At the beginning of the simulation, before switch SW1 closes, a surge of 0.5 V begins its propagation from source SRC1 along the line between nodes SR1 and SR2. At the moment this surge arrives to the open terminal SR2, the voltage at this terminal becomes 1 V. A reflected voltage surge of value 0.5 V starts the propagation to the source SRC1; the voltage at node SR1 increases its value to 1 V at the arrival of this surge.
- The initial voltage at nodes LD1 and LD2 is -1 V. This is the trapped charge value at the load side of switch SW1, and is originated after opening switch SW2, at $t = 0$.
- According to Eq. (6.8), and given that Z_L and Z_S have the same value, the peak values of surges V_L and V_S that are originated after closing switch SW1 is 1 V.
- The value of the voltage surge V_L after passing the bifurcation point is reduced to 0.666 V. This surge doubles at nodes LD1 and LD2. Since the voltage at these nodes is -1 V, the resulting voltage at the moment the first surge arrives is 0.333 V; see Figure 6.54.
- The subsequent propagations and reflections can cause a voltage higher than 1 V at nodes LD1 and LD2, as shown in the figure.

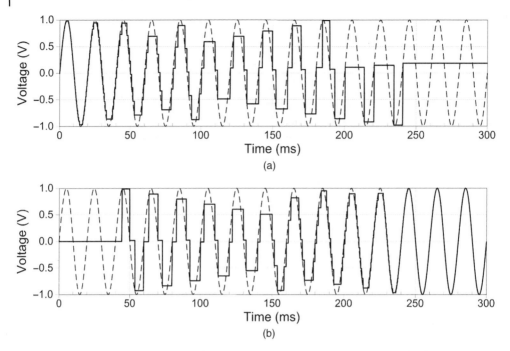

Figure 6.51 Case Study 6: Disconnector operations – Simulation results. (a) Opening, (b) Closing.

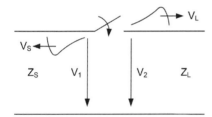

Figure 6.52 Case Study 7: Generation and propagation of VFFT in GIS.

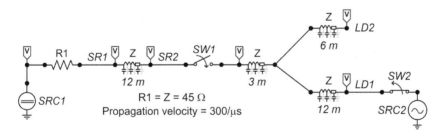

Figure 6.53 Case Study 7: ATPDraw implementation.

The maximum value of the voltages that can occur at nodes LD1 and LD2 will depend on the source resistance, R_1, and the surge impedance and length of each GIS section: with a lower value of R_1, voltage at LD1 and LD2 will increase and eventually exceed 2 V.

The level reached by overvoltages is random by nature. The maximum overvoltage produced by a disconnector breakdown depends on the geometry of the GIS, the measuring point, the voltage prior to the transient at the load side (trapped charge), and the intercontact voltage

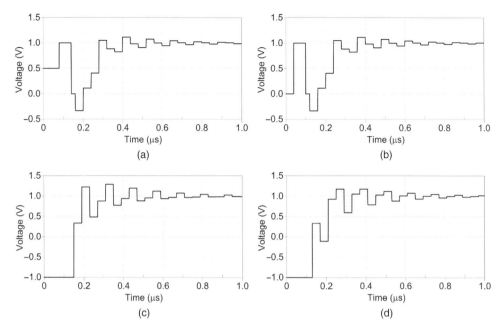

Figure 6.54 Case Study 7: Simulation results. (a) Voltage at Node SR1, (b) Voltage at Node SR2, (c) Voltage at Node LD1, (d) Voltage at Node LD2.

at the time of the breakdown. A very simple expression can be used to obtain the transient overvoltages as a function of time t and position s [77, 78]

$$V(t, s) = V_b * K(t, s) + V_q \tag{6.9}$$

where $K(t, s)$ is the normalized response of the GIS, V_b is the intercontact spark voltage, and V_q the voltage prior to the transient at the point of interest. As V_b and V_q are random variables, $V(t, s)$ is also random. This equation can be used to estimate worst-case values [78].

6.5.5.3 Case Study 8: Very Fast-Front Transients in a 765 kV GIS

The collapsing electric field during a breakdown produces travelling waves which propagate in both directions from the disturbance location. This propagation can be analysed assuming that propagation losses are negligible. Due to the very high frequencies generated by a dielectric breakdown, a simulation is restricted to calculations during the VFFT waveform period, usually 1 or 2 μs. If the simulation is performed with ATP, which uses a constant time-step size, then the value of this step size will depend on the shorter transit time in the GIS, and must be equal or smaller than one-half the shorter transit time.

Transient can usually be originated by any of the following causes: (i) a ramp voltage with a magnitude determined by the voltage across the switch; (ii) two ramp currents on opposite sides of the switch such that the voltage across the switch is equal to zero at the crest of the inputs; (iii) a switch closing operation after charging both sides of the switch to the desired value.

Low-voltage tests are a very useful tool for development and validation of GIS models. The case analysed in this section presents the simulation of internal transients during low voltage tests of a 765 kV GIS. Figure 6.55 shows the one-line diagram of a 765 kV GIS bay [46]. Detailed

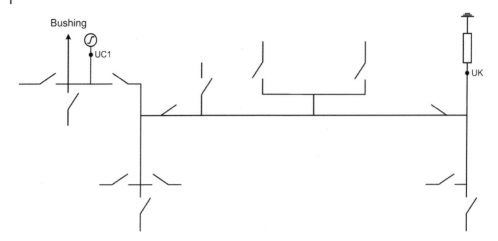

Figure 6.55 Case Study 8: One-line diagram of the test system [46].

Table 6.12 Case Study 8: Data of the 765 kV GIS [46].

Branch	Z (Ω)	Travel time (ns)	Branch	Z (Ω)	Travel time (ns)
UC1–J3	75	6.40	T17–T14	160	0.67
J3–J4	75	48.0	T14–T13	51	1.90
J4–T22	75	2.20	T13–T11	75	9.90
T22–T23	51	1.90	T11–T12	65	1.70
J4–D9	78	2.20	T11–J2	75	7.50
D9–D88	68	1.80	J2–T9	75	2.20
D88–D66	59	4.20	T9–T10	51	1.90
D44–D22	33	5.80	T10–T28	160	0.67
D22–D1	330	9.10	T28–J6	75	7.10
J3–T21	75	2.20	J6–UK	75	6.40
T21–T20	51	1.90	T28–T29	65	1.70
T20–T19	160	0.67	T28–J5	75	8.80
T19–T18	65	1.70	J5–T30	75	2.20
T19–T17	75	6.80	T30–T32	51	1.90
T17–T16	65	1.70	J2–J1	75	6.70
T17–J7	75	8.50	J1–T4	75	2.20
J7–T24	75	2.20	T4–T3	51	1.90
T24–T25	51	1.90	J1–T5	75	2.20
J7–T26	75	2.20	T5–T6	51	1.90
T26–T27	51	1.90			

data are given in Table 6.12. Models used to represent components of this case are those discussed above. These models were developed by using the following procedure [46]:

1) Low voltage tests on individual components were performed using waves with fronts of 4 and 20 ns.

2) Models based on physical dimensions were developed, assuming a propagation velocity equal to that of light.
3) Digital models were adjusted so simulation results were matched to measurements. The main adjustment was to decrease propagation velocity to 0.96 that of light.

Figure 6.56 displays the circuit implemented in ATPDraw to simulate this case. Given that a single-phase representation of GIS ducts is used, and the ducts are modelled as lossless lines, the implementation is straightforward. Two transients have been reproduced: (i) a ramp current is applied at $t = 0$; (ii) the ramp current source is also used but the transient starts after closing the switch at the instant the ramp reaches its maximum value. The waveforms obtained at two nodes with the first option (i.e. a ramp current that reaches a peak value of 4 A in 4 ns) are shown in Figure 6.57. Some differences can occur with the two options, but they are not significant.

It can be observed that waveforms for both cases are essentially the same, except for the first nanoseconds in the vicinity of the input node UC1. These simulation results were validated by comparison with low voltage measurements.

For slow switches the probability of a re-/prestrike with the greatest breakdown voltage (in the range 1.8–2 pu) is very small; however, due to the great number of re-/prestrikes which are produced with one operation this probability should not always be neglected. The value of the trapped charge is mainly dependent on the disconnect switch characteristics, the faster the switch, the greater the mean value that the trapped charge voltage can reach.

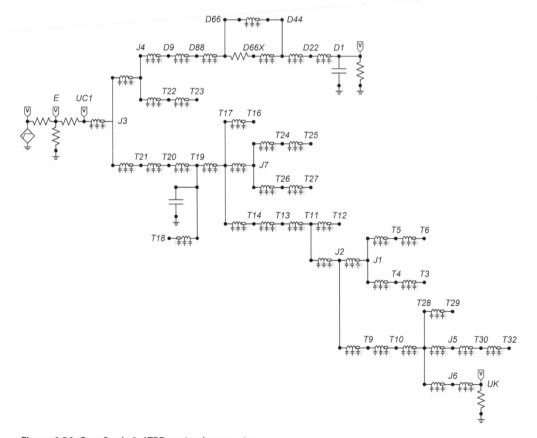

Figure 6.56 Case Study 8: ATPDraw implementation.

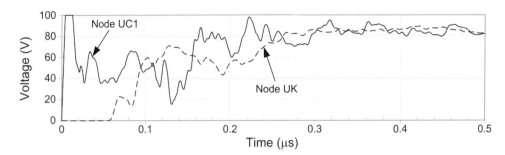

Figure 6.57 Case Study 8: Simulation results with a 4 ns ramp.

References

1 IEC 60071-1. (2019). Insulation co-ordination - Part 1: Definitions, principles and rules, Edition 9.0.
2 IEC 60071-2. (2018). Insulation co-ordination, Part 2: Application guide, Edition 4.0.
3 IEEE. (2010). C62.82.1–2010, IEEE Standard for Insulation Coordination – Definitions, Principles and Rules.
4 IEEE. (1999). Std 1313.2, IEEE Guide for the Application of Insulation Coordination.
5 Ragaller, K. (ed.) (1980). *Surges in High-Voltage Networks*. New York, NY-USA: Plenum Press.
6 Greenwood, A. (1991). *Electrical Transients in Power Systems*, 2e. Wiley.
7 Hileman, A.R. (1999). *Insulation Coordination for Power Systems*. Marcel Dekker.
8 Chowdhuri, P. (2003). *Electromagnetic Transients in Power Systems*, 2e. Taunton, UK: RS Press-John Wiley.
9 Martinez-Velasco, J.A. (ed.) (2009). *Power System Transients. Parameter Determination*. CRC Press.
10 Das, J.C. (2010). *Transients in Electrical Systems. Analysis, Recognition, and Mitigation*. New York, NY: McGraw-Hill.
11 Martinez-Velasco, J.A. (ed.) (2015). *Transient Analysis of Power Systems. Solution Techniques, Tools and Applications*. John Wiley-IEEE Press.
12 CIGRE WG C4.307. (2014). Transformer Energization in Power Systems: A Study Guide. CIGRE Brochure 568.
13 CIGRE WG 33.02. (1990). Guidelines for Representation of Network Elements when Calculating Transients. CIGRE Brochure 39.
14 Gole, A., Martinez-Velasco, J.A. and Keri, A. (eds.) (1998). Modeling and Analysis of Power System Transients Using Digital Programs. IEEE Special Publication TP-133-0. IEEE Catalog No. 99TP133–0.
15 IEC TR 60071–4. (2004). Insulation Co-ordination - Part 4: Computational Guide to Insulation Co-ordination and Modeling of Electrical Networks.
16 Iravani, R., Chaudhury, A.K.S., Hassan, I.D. et al. (1998). Modeling guidelines for low frequency transients. In: *Modeling and Analysis of System Transients Using Digital Programs* (eds. A. Gole, J.A. Martinez-Velasco and A. Keri). IEEE Special Publication TP-133-0, IEEE Catalog No. 99TP133-0.
17 CIGRE WG C4.307. (2014). Resonance and Ferroresonance in Power Networks. CIGRE Brochure 569.

18 Durbak, D.W., Gole, A.M., Camm, E.H. et al. (1998). Modeling guidelines for switching transients. In: *Modeling and Analysis of System Transients using Digital Systems* (eds. A. Gole, J.A. Martinez-Velasco and A.J.F. Keri). IEEE Special Publication, TP-133-0.

19 de León, F., Gómez, P., Martinez-Velasco, J.A., and Rioual, M. (2009). Transformers. In: *Power System Transients. Parameter Determination* (ed. J.A. Martinez-Velasco). CRC Press.

20 Can/Am Users Group (2000). *ATP Rule Book*. Can/Am Users Group.

21 Martinez-Velasco, J.A. and Popov, M. (2009). Circuit breakers. In: *Power System Transients. Parameter Determination* (ed. J.A. Martinez-Velasco). CRC Press.

22 CIGRE WG 13.02. (1995). Interruption of Small Inductive Currents. (ed. S. Berneryd). CIGRE Brochure 50.

23 CIGRE WG A3.07. (2004). Controlled Switching of HVAC Circuit Breakers. CIGRE Brochure 262.

24 Prikler, L., Ban, G., and Banfai, G. (1997). EMTP models for simulation of shunt reactor switching transients. *Electrical Power and Energy Systems* 19 (4): 235–240.

25 Durbak, D.W. (1987). The choice of EMTP surge arrester models. *EMTP Newsletter* 7 (3): 14–18.

26 Martinez, J.A. and Durbak, D. (2005). Parameter determination for modeling systems transients. Part V: Surge arresters. *IEEE Transactions on Power Delivery* 20 (3): 2073–2078.

27 Martinez-Velasco, J.A. and Castro-Aranda, F. (2009). Surge arresters. In: *Power System Transients. Parameter Determination* (ed. J.A. Martinez-Velasco). CRC Press.

28 Annakkage, U.D., Nair, N.K.C., Liang, Y. et al. (2012). Dynamic system equivalents: a survey of available techniques. *IEEE Transactions on Power Delivery* 27 (1): 411–420.

29 IEEE TF on Fast Front Transients (1996). Modeling guidelines for fast transients. *IEEE Transactions on Power Delivery* 11 (1): 493–506.

30 Imece, A.F., Durbak, D.W., Elahi, H. et al. (1998). Modeling guidelines for fast front transients. In: *Modeling and Analysis of System Transients using Digital Systems* (eds. A. Gole, J.A. Martinez-Velasco and A.J.F. Keri). IEEE Special Publication, TP-133-0.

31 Martinez-Velasco, J.A., Ramirez, A.I., and Dávila, M. (2009). Overhead lines. In: *Power System Transients. Parameter Determination* (ed. J.A. Martinez-Velasco). CRC Press.

32 Electrical Research Council (1982). *Transmission Line Reference Book, 345 kV and Above*, 2e. Palo Alto, California: Electric Power Research Institute.

33 CIGRE WG 33.01. (1991).Guide to Procedures for Estimating the Lightning Performance of Transmission Lines. CIGRE Brochure 63.

34 IEEE WG on Lightning Performance of Transmission Lines (1985). A simplified method for estimating lightning performance of transmission lines. *IEEE Transactions on Power Apparatus and Systems* 104 (4): 919–932.

35 CIGRE WG 33.01. (1992). Guide for the evaluation of the dielectric strength of external insulation. CIGRE Brochure 72.

36 Lee, K.C. (1983). Non-linear corona models in the Electromagnetic Transients Program. *IEEE Transactions on Power Apparatus and Systems* 102 (9): 2936–2942.

37 Eriksson, A.J. and Weck, K.H. (1988). Simplified procedures for determining representative substation impinging lightning overvoltages. CIGRE Session, Paper 33–16.

38 Elahi, H., Sublich, M., Anderson, M.E., and Nelson, B.D. (1990). Lightning overvoltage protection of the Paddock 362-145 kV gas insulated substation. *IEEE Transactions on Power Delivery* 5 (1): 144–149.

39 Gallagher, T.J. and Dudurych, I.M. (2004). Model of corona for an EMTP study of surge propagation along HV transmission lines. *IEE Proceedings - Generation, Transmission and Distribution* 151 (1): 61–66.

40 Wagner, C.F., Gross, I.W., and Lloyd, B.L. (1954). High-voltage impulse test on transmission lines. *AIEE Transactions* 73 (III-A): 196–210.

41 Martinez-Velasco, J.A. (2012). Basic methods for analysis of high frequency transients in power apparatus windings. In: *Electromagnetic Transients in Transformer and Rotating Machine Windings* (ed. C.Q. Su). IGI Global.

42 Popov, M., Gustavsen, B., and Martinez-Velasco, J.A. (2012). Transformer modelling for impulse voltage distribution and terminal transient analysis. In: *Electromagnetic Transients in Transformer and Rotating Machine Windings* (ed. C.Q. Su). IGI Global.

43 IEEE TF on Parameters of Lightning Strokes (2005). Parameters of lightning strokes: a review. *IEEE Transactions on Power Delivery* 20 (1): 346–358.

44 CIGRE WG 33/13–09. (1988). Monograph on GIS Very Fast Transients.

45 IEEE TF on Very Fast Transients (D. Povh, Chairman) (1996). Modelling and analysis guidelines for very fast transients. *IEEE Transactions on Power Delivery* 11 (4).

46 Martinez, J.A., Povh, D., Chowdhuri, P. et al. (1998). Modeling guidelines for very fast transients in gas insulated substations. In: *Modeling and Analysis of System Transients using Digital Systems* (eds. A. Gole, J.A. Martinez-Velasco and A.J.F. Keri). IEEE Special Publication, TP-133-0.

47 Martinez-Velasco, J.A. (2012). Very fast transients. In: *Power Systems*, 3e (ed. L.L. Grigsby). CRC Press.

48 Miri, A.M. and Stojkovic, Z. (2001). Transient electromagnetic phenomena in the secondary circuits of voltage- and current transformers in GIS (measurements and calculations). *IEEE Transactions on Power Delivery* 16 (4): 571–575.

49 Haznadar, Z., Carsimamovic, C., and Mahmutcehajic, R. (1992). More accurate modeling of gas insulated substation components in digital simulations of very fast electromagnetic transients. *IEEE Transactions on Power Delivery* 7 (1): 434–441.

50 Fujimoto, N., Stuckless, H.A., and Boggs, S.A. (1986). Calculation of disconnector induced overvoltages in gas-insulated substations. In: *Gaseous Dielectrics IV*. Pergamon Press.

51 Yanabu, S., Murase, H., Aoyagi, H. et al. (1990). Estimation of fast transient overvoltage in gas-insulated substation. *IEEE Transactions on Power Delivery* 5 (4): 1875–1882.

52 Glavitsch, H. (1980). Temporary overvoltages. In: *Surges in High-Voltage Networks* (ed. K. Ragaller), 115–129. Plenum Press.

53 EPRI Report EL-4202. (1985). Electromagnetic Transients Program (EMTP) Primer.

54 Martinez, J.A. and Mork, B. (2005). Transformer modeling for low- and mid-frequency transients – a review. *IEEE Transactions on Power Delivery* 20 (2): 1625–1632.

55 Martinez, J.A., Walling, R., Mork, B. et al. (2005). Parameter determination for modeling systems transients. Part III: Transformers. *IEEE Transactions on Power Delivery* 20 (3): 2051–2062.

56 Jazebi, S., Zirka, S.E., Lambert, M. et al. (2016). Duality derived transformer models for low-frequency electromagnetic transients – Part I: Topological models. *IEEE Transactions on Power Delivery* 31 (5): 2410–2419.

57 Mork, B.A., Gonzalez, F., Ishchenko, D. et al. (2007). Hybrid transformer model for transient simulation: Part I: Development and parameters. *IEEE Transactions on Power Delivery* 22 (1): 248–255.

58 Mork, B.A., Ishchenko, D., Gonzalez, F., and Cho, S.D. (2008). Parameter estimation methods for five-limb magnetic core model. *IEEE Transactions on Power Delivery* 23 (4): 2025–2032.

59 Cherry, E.C. (1949). The duality between interlinked electric and magnetic circuits and the formation of transformer equivalent circuits. *Proceedings of the Physical Society. Section B* 62: 101–111.

60 Slemon, G.R. (1953). Equivalent circuits for transformers and machines including non-linear effects. In: *Proceedings of IEE*, vol. 100, 129–143.

61 Corea-Araujo, J.A., Martinez-Velasco, J.A., González-Molina, F. et al. (2018). Validation of single-phase transformer model for ferroresonance analysis. *Electrical Engineering* 100 (3): 1339–1349.

62 CIGRE WG 13.02 (1973). Switching overvoltages in EHV and UHV systems with special reference to closing and reclosing of transmission lines. *Electra* 30: 70–122.

63 McGranaghan, M.F., Zavadil, R.M., Hensley, G. et al. (1992). Impact of utility switched capacitors on customer systems – Magnification at low voltage capacitors. *IEEE Transactions on Power Delivery* 7 (2): 862–868.

64 Pigini, A., Rizzi, G., Garbagnati, E. et al. (1989). Performance of large air gaps under lightning overvoltages: experimental study and analysis of accuracy of predetermination methods. *IEEE Transactions on Power Delivery* 4 (2): 1379–1392.

65 Mousa, A.M. (1994). The soil ionization gradient associated with discharge of high currents into concentrated electrodes. *IEEE Transactions on Power Delivery* 9 (3): 1669–1677.

66 Heidler, F., Cvetic, J.M., and Stanic, B.V. (1999). Calculation of lightning current parameters. *IEEE Transactions on Power Delivery* 14 (2): 399–404.

67 Brown, G.W. and Whitehead, E.R. (1969). Field and analytical studies of transmission line shielding: Part II. *IEEE Transactions on Power Apparatus Systems* 88 (3): 617–626.

68 Martinez, J.A. and Castro-Aranda, F. (2005). Lightning performance analysis of overhead transmission lines using the EMTP. *IEEE Transactions on Power Delivery* 20 (3): 2200–2210.

69 Martinez-Velasco, J.A. and Castro-Aranda, F. (2008). EMTP implementation of a Monte Carlo method for lightning performance analysis of transmission lines. *Ingeniare* 16 (2): 169–180.

70 Marsaglia, G. and Zaman, A. (1990). A random number generator for PC's. *Computer Physics Comminications* 60: 345–349.

71 Lyons, W.A., Uliasz, M., and Nelson, T.E. (1998). Large peak current cloud-to-ground lightning flashes during the summer months in the contiguous United States. *Monthly Weather Review* 126: 2217–2233.

72 Durbak, D.W. (1985). Zinc-oxide arrester model for fast surges. *EMTP Newsletter* 5 (1): 1–9.

73 IEEE Working Group on Surge Arrester Modeling (1992). Modeling of metal oxide surge arresters. *IEEE Transactions on Power Delivery* 7 (1): 302–309.

74 IEEE Std C62.22. (2013). IEEE guide for the application of metal-oxide surge arresters for alternating-current systems.

75 Ecklin, A., Schlicht, D., and Plessl, A. (1980). Overvoltages in GIS caused by the operation of isolators. In: *Surges in High-Voltage Networks* (ed. K. Ragaller), 115–129. Plenum Press.

76 Boggs, S.A., Chu, F.Y., Fujimoto, N. et al. (1982). Disconnect switch induced transients and trapped charge in gas-insulated substations. *IEEE Transactions on Power Apparatus and Systems* 101 (6): 3593–3602.

77 Boggs, S.A., Fujimoto, N., Collod, M. and Thuries, E. (1984). The modeling of statistical operating parameters and the computation of operation-induced surge waveforms for GIS disconnectors. CIGRE Session, Paper No. 13–15.

78 Fujimoto, N., Chu, F.Y., Harvey, S.M. et al. (1988). Developments in improved reliability for gas-insulated substations. CIGRE Session, Paper No. 23–11.

To Probe Further

ATP data files for the case studies presented in this chapter are provided in the companion website. The simulation results presented in the chapter are a small sample of the results that can be derived from the case studies, readers are encouraged to run the cases, check all results, and explore the performance of the test systems under different operating conditions or their response in front of different disturbances.

7

Simulation of Rotating Machine Dynamics
Juan A. Martinez-Velasco

7.1 Introduction

Several options are currently available in the Alternative Transients Program (ATP) for simulating rotating machine dynamics [1–8]. Built-in models do not only allow users to simulate machine dynamics, they also include other features such as automatic initialization, conversion of electrical parameters, and interface to control systems. The main goal of this chapter is to illustrate the range of applications that can be covered with the models presently available in ATP for representing rotating machines.

As for other power components, the representation of a rotating machine depends on the frequency range of the transient phenomena to be simulated. ATP models can basically be used to analyse network-machine interactions in low-frequency transients; some other studies, such as the simulation of inter-turn insulation failures, cannot be performed with the current models. Insulation failures are usually originated by steep-fronted surges, and ATP models are not valid for these frequency ranges; however, it is possible to develop accurate machine models using other ATP capabilities.

Section 7.2 summarizes the representation of rotating machines taking into account the frequency range of the transient phenomena. Section 7.3 provides a short background of ATP machine models and a summary of the models presently available through ATPDraw. Section 7.4 summarizes the capabilities and solution methods implemented in rotating machine models. Section 7.5 suggests a procedure to obtain the data input of a rotating machine in ATP format considering the frequency of the transients and the available machine data. Section 7.6 summarizes the capabilities and limitations of the current models. Sections 7.7 and 7.8 cover some of the most important applications of the ATP options, but considering only three-phase machines. Although it is possible to implement and simulate rotating machine models using TACS (Transient Analysis of Control Systems) or MODELS, only built-in models are applied in this chapter.

7.2 Representation of Rotating Machines in Transients Studies

Modelling of power components taking into account the frequency-dependence of parameters can practically be made by developing mathematical models which are accurate enough for a specific range of frequencies. Frequency ranges are classified into four groups with some overlapping [9] (see also Chapter 2): (i) Group I – low-frequency oscillations, from 0.1 Hz to 3 kHz; (ii) Group II – slow-front surges, from 50/60 Hz to 20 kHz; (iii) Group III – fast-front surges,

Transient Analysis of Power Systems: A Practical Approach, First Edition. Edited by Juan A. Martinez-Velasco.
© 2020 John Wiley & Sons Ltd. Published 2020 by John Wiley & Sons Ltd.
Companion Website: www.wiley.com/go/martinez/power_systems

from 10 kHz to 3 MHz; (iv) Group IV – very-fast-front surges, from 100 kHz to 50 MHz. The following comments summarize the modelling guidelines usually recommended for representing rotating machines in transients studies:

- ATP machine models can be used to simulate Group I transients;
- neither the representation of the mechanical part nor the saturation is needed for simulation of transients of Groups II–IV;
- the frequency-dependence of the subtransient inductance and losses (eddy current effects) has to be partially considered for Group II transients;
- capacitive effects are of concern for transient of Groups III and IV;
- representations for Group III and IV transients are required only if transients of the generator side of a generator-transformer set are investigated.

7.3 ATP Rotating Machines Models

7.3.1 Background

The first three-phase synchronous machine models were implemented in the mid-1970s [3, 10–12]; they incorporated a detailed representation of the mechanical and electrical parts, used a sophisticated solution method to solve machine-power system interface, and included interface to control systems. Although their implementation was raised by subsynchronous resonance (SSR) incidents [13–16], those models could also be used for other studies, such as loss of synchronism, load rejection, or transmission line reclosure. A simple and efficient representation of magnetic saturation was added in the late 1970s [17]. A new three-phase synchronous machine model based on a phase-domain interface method was implemented in 1996 [7]; its main advantages are a more stable machine-network interface and a more accurate representation of saturation.

Interests in the simulation of renewable energy sources motivated the demand for other machine models. The Universal Machine (UM) module was implemented in the BPA EMTP in 1980 [4]. This option allowed the representation of up to 12 different machine models and expanded the applications to the simulation of adjustable speed drives and wind power generation. The first UM release had limitations that were solved in subsequent versions [5].

Machine models currently implemented in the ATP can be used for simulating low-frequency transients; they are not adequate for simulation of fast-front transients (e.g. some switching motor operations can originate steep-front surges and cause large turn-to-turn winding stresses; lightning surges directly impacting a machine or transferred through transformers are also a source of high stresses and dielectric failures). Several models for analysing machine behaviour in fast-front transients and predicting distribution of inter-turn voltages caused by steep-fronted surges have been proposed to date [18–21]. The representation of a machine seen from its terminals in switching and steep-fronted surge studies has been the subject of other works; see, for instance, [22, 23].

Although the majority of ATP studies deal with three-phase synchronous and induction (asynchronous) machines, ATP models representing small and special machines have also been applied; see, for instance, [24] and [25].

7.3.2 Built-in Rotating Machine Models

The models implemented in ATP for representing rotating machine dynamics are the Type 56 (for induction machines), the Type 58 and Type 59 (for synchronous machines [3, 7]), the UM module [4, 5], and WindSyn-based models [26]. Table 7.1 lists the options currently available in

Table 7.1 Rotating machine models available in ATPDraw.

ATPDraw icon	Machine type	Capabilities
TEx SM 59 TPow	Type 59 three-phase synchronous machine	Three-phase model with detailed description of electrical and mechanical systems. The model accepts connection to control systems, incorporates steady-state initialization, and allows parallel operation of two or more machines. Solution method based on Park transformation.
TEx SM 58 TPow	Type 58 three-phase synchronous machine	Three-phase model with the same features as the Type 59 model. It uses a solution method on the phase domain.
SM ω	UM-1 three-phase synchronous machine	Three-phase model of a synchronous machine based on the Universal Machine module. Two options for data input.
IM ω	UM-3 three-phase squirrel-cage induction machine	Three-phase model of a short-circuited rotor induction machine based on the Universal Machine module.
IM ω	UM-4 three-phase wound-rotor induction machine	Three-phase model of a wound rotor induction machine based on the Universal Machine module.
SP ω	UM-6 single-phase induction machine	Single-phase induction machine based on the Universal Machine module
DC ω	UM-8 DC machine	DC machine with separate excitation based on the Universal Machine module
IM T	Type 56 three-phase induction machine	Induction machine model based on a phase-domain solution method.
Exfd SM WI Torque	Synchronous machine with manufacturer's data input (WindSyn type)	Synchronous machine model with manufacturer's data input based on WindSyn and the Universal Machine module. This model supports three types of synchronous machines.
IM WI Torque	Induction machine with manufacturer's data input (WindSyn type)	Induction machine model with manufacturer's data input based on WindSyn and the Universal Machine module. This model supports four types of induction machines.
Exfd UM/W IM-wnd Torque	Universal Machine with manufacturer's data (WindSyn type)	Universal machine model with manufacturer's data input based on WindSyn and the Universal Machine module. This option covers both synchronous and induction machines.

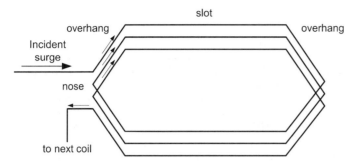

Figure 7.1 Rotating machine coil model for fast transients studies.

ATPDraw. The table displays the icons and provides a very short summary of capabilities (e.g. the ATP option on which it is based).

7.3.3 Rotating Machine Models for Fast Transients Simulation

The most accurate models proposed for representing rotating machines in fast transients are based on a multi-conductor line model [27]. Although there are some differences between the proposed models, all of them have several common features: each coil is segmented into several sections with uniform geometry, see Figure 7.1; mutual coupling between coil sections is usually neglected; a phase winding is modelled by cascading coil models, although this is not necessary, as inter-turn voltages are virtually independent of the connection at the far end of the coil. Elements of the impedance and admittance matrices of the multi-conductor line representation can be determined by means of simple procedures. Capacitances between turns and from turn to ground are calculated using the parallel-plate capacitor formula. Losses can also be added. Reference [20] proposes a simplified model to include resistance matrices, and assumes that these matrices are the same for slot and overhang regions.

7.4 Solution Methods

7.4.1 Introduction

The choice of a machine model can depend on the scenario to be analysed, the information available for simulating the rotating machines, or the method implemented to solve the interface machine-power system. The three-phase synchronous machine model can be simulated using several options. As for any other machine type, the choice within the UM module is restricted to the solution method. Two solution methods have been implemented, and the main aspect to be considered for selecting one or the other is the need to analyse a system with two or more machines running in parallel. This section presents a summary of the solution methods implemented in each model. The ultimate goal is to help users to select model and solution method.

7.4.2 Three-Phase Synchronous Machine Model

The three-phase synchronous machine models (Type-59 and Type-58 models) provide a detailed representation of both the electrical and the mechanical part of turbine–generator set [1–3, 7]. These two options can be applied in many case studies, such as transient stability studies, out-of-step synchronization, or SSR. Both options assume that the electrical part of the machine is based on the representation depicted in Figure 7.2a. The machine model may have: (i) three armature windings 'a', 'b', 'c', one per phase, connected to the network; (ii) one field winding 'f' which produces flux in the direct axis; (iii) one hypothetical winding 'kd' on the direct axis which represents the effects of the damper bars; (iv) one hypothetical winding 'g' on the quadrature axis representing the effects produced by eddy-currents; (v) one hypothetical winding 'kq' on the quadrature axis representing damper bar effects. Note that the convention about axes in Figure 7.2 is not the same that was shown in Figure 2.33.

Type 59 and Type 58 SM models may have any number of lumped masses in the shaft-rotor system, see Figure 7.2b. Each major element is considered to be a rigid mass connected to adjacent elements by massless springs. Each shaft mass, except generator and exciter masses, is allowed to have a constant mechanical power applied to it (in addition to the torque of the mechanical viscous damping and the spring connection to adjacent masses). The user specifies proportionality factors for each mass, with the actual constant power then determined internally by the ATP at the time of the sinusoidal steady-state initialization. The user is also allowed to represent prime mover dynamics by using TACS/MODELS to control the total mechanical power, see Figure 7.3. In this case, the required TACS/MODELS variable is a normalized multiplicative constant (equal to unity if it is to produce no effect) for scaling the otherwise-constant mechanical power. For output purposes, most machine parameters and

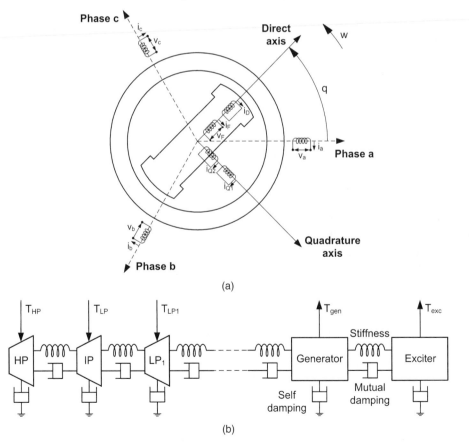

(a)

(b)

Figure 7.2 Schematic representation of a three-phase synchronous machine. (a) Electrical part, (b) Mechanical part.

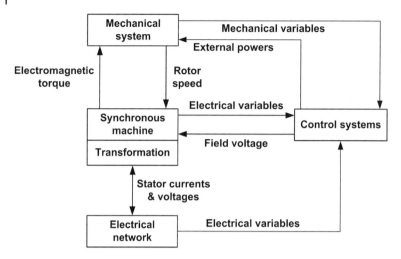

Figure 7.3 General structure of the Type 59 and Type 58 models.

variables of interest are available: electrical and mechanical parameters, initial steady state, electrical variables (winding currents, certain $dq0$ variables), mechanical variables (velocities and angles of rotor masses, inter-masses torques).

That representation has been widely used in many transient studies. However, equivalent circuits with a higher degree of complexity might be required. Table 7.2 shows the matrix of equivalent circuits with those model structures proposed in IEEE Std. 1110 [28]; see also [29]. Note that the table shows 12 combinations; however, only 7 are generally used. The selection of a model is usually based on the type of machine, the study to be performed, the user's experience, and the available information. Depending on the characteristic parameter source, the most complex models very often cannot be used due to lack of data.

The method implemented to solve the machine-network interface for the Type 59 synchronous machine model is based on reducing the machine equations to a three-phase Thevenin equivalent circuit. The reduced machine equations are added to the network equations and solved simultaneously. Using this method, there is no restriction on the number of machines which can be connected to one bus. However, since the prediction of several variables is required, the approach is more sensitive to the accumulation of errors. The solution at any time step is calculated using the following procedure [2]:

1) The machine angle is predicted.
2) The trapezoidal rule is applied to the branches of the equivalent circuits.
3) The resulting resistive networks are reduced to their Thevenin equivalent, one resistance in series with 1 V source. The values of some variables are predicted.
4) The resistive equivalent circuits, for $dq0$ quantities, are converted to phase quantities. In order to obtain a symmetric matrix, the equivalent resistances of d and q axis are averaged

Table 7.2 Synchronous machine models for transient studies.

	q-axis			
d-axis	No damper circuit	One damper circuit	Two damper circuits	Three damper circuits
Field circuit only	Model 1.0	Model 1.1	Not used	Not used
Field circuit + one damper circuit	Not used	Model 2.1	Model 2.2	Model 2.3
Field circuit + two damper circuits	Not used	Not used	Not used	Model 3.3

and the saliency terms are combined with the voltage sources into 1 V source. The values of i_d and i_q are predicted.

5) The reduced equations of each machine are added to the network equations. Instead of voltage sources in series with the resistive equivalent matrix, current sources in parallel with this resistive matrix are used. The equations of the complete network are solved.

6) The machine terminal voltages are extracted from this solution and converted to $dq0$ quantities. Next, the armature and rotor currents are calculated and used to find the electromagnetic torque. The mechanical equations are solved.

7) The predicted values are compared to those obtained at step 6); if the difference is larger than the accepted tolerance, return to step 6).

8) The process follows with step 1) to find the solution at the next time-step.

The Type 58 model has an input format fully compatible with the Type 59 model and uses a phase-domain formulation to solve the interface armature-power network; therefore, apart from the advantages and disadvantages inherent to this solution method, the limitations of this model are similar to those of the Type 59 model.

7.4.3 Universal Machine Module

The UM module was added to the BPA EMTP to provide a unique module for the study of various types of electric machines [4]. It can be used, combined with the rest of facilities available in the ATP, to represent 12 types of rotating machines [1]:

- *UM-1*. 3-phase ac synchronous machine
- *UM-2*. 2-phase ac synchronous machine
- *UM-3*. 3-phase ac induction machine – Cage rotor
- *UM-4*. 3-phase ac induction machine – Wound rotor
- *UM-5*. 2-phase ac induction machine – Cage rotor
- *UM-6*. 1-phase stator (synchronous or induction machine) ac machine – 1 phase rotor
- *UM-7*. 1-phase stator (synchronous or induction machine) ac machine – 2 phase rotor
- *UM-8*. DC machine – Separate excitation
- *UM-9*. DC machine – Series compound (long shunt) field
- *UM-10*. DC machine – Series field
- *UM-11*. DC machine – Parallel compound (short shunt) field
- *UM-12*. DC machine – Parallel field (self-excitation) armature winding on rotor to brush-commutator.

Since the first version of the UM module was implemented, several enhancements have been added to improve its efficiency and circumvent some limitations [5]. The present version of this module permits the use of the following capabilities:

a) Simulation of 12 different type of machines.

b) Simulation of mechanical systems which are arbitrary in size and configuration as equivalent electrical networks. The mechanical system is restricted to linear elements.

c) Simulation of multi-machine systems.

d) Multi-machine system sharing the same mechanical system.

e) Approximation of saturation effects.

f) Accepts data with parameters specified in SI (metric) units or in per-unit (pu) system.

g) Automatic initialization of the sinusoidal steady state.

h) Accepts three-phase synchronous machine parameters based on the standard short-circuit test, entered in Type 59 SM data format.

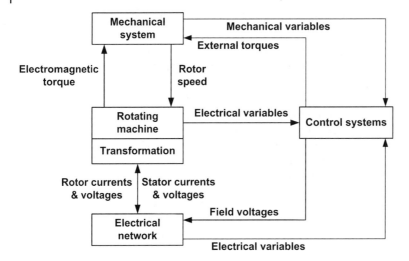

Figure 7.4 General structure of the Universal Machine module.

Figure 7.4 shows a schematic block diagram with the general interface structure of the UM. The following parts can be distinguished in this figure:

1) Electrical part, which includes the stator and the rotor of the machine.
2) Electrical network, which includes subnetworks connected to both stator and rotor.
3) Mechanical part, which must be transformed to an electrical equivalent.
4) Control systems.

The UM module does not have a built-in model for the mechanical part, so users must convert the mechanical system into an equivalent electrical network with lumped parameters. The electromagnetic torque appears as a current source in this network, which is treated by the ATP as part of the overall network. Table 7.3 describes the equivalence between mechanical and electrical components. The following rules are to be used for creating the equivalent mechanical network [2] (see Table 7.4):

- For each mass of the shaft, a node is created in the equivalent network with a capacitor to ground to represent the moment of inertia.
- If there is a mechanical torque acting on a mass, a current source is connected to the corresponding mass node.
- If there is damping proportional to speed on a mass, a resistor is located in parallel with the corresponding capacitor.
- If there are two or more masses, inductors that represent spring constants are used to connect adjacent shunt capacitors.
- If there is damping associated to this coupling, then the inductor is paralleled with a resistor.

Table 7.3 Equivalence between mechanical and electrical quantities.

	Mechanical			Electrical	
T	Torque acting on mass	[Nm/s]	i	Current into node	[A]
ω	Angular speed	[rad/s]	v	Node voltage	[V]
θ	Angular position	[rad]	q	Capacitor charge	[C]
J	Moment of inertia	[kg m^2]	C	Capacitance to ground	[F]
K	Spring constant	[Nm/rad]	$1/L$	Reciprocal of inductance	[1/H]
D	Damping coefficient	[Nms/rad]	G	Conductance	[S]

Table 7.4 Equivalence between mechanical and electrical components.

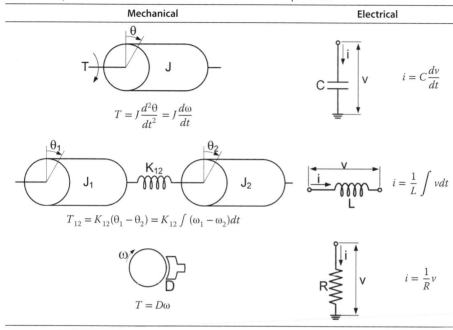

Mechanical	Electrical

$$T = J\frac{d^2\theta}{dt^2} = J\frac{d\omega}{dt}$$

$$i = C\frac{dv}{dt}$$

$$T_{12} = K_{12}(\theta_1 - \theta_2) = K_{12}\int(\omega_1 - \omega_2)dt$$

$$i = \frac{1}{L}\int v\,dt$$

$$T = D\omega$$

$$i = \frac{1}{R}v$$

Two interfacing methods for the solution of the machine equations with the rest of the network are available: Compensation and Prediction [1, 2].

Compensation based method: Each subnetwork connected to the UM is replaced by its Thevenin equivalent circuit at the terminals of the machine. These equivalent circuits are created at each time step without a knowledge of the UM variables, although they depend on the history of the UM variables. This interface method works as follows [2]:

1) Solve the complete network (armature side network, field side circuits, mechanical system) without the universal machines.
2) Extract the equivalent open-circuit voltages.
3) Predict the rotor speed.
4) Convert the Thevenin equivalent circuits from phase sequence to $dq0$ quantities.
5) Substitute the resulting equations into the difference equations of the armature side.
6) Substitute the Thevenin equivalent circuits of external networks connected to the field windings into their difference equations.
7) Solve the (linear) equations in $dq0$ quantities; obtain armature and field side currents.
8) Calculate the electromagnetic torque; use this torque as a current source in the mechanical network. Solve the equations of the mechanical network and obtain the rotor speed.
9) If the difference between predicted and calculated speeds is too large, return to step (4); otherwise, update the history terms.
10) Convert the armature currents from $dq0$ quantities to phase quantities.
11) Solve the complete network with the new terminal voltages.
12) Proceed to the next time-step.

As only one variable is predicted (i.e. the rotor angular speed), the iterations are confined to the machine equations and this approach results in a scheme free of any error amplification. On the other hand, mechanical variables change much more slowly than electrical variables and the prediction of the speed is usually good enough. Its major disadvantages come from the calculation of Thevenin equivalent circuits: this requires the subnetwork connected to the armature side to be linear, so there must not be more than one UM in the same subnetwork.

Prediction based method: Disadvantages of the compensation approach can be eliminated with an alternative interface method for the armature side connection. This new approach is based on the prediction of the armature side voltages at every time step from the knowledge of the variables calculated at previous time steps. Essentially, the UM is represented as voltage sources behind resistances. The interface method for the field side and the mechanical system is identical to that of the compensation-based method. The main differences between both methods are the following:

1) The complete network (armature and field circuits, mechanical system) is solved with UMs represented as voltage sources behind the armature resistances
2) Armature fluxes are predicted from the previous history of the same variables. The prediction is linear and takes place in the synchronously rotating reference frame. Armature-side voltages are calculated from these predicted fluxes

Regardless of the interface method, the machine equations are always solved in the reference frame, stationary with respect to the excitation side.

7.4.4   WindSyn-Based Models

WindSyn is a program created by Gabor Furst for conversion of manufacturer's machine data into electrical parameters [26]. This option uses the UM model and adds startup and machine control. ATPDraw splits the WindSyn machine handling in induction machines (WisInd.sup) and synchronous machines (WisSyn.sup), and adds two extra data parameters (i.e. Machine# and Kind) for internal use. The program recognizes up to three synchronous machine designs (salient poles with d-axis dampers; salient poles with d- and q-axis dampers, round rotor with d- and q-axis dampers) and four induction machine designs (wound rotor, single-cage rotor, double-cage rotor, and deep-bar rotor). The information required for any model includes the following common items: frequency, rated voltage, rated power, synchronous speed, rated-power factor, full-load efficiency, and moment of inertia.

7.5   Procedure to Edit Machine Data Input

The procedure to create the model of a rotating machine in ATP format may be split into three main steps: (i) choose the machine representation taking into account the transient phenomena to be simulated; (ii) perform the data conversion procedure depending on the data available and the chosen machine representation; (iii) select the ATP option that will represent the machine in the input file. The last two steps are closely related, and their order can be changed, while the first one might be made just following the guidelines proposed by the CIGRE WG [9]. The following comments discuss about the capabilities that should be used for simulating rotating machine dynamics.

1) *Choice of the machine representation.* Although it might be obvious that the rules to be followed are those suggested in the literature [9], for some cases, the choice of the best model might not be so simple due to overlapping between group frequency ranges. Therefore, at the end some experience is also useful to choose the machine representation.
2) *Data conversion procedure.* For a whole representation of a machine, it can be necessary to specify electrical and mechanical parameters, control system data, and initial conditions.

Mechanical data are only needed for Groups I and II transients, while the representation of the control system is only needed for Group I transients. An accurate representation of the mechanical part is important for some specific studies (e.g. torsional oscillations, transient torques, turbine-blade vibrations) [30]. For a more in-depth discussion about this subject, see [13–16, 31]. Synchronous machine electrical parameters may usually be obtained in one of the following forms: data supplied from manufacturer (conventional stability format data, standstill frequency response [SSFR]), data from field tests (e.g. on-line frequency response), computer calculation using the finite-element method [32]. A discussion about methods to obtain electrical and mechanical parameters can be found in [31]. Data from steady-state and short-circuit tests include reactances and time constants, armature resistance, and saturation effects. Several procedures have been proposed to pass from these data to the electrical parameters used in the transient solution of the machine [32–38]. Frequency response tests have received much attention during the last years. Several methods have been proposed to obtain parameters of d- and q-axis equivalent circuits, they are based on SSFR [39–41], and on-line frequency response [42, 43]. Some techniques have also been developed to account for saturation effects [29]. Although these tests and the corresponding procedures can also be used to obtain electrical parameters of an induction machine, data conversion procedures for these machines are performed from different specifications [29, 44, 45].

3) *Choice of the ATP option.* Three different groups of low-frequency models have been developed to solve the transient solution of a rotating machine [2]:

- *Models based on Park's transformation.* The electrical part of the machine is solved using a reference frame stationary with respect to field side; electrical parameters can be obtained by means of one of the methods mentioned above.
- *Models based on a phase-domain formulation.* The electrical part of the machine is solved in phase coordinates [7, 46, 47]; as for the previous formulation, electrical parameters can also be obtained using one of the procedures mentioned above.
- *Models based on data from frequency response tests.* The solution of the electrical part of the machine is based on the frequency response of the machine, and uses terminal impedance measurements. Such an option is not available in ATP.

7.6 Capabilities of Rotating Machine Models

The ATP representation of a rotating machine for simulating any transient is possible. For some high-frequency transients, a machine can be represented by simply using a lumped-capacitor component; for other scenarios, the simulation of high-frequency transients can be carried out by representing a machine using distributed-parameter components or some other ATP capabilities. The main drawbacks come from the fact that no data conversion procedure is implemented in the ATP to pass from machine data to ATP format input data. The rest of this section is dedicated to discussing the limitations of current ATP models for simulation of Group I transients.

The capabilities required in a transients program for an accurate simulation of any rotating machine transient might be listed as follows:

1) Several data conversion procedures are needed to pass from either manufacturer's data or field test data to internal parameters for both synchronous and induction machines, although specific procedures for each type of machine should also be available.
2) There should be no restriction for the number of coils on each rotor axis, using either $dq0$ domain or phase domain formulation; this could be especially important when machine

data come frequency response measurements. Discussions about equivalent circuits can be found in [31] and [48].

3) There should be no restriction on the number of masses of the mechanical part, and data conversion procedures should be implemented to pass from test data to mechanical parameters of each mass. Regardless of the model chosen to represent a machine, users have to know mechanical parameters, which are not always available as they have to be specified in ATP models. For a discussion about problems inherent in obtaining accurate mechanical parameters see [31] and [41].

4) It should be possible to specify different coupling between coils on a common rotor axis in all machine models. [36] and [49].

5) The effects of eddy current losses can be important in some transient studies. There are several ways to represent this phenomenon in machine windings; a strategy is to determine from frequency response measurements the parameters of additional short-circuited coils employed to represent this phenomenon [4]. Another way is to take advantage of an interface to control sections (TACS/MODELS) where this phenomenon could be represented; therefore, connection of all rotor coils to control systems is also important.

6) A multi-frequency steady-state initialization method is necessary if the machine is initially running under unbalanced conditions at its armature terminals.

Table 7.5 summarizes the most important features of the Type 59 SM model and the UM module. From the comparison of these capabilities, one can deduce that:

- the number of coils is limited to two per rotor axis for the Type 59 SM model, while there is no restriction for the UM module;
- Manual Initialization is not honoured by the Type 59 SM model;
- connection to control systems is only possible for the field coil in the Type-59 SM model, while it is possible to implement a control strategy in each rotor coil of the UM module;
- only a balanced steady-state initialization can be performed with the UM module; a Type 59 SM model can be initialized under unbalanced conditions;
- the Type 59 SM model assumes that a different coupling between coils on a common rotor axis is possible, while only one mutual flux links all UM coils on a common rotor axis.

Actually, some of these limitations are not very important since, except for very special cases, Manual Initialization does not have advantages in front of Automatic Initialization, and a different coupling between coils on one rotor axis can be represented with the UM module as all rotor coils of this module can be connected to external circuits.

The conclusions for the Type 59 SM model are also valid for the Type 58 SM model, although the solution (interface) method is different for each option.

Starting a simulation with a machine under unbalanced operating conditions is honoured by the Type 59 SM model, but it has its own limitations: when a three-phase machine is running under unbalanced conditions at its terminals, armature voltage and current harmonics are present. Therefore, only a multi-harmonic initialization method can perform a correct calculation, and this is not possible with the current ATP version.

Some other features, not shown in the table, are worth mentioning. Mechanical sharing is an interesting capability of the UM module, not available in the Type 59 SM model; it can be used for simulation of motor-generator sets. Both Type 59 SM model and UM module can be used for simulation of systems with machines running at different frequencies, provided that only single-frequency subnetworks are simulated; this feature could be used to simulate, for instance, a motor-generator set for which each machine is running at a different frequency. Several other limitations have been already recognized; see [2]. Remember that WindSyn-based options accept manufacturer's data.

Table 7.5 Capabilities of rotating machine models implemented in ATP.

Feature	Type-59 synchronous machine	Universal machine module
Electrical part	• A maximum of two coils per rotor axis is allowed. • Different coupling between rotor coils is possible. • Saturation effects can be represented.	• The machine may have any number of coils in each rotor axis. • Only one mutual flux links all windings on one axis. • Saturation effects can be represented.
Electrical parameters	The model accepts electrical parameters specified as internal parameters or based on standard tests.	The UM module accepts electrical parameters specified as internal parameters or based on standard tests for the Type 1 SM.
Mechanical part	The mechanical system may have any number of masses.	The mechanical system may have any number of masses.
Mechanical parameters	This option has a built-in model, and mechanical parameters can be specified in either Metric or English units.	There is no built-in model for the mechanical part; the user must convert the mechanical system into an electric network with lumped *RLC* parameters.
Initialization	• Only Automatic Initialization is honoured; initial conditions can be specified as voltages or power. constraints at the armature terminals. • Theoretically, the initialization can be performed for either balanced or unbalanced initial conditions.	• Manual and Automatic initialization are honoured; for the latter, initial conditions can be specified as voltages or power constraints at the armature terminals. • The initialization can be performed only for balanced conditions.

Solution method	• The electrical part is solved in Park's equations.
	• The machine equations are reduced to a three-phase Thevenin equivalent circuit; the reduced equations are added to the network equations and simultaneously solved.
	• There is no restriction on the number of machines running in parallel.
		• The electrical part is solved in transformed equations.
		• The module has two interface options:
		  ○ using Compensation, the equations of the networks connected to armature and rotor circuits are reduced to Thevenin equivalent circuits; these equivalents are added to machine equations; only one machine per node can be simulated;
		  ○ using Prediction, a machine is represented as voltage sources behind armature resistances, the interface is only used at the armature side, there is no restriction on the number of machines running in parallel.

7.7 Case Studies: Three-Phase Synchronous Machine

7.7.1 Overview

This section presents some case studies carried out with the built-in Type 59 SM model [1]. The cases simulate the response of a three-phase synchronous machine running as a stand-alone generator or as a component of a larger power system. The section illustrates the application of the Type 59 option in load rejection, transient stability, and SSR studies. The synchronous machine model is applied with different levels of complexity (e.g. single-mass versus multi-mass rotor representation). Since ATP models for representing rotating machines can only be used in low-frequency transients, the modelling guidelines applied to represent any other system component are those used in low-frequency transients. A description of the power system and a summary of modelling guidelines are provided in all case studies.

7.7.2 Case Study 1: Stand-Alone Three-Phase Synchronous Generator

Figure 7.5 shows the diagram of the test system: a stand-alone three-phase synchronous generator supplies a variable load represented by a custom-made module; see Chapter 9. The main goal of this case study is to analyse the transient response of the synchronous generator when the load is varied. The study is carried out without including any generator controller in the system model.

Figure 7.6 shows the circuit implemented in ATPDraw for testing the transient behaviour of the synchronous machine. The circuit has been created to simulate the response of the synchronous machine when there is a load variation, the machine is temporarily disconnected, or there is a fault at the machine terminals. The parameters of the synchronous generator are provided in Table 7.6. The window with the input of generator parameters is shown in Figure 7.7. A single-mass model is used for representing the mechanical system of the turbine-generator model, and only the moment of inertia is specified (in metric units).

Figure 7.8 depicts some simulation results corresponding to a load variation. The case has been simulated assuming the initial voltage at the generator terminals is the rated one. The way in which the load varies is more or less evident from Figure 7.8b, which displays the stator current. The plot of the rotor speed deviation proves that the generator comes back to the original speed after a transient period in which the speed decreases as a consequence of the load increment. Since the load is not the same at the end of the simulation and the model does not include controllers, the operating conditions (i.e. voltages, currents and rotor speed) are not the same that at the beginning of the simulation, although differences are not too large.

7.7.3 Case Study 2: Load Rejection

Load rejection is the condition created by a sudden load loss in the system. This event can cause temporary overvoltages, which can reach values that are a function of the rejected load, the system topology, and the characteristics of the sources (e.g. speed and voltage regulators of generators). A distinction should be made between system configurations when large loads

Figure 7.5 Case Study 1: Diagram of the test system.

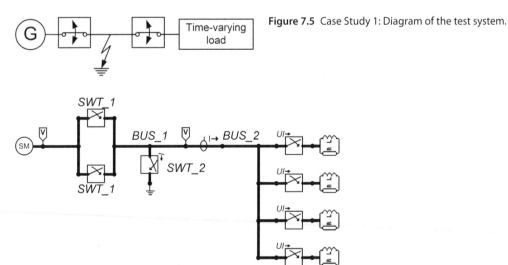

Figure 7.6 Case Study 1: ATPDraw implementation of the test system.

Table 7.6 Case Study 1: Parameters of the synchronous generator.

Parameter	Value	Units
Phase-to-line rated voltage	11	kV (rms)
Three-phase rated power	60	MVA
Rated frequency	50	Hz
Number of poles	4	—
Field current	20	A
Armature resistance	0.003	pu
Armature leakage reactance	0.132	pu
d-axis synchronous reactance	1.825	pu
q-axis synchronous reactance	1.571	pu
d-axis transient reactance	0.172	pu
q-axis transient reactance	0.181	pu
d-axis subtransient reactance	0.143	pu
q-axis subtransient reactance	0.158	pu
Open-circuit d-axis transient time constant	1.019	s
Open-circuit q-axis transient time constant	0.632	s
Open-circuit d-axis subtransient time constant	0.031	s
Open-circuit q-axis subtransient time constant	0.027	s
Zero-sequence reactance	0.132	pu
Canay's characteristic reactance	0.132	pu
Moment of inertia of rotor mass	0.013	Million kg m^2

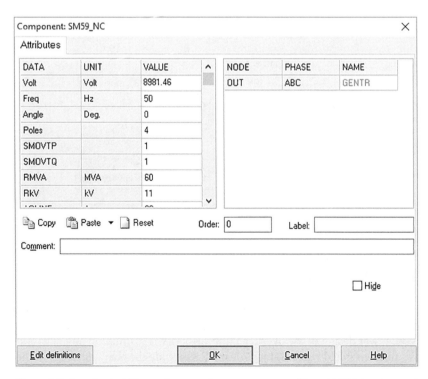

Figure 7.7 Case Study 1: Three-phase synchronous generator (Type 59 SM) data input window.

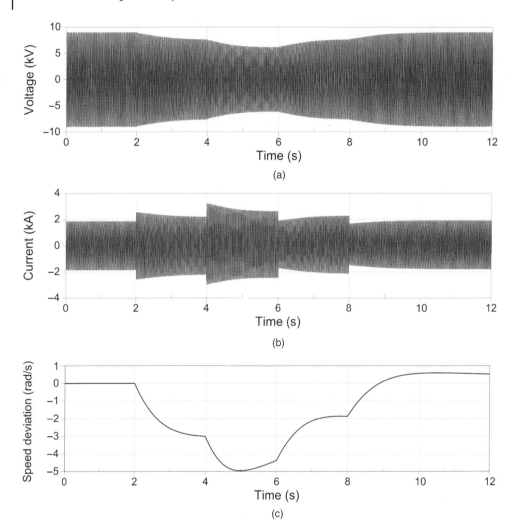

Figure 7.8 Case Study 1: Load variation at the generator terminals. (a) Generator terminal voltages, (b) Load current, (c) Rotor speed deviation.

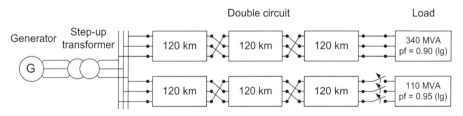

Figure 7.9 Case Study 2: Diagram of the test system.

are rejected; in general, the highest overvoltages will occur in systems with long lines and low short-circuit power at the generation side of the lines. Figure 7.9 shows the diagram of the test system: a synchronous generator supplies power to a load by means of a long overhead transmission line; the line is a double circuit with different load condition at each circuit terminals.

Figure 7.10 shows the tower design of the overhead line, a 400-kV transmission line with two conductors per phase and two shield wires. Tables 7.7 and 7.8 provide the characteristics of the

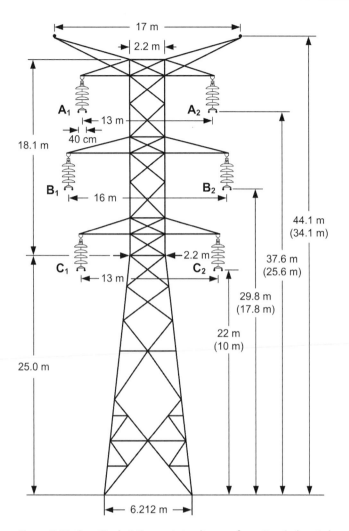

Figure 7.10 Case Study 2: Transmission line configuration (values in brackets are midspan heights).

Table 7.7 Case Study 2: Characteristics of conductors and shield wires of the transmission line.

	Conductor type	Diameter (mm)	DC resistance (Ω/km)
Phase conductors	CONDOR	27.72	0.0718
Shield wires	AW7	9.78	1.4630

overhead line conductors, and the parameters of the generator and the step-up transformer, respectively. Details about the loads are provided in Figure 7.9.

The circuit implemented in ATPDraw to simulate this system without any generator controller is shown in Figure 7.11. Details about the implementation of the step-up transformer and the overhead transmission line are shown in Figures 7.12 and 7.13. The menu window for the synchronous generator is similar to that of Case Study 1 (see Figure 7.7) with the new machine parameters. A single-mass model is again used to represent the mechanical system

Table 7.8 Case Study 2: Parameter of the synchronous generator and the transformer.

	Synchronous generator	
Parameter	**Value**	**Units**
Phase-to-line rated voltage	19	kV (rms)
Three-phase rated power	450	MVA
Rated frequency	50	Hz
Number of poles	2	—
Field current	500	A
Armature resistance	0.0035	pu
Armature leakage reactance	0.1985	pu
d-axis synchronous reactance	1.6025	pu
q-axis synchronous reactance	1.4024	pu
d-axis transient reactance	0.4095	pu
q-axis transient reactance	0.4683	pu
d-axis subtransient reactance	0.3168	pu
q-axis subtransient reactance	0.2645	pu
Short-circuit d-axis transient time constant	2.1735	s
Short-circuit q-axis transient time constant	0.8632	s
Short-circuit d-axis subtransient time constant	0.0442	s
Short-circuit q-axis subtransient time constant	0.0313	s
Zero-sequence reactance	0.1785	pu
Canay's characteristic reactance	0.1985	pu
Moment of inertia of rotor mass	0.1614	Million kg m^2

	Step-up transformer	
Line-to-line rated voltages	19/400	kV
Rated power	500	MVA
Winding connections	DY 11	—
Core design	Triplex core	—
Short-circuit test – voltage	6.58%	%
Short-circuit test – losses	500	kW
No-load test – saturation curve	See Figure 7.12	—

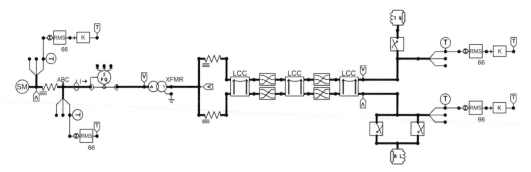

Figure 7.11 Case Study 2: ATPDraw implementation of the test system.

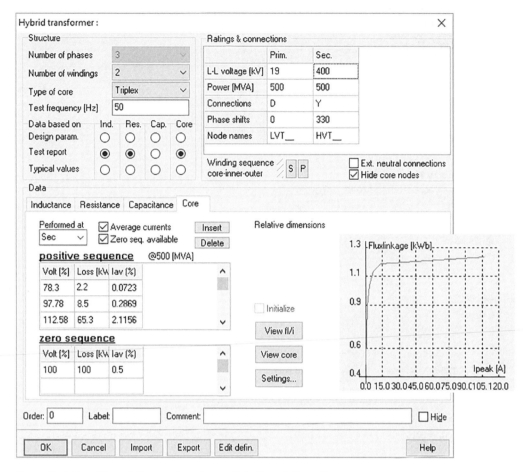

Figure 7.12 Case Study 2: Step-up transformer (hybrid model) window.

of the turbine-generator set. The transformer is represented with the 'Hybrid' option, while the transmission line sections are represented by means of a constant and lumped parameter model (i.e. Bergeron option). Custom-made models are used to represent the loads and measure the active and reactive powers supplied by the synchronous generator. The TACS device 66 is used to obtain the rms values of voltages and currents.

Figure 7.14 shows some simulation results that correspond to a temporary disconnection of Load 2 during two seconds (Figure 7.11). The initial voltage at the generator terminals is the rated one. The main conclusions derived from these results can be summarized as follows:

- The voltages at the generator and load terminals significantly increase after the disconnection of Load 2; the voltage at the node to which Load 2 is connected, reaches a value close to 1.4 pu. After the load is reconnected, all voltages come to their original values.
- The value of the active power measured at the generator terminals decreases immediately after load loss, but it changes before load reconnection; this is due to the values reached by voltages during the load loss and the fact that loads are represented as constant impedances.
- Since the reactive power generated by the line exceeds that demanded by loads, the reactive power at the generator terminals is always negative, and it increases during the load loss. The consequence is a generator terminal voltage higher than the constant internal voltage.

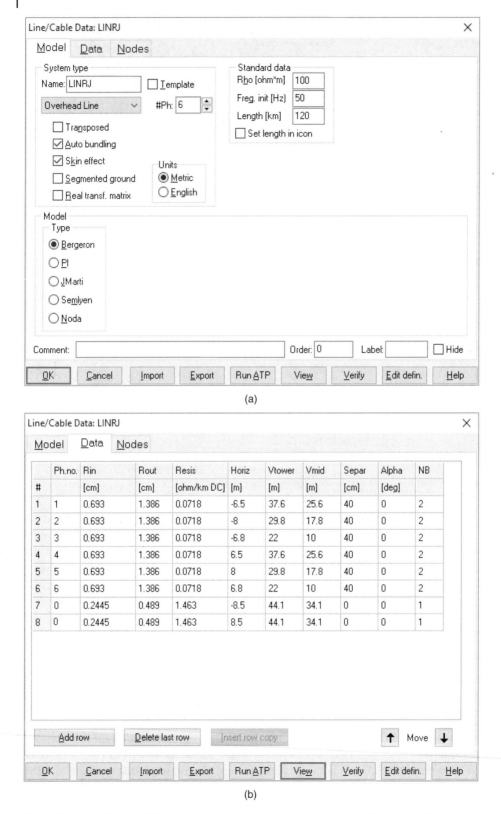

Figure 7.13 Case Study 2: LCC menus for the overhead transmission line. (a) Selection of line model and some line specifications, (b) Conductor geometry and physical characteristics.

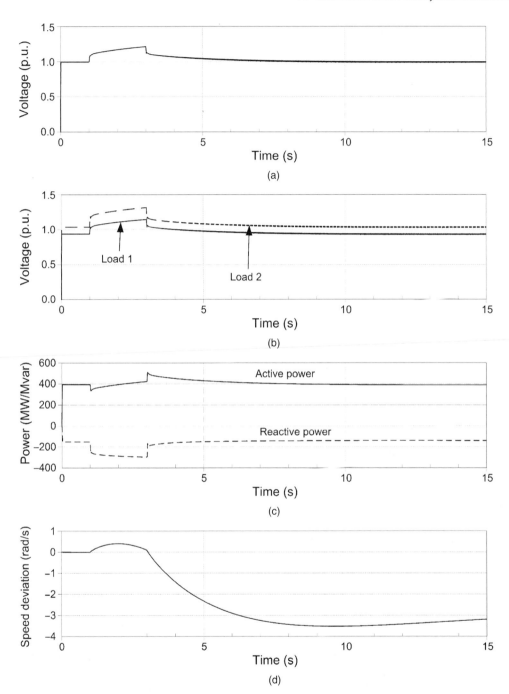

Figure 7.14 Case Study 2: Simulation results – Load rejection (temporary load loss). (a) Voltage at the generator terminals, (b) Voltages at the load terminals, (c) Active and reactive power at the generator terminals, (d) Generator speed deviation.

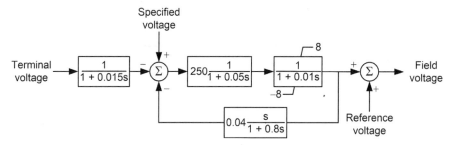

Figure 7.15 Case Study 2: Diagram of the excitation controller.

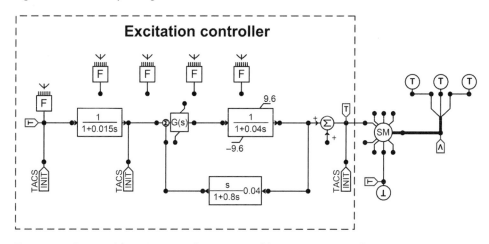

Figure 7.16 Case Study 2: ATPDraw implementation of the excitation controller.

- The rotor speed deviation initially exhibits a pattern opposite to that of the active power (i.e. it is positive after the load loss), and becomes negative after the load reconnection. According to the plot, the rotor speed slowly comes back to their original value.

A second study aimed at assessing the effect of the excitation control has been carried out. Figure 7.15 shows the diagram of the controller based on a model proposed by IEEE; see [50]. Figure 7.16 displays the ATPDraw implementation of the controller; note how the icon of the synchronous machine changes when a controller is added to the machine.

Figure 7.17 displays the simulation results with the control of the excitation:

- Voltage variations are as large as with the previous scenario; however, the voltage at the generator terminals is 1 pu before load disconnection and after load reconnection.
- The field voltage exhibits quick variations at the time the load is lost and the time the load is reconnected. This voltage reaches values close to 3 pu.
- The speed deviation of the machine rotor is larger during the disconnection period than in the previous study and, as with the previous study, it slowly recovers its initial value. These results support the need of a speed control (i.e. a governor) to recover the synchronous speed of the generator rotor.

Since the voltages increase during the load-loss period, the decrement of the active power measured at the generator terminals (not shown in the figure) is not as large as the actual decrement of the line active power, and its value remains below the initial value during the transient period. The behaviour of the reactive power is very similar to that of the previous study, although the values are different, and it recovers its initial value as soon as Load 2 is reconnected.

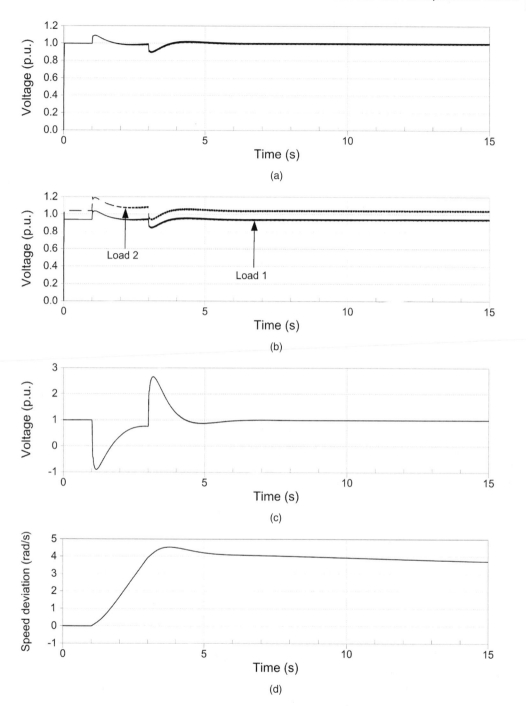

Figure 7.17 Case Study 2: Simulation results – load rejection (with excitation control). (a) Voltage at the generator terminals, (b) Voltages at the load terminals, (c) Field winding voltage, (d) Generator speed deviation.

7.7.4 Case Study 3: Transient Stability

Power systems never operate at steady state: the load continuously changes and the generation continuously responds to the load change to maintain the system frequency within acceptable levels. In addition, the power system is subject to disturbances due to faults and manoeuvres. Once a fault is detected by the protection system, the faulted section must be isolated to prevent the disturbance from spreading into the rest of the network. A disturbance causes a mismatch of power generation and consumption, which, in turn, results in disturbing the system frequency, voltages, and the speed of generators. A load rejection event is a particular case of disturbance.

A stable power system is capable of returning to a new steady-state operation with satisfactory voltage levels and system frequency. The stability of power systems can be classified based on the following considerations [51]: (i) the physical nature of the resulting mode of instability as indicated by the main variable in which instability can be observed; (ii) the nature of the disturbance, which influences the method of calculation and prediction of stability; (iii) the devices, processes, and the timespan that must be taken into consideration in order to assess stability. The response of a power system, and therefore its stability, depends on the operating conditions at the time a disturbance occurs, the design and performance of the protection system, or the response of the various controllers of the generators affected by a given disturbance. It is important to mention that a power system can be subject to different forms of instability. For more details, see [51].

The application of the ATP to the analysis of the so-called rotor angle stability is illustrated with two cases: a simple generator-infinite bus system and a multi-machine system.

Figure 7.18 shows the configuration of the first test system: a synchronous generator is connected to a power system, represented by an infinite bus, by means of a step-up transformer, a short cable, and two transmission lines. The parameters of the generator and the transformer are provided in Table 7.9.

The circuit implemented in ATPDraw to analyse the transient stability of this system is depicted in Figure 7.19. This figure includes some information about the various test system components and the tasks carried out by some ATP capabilities. Note that FIX SOURCE option (see Chapters 3 and 4) is used to obtain the initial steady state of the test system. The generator model does not include any type of control. The cable and the two transmission lines are represented by means of mutually coupled RL elements, characterized by their zero and positive sequence impedances; see figure.

Figure 7.20 shows some simulation results that correspond to a three-phase fault at a location in Line 2 close to the point of coupling with the cable (see Figures 7.18 and 7.19). The fault is permanent, so the switches located at both terminals of the faulted line will open. In the simulated case, the response of the protection system is delayed 120 ms (i.e. 10 cycles of the power frequency). The operating conditions in the test system at the time the fault occurs

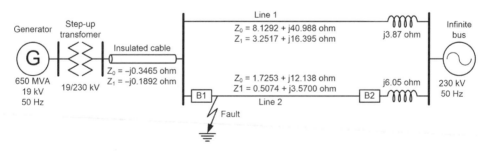

Figure 7.18 Case Study 3A: Test system configuration.

Table 7.9 Case Study 3A: Parameters of the synchronous generator and the transformer.

Synchronous generator		
Parameter	Value	Units
Phase-to-line rated voltage	19	kV (rms)
Three-phase rated power	650	MVA
Rated frequency	50	Hz
Number of poles	2	—
Field current	1250	A
Armature resistance	0.005	pu
Armature leakage reactance	0.194	pu
d-axis synchronous reactance	2.250	pu
q-axis synchronous reactance	2.200	pu
d-axis transient reactance	0.350	pu
q-axis transient reactance	—	pu
d-axis subtransient reactance	0.260	pu
q-axis subtransient reactance	0.223	pu
Open-circuit d-axis transient time constant	6.740	s
Open-circuit q-axis transient time constant	—	s
Open-circuit d-axis subtransient time constant	0.023	s
Open-circuit q-axis subtransient time constant	0.157	s
Zero-sequence reactance	0.194	pu
Canay's characteristic reactance	0.194	pu
Moment of inertia of rotor mass	0.752	Million lb ft^2

Step-up transformer		
Line-to-line rated voltages	19/226.6	kV
Rated power	—	MVA
Winding connections	DY 11	—
Core design	Triplex core	—
Primary winding – impedance	$0.01487 + j0.11900$	Ω
Secondary winding – impedance	$0.70133 + j5.61060$	Ω
No-load test – saturation curve	No saturation	—

are fixed by the FIX SOURCE capability; in this case, the voltage at the generator terminals (phase-to-ground peak value) and the three-phase active power injected by the generator into the system are specified. The voltage is 15.513 kV, and corresponds to the rated peak value, while the active power is 500 MW.

In transient stability studies, two variables of concern are the rotor angle and speed of the generators affected by the disturbance. As shown in the figure, both the angle and the speed deviation of the synchronous generator exhibit the typical oscillation of a stable system: during the fault condition, the rotor speed increases linearly; this increment does not stop at the moment the fault is cleared (i.e. when the faulted line is separated from the system), but if

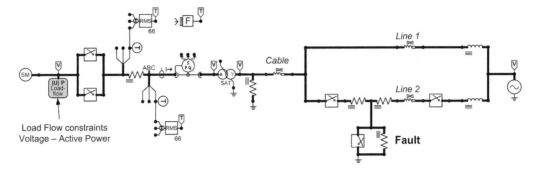

Figure 7.19 Case Study 3A: ATPDraw implementation of the test system.

the system is stable, the increment will stop and reverse the sense. The rotor speed deviation oscillates during a certain period and reaches a final zero value; this means that the generator is synchronized again to the power system and the system is stable.

However, this does not mean that other system and generator variables will reach the same values they had at the beginning of the transient process. As controllers are not included in the synchronous generator model, the internal generator voltage and the infinite bus voltage will remain the same during the simulation. Since the impedance between the two nodes is now larger (impedance of Line 2 is no longer in parallel with that of Line 1), the steady-state current between the two nodes will be smaller after Line 2 has been separated from the system. This will affect not only the stator currents (see Figure 7.20c) but also the terminal voltage of the generator (which is steadily increasing as shown in Figure 7.20a), the rotor angle, or the active and reactive powers injected into the system (see Figure 7.20b).

To check the system behaviour, two new disturbances have been simulated: the faults are again permanent and the protection system separates the faulted line after 120, 180 and 185 ms, respectively. Figure 7.21 compares the speed deviation of the machine rotor that results from the previous case and the two new scenarios. From these results, it is evident that adequately setting the protection system is important to avoid an unstable behaviour.

A new transient stability study has been carried out with a multi-machine power system including a representation of voltage and speed controllers. Table 7.10 provides the main parameters of the power system components and the load values. Figure 7.22 shows the ATP-Draw implementation of the system. With this system model, the response to both temporary and permanent bolted faults in Line 7–8, at a location close to Bus 7, can be simulated. Although this case analyses the transient response caused by a permanent fault, and the response of the breakers located at the two ends of Line 7–8 is correct, the response to a failure of the protection system can also be simulated.

Some simplifications are introduced in all system components; see Table 7.10. The transformer models are lossless and unsaturable, and the transmission lines are represented by means of three uncoupled PI circuits.

The system will be tested with models of voltage and speed controllers in the three generators. Figure 7.23 shows the diagrams and the parameters of the two controllers; they are the same for all generators.

Figure 7.24 shows some simulation results corresponding to a temporary fault of duration of 120 ms (i.e. six cycles of power frequency). These results prove that the system is stable for this disturbance: mechanical powers from the turbines exhibit long-duration oscillations, but their values are within reasonable limits and come back to unity, as field voltages; voltages at generator terminals remain at their original values after a transient period; the rotor angular

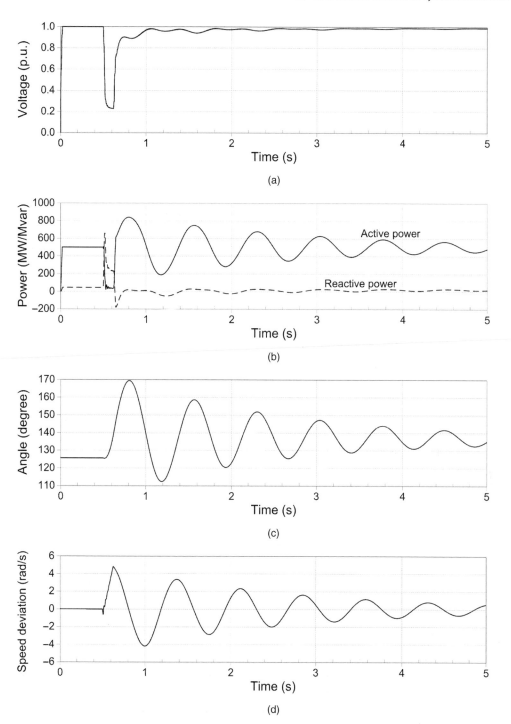

Figure 7.20 Case Study 3A: Simulation results – transient stability (without excitation control): fault clearing: 120 ms. (a) Voltage at the generator terminals, (b) Active and reactive power at the generator terminals, (c) Rotor angle, (d) Generator speed deviation.

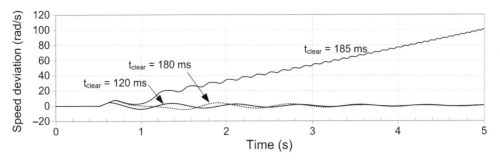

Figure 7.21 Case Study 3A: Simulation results – transient stability: different fault durations.

speeds (not shown in the figure) have a limited deviation and return to the synchronous speeds. Active powers measured at generator and load terminals also exhibit some oscillations, mainly due to the oscillations of currents (not shown in the figure) although they also come back to their original values.

The plots of the figure confirm the different nature of electromagnetic and electromechanical transient: it is evident that the oscillations of the mechanical variables (i.e. mechanical powers) have a frequency much lower than those of voltages, currents and powers.

Protective devices in actual transmission systems have a rather fast response so it should be expected that for this fault duration the line would have been separated from the system. However, this is a different disturbance since it will affect the system configuration and cause a load-flow reconfiguration and different share of active power between generators. Therefore, the system has been simulated again assuming the protection system separates the fault line. The new results, not displayed here, and for which the protection system separates the faulted line 120 ms after the fault occurrence, again show a stable system, since rotor speed deviations are limited and the speeds come back to the synchronous values, although they exhibit slow oscillations (i.e. several seconds). Interestingly, the system exhibits a large stability margin in front of this disturbance since only a very late operation of the protection system could lead to an unstable response. It is also important to mention that steady-state calculations with the Line 7–8 disconnected prove that the system can operate without any problem.

7.7.5 Case Study 4: Subsynchronous Resonance

The mechanical shaft system of a turbine-generator set has a natural resonant frequency that may fall below the power system (synchronous) frequency. This could mean that as the shaft is being accelerated from standstill to normal operating angular speed, there is a point at which the machine experiences more than the usual degree of vibration.

SSR is a phenomenon in which the resonant frequency of the turbine-generator shaft coincides with a natural resonant frequency of the electrical system such that there is a sustained exchange of energy between the mechanical shaft and the electrical system [13, 15, 16].

A very common scenario for SSR will usually occur in long-distance transmission systems with series compensated line: an unstable exchange of energy between the mechanical shaft system of a turbine-generator set and the series capacitor compensated transmission line to which it is connected. This exchange of energy results in torsional stress on the turbine generator shaft that can lead to severe damage. In extreme cases, the shaft can actually fracture. The analysis of the SSR phenomenon can be performed by using several methods: frequency scanning, eigenvalue analysis in the frequency domain, and time-domain transient analysis using a tool such as ATP.

Table 7.10 Case Study 3B: Parameters of system components and loads.

Synchronous generators				
	Value			
Parameter	**Generator 1**	**Generator 2**	**Generator 3**	**Units**
Phase-to-line rated voltage	16.5	18	13.8	kV (rms)
Three-phase rated power	247.5	192	128	MVA
Rated frequency	50	50	50	Hz
Number of poles	4	2	2	—
Field current	450	410	325	A
Armature resistance	0.000 01	0.000 01	0.000 01	pu
Armature leakage reactance	0.083 16	0.100 03	0.094 98	pu
d-axis synchronous reactance	1.361 35	1.719 94	1.681 00	pu
q-axis synchronous reactance	0.239 83	1.659 84	1.610 00	pu
d-axis transient reactance	0.150 48	0.230 02	0.232 06	pu
q-axis transient reactance	0.239 83	0.378 05	0.320 00	pu
Open-circuit d-axis transient time constant	8.960 00	6.000 00	5.890 00	s
Open-circuit q-axis transient time constant	—	0.535 00	0.600 00	s
Zero-sequence reactance	0.150 00	0.150 00	0.150 00	pu
Canay's characteristic reactance	0.083 16	0.100 03	0.094 98	pu
Moment of inertia of rotor mass	1.315 73	1.213 70	1.100 51	Million lb ft^2

Step-up transformers				
	Value			
Parameter	**TR-1**	**TR-2**	**TR-3**	**Units**
Line-to-line rated voltages	16.5/230	18/230	13.8/230	kV
Winding connections	DY 11	DY 11	DY 11	—
Core design	Triplex core	Triplex core	Triplex core	—
Primary winding – impedance	j0.23522	j0.30375	j0.16739	Ω
Secondary winding – impedance	j15.2350	j16.5310	j15.4990	Ω

Transmission lines		
	Parameters	
Line	**Series impedance/Shunt capacitance**	**Units**
Bus4 – Bus5	$Z = 5.2900 + j44.965\ Y = j166.35$	Ω/µS
Bus4 – Bus6	$Z = 8.9930 + j48.668\ Y = j149.33$	Ω/µS
Bus7 – Bus5	$Z = 16.928 + j85.169\ Y = j289.22$	Ω/µS
Bus7 – Bus8	$Z = 4.4965 + j38.088\ Y = j140.83$	Ω/µS
Bus9 – Bus6	$Z = 20.631 + j89.930\ Y = j338.37$	Ω/µS
Bus9 – Bus8	$Z = 6.2951 + j53.323\ Y = j197.54$	Ω/µS

Loads	
Load 1	230 kV, 150 MVA, pf $= 0.95$ (lg)
Load 2	230 kV, 150 MVA, pf $= 0.95$ (lg)
Load 3	230 kV, 150 MVA, pf $= 0.95$ (lg)

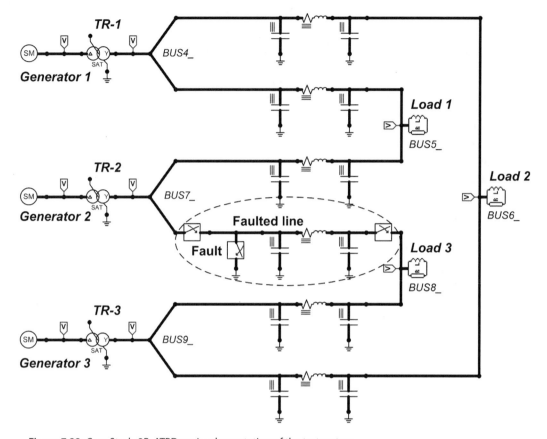

Figure 7.22 Case Study 3B: ATPDraw implementation of the test system.

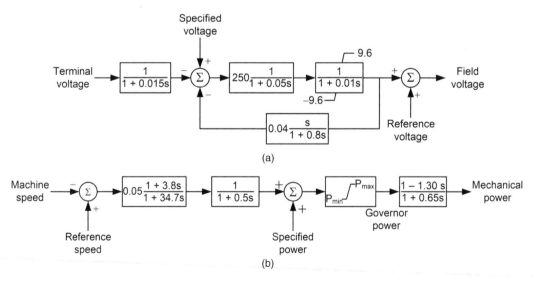

Figure 7.23 Case Study 3B: Diagrams of generator controllers. (a) Control of excitation (voltage control), (b) Governor (speed control).

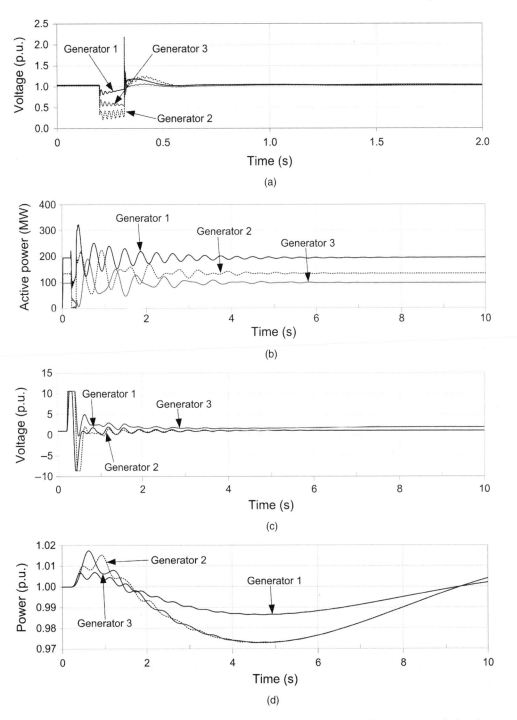

Figure 7.24 Case Study 3B: Simulation results – transient stability with controllers: temporary fault – clearing after 120 ms. (a) Generator terminal voltages, (b) Active powers at the generator terminals, (c) Field voltages, (d) Mechanical input powers.

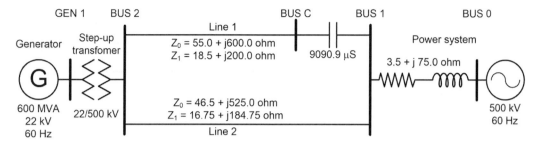

Figure 7.25 Case Study4: Test system configuration for subsynchronous resonance studies (Second IEEE Benchmark).

Two common SSR modes have been detected in series-compensated systems: self-excitation and torque amplification [13, 15, 16, 52]: The first mode occurs when small disturbances lead to unstable build-ups of shaft torque amplifications over a long period of time. Torque amplification occurs as a consequence of large disturbances that can lead to a stable system but causes high-level transient shaft torques and significant fatigue loss-of-life. The case study presented here is based on the Second IEEE Benchmark System for SSR Studies [16].

Figure 7.25 presents the diagram of this test system. Table 7.11 provides the parameters of the synchronous generator and the step-up transformer. Line parameters are shown in the figure. A 55% compensation level has been chosen for the series-compensated line. Note that the mechanical shaft system is modelled as a multi-mass system. In addition, the electrical system assumes a saturable model: the field current value to obtain rated voltage on the air gap line is 1970 A, while the current to obtain a 1.2 pu voltage on the saturation characteristic is 2600 A [16, 52].

To create the conditions for obtaining SSR, the following aspects have been considered [52]:

- The phase angle of the generator terminal voltage has been fitted to obtain a no-load condition (i.e. a zero-power injection into the power system).
- The natural frequencies of the mechanical shaft system can be obtained from the solution of an eigenvalue problem. For the synchronous generator under study these frequencies are 0, 24.65, 32.39, and 51.10 Hz [16, 52].
- The natural frequency of the series *LC* circuit (upper line in Figure 7.25) is given by

$$f_n = f \cdot \sqrt{\frac{Comp}{100}} \tag{7.1}$$

where f is the power frequency of the test system (i.e. 60 Hz for this system) and *Comp* is the compensation percentage of the line impedance.
- With the 55% compensation level selected for the series compensated line, resonance can occur at 24.65 Hz, the first natural frequency; see [16] and [52].
- The energy exchange occurs in the machine air gap at the complementary frequency, in this case 35.35 Hz. According to the study presented in [52], the system will be stable at this mode of oscillation with a damping of 0.40 rad/s, which is much larger than the natural damping.

Figure 7.26 depicts the circuit implemented in ATPDraw for simulating the SSR phenomenon. The circuit includes a power monitor to check the initial operating conditions and some connectivity requirements at both sides of the step-up transformer. To achieve the compensation level obtained above, a three-phase capacitor bank with 9090.9 µS per phase is connected in series with the upper line. The mechanical parameters were converted to Metric units.

Table 7.11 Case Study 4: Parameters of system components and loads.

Synchronous generator		
Parameter	**Value**	**Units**
Phase-to-line rated voltage	22	kV (rms)
Three-phase rated power	600	MVA
Rated frequency	60	Hz
Number of poles	2	—
Field current – rated voltage	1970	A
Field current – $1.2 \times$ rated voltage	2600	A
Armature resistance	0.0045	pu
Armature leakage reactance	0.14	pu
d-axis synchronous reactance	1.65	pu
q-axis synchronous reactance	1.59	pu
d-axis transient reactance	0.25	pu
q-axis transient reactance	0.46	pu
d-axis subtransient reactance	0.20	pu
q-axis subtransient reactance	0.20	pu
Open-circuit d-axis transient time constant	4.50	s
Open-circuit q-axis transient time constant	0.55	s
Open-circuit d-axis subtransient time constant	0.04	s
Open-circuit q-axis subtransient time constant	0.09	s
Zero-sequence reactance	0.14	pu
Moment of inertia – exciter	0.001 383	Million lb ft^2
Moment of inertia – generator	0.176 204	Million lb ft^2
Moment of inertia – low pressure turbine	0.310 729	Million lb ft^2
Moment of inertia – high pressure turbine	0.049 912	Million lb ft^2
Damping – exciter	4.3	lb ft s/rad
Damping – generator	547.9	lb ft s/rad
Damping – low pressure turbine	966.2	lb ft s/rad
Damping – high pressure turbine	155.2	lb ft s/rad
Spring constant – exciter-generator	4.39	Million lb ft/rad
Spring constant – generator-low pressure turbine	97.97	Million lb ft/rad
Spring constant – low pressure turbine-high pressure turbine	50.12	Million lb ft/rad

Step-up transformers		
Line-to-line rated voltages	22/500	kV
Winding connections	YD 11	—
Core design	Triplex core	—
High-voltage winding – impedance	$0.250\,00 + j25.000$	Ω
Low-voltage winding – impedance	$0.001\,45 + j0.1452$	Ω

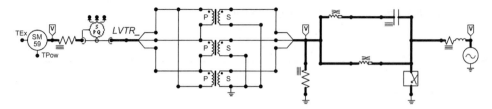

Figure 7.26 Case Study 4: ATPDraw implementation of the test system – Second IEEE Benchmark.

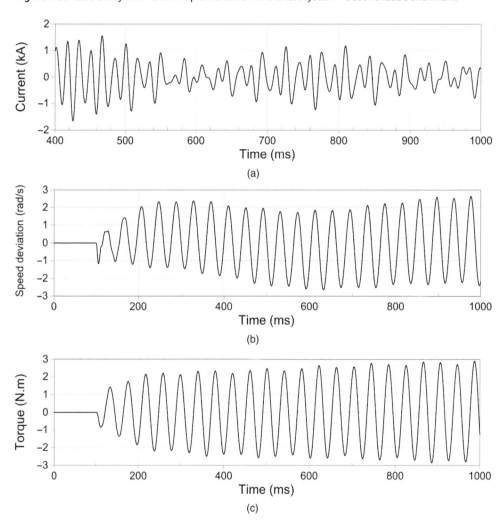

Figure 7.27 Case Study 4: Simulation results – self-excitation case (fault cleared after 0.2 ms). (a) Currents injected by the generator (after fault clearing), (b) Rotor speed deviation, (c) GEN-LP shaft torque.

Figure 7.27 shows some simulation results corresponding to a three-phase bolted fault at the right side of the parallel lines, see Figure 7.26. The duration of the fault is 0.2 ms. This will not cause any stability problem but will fire an SSR phenomenon as shown in the plots:

- The currents (and powers) injected by the generator into the power system are negligible before the fault; however, after fault clearing, they remain oscillating, and reach non-small values; see Figure 7.27a.

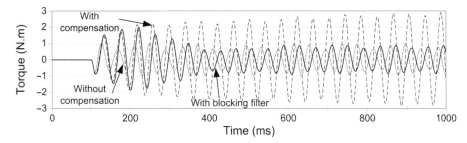

Figure 7.28 Case Study 4: Comparison of results.

- The rotor speed deviation remains oscillating, but within a narrow, limited range of values; see Figure 7.27b.
- The oscillations of the GEN-LP torque steadily grow, which can cause fatigue loss-of-life.

Active and reactive powers (not shown in the figure) remain oscillating within values that are larger than those at the beginning of the transient. Both active and reactive flows are bidirectional and oscillate at the predicted frequency 24.65 Hz.

Several countermeasures have been proposed and implemented to avoid the SSR phenomenon [13, 52, 53]. The list includes excitation dampers, dynamic stabilizers, static blocking filters, line filters, bypass damping filters, dynamic filters, torsional motion relay, or armature current SSR relays. In addition, operators can install monitors to detect and measure vibrations in the mechanical shaft system and decide a procedure to avoid damages.

Figure 7.28 shows the result derived from the application of a static blocking filter. The parameters of the filter are those proposed in [51]. The figure compares three scenarios: without series compensation, with series compensation, and with series compensation and static blocking filter. The figure shows how a damped shaft torque is achieved with the static blocking filter: after an initial period in which the fault causes some self-excitation and an increased shaft torque, the torque oscillations are later damped and reach values similar to those derived from a case without series compensation. The simulation of the test system without series compensation confirms that the SSR phenomenon is due to the presence of a series-compensated line. In all cases, the phase angle of the generator terminal voltage has been fitted to achieve a negligible initial power.

7.8 Case Studies: Three-Phase Induction Machine

7.8.1 Overview

This section presents some case studies based on the UM module which is used for simulating three-phase induction machines running either as motors or generators. The cases are dedicated to duplicate the behaviour of three-phase induction machines with short-circuited or squirrel-cage induction machine (SCIM), and wound rotor or doubly-fed induction machine (DFIM). The UM options for simulating each machine design are respectively known as UM-3 for SCIM, and UM-4 for DFIM. The case studies cover from very simple test systems, aimed at presenting the options available in ATP for simulating three-phase IMs operating under steady-state conditions or the startup of IMs, to more complex scenarios in which the IMs are part of a larger system.

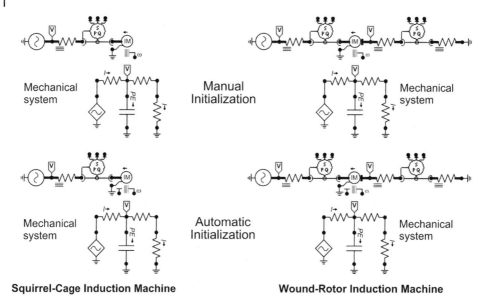

Mechanical system	Manual Initialization
Mechanical system	Automatic Initialization

Squirrel-Cage Induction Machine **Wound-Rotor Induction Machine**

Figure 7.29 Case Study 5: ATPDraw implementation of the test systems.

7.8.2 Case Study 5: Induction Machine Test

Figure 7.29 depicts the circuits implemented in ATPDraw for simulating this case study. Each part of the figure presents SCIM and DFIM models directly connected to ideal voltage sources respectively. The ratings and parameters of the SCIMs are as follows:

- Ratings: Voltage = 400 V, Stator connection = Y, 2 pole pairs, Non-saturable model
- Electrical system: d/q-axis magnetization inductance = 25.47 mH

  Stator zero sequence impedance $R = 0.0630\,\Omega$, $L = 0.1717\,\text{mH}$

  Stator d/q-axis impedance $R = 0.0630\,\Omega$, $L = 0.3535\,\text{mH}$

  Rotor d/q-axis impedance $R = 0.1020\,\Omega$, $L = 0.1020\,\text{mH}$

- Mechanical system: Moment of inertia = 3 kg m^2, Damping = 0.01 Nm/(rad/s)

Each case includes an additional circuit needed to represent the mechanical system of the machines. Custom-made modules for measuring active and reactive powers (flowing to the stator of both machines, and in the rotor of a DFIM) are included in the circuits implemented in ATPDraw. Some additional details and explanations are provided in the figure. The circuits can be used to obtain the steady-state operation and simulate the transient response of the two machines during the startup. The two machines can run either as motor or generator; the differences are in the initial slip (positive for a motor; negative for a generator) and the sign of the current source that represents the electromagnetic torque (positive for a motor; negative for a generator).

Two options have been implemented in the UM module for solving this case study.

- *The Automatic Initialization option.* It is required to obtain the steady state of an induction machine (running either as motor or generator). Any later transient response can be based in Compensation or Prediction; see Figure 7.30. Compensation is the usual choice when only

Figure 7.30 Specification of the UM initialization option ('Settings' menu).

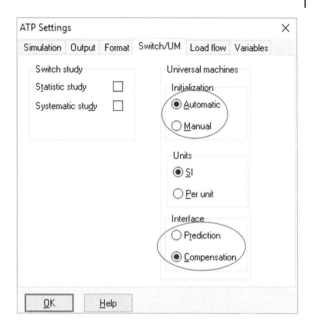

one machine is to be represented; Prediction must be selected when there is more than one machine running in parallel (see Section 7.4.3).

- *The Manual Initialization option.* It can be selected when all the machines to be simulated are initially 'dead'. The goal of the second part of this case study is to simulate the startup of an IM in which all its variables are initially zero.

When using ATPDraw, the selection of Manual or Automatic Initialization, Compensation or Prediction is made in the Settings window; see Figure 7.30. An example of the input menu for a SCIM (represented by the UM-3 type) is shown in Figure 7.31.

Figure 7.32 provides some simulation results that correspond to the steady-state operation and the startup of a SCIM running as a motor. The current source connected to the mechanical network in the circuits using Manual Initialization represent an external torque that is not included in the initialization; in this case, this current source is activated at $t = 1$ second; see right-side plots. One can observe the large values reached by currents, torque and powers during the initial period of the transient caused by the startup. Note also how the rotor speed decreases after the external torque has been applied.

For this simple case study, the results obtained with both types of induction machines would be the same; the differences are mainly due to the different parameters used to represent the rotor of each machine type.

The main differences between the results obtained when any IM design runs as motor or generator is the sign of the active power and the electromagnetic torque, and the value of the rotor speed with respect the synchronous speed: (i) the active power flows from the source to the machine stator when the machine runs as a motor, and from the machine stator to the source when it runs as a generator; (ii) in all cases, the reactive power measured at the stator side of the both machine types is flowing from the source to the machine stator, no matter if the machine runs as a motor or generator; (iii) the electromagnetic torque value is negative for a motor and positive for a generator; (iv) the rotor speed is below the synchronous speed for a motor, and above for a generator.

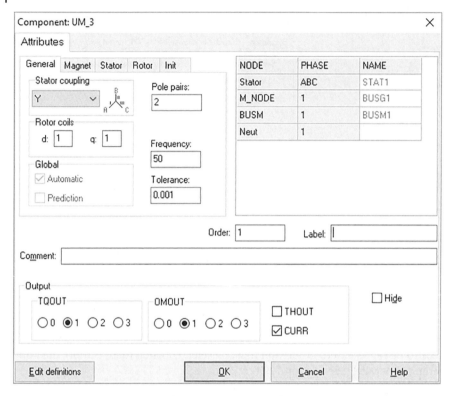

Figure 7.31 Case Study 5: UM-3 Induction Machine (SCIM design) window.

The frequency of the rotor currents (not shown in the figure) depends on the slip. The values of active and reactive powers flowing in the rotor of a DFIM are very low for the test system configuration analysed here.

In order to understand the differences between the two machine designs running as motors or generators, a parametric study of both machines under steady-state operation has been carried out. The goal of the study is to show the relationship between stator powers and slip. Figure 7.33 shows the variation of the active and reactive powers measured at the stator terminals of the SCIM model tested in this case study as a function of the slip. Note that:

- The active power is positive (i.e. flows from the source to the machine) when the slip is positive (i.e. the rotor speed is below the synchronous speed) and the machine runs as a motor, and negative (i.e. flows from the machine to the source) when the slip is negative (i.e. the rotor speed is above the synchronous speed) and the machine runs as a generator. The magnitude of the active power linearly varies with the slip.
- The reactive power is always positive (i.e. it flows from the source to the machine), is larger for a motor than for a generator, and does not exhibit a linear variation. Its magnitude increases as the slip approaches to zero and the machine runs as a generator, and decreases as the slip approaches to zero when it runs as a motor.

An additional study has been carried out with the DFIM model; the goal is to assess the startup transient of the machine when a variable resistor is inserted in series with the rotor windings. It is very evident from the results displayed in Figure 7.34 that the rotor speed will need a longer period of time to reach its final value as the value of the resistance increases.

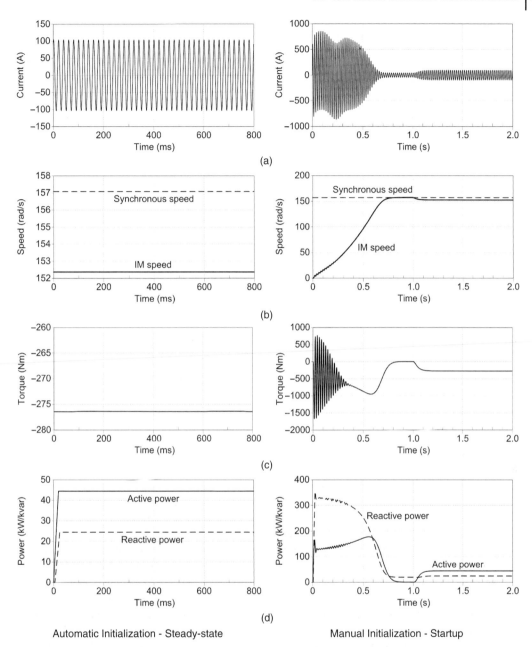

Automatic Initialization - Steady-state   Manual Initialization - Startup

Figure 7.32 Case Study 5: Simulation results with a SCIM running as a motor. (a) Stator currents, (b) Rotor speed, (c) Torque, (d) Active and reactive powers.

7.8.3 Case Study 6: Transient Response of the Induction Machine

This study includes three cases dedicated to the analysis of the response of induction machines running either as motors or generators in front of several types of disturbances. The design of the induction machine is in all cases of short-circuited rotor (i.e. SCIM type).

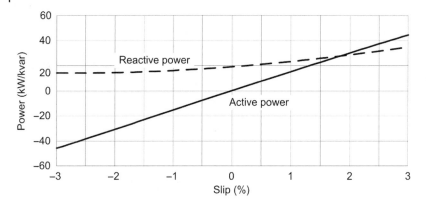

Figure 7.33 Case Study 5: Parametric study – variation of the active and reactive power with the slip.

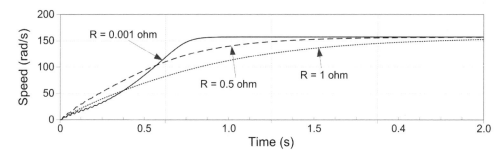

Figure 7.34 Case Study 5: Startup of a DIFM with different values of the series rotor resistance.

7.8.3.1 First Case

Figure 7.35 shows the circuit implemented in ATPDraw for simulating the test system. The IM is running as a generator, with a capacitor bank permanently connected at its terminals. The system has two loads, represented by the custom-made module already used in previous case studies. The ratings and main parameters of this test system are provided in Table 7.12. This case can be used to analyse the behaviour of the system and the island that will appear if the generator is (permanently or temporarily) disconnected.

Figure 7.36 shows results corresponding to a permanent disconnection of the induction generator. After the disconnection, the generator remains connected to the load. The voltages at the generator terminals depend on the initial slip: the larger this value, the higher the probability of keeping the voltages at an acceptable value. The angular speed of the rotor drops to a rather low value and reaches a value at which it remains, since the external torque to the generator is assumed constant. The measured active and reactive powers are different for the load and the generator, although in this case, the value of the reactive power depends on the capacitor bank. Since the external mechanical torque is assumed constant, the electromagnetic torque will not be constant and exhibit an opposite behaviour during the transient that follows the disconnection. Although the island can continue working, the frequency of voltages and currents within the island is no longer the power frequency: the equilibrium is reached with a frequency much lower that the power frequency, which is a non-desired scenario.

7.8.3.2 Second Case

Figure 7.37 depicts the circuit implemented in ATPDraw for simulating the second part of this case study: a synchronous generator supplies two medium-voltage induction motors. The main

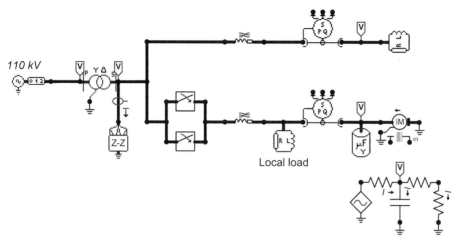

Figure 7.35 Case Study 6A: ATPDraw implementation of the test system. Disconnection of the induction machine running as generator.

Table 7.12 Case Study 6A: Parameters of system components.

Component	Parameters
High-voltage system	110 kV, 1200 MVA, $Z_0 = 0.11 + j1.1$ pu, $Z_{1/2} = 0.10 + j1.0$ pu
Substation transformer	110/12 kV, 4 MVA, Yd1
	Short-circuit impedance = 8%, $X/R = 8$, No saturable
	Grounding: Zig-zag transformer (with a current limit of 500 A)
Cables	Lengths = 4 km (top cable), 6 km (bottom cable)
	$R_0 = 0.10\,\Omega/\text{km}$, $L_0 = 1.5\,\text{mH/km}$; $R_1 = 0.01\,\Omega/\text{km}$, L1 = 0.5 mH/km
Loads	Top load: 12 kV, 2 MVA, pf = 0.95 (lg)
	Bottom load: 12 kV, 2 MVA, pf = 0.95 (lg)
Capacitor bank	12 kV, 2 Mvar
Induction machine	Squirrel-cage, Stator connection = Y, 2 pole pairs, No saturable
	d-axis magnetization inductance/q-axis magnetization inductance = 1.834 H
	Stator: d-axis impedance/q-axis impedance $R = 0.842\,\Omega$, $L = 0.0120$ H
	Rotor: d-axis impedance/q-axis impedance $R = 0.641\,\Omega$, $L = 0.0641$ H
	Moment of inertia = 8 kg m^2; damping = 0.01 Nm/(rad/s)

parameters of the three machines and the two cables are presented in Table 7.13. The two induction motors have the same ratings and parameters. The goal is to analyse the behaviour of the two motors when Motor 1 (i.e. the top machine in Figure 7.37) is suddenly connected, and Motor 2 (i.e. the bottom machine) is already running. The following transient is simulated:

1. Motor 1 is activated without mechanical load at the beginning of the simulation: the switch that connects Motor 1 to the generator is closed at $t = 0$. The Automatic Initialization option is selected, being the initial slip 99.9%. Motor 2 is already running with a 3% slip.

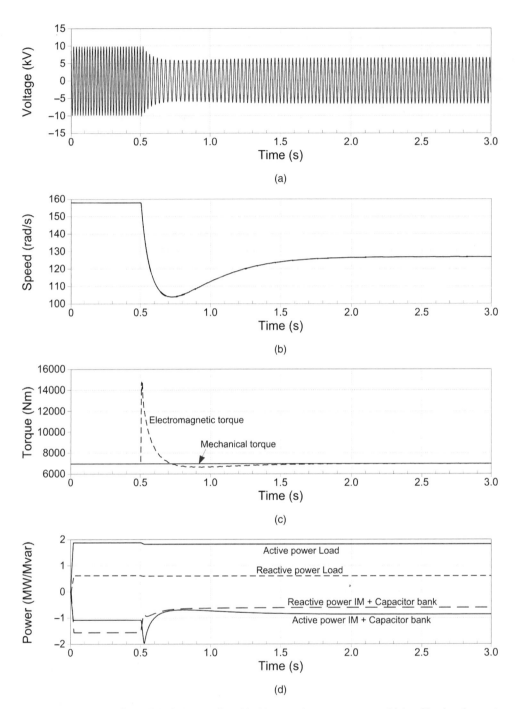

Figure 7.36 Case Study 6A: Simulation results – Machine running as generator with local load and capacitor bank – permanent disconnection from the grid (slip = −0.5%). (a) Voltages at the induction machine terminals, (b) Induction machine rotor speed, (c) Electromagnetic and mechanical torques, (d) Active and reactive powers at the load and induction generator terminals.

Table 7.13 Case Study 6B: Parameters of system components.

Synchronous generator		
Parameter	Value	Units
Phase-to-line rated voltage	3.3	kV (rms)
Three-phase rated power	5	MVA
Number of poles	2	—
Field current – rated voltage	100	A
Armature resistance	0.002	pu
Armature leakage reactance	0.20	pu
d-axis synchronous reactance	1.50	pu
q-axis synchronous reactance	1.45	pu
d-axis transient reactance	0.38	pu
q-axis transient reactance	0.50	pu
d-axis subtransient reactance	0.25	pu
q-axis subtransient reactance	0.25	pu
Open-circuit d-axis time constant	0.30	s
Open-circuit q-axis time constant	0.10	s
Open-circuit d-axis subtransient time constant	0.030	s
Open-circuit q-axis subtransient time constant	0.025	s
Zero-sequence reactance	0.2	pu
Canay's characteristic reactance	0.2	pu
Moment of inertia	0.001	Million kg m^2
Damping – exciter	5	N m s/rad

Insulated cables	
Cable Motor 1	Length $= 1$ km; $R_0 = 0.09\,\Omega$/km, $L_0 = 0.68$ mH/km; $R_1 = 0.01\,\Omega$/km, $L_1 = 0.32$ mH/km
Cable Motor 2	Length $= 2$ km; $R_0 = 0.09\,\Omega$/km, $L_0 = 0.68$ mH/km; $R_1 = 0.01\,\Omega$/km, $L_1 = 0.32$ mH/km

Induction motors		
General specifications	Squirrel-cage, Stator connection $=$ Y, 2 pole pairs, No saturable	
Electrical parameters	d-axis magnetization inductance/q-axis magnetization inductance $= 0.1807$ H	
	Stator:	d-axis impedance/q-axis impedance	$R = 0.4892\,\Omega, L = 5.151$ mH
	Rotor:	$d1$-axis impedance/$d2$-axis impedance	$R = 0.5871\,\Omega, L = 5.149$ mH
		$q1$-axis impedance/$q2$-axis impedance	$R = 2.0629\,\Omega, L = 16.865$ mH
Mechanical parameters	Moment of inertia $= 5$ kg m^2; damping $= 0.01$ Nm/(rad/s)	

2. At $t = 1$ seconds, the mechanical torque of Motor 1 is increased. This torque is represented by means of a dc current source in parallel to the current source that represents the mechanical torque during the motor startup; see Figure 7.37. The new mechanical torque value is 3900 Nm; the goal is to obtain similar operating conditions with both motors.

Figure 7.38 shows some simulation results that look reasonable:

- The rotor speed of Motor 1 reaches a value very close to the synchronous speed and, after its mechanical torque has been increased, it decreases to a speed very similar to that of Motor 2.

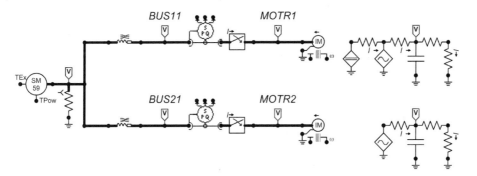

Figure 7.37 Case Study 6B: ATPDraw implementation of the test system.

- After the startup transient, the torque for Motor 1 becomes negligible, but once the external torque is varied, the final value of the two motor torques are very similar.
- The differences observed in the power values measured at the two motor terminals is due to the different voltage drop caused in the cables that connect the two motors to the generator; the voltages at the terminals of the two motors are very similar, but the currents are not, because of the different cable lengths.
- The generator speed deviation (as well as the variation of the terminal voltages, not shown) suggests that generator controllers should be used to obtain a correct response of the system.

7.8.3.3 Third Case

Figure 7.39 shows the circuit implemented in ATPDraw to analyse the test system that corresponds to the third part of this case study. The goal is to simulate the response of the induction motors that the system is feeding when a fault occurs in a location 2 km away from the MV terminals of the substation transformer. The main parameters of this system are provided in Table 7.14. The implemented circuit includes several custom-made models used for representing the network equivalent, the substation transformer and its grounding impedance, all overhead lines, distribution transformers, and power monitors. All motors have the same ratings and parameters, and their mechanical systems have been represented by means of a compressed group (see Figure 7.39).

The system has been tested assuming two different types of fault and different fault durations. The initial conditions for the motors are as follows: (i) IM1 slip: 2.0%; (ii) IM2 slip: 1.5%; (iii) IM3 slip: 1.5%; (iv) IM4 slip: 2.0%; (v) IM5 slip: 1.0%. Figures 7.40 and 7.41 show the angular speeds of the five motors in front of a single-phase-to-ground and a three-phase fault, respectively. All the simulation results derived with a single-phase-to-ground fault exhibit a stable response of all motors, irrespective of the fault duration and the mechanical parameters; however, the performance of some machines can be unstable for a three-phase fault depending on the fault duration. One important reason for this behaviour is the fact that with a single-phase-to-ground fault, the dip and swells that can be noticed at the MV side of distribution transformers are not propagated to the LV side. Figure 7.42 shows the voltages at both sides of the transformer that feeds IM1: the voltage decrement noticed during the fault is significant at the MV side but it is hardly noticed at the LV side.

The behaviour in front of a three-phase fault is, however, very different. Figure 7.41a shows a stable performance: all rotor speeds recover their original values once the fault has been cleared. However, if only the fault duration is varied (i.e. increased), then the rotor speed of IM1, whose distance to the fault location is the shortest one, cannot recover its original value (see Figure 7.41b).

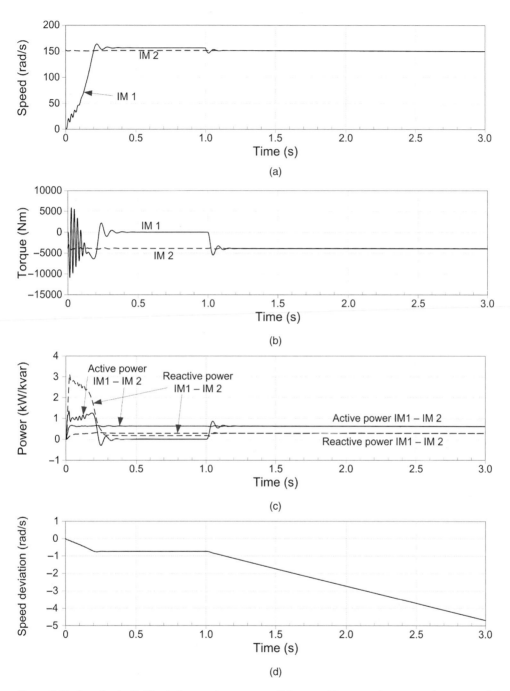

Figure 7.38 Case Study 6B: Simulation results – startup of Motor 1 – The motor is connected at $t = 0$ and the mechanical torque is increased at $t = 1$ second. (a) Angular speeds, (b) Electromagnetic torques, (c) Active and reactive powers, (d) Synchronous generator rotor speed deviation.

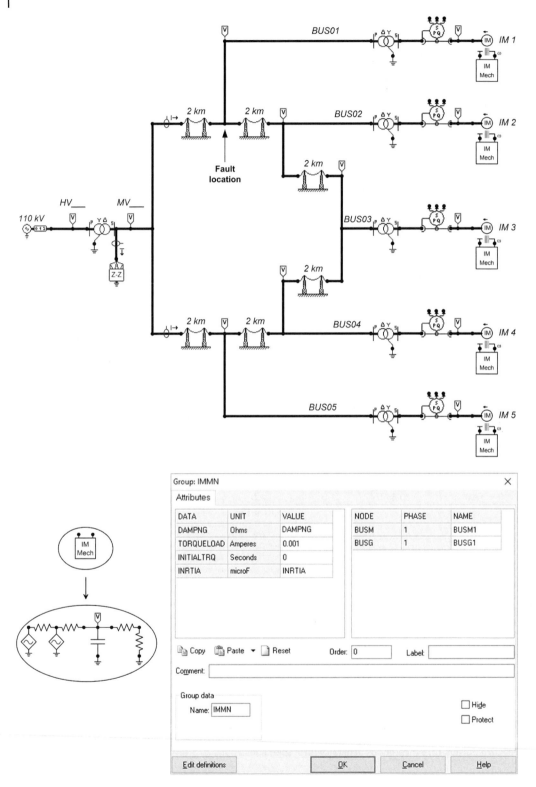

Figure 7.39 Case Study 6C: ATPDraw implementation of the test system.

Table 7.14 Case Study 6C: Parameters of system components.

Component	Parameters
High-voltage system	110 kV, 1500 MVA, $Z_0 = 0.11 + j1.1$ pu, $Z_{1/2} = 0.10 + j1.0$ pu
Substation transformer	110/25 kV, 10 MVA, Yd1
	Short-circuit impedance = 6%, $X/R = 10$, No saturable
	Grounding: Zig-zag transformer (with a current limit of 577 A)
Overhead lines	Length = 2 km, $Z_0 = 0.764 + j1.564 \, \Omega/\text{km}$, $Z_{1/2} = 0.614 + j0.391 \, \Omega/\text{km}$
Distribution transformers	25/0.415 kV, 1 MVA, Dy11
	Short-circuit impedance = 4%, $X/R = 10$, No saturable
Induction motors	Squirrel-cage, Stator connection = Dlead, 4 pole pairs, No saturable
	d-axis magnetization inductance/q-axis magnetization inductance = 7.17 mH
	Stator: d-axis impedance/q-axis impedance $R = 0.0108 \, \Omega$, $L = 0.224$ H
	Stator: d-axis impedance/q-axis impedance $R = 0.0236 \, \Omega$, $L = 0.224$ H
	Moment of inertia = 8 kg m2; Damping = 0.01 Nm/(rad/s)

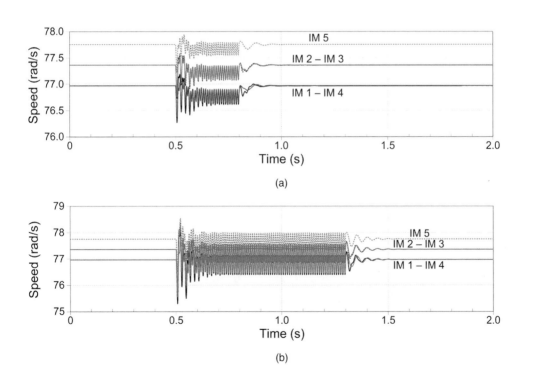

Figure 7.40 Case Study 6C: Simulation results – angular speeds – single-line-to-ground fault. (a) Fault duration: 0.3 seconds ($C = 20$ F; $R = 50 \, \Omega$), (b) Fault duration: 0.8 seconds ($C = 8$ F; $R = 100 \, \Omega$).

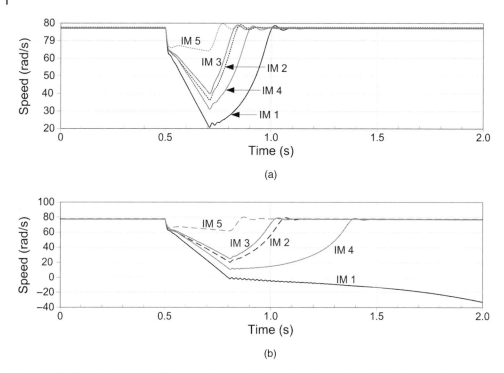

(a)

(b)

Figure 7.41 Case Study 6C: Simulation results – angular speeds – three-phase fault – $C = 20$ F; $R = 50\,\Omega$. (a) Fault duration: 0.2 seconds, (b) Fault duration: 0.3 seconds.

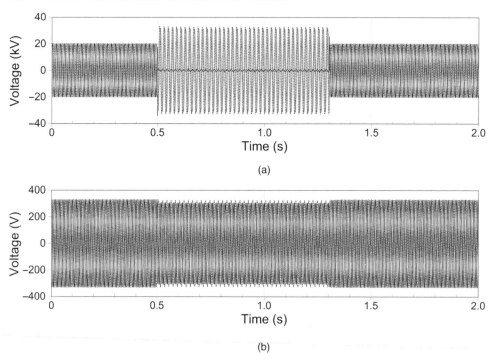

(a)

(b)

Figure 7.42 Case Study 6C: Simulation results – voltages at the two terminals of the transformer that supplies IM2 – single-phase-to-ground fault. (a) MV side, (b) LV side.

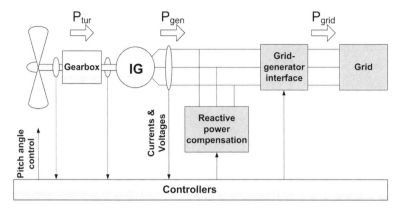

Figure 7.43 Case Study 7: Scheme of a wind energy conversion system.

7.8.4 Case Study 7: SCIM-Based Wind Power Generation

Wind is a cheap, clean and renewable energy source. However, the erratic behaviour of this source adds some difficulties in the operation of a wind energy conversion system (WECS): the transient response as a result of a change in the electric grid conditions or in the wind power input can be accurately reproduced if the wind dynamics are well known and an accurate model is used for each component of the WECS [54, 55]. This case study presents the ATP implementation of a two-blade horizontal-axis turbine-generator set based on an SCIM design. Although this design is a low-cost and robust option, the most popular WECS designs are presently based on a DFIM.

Figure 7.43 shows the general diagram of a WECS. The main components of this system are the grid, the wind turbine, the gearbox, the generator, an interface between the generator and the grid, a reactive power compensator, and the controllers.

Control systems can be used to control the turbine, the reactive power compensator and any interface between the generator and the grid (e.g. an interface based on a static converter). When an SCIM is used without any converter-based interface, the reactive power needed to provide the excitation of the generator can be obtained from the grid, from a local compensator, or from both.

The implementation of a system like that shown in the figure is a complex task. Some important simplifications have been introduced in the approach implemented here to represent the wind dynamics and obtain the torques that act on the turbine blades. According to the Betz theory [54, 55], the power available from the wind can be approached by a cubic form, kv^3, where k is a factor that depends on the cross-sectional area of the system and v is the wind speed. Only a part of this power is extracted by the turbine, the ratio of the power extracted to the actual power available is known as *power coefficient*, C_p. For a particular system, this coefficient can be obtained from the so-called *tip speed ratio*, λ; that is, the ratio of the tangential speed of a blade tip to the undisturbed wind speed.

The test system analysed in this case study consists of a horizontal-axis turbine, a squirrel-cage rotor induction generator and a capacitor bank. The generator is directly connected to the grid, which is represented by an infinite bus in series with the equivalent impedance. This approach is acceptable if the interest is concentrated in low-frequency transients; that is, those transient phenomena that could mainly affect the behaviour of the turbine-generator set. Active blade pitch control is not covered in this case study. The turbine has two blades and its shaft is connected to the rotor generator by means of a gearbox.

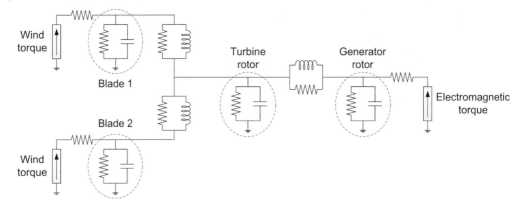

Figure 7.44 Case Study 7: Equivalent electrical circuit of the mechanical part.

Figure 7.44 depicts the equivalent electrical network of the complete mechanical system. According to this figure:

- each blade of the turbine is represented by means of a mass with damping, the connection to the shaft is made assuming stiffness and damping;
- turbine and generator shafts are represented by means of a mass with damping, the connection between them is made assuming stiffness and damping;
- a current source controlled from TACS represents the torque acting on each blade.

The electromagnetic torque from the generator is represented by means of a current source, whose value is calculated in the UM module. The gearbox can be modelled by means of an ideal transformer placed between turbine and generator parameters. This transformer is not required if turbine parameters and variables are referred to the generator side. The main parameters of this test system, as modelled in Figure 7.44, are provided in Table 7.15.

If the wind speed at the tower height has been specified, the torque acting on each turbine blade is calculated by means of the following procedure:

1) The blade positions are obtained by integrating the instantaneous blade speeds.
2) The instantaneous value of the tip-speed ratio in each blade is calculated from their speed and position. Figure 7.45 illustrates the approach used in this case study. Wind speed at a height h over ground is approached using the following form:

$$v = v_0 \left(\frac{h}{h_0} \right)^n \qquad n = 0.168 \tag{7.2}$$

where v_0 is the wind speed at the tower-axis height, h_0.
The position of each blade tip is calculated (see Figure 7.45) as follows:

$$h_1 = h_0 + R \sin \theta_1 \qquad h_2 = h_0 + R \sin \theta_2 \tag{7.3}$$

where R is the blade radius.
3) The power coefficient, C_p, is obtained from the tip-speed ratio, λ, using a curve like that shown in Figure 7.46. This coefficient is a measure of the wind turbine efficiency; it is the ratio of actual electric power produced by a wind turbine divided by the total wind power at a specific wind speed, and is assumed to be proportional to the pitch angle.
4) Finally, the torque acting on each blade is calculated by means of the following expression:

$$T = \frac{1}{2} \frac{(\pi R^2) \cdot \rho \cdot v_0^3 \cdot C_p}{\omega} \tag{7.4}$$

where ρ is the air density and ω is the blade speed.

Table 7.15 Case Study 7: Parameters of system components.

Component	Parameters
Supply system	3.6 kV, 50 MVA, 60 Hz, $Z_0 = 0.11 + \mathrm{j}1.1$ pu, $Z_{1/2} = 0.10 + \mathrm{j}0.95$ pu
Capacitor bank	3.6 kV, 1.5 Mvar
Wind energy conversion system	*Generator*: Three-phase SCIM, Wye connection, 4 poles, No saturable

Generator: Three-phase SCIM, Wye connection, 4 poles, No saturable

Electrical parameters:	$R_s = 0.412\,\Omega$; $L_{la} = 1.20\,\mathrm{mH}$
	$L_m = 23.58\,\mathrm{mH}$
	$R_r = 0.110\,\Omega$; $L_{lr} = 1.20\,\mathrm{mH}$
Mechanical parameters:	Moment of inertia $= 98.0\,\mathrm{kg\,m^2}$
	Damping $= 1.099\,\mathrm{Nms/rad}$

Gearbox: Ratio $= 1 : 100$
Turbine: Horizontal-axis, two blades, tower height $= 40\,\mathrm{m}$, blade radius $= 20\,\mathrm{m}$
(Mechanical parameters are reduced to generator side)

Blades:	Moment of inertia $= 1300\,\mathrm{kg\,m^2}$
	Damping coefficient $= 1/3.2485\,\mathrm{Nms/rad}$
Connection blades-shaft:	Stiffness coefficient $= 1/16.5\mathrm{E}\text{-}6\,\mathrm{Nm/rad}$
	Mutual damping coefficient $= 1/0.46\mathrm{E}\text{-}3\,\mathrm{Nms/rad}$
Shaft:	Moment of inertia $= 68\,\mathrm{kg\,m^2}$
	Damping coefficient $= 1/2.5266\,\mathrm{Nms/rad}$
Connection turbine-generator:	Stiffness coefficient $= 1/933.7\mathrm{E}\text{-}6\,\mathrm{Nm/rad}$
	Mutual damping coefficient $= 1/15.5\mathrm{E}\text{-}3\,\mathrm{Nms/rad}$

Figure 7.45 Case Study 7: Wind speed and blade positions.

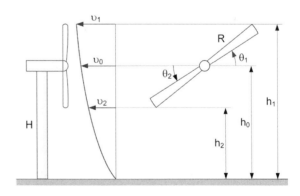

Wind gusts can be represented using the following approach:

$$v = v_0 \cdot [1 + k(1 - \cos \omega_g t)] \qquad (\omega_g = 2\pi f_g) \tag{7.5}$$

where v_0 is the average wind speed, k is a factor that measures the gust magnitude, and f_g is the frequency of the gusts.

Figure 7.47 shows the circuit implemented in ATPDraw to analyse this test system. The model includes the MV supply system, a capacitor bank for reactive power compensation, and the wind generator. The mechanical system model is that presented in Figure 7.44. Note that there are two nodes (i.e. BUSW1 and BUSW2) to which current sources representing wind power torques are connected, plus one node (i.e. BUSMT) to which the generator electromagnetic torque is applied. The TACS section calculates the complementary torque needed to represent the

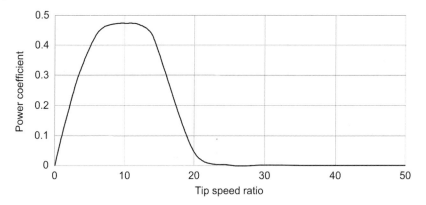

Figure 7.46 Case Study 7: Power coefficient curve.

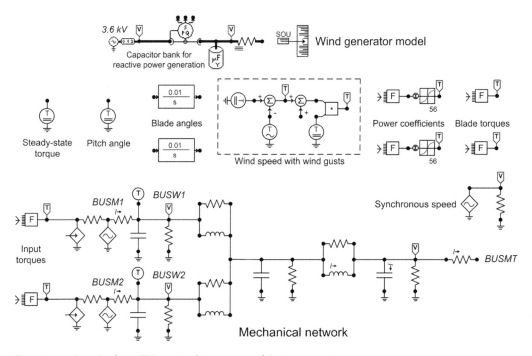

Figure 7.47 Case Study 7: ATPDraw implementation of the test system.

impact of the wind on the two turbine blades during their rotation. The version depicted in the figure includes wind gust calculation. The induction generator is represented by means of a user-specified code. The main reason is that the turbine uses an option available in the UM module for representing machines with more than one mechanical input; see the ATP Rule Book [1].

Since the torque caused by the wind on each turbine blade varies with the angular position of the blade, the torques on the blade tips are not constant, even with a constant wind speed.

The simulation of a transient phenomenon is usually performed assuming that the system is initially running at steady-state conditions. The strategy used here to obtain the wind speed to be specified in the TACS section and that corresponding to the desired generator slip can be summarized as follows:

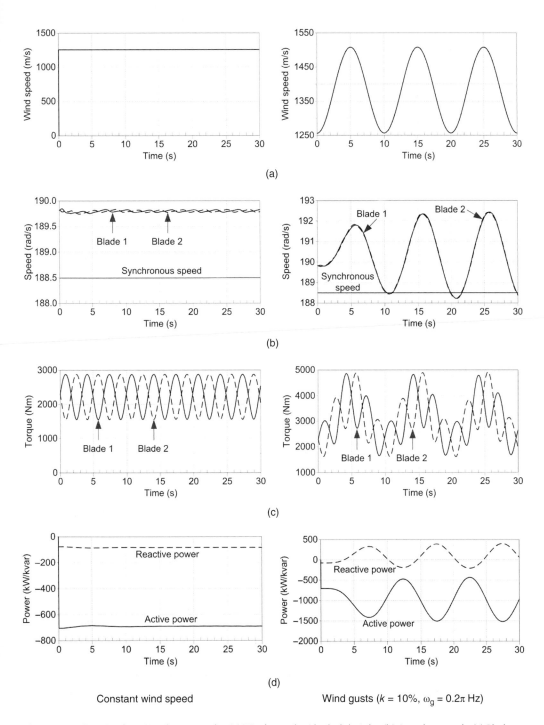

Figure 7.48 Case Study 7: Simulation results. (a) Wind speed at the hub height, (b) Angular speeds, (c) Blade torques, (d) Active and reactive powers.

- Obtain the steady-state operating conditions of the induction generator that correspond to the initial wind speed without including the TACS section needed to estimate the wind torques. The Automatic Initialization option will provide the torque applied to each turbine blade.
- Estimate by trial and error the wind speed (to be specified in the TACS section) that will produce the mechanical torque estimated in the previous step. Remember that mechanical parameters are reduced to the generator side and the gearbox ratio is 100.
- Place in parallel with each current source that represents a wind torque a TACS-controlled current source to represent the deviation of each wind torque during the transient simulation. Since this source is not honoured by the ATP at the initial step, specify a zero vale for t_{start}.

Figure 7.48 shows some simulation results obtained with a constant wind speed and wind gusts. The plots of the figure compare results from the two scenarios assuming the initial slip of the generator is, in both cases, 0.7%. The values of the electromagnetic torque and the wind torques to the turbine blades are respectively 4194.2 and 2297.7 Nm. This later value is needed to begin the simulation of a transient case. Remember that mechanical parameters and variables are reduced to the turbine rotor side (e.g. the angular speed is 100 times faster at the rotor side than at the blade side, so it is the wind speed to be used). The actual wind speed in this case is 12.56 m/s, which is a rather high value.

The values of the variables depicted in the various plots of the figure can help to understand the design of this wind turbine-generator set. The results compare those derived when the wind is constant and those obtained with wind gusts. The gusts are assumed to behave according a sinusoidal variation with a frequency of 0.1 Hz. The magnitude assumed for gusts is a 20% of the wind speed used for the simulation of the previous scenario (i.e. left plots in Figure 7.48). The 30-second simulation exhibits very stable values once the initial transient (required by the model to get accommodated to the variation suffered by blade torques during a rotation) is over. Due to the parameters of the turbine-generator set and the operation conditions with which the gusts occur, a 20% of wind variation can significantly affect some variables. For instance, active power and electromagnetic torque exhibit oscillations of about 100% with respect to the values obtained without gusts, and wind torques on blades are even above this percentage. On the other hand, the oscillations of the angular speeds (i.e. blades and generator rotor) oscillate less than 2% with respect the angular speed without gusts. As for the sign of the active power, take into account that it is measured flowing from the grid to the WECS: the negative sign means that the power is actually supplied by the WECS.

References

1 Can/Am Users Group (2000). *ATP Rule Book*. Can/Am Users Group.

2 Dommel, H.W. (1986). *Electromagnetic Transients Program. Reference Manual (EMTP Theory Book)*. Portland: Bonneville Power Administration.

3 Brandwajn, V. and Dommel, H.W. (1977). A new method for interfacing generator models with an electromagnetic transients program. In: *Proceedings of the IEEE PICA*, 260–265.

4 Lauw, H.K. and Meyer, W.S. (1982). Universal machine modeling for the representation of rotating machinery in an electromagnetic transients program. *IEEE Transactions on Power Apparatus Systems* 101 (6): 1342–1352.

5 Lauw, H.K. (1985). Interfacing for universal multi-machine system modeling in an electromagnetic transients program. *IEEE Transactions on Power Apparatus Systems* 104 (9): 2367–2373.

6 Shirmohammadi, D. (1985). Universal machine modeling in electromagnetic transients program. *EMTP Newsletter.* 5 (2): 5–27.

7 Cao, X., Kurita, A., Mitsuma, H. et al. (1999). Improvements of numerical stability of electromagnetic transient simulation by use of phase-domain synchronous machine models. *Electrical Engineering Japan* 128 (3): 53–62.

8 Martinez-Velasco, J.A. (1997). Modeling and simulation of rotating machines using the ATP. European EMTP Users Group Meeting in Barcelona, Spain (November 1997).

9 CIGRE WG 33.02 (1990). *Guidelines for Representation of Network Elements when Calculating Transients*, 39. CIGRE Brochure.

10 Gross, G. and Hall, M.C. (1978). Synchronous machine and torsional dynamics simulation in the computation of electromagnetic transients. *IEEE Trans. Power Appar. Syst.* 97 (4): 1074–1086.

11 Olive, D.W. (1981). Modeling synchronous machines for digital studies, Chapter V. In: *Digital Simulation of Electrical Transient Phenomena* (ed. A.G. Phadke), 30–39. IEEE Tutorial Course, Publication No. 81EHO173-5-PWR.

12 Baker, D.H. (1981). Synchronous machine modeling in EMTP, Chapter VI. In: *Digital Simulation of Electrical Transient Phenomena* (ed. A.G. Phadke), 40–46. IEEE Tutorial Course, Publication No. 81EHO173-5-PWR.

13 Bowler, C.E.J., Lawson, R.A. and Baker, D.II. (1981) Symposium on Countermeasures for Subsynchronous Resonance. IEEE Power Engineering Society, Publication 81TH0086-9-PWR.

14 Phadke, A. (ed.). *Digital Simulation of Electrical Transient Phenomena*. IEEE Tutorial Course Course Text 81 EHO173-5-PWR.

15 IEEE Committee Report (1977). First benchmark model for computer simulation of subsynchronous resonance. *IEEE Transactions on Power Apparatus Systems* 96 (5): 1565–1572.

16 IEEE Committee Report (1985). Second benchmark model for computer simulation of subsynchronous resonance. *IEEE Transactions on Power Apparatus Systems* 104 (5): 1057–1066.

17 Brandwajn, V. (1980). Representation of magnetic saturation in the synchronous machine model in an electromagnetic transients program. *IEEE Transactions on Power Apparatus Systems* 99 (5): 1996–2002.

18 Bacvarov, D.C. and Sarma, D.K. (1986). Risk of winding insulation breakdown in large ac motors caused by steep switching surges. Part I: computed switching surges. *IEEE Transactions on Energy Conversions* 1 (1): 130–139.

19 McLaren, P.G. and Abdel-Rahman, M.H. (1988). Modeling of large AC motor coils for steep-fronted surge studies. *IEEE Transactions on Industry Applications* 24 (3): 422–426.

20 Guardado, J.L. and Cornick, K.J. (1989). A computer model for calculating steep-fronted surge distribution in machine windings. *IEEE Transactions on Energy Conversions* 4 (1): 95–101.

21 Narang, A., Gupta, B.K., Dick, E.P., and Sharma, D.K. (1989). Measurement and analysis of surge distribution in motor stator windings. *IEEE Transactions on Energy Conversions* 4 (1): 126–134.

22 Dick, E.P., Gupta, B.K., Pillai, P. et al. (1988). Equivalent circuits for simulating switching surges at motor terminals. *IEEE Transactions on Energy Conversions* 3 (3): 696–704.

23 Dick, E.P., Cheung, R.W., and Porter, J.W. (1991). Generator models for overvoltages simulations. *IEEE Transactions on Power Delivery* 6 (2): 728–735.

24 Knudsen, H. (1995). Extended Park's transformation for 2x3-phase synchronous machine and converter phasor model with representation of harmonics. *IEEE Transactions on Energy Conversions* 10 (1): 126–132.

25 Domijan, A. and Yin, Y. (1994). Single-phase induction machine simulation using the electromagnetic transients program: theory and test cases. *IEEE Transactions on Energy Conversions* 9 (3): 535–542.

26 Furst, G. (2002). Demonstration of the new WindSyn Program. EEUG Meeting in Sopron, Hungary (December 2002).

27 Martinez-Velasco, J.A. (2013). Basic methods for analysis of high frequency transients in power apparatus windings, Chapter 2. In: *Electromagnetic Transients in Transformer and Rotating Machine Windings* (ed. C.Q. Su), 45–110. IGI Global.

28 IEEE (1991). IEEE guide for synchronous generator modeling practices in stability studies. *IEEE Standard* 1110.

29 Martinez, J.A., Johnson, B., and Grande-Morán, C. (2005). Parameter determination for modeling systems transients. Part IV: rotating machines. *IEEE Transactions on Power Delivery* 20 (3): 2063–2072.

30 IEEE Task Force on Slow Transients (M.R. Iravani, Chairman) (1995). Modelling and analysis guidelines for slow transients. Part I: torsional oscillations; transient torques; turbine blade vibrations; fast bus transfer. *IEEE Transactions on Power Delivery* 10 (4): 1950–1955.

31 Anderson, P.M., Agrawal, B.L., and Van Ness, J.E. (1990). *Subsynchronous Resonance in Power Systems*. IEEE Press.

32 (1983). *Symposium on Synchronous Machines Modelling for Power System Studies*. IEEE Power Engineering Society, Publication 83THO101-6-PWR.

33 IEEE (1983). IEEE guide: test procedures for synchronous machines. *IEEE Standard* 115.

34 IEEE Committee Report (1980). Supplementary definitions and associated test methods for obtaining parameters for synchronous machine stability study simulations. *IEEE Transactions on Power Apparatus Systems* 99 (4): 1625–1633.

35 IEEE Committee Report (1986). Current usage and suggested practices in power system stability simulations for synchronous machines. *IEEE Transactions on Energy Conversions* 1 (1): 77–93.

36 Canay, I.M. (1983). Determination of model parameters of synchronous machines. *IEE Proceedings* 130, Pt. B (2): 86–94.

37 Alvarado, F.L. and Cañizares, C. (1989). Synchronous machine parameters from sudden-short tests by back-solving. *IEEE Transactions on Energy Conversions* 4 (2): 224–236.

38 Canay, I.M. (1993). Modelling of alternating-current machines having multiple rotor circuits. *IEEE Transactions on Energy Conversions* 8 (2): 280–296.

39 IEEE (1987). IEEE standard procedures for obtaining synchronous machine parameters by standstill frequency response testing. *IEEE Standard* 115A.

40 Martinez, J.A. and Gustavsen, B. (2009). Parameter estimation from frequency response measurements. IEEE PES General Meeting in Calgary, Canada (July 2009).

41 Karaagac, U., Mahseredjian, J., and Martinez-Velasco, J.A. (2009). Synchronous machines, Chapter 5. In: *Power System Transients. Parameter Determination* (ed. J.A. Martinez-Velasco), 251–350. CRC Press.

42 Dandeno, P.L., Kundur, P., Poray, A.T., and Zein El-din, H.M. (1981). Adaptation and validation of turbogenerator model parameters through on-line frequency response measurements. *IEEE Transactions on Power Apparatus Systems* 100 (4): 1656–1645.

43 Dandeno, P.L., Kundur, P., Poray, A.T., and Coultes, M.E. (1981). Validation of turbogenerator stability models by comparison with power system tests. *IEEE Transactions on Power Apparatus Systems* 100 (4): 1637–1645.

44 IEEE (1978). IEEE guide: test procedure for polyphase induction motors and generators. *IEEE Standard* 112.

45 Rogers, G.J. and Shirmohammadi, D. (1987). Induction machine modelling for electromagnetic transient program. *IEEE Transactions on Energy Conversions* 2 (4): 622–628.

46 Marti, J.R. and Louie, K.W. (1997). A phase-domain synchronous generator model including saturation effects. *IEEE Transactions on Power Systems* 12 (1): 222–229.

47 Oguz Soysal, A. (1993). A method for wide frequency range modeling of power transformers and rotating machines. *IEEE Transactions on Power Delivery* 8 (4): 1802–1810.

48 Atarod, V., Dandeno, P.L., and Iravani, M.R. (1992). Impact of synchronous machine constants and models on the analysis of torsional dynamics. *IEEE Transactions on Power Systems* 7 (4): 1456–1463.

49 Canay, I.M. (1969). Causes of discrepancies on calculation of rotor quantities and exact equivalent diagrams of the synchronous machine. *IEEE Transactions on Power Apparatus Systems* 88 (7): 1114–1120.

50 IEEE Std 421.5-2005. (2005). IEEE Recommended Practice for Excitation System Models for Power System Stability Studies.

51 Kundur, P. (1994). *Power System Stability and Control*. New York: McGraw-Hill.

52 EPRI Report EL-4202. (1985). Electromagnetic Transients Program (EMTP) Primer.

53 EPRI Report EL-4651. (1989). Electromagnetic Transients Program (EMTP) Volume 3: Workbook III.

54 Freris, L.L. (ed.) (1990). *Wind Energy Conversion Systems*. Prentice-Hall.

55 Ackermann, T. (ed.) (2012). *Wind Power in Power Systems*, 2e. Wiley.

To Probe Further

The simulation results presented in the chapter are a very small sample of the results that can be derived from the case studies. ATP data files are available in the companion website. Readers are encouraged to run the cases, check all results, and explore the performance of the test systems under different operating conditions or their response in front of different disturbances.

8

Power Electronics Applications
Juan A. Martinez-Velasco and Jacinto Martin-Arnedo

8.1   Introduction

Power electronics systems are spread to all voltage levels, from extra high voltage (EHV) transmission to low-voltage (LV) circuits in end-user facilities [1–4]. Power electronics applications include, among others, high-voltage dc (HVDC) links, adjustable-speed drive technologies, energy storage and renewable energy integration, Flexible AC Transmission Systems (FACTS) and Custom Power devices. Solid-state technologies have drastically changed some applications (e.g. rotating machine drives, integration of renewable energy sources) due to the flexibility they add in the control of power variables. As these applications are increasing, there is an increasing demand for a better understanding of their behaviour and dynamics. The need for power electronics modelling and simulation is driven by both existing and new applications. Computer simulation appears as an adequate solution for evaluating converter designs and testing control strategies. Objectives of the simulation include: verification of an application design, prediction of a system performance, identification of potential problems, or evaluation of possible problem solutions. Simulation can be especially important for concept validation and design iteration during the development of a new product.

The theories of power electronics are not discussed in this chapter, instead the focus is on the simulation of the interaction between the power electronics (sub)systems and the power system to which they are connected. The switching of power electronics devices introduces nonlinearities and reproducing their behaviour with the desired degree of accuracy represents a significant challenge. In many cases, approximations in the converter modelling can be made without degrading accuracy too significantly. In other cases, simpler idealized switch models can be used for modelling power semiconductor devices [5–7]. However, in some studies, detailed modelling of both the converter topology and switching devices is required to accurately assess switching transients, device stresses and losses, although sufficient data to appropriately model the semiconductor devices and related control strategies often lacks.

Power electronics simulation can be divided into two basic categories, depending on the study goals [6, 7]. The first category covers steady-state evaluations; the focus is on the power system response to the harmonics injected from a power electronics subsystem. Examples of this type include steady-state harmonic propagation in transmission and distribution systems, harmonic resonances, voltage and current distortion, or filtering design. In this type of study, the harmonic current injection can often be assumed to be independent of the voltage variations at a point of common coupling (PCC), so the power electronics subsystem can be greatly reduced to a shunt circuit equivalent. The second category covers a more complex range of practical problems. In many applications, operation of a power electronics subsystem depends on the operation state of the connected system. To evaluate the dynamic and transient performance

Transient Analysis of Power Systems: A Practical Approach, First Edition. Edited by Juan A. Martinez-Velasco.
© 2020 John Wiley & Sons Ltd. Published 2020 by John Wiley & Sons Ltd.
Companion Website: www.wiley.com/go/martinez/power_systems

of a system with power electronics interfaces, the monitoring and control loops of the system, including detailed signal processing and device firing need to be modelled. Examples of this type of study can be the impact of an active power conditioning system, the transient response of an adjustable speed drive, or the performance of a wind energy conversion system.

This chapter presents a survey of the work performed in the time-domain simulation of power electronics systems using the Alternative Transients Program (ATP). The chapter discusses the modelling guidelines to be applied when simulating power electronics devices as either stand-alone systems or as part of a larger power system. The chapter targets applications of semiconductor devices commonly used in power systems (e.g. power diodes, thyristors, gate turn-off thyristors (GTOs), insulated-gate bipolar transistor (IGBTs)). Sections 8.2 and 8.3 discuss the approaches that can be used for representing power electronics converters and semiconductor devices respectively. Section 8.4 provides a very short summary of the ATP solutions methods. The ATP capabilities that can be used for representing the different blocks that form power electronics systems and some guidelines for their application are detailed in Section 8.5. Section 8.6 summarizes the current power electronics applications in transmission, distribution, and distributed energy resources (DER). The last two sections present some illustrative case studies; Section 8.7 begins with some simple and introductory examples, while Section 8.8 details some case studies of increasing complexity.

8.2 Converter Models

This section presents a short discussion of the two main approaches commonly used for representing power electronics systems in dynamic and transient studies [8].

8.2.1 Switching Models

The degree of detail in the converter model often depends on the relationship between the frequency of interest in the simulation results and the switching frequencies in the converter. Similarly, the degree of detail in semiconductor models depends on the relationship between the time periods of the frequencies of interest and the transition times of the power semiconductor devices. The latter are usually much shorter than the intervals between converter switching operations. There are transients associated with these turn-on/turn-off transitions, so if these transients are of interest, then more detailed device models are required. In addition, when the slowest transition time of a power semiconductor device (which is often at turn-off) approaches the period between switch operations, more detailed device turn-on and turn-off models are required. Use of these models also requires more detail in modelling the parasitic inductances and capacitances in the converter. On the other hand, simpler or aggregate device models can be used in many cases, and the converter can be reduced to a simple equivalent. In such cases, it is often appropriate to model series and parallel connected power electronic devices as one or two equivalent devices. In addition, it is sometimes sufficient to represent a converter made up of many converter modules as a simpler converter. For example, a 48-pulse voltage source converter could be represented with a simpler, lower pulse-order model if the response is sufficient for the studies to be performed.

8.2.2 Dynamic Average Models

For system-level transient studies, the dynamic behaviour of a power electronics (sub)system may be predicted with sufficient accuracy using the so-called dynamic average models, which approximate the slower system dynamics whilst neglecting the details of fast switching [9, 10]. The objective of these models is to replace the discontinuous switches (or group of switches that operate together as a power conversion cell) with continuous elements and dependent sources that represent the averaged behaviour of the switching cell within a prototypical switching interval. Average models can use much larger integration time steps: they typically execute orders of magnitude faster than the corresponding detailed switching models. The converter is often represented as either a dependent current source or a dependent voltage source. These models are used for steady-state operation and to study the response of slower converter control schemes for studies where large simulation time steps are often preferred. Examples include fundamental component and harmonic component models for converters under steady-state and slow-transient conditions, where the transient response of the converter and its faster controls are not the focus of the simulation.

8.3 Power Semiconductor Models

8.3.1 Introduction

The list of the most common power semiconductor devices includes power diodes, thyristors, GTOs, and IGBTs. The list can be expanded by adding gate commutated thyristors (GCT/IGCT), MOS-controlled thyristors (MCT), MOS turn-off thyristors (MTO), and static induction transistor/static induction thyristor (SIT/SiTh) [1, 8]. All these devices have specific turn-on and turn-off characteristics that are visible in the voltage and current characteristics.

Each device has conduction losses while the device is turned on. In some devices, the conduction losses can be modelled as a steady-state voltage drop or a nonlinear temperature-dependent resistance. Although models representing these characteristics can be simulated with ATP, some degree of approximation will be necessary in most studies. In some cases, resistances, inductances, and dependent voltage sources are sufficient to represent device behaviour. In other cases, more detailed equations will be needed to achieve accurate-enough device models.

8.3.2 Ideal Device Models

When a power electronics converter is connected to a system where the timescale of the response is much longer than the device turn-on and turn-off times, ideal switch models can be used. In this case, the device is assumed to open or close in one time step as the simulation progresses. The behaviour of ideal models can be summarized as follows (see references [1, 8]): (i) when the device is off, it behaves as an open circuit; (ii) when the device is on, it behaves as a short circuit; (iii) the device turns on at the next time step after a firing command; (iv) the device turns off one time step after a firing command, or for diodes and thyristors one time step after next current-zero crossing. These models can be applied when the periods of the frequencies of interest are much longer than turn-on and turn-off times, and converter losses, as well as voltage and current stresses, are not important [8].

8.3.3 More Detailed Device Models

They can be required in other circumstances, usually when the transient response of the converter and the immediate converter subsystems are of interest. Examples of these situations include studies for switching and conduction loss prediction; simulations to evaluate voltage and current stresses on the power electronics devices; simulations of converters with high-switching frequencies and slow devices; electromagnetic interference studies; thermal analysis; design of device protection [8]. In these cases, a reasonably accurate switch model is critical to the performance of the study. In general, these scenarios are of interest to the converter designer. More detailed switch models are also important for studies on the impact of device switching transients on machine or transformer insulation; the device turn-on and turn-off for IGBT-based voltage source converters produce repetitive steep wavefront transients that can interact with cable impedances and lead to insulation failures in motors [11, 12]. As voltage-source converter (VSCs) move to custom power and distributed generation applications, it could become necessary to perform insulation coordination studies for these applications, especially for stresses experienced by transformers. More accurate device models must include device turn-on/turn-off behaviour and conduction behaviour while the device is on or off, parasitic inductances and capacitances, wire and lead resistance, snubber circuit characteristics, and accurate gate circuit models. If more accurate models of the fastest devices, such as IGBTs and MOSFETs, are required, then simulation time steps shorter or much shorter than those used with ideal switch models will be needed. The degree of detail will vary with the device in question, the case under study, and the capabilities of the software tool. Very detailed power semiconductor device models are available as pre-compiled libraries for some software tools [13, 14]. These programs can utilize variable time-step numerical integration methods that are better suited to the accurate modelling of turn-on and turn-off behaviour of power semiconductor devices than a fixed time-step method. Loss estimates and voltage and current transients obtained from simulations based on these models are usually quite good and match up with experimental results closely. The most significant trade-off involved in using these models is the simulation time.

8.3.4 Approximate Models

Accurate-enough device representations can be implemented through some simple approximations without using a highly detailed switch model [15–17]. A very simple approximation to represent the device forward voltage drop is implemented by adding a resistance or a constant voltage source in series with a switch model. Figure 8.1 shows such a model for a diode. However, adding either element has limitations: a linear resistance is not an accurate representation of the forward voltage drop when the resistance varies with current, so this type of model will tend to be accurate only over a narrow range of currents. The series voltage source also poses some problems if the forward voltage drop varies with current. The voltage source should only represent a voltage drop as a loss component; it must not be able to supply power to the circuit [18]. A better approximation for the on-state model would be to use a controlled (voltage or current) source to represent the switch, since the user can include turn-on and turn-off times from a look-up table to control the source behaviour. The challenge with this representation is creating an approximation that is valid over the voltage and current operating ranges of interest. A better approximation is to model the device using a nonlinear passive circuit element, either in combination with an ideal switch or as a stand-alone element that has large impedance while the device is off. This nonlinear circuit element incorporates feedback of circuit conditions and possible gate signals. ATP allows users to insert a nonlinear circuit element based

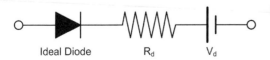

Ideal Diode R_d V_d

Figure 8.1 Forward model of a diode.

on a user-specified data. The nonlinear element approximates the device characteristic more appropriately, especially the on-state voltage drop. Some modifications could also be made to include turn-on and turn-off. Data for detailed models can be obtained from manufacturer's specifications. However, this is not an easy task, and some approximations might be required. If data is lacking, it may be possible to develop a reasonable approximation based on converter voltage and current ratings, although this approximation can lead to errors.

8.4 Solution Methods for Power Electronics Studies

8.4.1 Introduction

Specialized simulation tools offer the best accuracy for simulating power electronics when a detailed topology and complex operation controls need to be simulated and the main interest of the study is in the power electronics (sub)system. When the scenario requires the simulation of a larger zone, a tool such as ATP becomes very useful because it offers capabilities for characterizing power system components, including power electronics switches, with reasonably simplified accuracy. Two basic solution methods can be considered in power electronics studies: frequency-domain solution and time-domain solution [6, 7]. Time-domain simulation tools can only simulate circuit phenomena at discrete frequencies or at discrete intervals of time. This leads to discretization errors. Compared with the time-domain calculation, a frequency-domain simulation is more robust because a circuit solution is found at each individual frequency and these errors are not accumulated. Tools based on a frequency-domain solution method often treat the nonlinearity of a system as known current sources. For a harmonic evaluation, the frequency-domain solution usually requires less computation time compared with a time-domain solution. However, most frequency-domain solution programs have difficulties in handling system dynamics, control interfaces, and fast transients. A time-domain solution is based on the integration over a discrete time interval, and its stability and accuracy are closely related to the time-step size. Because discretization errors can accumulate, the solution may diverge from the true solution if an improper time step is selected. A time-domain simulation has advantages over a frequency-domain simulation in handling system dynamics, power electronic models, nonlinearities, and transient events. Only time-domain simulation is used in this chapter.

8.4.2 Time-Domain Transient Solution

The solution technique implemented in the ATP is based on nodal analysis [19]: an implicit trapezoidal rule is used for discretization and formulating the network nodal equations; the nodal voltages are unknown and calculated at every time-step, which is fixed. As the conductance matrix is usually sparse, specially designed techniques such as optimal reordering schemes and partial LU factorization are often used to solve this linear system efficiently instead of inverting the matrix directly [19]. Although the trapezoidal rule has order accuracy lower than other algorithms, it is numerically stable; this rule belongs to a class of implicit integration schemes that are adequate for solving stiff systems. Opening and closing of a switching element (diode, thyristor, transistor, etc.) represents an event that causes a topological change, and numerical oscillations can result in cases with inductive nodes or capacitive loops [19]. In power electronics applications a very common way for solving numerical oscillations is the use of snubber circuits in parallel with switches [6, 7]. An advantage of this solution is that snubbers do physically exist in many real cases [1]; however, they can also introduce some error in the solution. Other solutions have been proposed and implemented in other tools in order to suppress these oscillations; they are based on an interpolation technique [20–22] or a CDA (Critical Damping Adjustment) procedure [23, 24]. These algorithms modify the basic algorithm only at discontinuities.

8.4.3 Initialization

The automatic calculation of the initial steady-state solution of a system with power converters is not possible with the current ATP version. A power converter is a nonlinear device and a harmonic source; ATP steady-state initialization honours only linear components although it can solve circuits with more than one frequency during the initialization phase. At least two approaches can be used to overcome this drawback: (i) the simulation is started with the system initially deactivated, all semiconductors are blocked and control units do not operate, so the steady-state solution is reached after some simulation time, once the transients settle down; (ii) the initial state is manually calculated and specified using ATP options to override the steady state initialization; this solution can be used with small and simple systems.

8.5 ATP Simulation of Power Electronics Systems

8.5.1 Introduction

A power electronics converter can be seen as a set of switches that control the transfer of energy between two sources or between a source and a load. Different topologies and control strategies can be used taking into account the types of sources between which the energy transfer is made. The simulation of a power system with one or several power electronics (sub)systems requires of the following capabilities: (i) components for representing all types of sources, including detailed rotating machine models, and power delivery components (e.g. transformers, lines, cables); (ii) components for representing semiconductor devices; (iii) capabilities for representing control strategies.

A special application of a power electronics converter is the control of a rotating machine. Figure 8.2 shows the schematic diagram of an adjustable speed drive. The diagram has four blocks: the supply power system, the power converter, the rotating machine, and the control system, including sensors for monitoring variables [25, 26].

Guidelines for modelling every block might be summarized as follows:

1. *Supply power system.* The level of representation of the supply system depends on the objectives of the study. If the simulation aims to study the behaviour of the drive, testing the converter and/or its control strategy, the supply system can be represented by a simplified equivalent (i.e. ideal source plus equivalent impedance). If the goal is to evaluate the effects of the drive on the supply system, a more detailed description of the system could be needed.

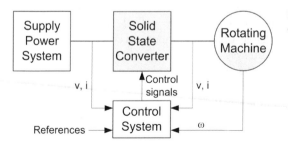

Figure 8.2 Schematic diagram of a solid-state electrical drive.

2. *Solid-state converter.* As discussed in Section 8.2, the converter can be simulated using two different approaches: dynamic average and switching models. With a dynamic average model, the converter is not represented, instead the machine terminals are directly fed from a voltage or a current regulator. In general, field and armature time constants are preserved in the rotating machine model, as well as the mechanical time constants. This approach is a good solution when the main interest is the study of a control strategy.

3. *Rotating machine.* If the study involves low-frequency transients, then ATP built-in models of rotating machines should be used. If the transients, including those related to electrical variables of the machine, are within a high-frequency range (i.e. several dozens of kHz), then some alternative options should be considered; for instance, implementing a custom-made rotating machine model taking advantages of ATP capabilities.

4. *Control system.* A control section will usually consist of three major parts dedicated respectively to monitoring and sampling, signal processing, and generation of gate signals. Remember that two options are available in the ATP for representing a control strategy: TACS (Transient Analysis of Control Systems) and MODELS; see Chapter 4. A common feature to both of them is the time-step delay that exists when variables are passed from the control section to the power system/converter. In general, this does not represent a serious problem; however, the choice of the time-step size must be done taking into account this fact and the sampling resolution of the simulated system. A control section, using either TACS or MODELS, is enough to simulate a drive when a macro-model or a block diagram approach is used to represent its behaviour [27]. Both options occupy the same location in a data file; the choice depends on the application case, the type of control, and the user's experience. Table 8.1 shows a summary of their main features.

A detailed discussion about the capabilities available in ATP for representing each block is provided in the next subsections.

8.5.2 Switching Devices

8.5.2.1 Built-in Semiconductor Models

Several types of built-in switches are available in ATP for representing semiconductor devices (see Figure 8.3) [28]:

- Type 11 switch, without a control signal it is used to represent a diode;
- Type 11 switch, with a gate signal it is used to represent a thyristor;
- Type 12 switch, used to simulate a triac;
- Type 13 switch, used to model bidirectional controlled devices.

Table 8.1 Comparison of TACS and MODELS features.

TACS	MODELS
<ul><li>Formatted grammar</li><li>Transfer-function representation</li><li>Representation of physical devices (i.e. frequency meters)</li><li>z-transform</li><li>Logical and algebraical operations by using FORTRAN-like expressions</li></ul>	<ul><li>Non-formatted grammar and regular syntax for the use of variables, arrays, and functions</li><li>Distinction between description and use of a model</li><li>Decomposition of a large model into submodels, with a hierarchical structure</li><li>Self-documenting nature</li><li>Conditional branching and loops (IF, WHILE, FOR statements)</li><li>User-established initialization</li><li>Connection to external routines</li></ul>

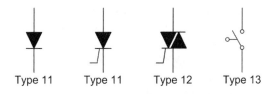

Type 11   Type 11   Type 12   Type 13

Figure 8.3 Built-in ideal semiconductor devices.

The Type 11 switch model can be used with or without gate signal. If no gate signal is applied to the switch, the switch model will behave as an ideal diode, with a turn-on time of one-time step when it is forward biased (voltage greater than setting for ignition voltage). The switch model will turn off at the next time step after current through the switch passes zero, and can exhibit the same problems with numerical oscillations that conventional switch models turning off at a natural current zero. This switch model can also behave as an ideal thyristor when a gate pulse is applied. This model implements a latching device, so a short pulse is sufficient to turn the device on, and it will stay on as long as the conduction conditions are met. The switch will normally turn off like the diode model.

The Type 13 controlled switch model is used to represent self-commutated switches. This switch model is able to turn on or turn off immediately at the next time step following application of a gate command, thus interrupting current. Since the interruption in current always occurs at a simulation time step, this switch does not create numerical oscillation problems. This switch can carry current in both directions.

8.5.2.2 Custom-made Semiconductor Models

The diode model is a two-terminal uncontrollable device while the others are controllable devices. Representing the reverse recovery characteristic, leakage current, and forward voltage drop of a diode is not necessary in many application studies. Some details of the device characteristic can be reduced and a simplified model is acceptable: instead of representing a diode using its switching characteristic, a simplified characteristic can be used; see Figure 8.4 [6, 7].

The switching characteristics of an actual and an idealized thyristor are shown in Figure 8.5 [6, 7]. To represent the simplified characteristic, the turn-on control is added on the simplified diode model. If the control is applied continuously, this switch simulates the diode which allows unidirectional current flow when the switch is forward biased. Delaying the gate pulse allows control over the turn-on instant of the forward biased switch.

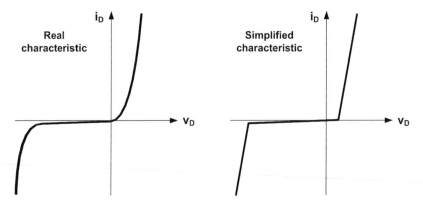

Figure 8.4 Actual and simplified characteristic of a diode.

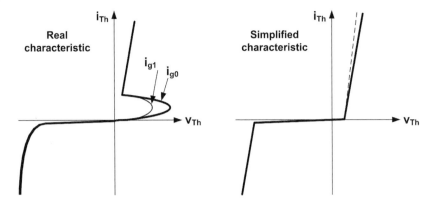

Figure 8.5 Actual and simplified characteristic of a thyristor device.

Figure 8.6 Semiconductor cells using built-in switches.

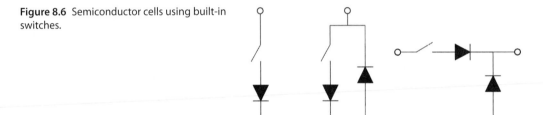

If the control system is excluded from the study, there is very little difference between modelling a GTO, IGBT or any other three-terminal and controllable switch. These devices can all be represented by a simplified switch with gate turn-on and turn-off controls. The different switching characteristics can be realized by applying different firing strategies. However, additional errors can occur because, with ATP, the gating pulses can only be applied at integer time-step multiples, which may not be the case in reality.

Additional options can be considered for representing semiconductor devices with ATP:

1) Since built-in switch models do not cover all power semiconductor models, users can consider the association of the elementary switches listed above to construct other ideal semiconductor models. Figure 8.6 shows the so-called T and TD-cells. The T-cell can be used to represent a transistor. The figure shows a TD that uses the Type 13 switch plus a specific firing sequence; see [29].
2) Instead of switches, the representation of semiconductor devices can be made by means of nonlinear resistors [30]. Several strategies can be used; the model can be made by either using a very accurate value, with resistances calculated from a control unit, or a binary model: the resistor value is *high* at an open state and *low* at a closed state. With this approach, a larger time-step size can be used and numerical oscillations are generally avoided. Figure 8.7 shows the large-signal transistor model proposed in [31].
3) In some applications, mainly those with high-frequency transients, the preceding options are not accurate enough and more sophisticated models have to be used. Successful models of an IGBT, based on a rigorous mathematical description, have been implemented in ATP. They take advantage of capabilities for representing control strategies; see [13, 32].

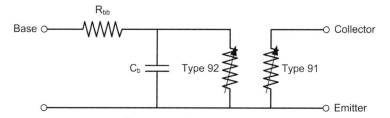

Figure 8.7 Large-signal transistor model.

4) The so-called Type 94 branch can also be used to represent single or multi-branch semi-conductors. Interfaces based on both compensation and prediction are available with this option [33]. Using compensation, at each time step of the simulation, the model represented by the Type 94 branch receives from the circuit to which it is connected, a Thevenin equivalent as seen from the component, as well as branch voltages and currents that exist at the previous time step of the simulation. A dedicated MODELS section will calculate actual voltages and currents of each branch of the semiconductor. Their values will then be included in the solution of the electrical circuit using superposition and without performing any iteration. A Type 94 branch is excluded from the steady-state initialization. However, other branches can be connected in parallel for including a simplified representation during this initialization. Voltage and current of the branch, which is calculated at time zero of the simulation, can be used by the model of the component for proper initialization. The network is seen from a Type 94 component as linear, so no other nonlinear element must be connected to the network, unless they are separated from the local subnetwork using stub lines. This is an important limitation in many applications.

A diode model aimed at representing the correct operation during a transient process must represent the so-called forward and reverse recoveries. During a rapid switch from off-state to on-state, a high forward voltage builds up across the diode due to the initial low conductance. This process is known as forward recovery. The reverse recovery is a transient mode that occurs when a forward conducting diode is turned off quickly and the internally stored charges are evacuated during a reverse current period with a high reverse voltage. An extension of the basic charge control model, derived from the semiconductor charge transport equations, considering both processes, was presented and validated in [34, 35]. A complete diode model taking into account both forward and reverse recovery based on the Type 94 branch was presented in [36].

8.5.3   Power Electronics Systems

If every individual switching device is represented, a power system model with power electronics (sub)systems can easily reach a complication level that is difficult for implementation. However, for many applications, it is not necessary to represent all individual devices. Usually, what need to be simulated are the terminal characteristics of a subsystem and how it interfaces with the connected system. The following recommendations can be used to reduce the modelling complexity [6, 7]:

- Use one or a few equivalent devices to represent series and parallel combination of a group of devices.
- Use the simplest device model which is appropriate for the application.
- Represent a power electronics (sub)system by an equivalent source injection whenever it is acceptable.

- Represent only the front end of the drive system when the major concern is utility interfacing.
- Include the system dynamic and controls only when necessary.
- Use modular approach for large scale models.
- Represent power electronic loads with similar characteristics by a single equivalent load.

Important issues that should be addressed are [6, 7]: harmonic cancellation when multiple loads are represented by their lumped equivalent; existing system distortion; appropriate source topology for power electronics (sub)system representation; system imbalance; effects of a dc link or the inverter side connection on the front end interface with the power supply system; current or voltage sharing among the parallel or series switching devices; switching loss prediction.

As already mentioned, the trapezoidal integration method is inherently prone to spurious oscillations in capacitive and inductive circuits when subjected to sudden changes (e.g. step change in voltage, current injection, switching). One of the following measures, or a combination of them, can be taken to prevent numerical problems [6, 7]: (i) select a smaller time step, (ii) use artificial snubber circuits, (iii) introduce a small smoothing reactor for dc links, (iv) introduce proper stray capacitances in the system model, (v) provide damping. This problem is usually solved with the use of snubber circuits in parallel with semiconductor devices. The parameter values of the snubber depend on the time-step size, the system configuration (capacitors and inductors in the system), and the load current level.

8.5.4 Power Systems

The model of the power supply system does not necessarily have to cover a large electrical and geographic radius. In many scenarios, the system model can be simplified. The proper level of system reduction depends on the study objectives.

- If the purpose is to characterize the harmonics generated by a particular type of power electronics application, the power system model can be significantly reduced. When a pre-existing voltage distortion level at a power electronics interfacing bus is low, the rest of the power system can be satisfactorily represented by one or a set of first-order equivalents connected to the bus at a higher system voltage level.
- If the objective is to evaluate effects of the power electronics on a connected utility system, the model shall be extended to cover all sensitive loads (i.e. rotating machines and all other major power electronics) within a concerned electrical radius. Special attention is needed if an unbalanced system condition is involved.
- A large power system model might be required for harmonic propagation and resonance studies. The main system components and dominant topology need to be kept in the power system model. Filter banks, nonlinear passive circuit components, and all other harmonic injection sources should be represented. Frequency-dependent characteristics of the system components might need to be considered.

8.5.5 Control Systems

The system control is one of the most important aspects of a power electronics simulation since the proper performance of a switching device is realized via appropriate gate controls. In a time-domain simulation, the highest resolution for a signal sampling is determined by a selected time step. In general, this presents no problem for analogue controls. When simulating digital controls, if the selected time step is too large and the simulated sampling resolution is significantly different from the real system sampling resolution, errors can be introduced and even lead to instability.

When modelling a control response, it is important to keep in mind the time delay introduced by the program between the primary system and the control interface. This may not cause problems in some simulations. However, if the modelled control logic makes this time delay caused error to accumulate over a period of time, it can eventually result in the solution divergence. The problem can be corrected in most cases by reducing the time-step size or avoiding the possible accumulation mechanism in the control model.

Different methods may be used to synchronize power electronics gating signals with required system references. In many cases, a real phase-locked-loop (PLL) can be greatly simplified to reduce the modelling system complexity. However, when the system contains significant waveform distortions, either harmonics or transient disturbances, a practical PLL with all signal filters should be carefully implemented in order to accurately predict control response. This is particularly important when the objective of the simulation is to verify control design and to evaluate the response of a power electronics application to primary system dynamics.

All power electronics devices have their limit in switching frequency. When a load commutation or a standard pulse width modulation (PWM) scheme is simulated, the highest switching frequency in the simulation is controlled by the system frequency or by a carrier frequency. Even considering a variable carrier frequency, the number of switching events per fundamental frequency cycles is known and the highest switching frequency can be made under a physical limit of the simulated device. However, if the device firing is determined by a simple comparison between the system control reference and the system output, a device switching may take place in simulation whenever a comparison difference is detected. Therefore, the switching frequency becomes highly dependent on the time-step size, and the average switching frequency becomes unpredictable. When using this type of firing logic, users should always take additional measures (e.g. introducing a hysteresis loop) to ensure that the modelled device is working under its physical switching capability.

8.5.6 Rotating Machines

Depending on the case study, ATP users can consider two options for simulating a rotating machine: built-in and custom-made models. Obviously, the second option will only be used when the rotating machine could not be represented by means of a built-in model.

8.5.6.1 Built-in Rotating Machine Models
A detailed discussion about the models implemented in ATP for representing rotating machines was provided in Chapter 7. By default, ATP users will take advantage of the Type 59 and Type 58 models for representing synchronous machines, and the UM module for representing any other machine design, although the three-phase synchronous machine is also available in this option. The UM module has generally been used for representing three-phase ac machines; however, some cases illustrating the simulation of other types of machines (single-phase ac machines, dc machines) have also been presented [37–40].

8.5.6.2 Custom-made Rotating Machine Models
The UM module does not cover every type of machine. TACS and MODELS capabilities can be used for representing the dynamics of a rotating machine [41, 42]. This option is a good solution when a macro-model approach can suffice or is needed. Although disadvantages of this approach are obvious (e.g. users have to develop their own models), it can be a good solution for detailed or special representations [43]. A complex case using MODELS for representing the rotating machine and the converter was reported in [44]; in this case, the justification was the need for the representation of a special machine not available as a built-in model.

8.5.7 Simulation Errors

Errors in power electronics simulations can come through the following sources [6, 7]: (i) switching device approximation and system reduction; (ii) circuit elements added for mitigating or avoiding numerical oscillations; (iii) control system simplification; (iv) time-step related truncation; (v) interfacing time delay introduced by the solution method; (vi) incorrect system initial conditions. Some errors resulting from the system simplification and the measures for controlling numerical oscillations might be acceptable; other errors can be controlled by reducing the time-step size. As a rule of thumb, the size should be between 1/20 and 1/5 of the period of the highest frequency. Errors caused by incorrect system initial conditions can be reduced by just letting the simulation run for a long-enough period of time to reach a corrected initial condition. This may take more computing time, but time is saved in model construction, especially if the program allows restarting (as ATP does). Some tricks can also be used to accelerate the system simulation and quickly reach the correct initial condition; they will generally be based on users' experience.

8.6 Power Electronics Applications in Transmission, Distribution, Generation and Storage Systems

8.6.1 Overview

Generation of electricity has customarily been done at high voltages. For long-distance transmission systems, the voltages are stepped up to high levels using step-up power transformers. The voltages are stepped down to lower levels, again using transformers, for distribution over relatively short distances to reach loads. This electric power flow transmission may be inefficient because transmission lines carry both real power and reactive power, and a number of lines may reach their rating limits before the rest of the system, as a consequence of which previously under-loaded lines that will have to pick up the load may become overloaded.

The flow of electricity in a particular transmission line can be controlled by means of a power flow controller, which regulates the parameters that affect the flow of power. Solutions based on power electronics have given system operators the ability to influence these parameters, thereby allowing them to control power flow throughout the network. The demand for electrical energy increases continuously, but the construction of new transmission lines is becoming increasingly difficult because of various reasons (e.g. regulatory and environmental constraints, public policies, escalating costs). Power electronics can provide a solution by using the existing lines in a more efficient way to carry maximum real power with a minimal reactive power. Power electronics enable the development of advanced compensators that can rapidly control the flow of real and reactive powers, control the voltage profile of the network, and offer improved stability.

AC transmission has been the dominant form in electric power systems. Alternatively, it is also possible to transmit electric power by converting the generated ac power to dc and back to ac at the end of the line for distribution. In the HVDC scheme, the dc line needs to carry only real power, thereby eliminating line losses due to reactive power flow and increasing line utilization when compared with ac transmission systems. However, for an economically viable design, the line needs to be of a minimum length, since the two conversion stages – ac-dc and dc-ac – cost significantly more when compared with an ac transmission system where no such conversions are required. Another area in which power electronics is playing a crucial role is in interfacing of renewable energy sources, such as wind and solar, into the grid. This also enables micro-grids and custom power applications at distribution levels. Figure 8.8 shows a schematic diagram of the major areas where power electronics is used in power systems [45].

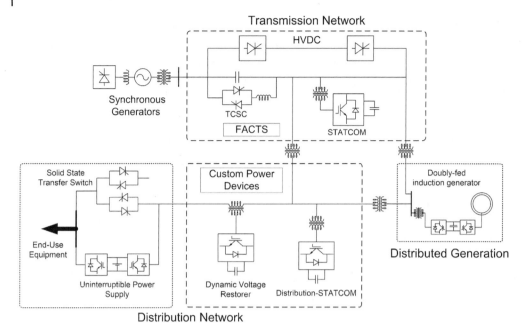

Figure 8.8 Power electronics in power systems.

8.6.2 Transmission Systems

Control of the power flow in an electric power system involves control of the magnitude and phase angle of the voltage at certain points in the system. This was accomplished in the past by using a synchronous condenser. Power electronics technologies have enabled the replacement of the bulky, slow, and high-maintenance compensators with compact, fast, and high-performance compensators. Actually, the concept of a synchronous condenser has been extended with the use of a static synchronous compensator (STATCOM) that connects an electronically-generated sinusoidal voltage (with some harmonic components) in shunt with the transmission line through a tie inductor. The VSC used in a STATCOM is the controllable voltage source that enables such an undertaking.

The VSC-based concept has been further developed to include static synchronous series compensator (SSSC), unified power flow controller (UPFC), back-to-back STATCOMs, also known as VSC-based high voltage direct current (VSC-HVDC), and back-to-back SSSCs, also known as the interline power flow controller (IPFC), for transmission applications [2, 46–48]. The semiconductor switches that are used in the implementation of a VSC are fully controllable, and the switches can be turned on and off at the desired time.

The concept FACTS was adopted, and defined as alternating current transmission systems incorporating power electronic based and other static controllers to enhance controllability and increased power transfer capability. The key to independent control of real and reactive power flows in a transmission line is to control both the magnitude and phase angle of the transmission line voltage simultaneously. This can be achieved with either shunt-series or shunt-shunt configurations.

8.6.3 Distribution Systems

The primary source of electrical energy for consumer equipment is the power distribution utility. Utilities provide power to large consumers at voltages below 33 kV. The voltage supply is stepped down for domestic use. The main concern of utilities is to provide continuity of power supply at an acceptable voltage and frequency, with no sustained interruptions or 'outages'. The level of reliability as measured by the number of outages that an average consumer is subjected to in a year may not always be appropriate. This is because very short-duration disturbances like voltage dips can cause significant disruptions in sensitive loads. Another concern is the unsatisfactory operation, or even damage, which may be caused to consumer equipment by temporary or steady-state distortions in the supply voltage waveforms; the quality of power is an important concern in distribution systems. Power-quality problems may arise due to disturbances on the utility side such as lightning and switching events, and equipment failures. They may also be caused due to energization of large loads and by the presence of nonlinear and time-varying loads [49]. Uninterruptible Power Supply (UPS) systems, stand-by generators and alternative tie points to the utility are possible solutions to overcome long-duration disruptions in power supply. Some power-quality problems such as harmonic distortions and dynamic overvoltages can be addressed by having filters and overvoltage suppressors. These may be seen as local or load-side solutions, and may be established by end users. However, most voltage variations and interruptions result from events that involve the utility system, or are transferred from other loads via the utility system. Therefore, distribution utilities have to offer solutions, which ensure supply with the desired quality. This concept is known as Custom Power (CP) [50, 51]. CP solutions involve the application of power electronics controllers to distribution systems at the supply end of industrial/commercial customers and industrial parks. They can be used for network reconfiguration (e.g. static transfer switch, static circuit breaker, current limiter), or for load and supply compensation (e.g. distribution static compensator (DSTATCOM), dynamic voltage restorer (DVR), unified power quality conditioner (UPQC)) [50, 51].

8.6.4   DER Systems

Small-scale distributed generation (DG) and distributed storage (DS) systems are usually connected to the power system at low- and medium-voltage levels. DG and DS systems, usually referred to as DER systems, can significantly affect the power system performance. This justifies the need for advanced modelling and simulation of DERs to reveal their impacts on the power system. Broadly speaking, DER systems can be classified under two categories [4]:

- Machine-based DER systems, which employ a synchronous or asynchronous generator to convert mechanical energy to electricity. The machine is directly coupled with the power system and bound to operate at the power-system frequency. Some examples include biomass- and diesel-fuelled generators, generators driven by gas-fuelled internal combustion engines, small hydroelectric generators, and fixed-speed wind energy systems.
- Electronically-interfaced DER systems, which may or may not utilize a rotating machine to generate electricity out of the prime energy resource. Since energy resources can generate either dc power or ac power at a different or even variable frequency with respect to the power-system frequency, the generation component is coupled with the power system via a power electronics interface. Examples are photovoltaic (PV) solar energy systems, fuel-cell systems, wind-energy systems, battery-energy storage systems, supercapacitor-energy storage systems, and flywheel-energy storage systems.

Figure 8.9 shows a generic schematic diagram of an electronically-interfaced DER system, which consists of an energy resource and a power electronics interface, interconnected via a dc link. The energy resource exchanges the power with the interface, which consists of a dc-ac VSC [4], a dc-side capacitor, an ac-side filter to mitigate harmonic current injection by the converter, and a start-up/shut-down switchgear. This interface exchanges real and reactive powers with the host grid.

The converter is controlled by means of a PWM strategy. A dedicated closed loop can control the magnitude and phase angle of the ac-side current. This, in turn, enables the regulation of P_g and Q_g at their corresponding setpoints. It is also possible to directly control P_g and Q_g by the magnitude and phase angle of v_t; this strategy, known as *voltage-mode control*, renders the converter vulnerable to overcurrents (during either transient or steady-state operation) as well as to network faults. By contrast, the *current-mode control* can avoid such vulnerability by allowing for the imposition of a safe upper limit on the magnitude of i_t. The capability to control P_g and Q_g permits several control scenarios for the DER system [4].

Figure 8.9 indicates that the DER system may alternatively employ a *conditioned energy resource* [4, 52]. As illustrated in Figure 8.10, an intermediate electronic power conditioner must precede the energy resource: if the energy resource generates dc power, a dc-dc electronic power conditioner is utilized; if the energy resource generates single- or three-phase ac power, the electronic power conditioner is of ac-dc type. In both systems, the output dc terminals of the conditioner define the dc port of the conditioned energy resource. The electronic power conditioner may be based on a variety of configurations. The conditioning of an energy resource is usually exercised to enable the integration of energy resources with low-voltage

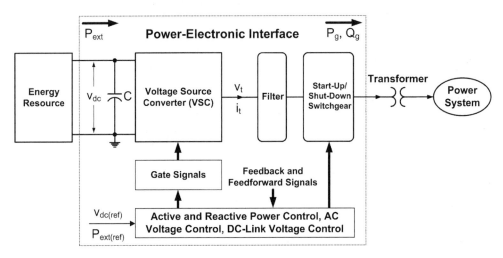

Figure 8.9 Generic schematic diagram of an electronically-interfaced DER system.

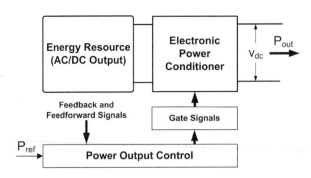

Figure 8.10 Schematic diagram of a conditioned energy resource.

outputs (e.g. in fuel cells and super capacitors), enable output power control or maximum power point tracker MPPT in variable-speed wind energy conversion system (WECS) and two-stage PV energy systems, and allow for optimal operation of the power electronics interface of the DER system (e.g. in two-stage PV systems). The use of the electronic power conditioner, however, results in a somewhat lower efficiency for the DER system, due to the additional power losses.

8.7 Introduction to the Simulation of Power Electronics Systems

8.7.1 Overview

The built-in models that can be used for representing semiconductor devices were listed in Section 8.5.2:

- The Type 11 switch can carry current in only one direction and will turn off like a diode. The user can also specify a minimum ignition voltage (with a default value of zero), a minimum holding current for maintaining continuous conduction, and a de-ionization time. The minimum ignition voltage can be useful to ensure the device does not turn on before the forward voltage exceeds the device's forward voltage rating; however, the voltage drop across the switch drops to zero after it turns on, so this may not adequately represent the on-state forward voltage drop. Since this switch model will turn off at the next time step after current through the switch passes zero, it can exhibit the same problems with numerical oscillations as conventional switch models turning off after a natural current zero.
- The Type 11 switch model can be used with or without gate pulses applied. If no gate signal is applied to the switch, the model will behave as an ideal diode; however, this switch model can also behave as an ideal thyristor when a gate pulse is applied. This model implements a latching device, so a short pulse is sufficient to turn the device on, and it will stay on as long as the conduction conditions are met. The switch will normally turn off like the diode model, and will exhibit the same problems with numerical oscillations as the pure diode model.
- The Type 13 switch model can be used to represent a self-commutated switch. This model is able to turn on or turn off immediately at the next time step following application of a gate command, thus interrupting current. This switch model will carry current in both directions; that is, this switch is a bidirectional controlled device that supports both direct and reversal currents.

A fourth model, the Type 12 switch model, can be used to simulate a triac; it is rarely used and will not be used in any case studied in this chapter.

This section is aimed at presenting some common numerical problems that can arise when using ATP built-in models for implementing power electronics systems/converters and how to solve them. The section analyses case studies of increasing complexity: it begins with very simple scenarios that involve a single-switch model and finishes with more complex cases in which a full converter is analysed. In all cases, control strategies are very simple; the goal is to illustrate problems that can occur during on-off commutations and commutations between switches. The solutions to numerical problems that can occur in ATP simulations are basically the following ones:

- The use of snubber circuits in parallel to switches (semiconductors) is probably the most common solution since snubber circuits are also used in actual power electronics systems to avoid or limit overvoltages across semiconductors [1]. See Chapter 3.

- The use of parallel damping with inductors and series damping with capacitors is another solution that can limit or mitigate numerical oscillations, as explained in Chapter 3.
- The use of smaller time steps is a solution that can always improve simulation results. When simulating power electronics systems, the time step will depend on the supply and switching frequencies, but it will always be in the order of 1 μs. Larger and smaller values can also be used or required, as illustrated in the cases studied in this and the next section.
- Another scenario that can often occur is the creation of loops of closed switches. This can basically be solved by inserting small resistances and inductances in series with switch models.

8.7.2 One-Switch Case Studies

Figure 8.11 shows the test circuit – a 50-Hz ideal voltage source feeds a load represented as an *RL* branch – with three different semiconductor devices. The load voltage and current are controlled by means of a semiconductor device. When a gate signal is required, the signal is a constant-period constant-width pulse. The main goals are to check the performance of each switch model and solve the numerical problems that could occur.

The parameters used in this case are as follows:

- Source: Voltage = 1 V (peak value), Frequency = 50 Hz, Phase angle = −90°
- Load: $R = 1\,\Omega, L = 3\,\text{mH}$
- Control signal: Pulse type, Period = 20 ms, Width = 5 ms

Figures 8.12–8.14 show the ATPDraw implementation of the three cases and some simulation results obtained with each switch model. The time step used with the three switch models is 1 μs. Numerical oscillations are expected with the three switch models, so a parallel snubber circuit has been used in all cases. However, it was found that the best cure for those oscillations also depends on the load inductance. Additionally, with the parameters shown above, a parallel damping was used in the three scenarios; see figures. Snubber parameters with the three switch models were: $R = 1000\,\Omega, C = 5\,\mu\text{F}$. The factor K_p to be specified in the inductor menu was 5.

From the analysis of these results and the comparison between them, one can easily deduce the performance of each switch model and the different behaviours. All switch models are assumed ideal and their behaviour is as expected.

The most interesting difference between the two controlled switch models is in the effect that the control signal has on the behaviour of each model:

- the Type 11 switch behaves as a thyristor; once the signal is applied to its gate, the latching mechanism keeps the device on, and it is deactivated by a natural line-commutated zero;
- the Type 13 switch is a bidirectional and self-commutated switch that can behave as a GTO with a series diode; the semiconductor turns on when the signal is positive and turns off when it changes the sign or becomes zero.

Note the differences between voltages and currents in the test circuit shown in Figures 8.13 and 8.14. Remember that the only difference between the two circuits is the semiconductor type.

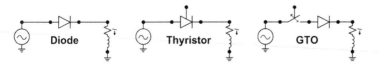

Figure 8.11 Introductory case study: one-switch scenarios.

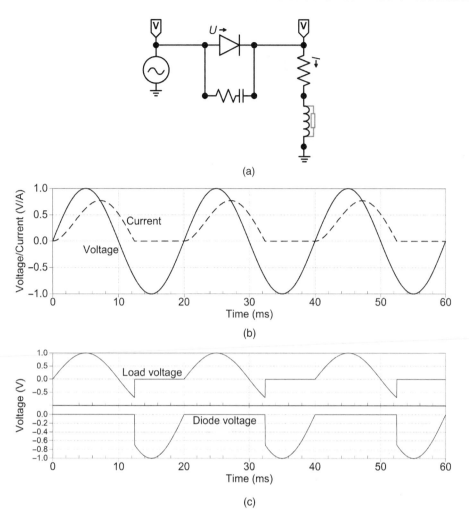

Figure 8.12 One-switch case study: simulation results – diode performance. (a) ATPDraw implementation, (b) Source voltage and load current, (c) Load and diode voltages.

8.7.3 Two-Switches Case Studies

Figure 8.15 shows the new circuit configurations: two 50-Hz ideal voltage sources supply an *RL* load, whose voltage and current are controlled by means of two semiconductor devices. The control signals when a gate signal is required are again constant-period constant-width pulses. As for the previous case study, the goal is again to check the performance of each switch model and solve the numerical problems that could occur with the new circuit configuration. The parameters used in this case are as follows:

- Sources: Voltages = 280 V (peak value), Frequency = 50 Hz, Phase angles = ±90°
- Load: $R = 5\,\Omega$, $L = 10\,\text{mH}$
- Control signals: Pulse type, Peak value = 1, Period = 20 ms, Width = 3 ms

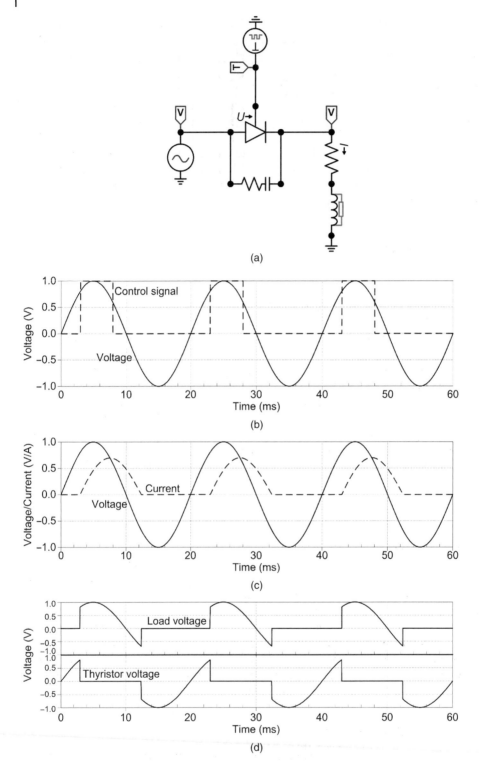

(a)

(b)

(c)

(d)

Figure 8.13 One-switch case study: simulation results – thyristor performance. (a) ATPDraw implementation, (b) Source voltage and gate signal, (c) Source voltage and load current, (d) Load and thyristor voltages.

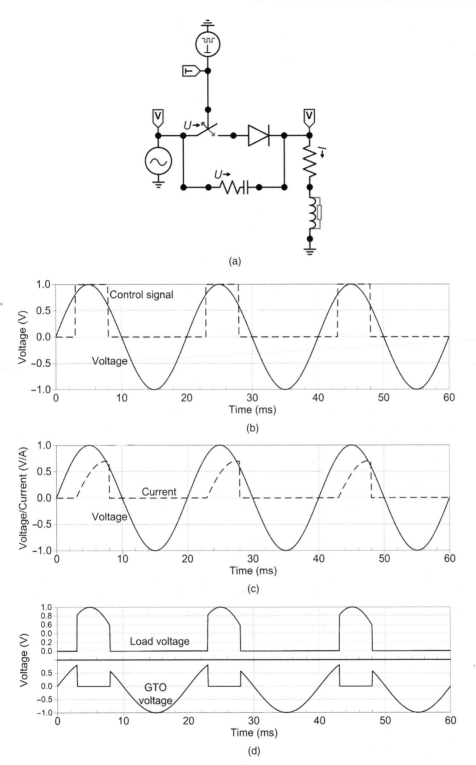

Figure 8.14 One-switch case study: simulation results – GTO performance. (a) ATPDraw implementation, (b) Source voltage and gate signal, (c) Source voltage and load current, (d) Load and GTO voltages.

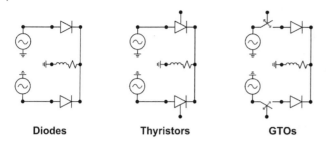

Diodes   **Thyristors**   **GTOs**

Figure 8.15 Introductory case study: two-switches configurations.

Figures 8.16–8.18 show the ATPDraw implementation of the three cases and some simulation results. The time step used with the three switch models was again 1 μs.

Note that, as for the previous case study, a parallel snubber circuit has been placed in parallel to all semiconductor models. The circuit parameters are now $R = 10\,000\,\Omega$, $C = 5\,\mu F$.

The fact that the load current comes from two different sources means that there will be a commutation between the sources. To avoid the creation of a loop of closed switches, a small resistor has to be used in series with each semiconductor model. With thyristors and GTOs, the series resistors are not required unless the load current was continuous (i.e. without zero-current periods). In the case of GTOs, this will also avoid the problem if there was overlapping between the two control signals. The results with the new circuit configuration are again as expected.

With the load parameters used in this case study, a continuous current results when the load current is controlled by means of diodes and thyristors. However, such continuity is not achieved with GTOs due to the pulse duration, which is not long enough.

The differences between results with thyristors and GTOs are again evident: the currents through thyristors remain after the control signals are deactivated due to the latching mechanism (i.e. the currents become zero due to line-commutation); however, the change in the control signals has a different consequence with GTOs since the self-commutation mechanism

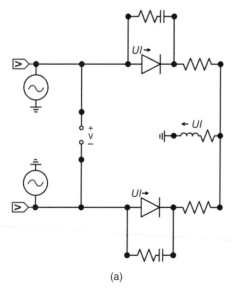

(a)

Figure 8.16 Two switches: ATPDraw implementation and simulation results – diode performance. (a) ATPDraw implementation, (b) Source voltage and current – top circuit, (c) Source voltage and current – bottom circuit, (d) Voltages across diodes during non-conducting periods, (e) Load voltage and current.

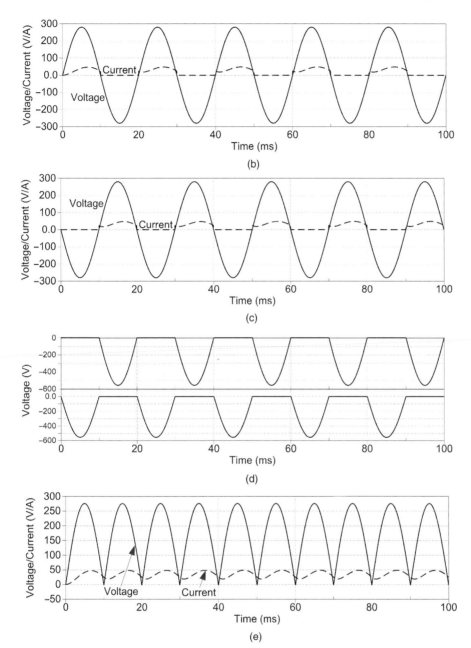

Figure 8.16 *(Continued)*

operates now and the devices become closed after the control signals are deactivated. In the case of thyristors, the pulse duration is irrelevant, in the case of GTOs that duration is crucial.

8.7.4 Application of the GIFU Request

The ATP has a special option that can be useful when simulating some power electronics systems. Figure 8.19 shows the circuit that is used in this section to test the so-called GIFU request.

It is a simple buck-boost circuit that can be used to obtain a unidirectional voltage of different polarity and value to those of the supply dc source (i.e. with a positive polarity dc source, the voltage polarity of the resistive load will be negative). The circuit will also work without the diode, but the resulting voltage will oscillate (between positive and negative values) with a frequency that will depend on the control signal frequency.

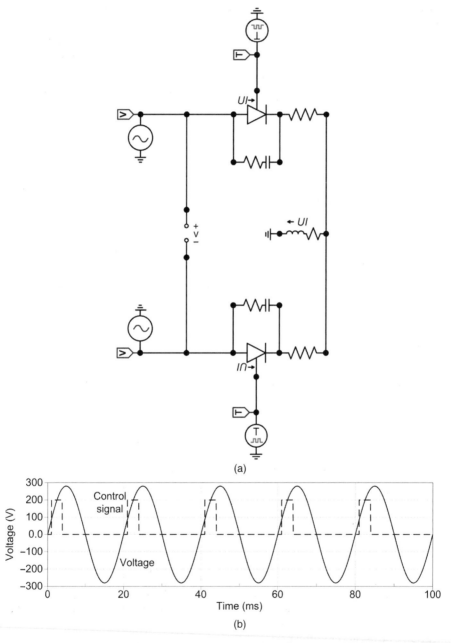

(a)

(b)

Figure 8.17 Two switches: ATPDraw implementation and simulation results – thyristor performance. (a) ATPDraw implementation, (b) Source voltage and gate signal – top circuit, (c) Load and (bottom) source voltages, (d) Load and (top) thyristor currents, (e) Load voltage and current.

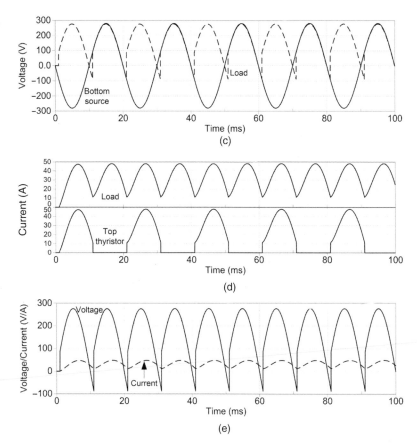

Figure 8.17 (*Continued*)

With the configuration shown in Figure 8.19, the resistive load voltage and currents will be discontinuous, and oscillate at the frequency of the control signal. To obtain a continuous voltage a shunt capacitor has to be used in parallel to the load resistance. On the other hand, the current through the inductance L_1 will be continuous, as shown below.

The way in which an opposite polarity voltage can be created between the terminals of the load resistance is very simple: (i) when the switch is closed, a current will begin circulating through the circuit $R_1 - L_1$; (ii) when the switch is opened, the current will continue circulating through the circuit $L_1 - R_l -$ Diode.

Note that if the circuit is initially dead, the current through L_1 will initially be zero, but after the first commutation, the current will no longer be zero. Although a current decay will be caused by the load resistance, with every new switch commutation the current may increase. Take into account that the current caused by the switch closing will produce a current variation that will depend on the values of R_1 and L_1, as well as the pulse duration; this means that only with a long enough pulse duration the current can increase with every new commutation. If the pulse duration is rather short (as in the case simulated here), the current will increase a certain value during any new commutation, will decrease every time the switch opens due to the damping caused by the load resistance, and will increase again after the switch closes. This will continue until the current through the circuit $R_1 - L_1$ will reach its maximum value, which is given by the dc source voltage and the resistance R_1.

The parameters used in this case are as follows:

- DC source: Voltage = 20 V (positive polarity)
- Circuit parameters: $R_1 = 1\,\Omega$, $L_1 = 1\,\text{mH}$, $R_l = 1\,\Omega$
- Control signal: Period = 0.1 ms, Duration = 0.05 ms

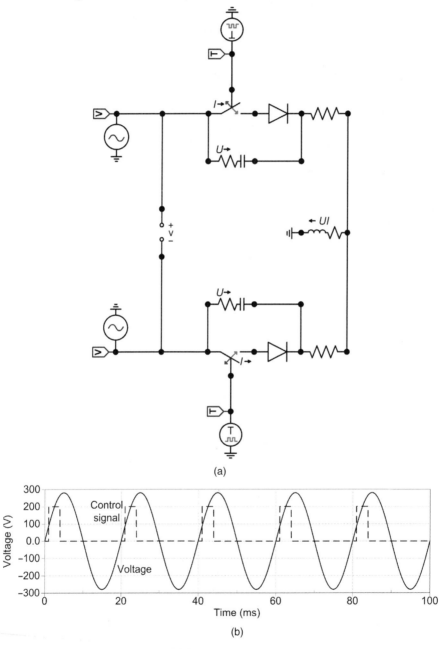

(a)

(b)

Figure 8.18 Two switches: ATPDraw implementation and simulation results – GTO performance. (a) ATPDraw implementation, (b) Source voltage and control signal, (c) Source voltage and load current, (d) Currents through GTOs, (e) Load voltage and current.

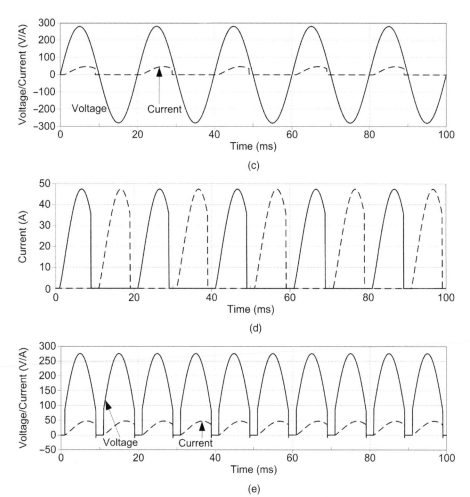

(c)

(d)

(e)

Figure 8.18 (Continued)

Figure 8.19 Buck-boost circuit for testing the GIFU option.

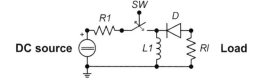

The transient that will follow the switch opening will be wrong and the commutation from the circuit dc source – R_1 – Switch – L_1 to the circuit L_1 – R_{load} – Diode will not be correct, unless the GIFU request was used. This request was implemented in 1996. The details about the option and some illustrative examples were presented in [53]. Readers interested in more details can also consult the Can-Am News [54].

Figure 8.20 shows the simulation results obtained with the GIFU option. They exhibit the behaviour discussed above: since the dc voltage source polarity is positive, the voltage polarity between the load resistance terminals is negative and discontinuous (since no shunt capacitor

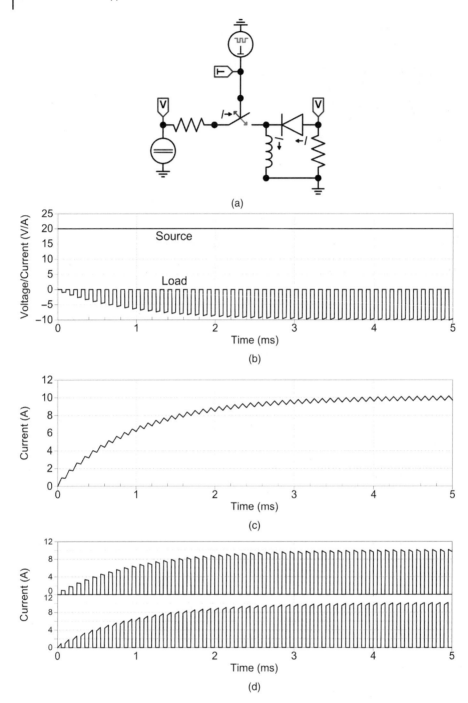

Figure 8.20 ATPDraw implementation and simulation results: buck-boost circuit. (a) ATPDraw implementation, (b) Source and load voltages, (c) Inductor current, (d) Currents through the switch and the diode.

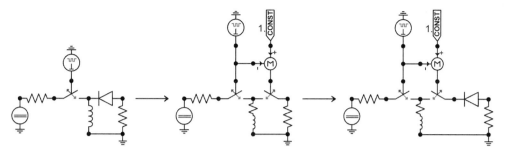

Figure 8.21 ATPDraw implementation of equivalent circuits without the GIFU option.

has been used), with a frequency equal to that of the control signal. On the other hand, the currents through the switch and the diode (and consequently through the load resistance) are discontinuous, but that through the inductor is continuous.

The GIFU request can be used with the Type 11 and the Type 13 switch models, and will work every time a diode is involved. Therefore, it is an option with a limited range of applications.

It is also worth mentioning that there are alternative solutions to some problems that can be solved with the GIFU option. Figure 8.21 shows circuit configurations that will produce the same solution but without using the GIFU request. In the circuits with two Type 13 switches, the control signal required for the second switch is the complementary signal of the first switch.

8.7.5 Simulation of Power Electronics Converters

Two simple examples of converters are studied in this section to present some problems that can occur when more than two switch devices are involved. The first is a single-phase inverter, the second one is a three-phase (line-commutated) uncontrolled rectifier.

8.7.5.1 Single-phase Inverter

Figure 8.22 shows the configuration of the first test circuit: a single-phase inverter that uses a dc voltage source to supply a load represented by an *RL* branch with an ac current of the desired frequency. The goal is to control the switches to obtain a 50-Hz current through the load branch. The switches are modelled as IGBTs with antiparallel diodes, as shown in Figure 8.23. An ungrounded dc voltage source is used to represent the supply source. The parameters of this circuit are: dc source voltage = 325 V; $R_{\text{load}} = 5\,\Omega$, $L_{\text{load}} = 10\,\text{mH}$. The control unit is based on a simple modulation strategy implemented in TACS. Figure 8.24 summarizes the main principles on which this strategy is based:

Figure 8.22 Configuration of the single-phase test inverter.

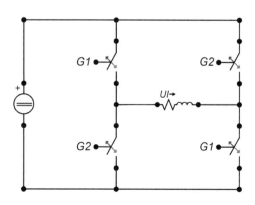

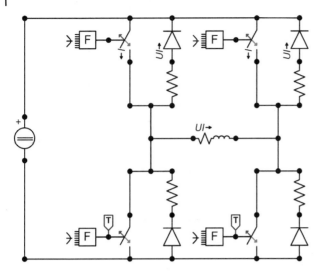

Figure 8.23 ATPDraw implementation of a single-phase inverter.

- a pulse train signal of period 0.1 ms and duration 0.05 ms, and whose magnitude oscillates between 0 and 1, is converted into a saw-tooth signal of the same period and whose magnitude oscillates between 0.2 and −0.2;
- this signal is compared to a sinusoidally oscillating signal of magnitude 0.2 and frequency 50 Hz;
- the intersection between the two signals will provide the gate signals, G1 and G2 (see Figure 8.22).

The resistances in series with diodes are needed to avoid loops of closed switches. This configuration cannot work unless GIFU option was requested in all Type-13 switches.

Figure 8.25 shows some simulation results with two different control signals (see Figure 8.24): the period of the signal is in both cases 100 μs, but the pulse duration is different. The current through the load branch, shown in Figure 8.25c, exhibits the desired 50-Hz waveform, and it is the result of adding the two currents depicted in Figure 8.25a. Note that the load voltage is neither sinusoidal nor continuous irrespective of the pulse width; however, when the pulse is one half of the period the load current becomes sinusoidal.

Although the case studies presented in this book are basically modelled with ideal switches, the ATP built-in models of the diode and the thyristor allow users to specify some additional features of the two semiconductors. One of these features is the ignition voltage. As already mentioned, the results presented in Figure 8.25 were derived by requesting the GIFU option in Type 13 switches, and assuming a 0 V ignition voltage for all diodes. If this voltage value is positive for all diodes, then the GIFU option is not required, and the simulation results will be the same.

A detailed study of the simulation results reveals that there are no currents through the antiparallel diodes. This means that the current through one IGBT (e.g. G1) commutates to other (e.g. G2) and diodes are not required. The simulation of the test circuit without diodes will provide the results shown in Figure 8.25. Take into account that, since the new test circuit configuration does only include IGBTs (but not diodes), the GIFU option is no longer needed.

8.7.5.2 Three-phase Line-Commutated Diode Bridge Rectifier
Figure 8.26 shows the configuration of this test circuit. The supply system is represented by means of three balanced ideal voltage sources in series with its equivalent impedance, modelled

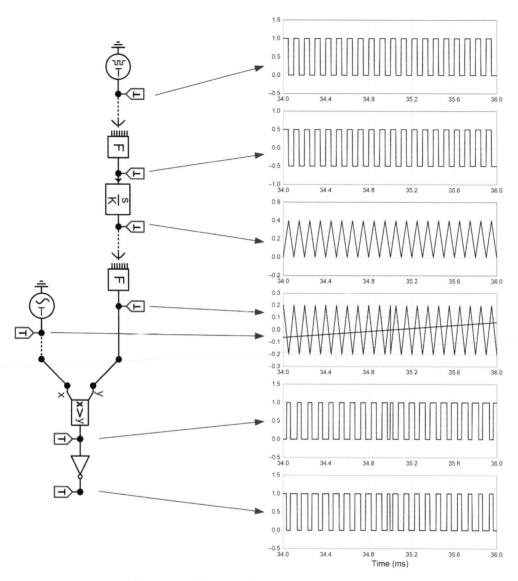

Figure 8.24 ATPDraw implementation of the control strategy.

as three symmetrical uncoupled *RL* branches. The load at the dc side is represented as an *RL* branch paralleled by a capacitor that should reduce the voltage ripple between load terminals. The system model can be further improved by adding a branch to represent the connection between the rectifier and the load.

Assume the circuit parameters listed below:

- AC source: Voltage = 400 V (phase-to-phase rms value), Frequency = 50 Hz
- Source impedance: $R_s = 0.1\,\Omega$, $L_s = 0.5\,\text{mH}$
- DC capacitor: $C = \text{Between 100 and 800 μF}$
- Connection branch: $R_c = 0.01\,\Omega$, $L_c = 0.5\,\text{mH}$
- DC load: $R_l = 50\,\Omega$, $L_l = \text{Between 5 and 20 mH}$

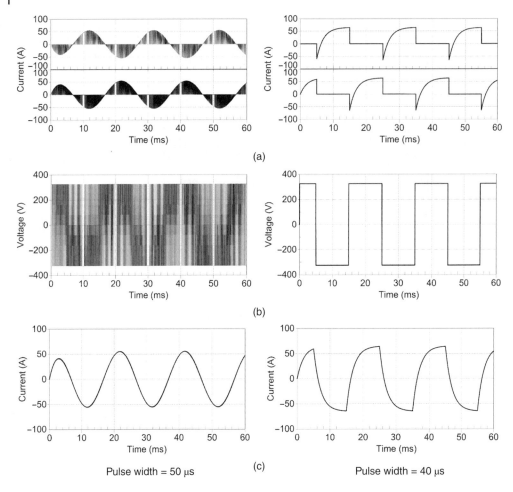

Figure 8.25 Simulation results – single-phase inverter. (a) Currents through IGBTs (G1 and G2), (b) Load voltage, (c) Load current.

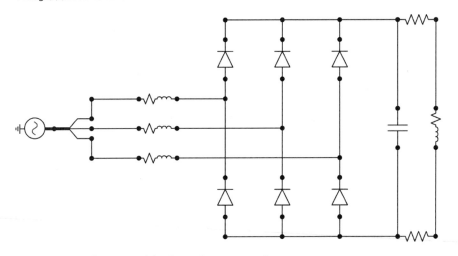

Figure 8.26 Configuration of the three-phase test rectifier.

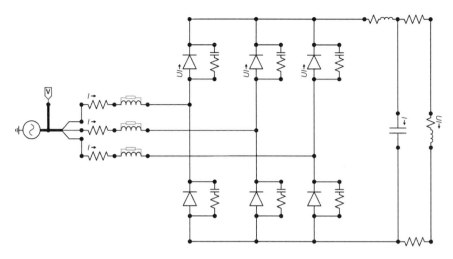

Figure 8.27 ATPDraw implementation of the three-phase rectifier.

The implementation and simulation of this power electronics configuration poses several challenges since the simulation results can be correct with some combination of parameters and wrong with some others. In any case, a snubber circuit in parallel to each diode is required.

Consider, for instance, a parallel capacitor with 500 μF. If the load inductance is $L_l = 5$ mH, the results are correct, but if the capacitor is removed, then results are no longer correct. However, if the inductance value is increased to 20 mH, then results are correct irrespective of the capacitor. This problem can be solved by adding parallel damping to the source inductance.

Figure 8.27 shows the final ATPDraw implementation of the test circuit: parallel snubber circuits and parallel damping have been added to avoid numerical oscillations. Figure 8.28 shows some simulation results when a 500 μF capacitor is in parallel to the dc load, which is represented by an RL branch with $R_l = 50 \Omega$ and $L_l = 5$ mH. Notice that during the activation period, the presence of the capacitor causes high currents, whose peaks depend on the phase and the capacitance value: the higher the capacitance, the higher the peak current values.

8.7.6 Discussion

This section has presented some scenarios for which numerical problems can occur during the simulation of power electronics systems. The first aspect that the user should analyse is the implemented model: numerical problems can usually occur due to improper modelling. In fact, the idealization followed when using ATP built-in models should be seen as a first cause of numerical problems; such ideal behaviour can be beyond acceptable representations. However, users can be interested in using such idealized models if they can be accurate enough for simulation purposes. Then, some of the solutions discussed in this section should be considered.

Although it has been shown that snubber circuits are not always necessary, they should be considered a by-default solution: sometime numerical problems can occur just by changing some circuit parameters, without modifying the circuit configuration. Another solution that can cure some unacceptable results, is the reduction of the time step: by default, it should be always equal or smaller than 10 μs. Of course, shorter time step could also be justified because of the switching frequency of the control strategy. It has been mentioned that when commutation between switches involves diodes, ATP users can consider the application of the GIFU option. The cases included in this section have also shown that there are some scenarios for which an alternative solution to the GIFU option can be implemented.

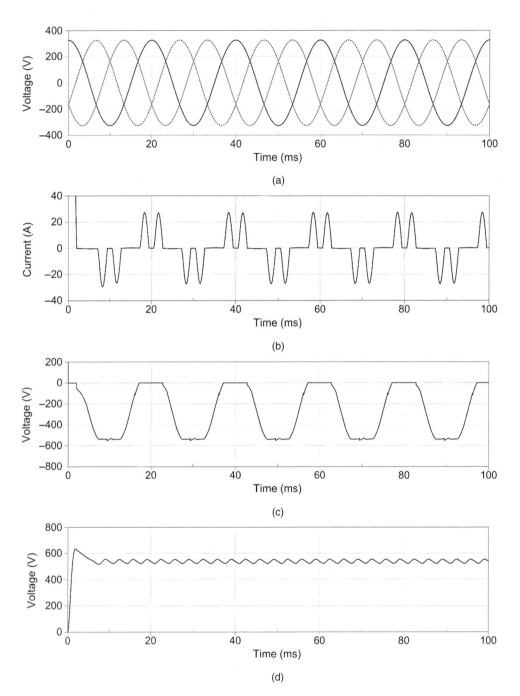

Figure 8.28 Simulation results – three-phase rectifier. (a) Source voltages, (b) Source current – phase a, (c) Diode voltage – phase a, (d) Load voltage.

8.8 Case Studies

8.8.1 Introduction

This section presents a collection of power electronics applications that have been ordered according to their complexity. The case studies range from a very simple three-phase controlled rectifier to the very complex model of a three-phase modular multilevel solid state transformer (SST). The list of case studies is the following one:

1. Three-phase controlled rectifier.
2. Three-phase adjustable speed ac drive.
3. Single-phase digitally-controlled static var compensator (SVC).
4. UPFC.
5. SST.

As discussed in previous sections, the implementation of power electronics applications involves the modelling of both the power and control sections. In addition, users of transients tools can consider two options for representing power electronics based system: detailed switching models and dynamic average models. When the complexity of the system is too high (e.g. the model must include several converters and hundreds of semiconductors), the use of a dynamic average model can significantly reduce the computing time and, at the same time, keep the accuracy of simulation results by neglecting or averaging the effect of fast switching transients. As for the representation of the control strategies, ATP users can also choose between two options, TACS and MODELS, and even mix both languages in the same case study.

The case studies presented in this section illustrate the application of all these alternatives; most cases use TACS except Case Study 3 which aims at duplicating a digitally-controlled SVC, for which MODELS is a more adequate option. Except for the study of the UPFC, the power part of all the power electronics converters is represented using detailed switching models. The application of an average model is illustrated here with the modelling of a UPFC model.

8.8.2 Case Study 1: Three-phase Controlled Rectifier

Figure 8.29 shows the ATPDraw implementation of this case study: a thyristor-based three-phase rectifier supplies an RL branch paralleled by a capacitor installed to reduce the voltage ripple.

The approach used to model this system follows the guidelines recommended for (low-frequency) power electronics applications: (i) semiconductors (i.e. thyristors) are modelled as ideal controlled switches; (ii) a snubber circuit is installed in parallel to each thyristor to avoid numerical oscillations; (iii) the supply source is represented as a three-phase ideal source in series with its equivalent balanced uncoupled RL impedance; (iv) any other component (e.g. the dc load) is represented by means of lumped-parameter branches; (v) the controller of this rectifier is modelled using TACS capabilities. Details about the control strategy are provided below.

The main characteristics of the test system are as follows:

- Voltage source: Voltage = 400 V (phase-to-phase rms value), Frequency = 50 Hz
- Impedance: $R_{grid} = 1\,\Omega$, $L_{grid} = 1\,\text{mH}$
- Load: $R_{load} = 10\,\Omega$, $L_{load} = 5\,\text{mH}$, $C_{load} = 500\,\mu\text{F}$
- Snubbers: $R_{snb} = 200\,\Omega$, $C_{snb} = 0.1\,\mu\text{F}$

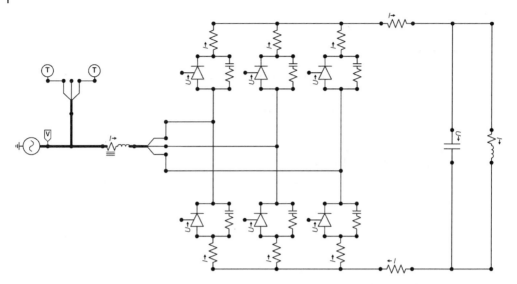

Figure 8.29 Case Study 1: ATPDraw implementation of the test system – controlled rectifier.

Figure 8.30 shows the TACS implementation of the open-loop control strategy for this case study. The controller has two inputs (i.e. phase-to-ground source voltages of phases 'a' and 'c') and three specifications: (i) source frequency; (ii) phase angle; (iii) delay for defining the duration of thyristor gate pulses. The difference between the source voltages is used to synchronize a unity ramp: the ramp begins climbing at the time this difference crosses zero from negative to positive, and becomes zero at the zero crossing of the voltage from positive to negative. The time required to reach unity is half period of the source frequency (i.e. 10 ms for a 50 Hz frequency). See simulation results depicted below.

The specified phase angle, α, for firing thyristors is normalized: 1 for $\alpha = 180°$, 0.5 for $\alpha = 90°$, 0 for $\alpha = 0°$. This normalized value is compared to the ramp: the initiation of the reference gate pulse is obtained at the time both signals are equal; the duration of this pulse is defined with the specified delay. The other gate pulses are delayed one sixth of the power frequency with respect to each other; see Figure 8.30.

Figure 8.31 displays some simulation results obtained with a phase angle of 90°. The unity ramp is generated from the difference $(v_a - v_c)$, as shown in Figure 8.31b. This ramp is compared to the normalized phase angle (see Figure 8.30), as shown in Figure 8.31c, and the reference gate pulse is obtained. The pulses to fire the six thyristors are next generated. Figure 8.31d depicts the waveform of the voltage across the phase a thyristor.

The waveform of the voltage across the dc-side capacitor and, consequently, across the dc load branch, as well as those of the currents supplied by the source and the current through the load branch depend on the specified parameters: with the parameters selected for the present case, the source currents and the current supplied from the rectifier are discontinuous (i.e. there is a time period during which these currents are zero), but the current through the *RL* load branch is continuous; this is an effect of the capacitor. As for the dc-side voltage, its waveform exhibits an average value below 150 V and some significant ripple; see Figure 8.31f.

Both the ripple and the average value can be controlled by means of the capacitance and the phase angle.

There is an important aspect that has already been mentioned in other case studies included in this chapter. ATP cannot perform a full steady-state initialization for test systems with power electronics devices/converters; although some steady-state initialization is possible when these devices/converters are present and some tricks can also be used to speed-up the

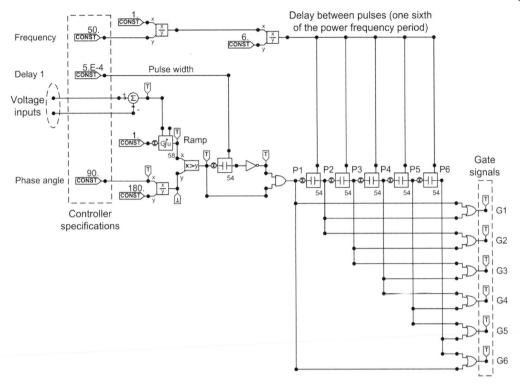

Figure 8.30 Case Study 1: ATPDraw implementation of the controller using TACS.

initial calculation. In this case study, the actual initialization does not begin until the first pulse gates are generated and the first semiconductors become closed; see, for instance, the plots of the thyristor voltage, grid current, and dc load voltage.

8.8.3 Case Study 2: Three-phase Adjustable Speed AC Drive

Figure 8.32 shows the ATPDraw implementation of this case study: an induction motor is supplied from a variable-frequency converter and a LV network. The figure shows both the electrical circuit and the control section, which has been implemented using TACS capabilities. The converter consists of an uncontrolled rectifier, a dc link, and a variable-frequency inverter. The frequency of the inverter is controlled by means of a closed-loop control strategy whose main principles are summarized below.

The modelling guidelines for this test case are again those discussed in the previous sections and to be applied in low-frequency applications. As shown in the simulation results, the output frequency of the inverter will always be below the source frequency.

- The supply system is represented as an ideal voltage source in series with its equivalent impedance. The source is ungrounded and a damping resistance is installed in parallel to each source inductance. The selected damping factor, K_p, is 10.
- The semiconductors for both the rectifier and the inverter are ideal. Snubber circuits are used only with diodes.
- The impedance between the converter and motor terminals is represented by uncoupled lumped *RL* branches.
- The induction motor is represented with the option UM 3 of the Universal Machine module.

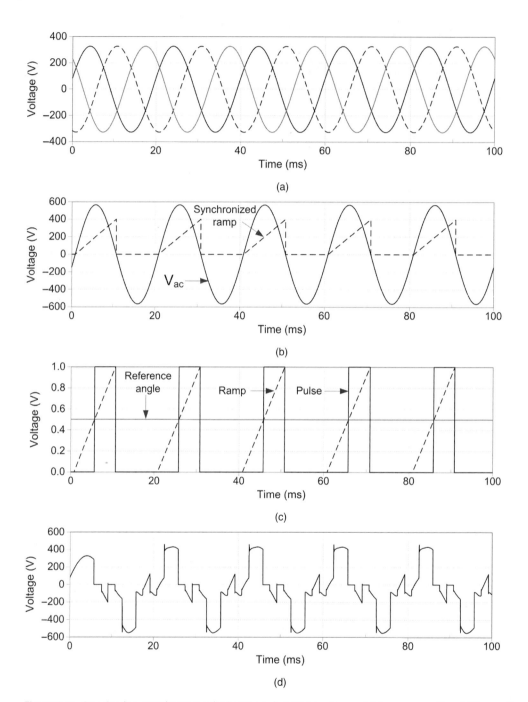

Figure 8.31 Case Study 1: simulation results (continued). (a) Grid voltages, (b) Ramp generation, (c) Generation of the reference pulse, (d) Upper-side thyristor voltage – phase a. (e) Grid current – phase a, (f) Load voltage.

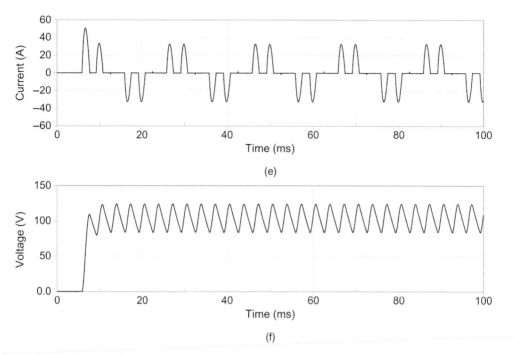

Figure 8.31 (Continued)

Table 8.2 presents the main parameters of this test system.

The induction motor can be controlled as a dc motor with independent flux and torque control using a feedforward loop for the flux and a feedback mode for the torque. Figure 8.33 displays the main blocks of the controller, which assumes there is proportionality between the d-axis current and the rotor flux, on one hand, and between the q-axis stator current and the electromagnetic torque, on the other hand. The regulator is a PI block with limits for the reference torque.

The case analysed here includes the simulation of the motor startup and a variation of the external mechanical torque. The specified angular speed of the rotor is 150 rad/s: see Figure 8.32.

Figure 8.34 shows some simulation results: the startup is carried out using Automatic Initialization, and the external torque is increased at 1 second during a period of 0.3 second.

- Figure 8.34a depicts the variation of the angular speed of the motor rotor: after reaching the specified value, the speed exhibits a quick decrement speed as a consequence of the external torque. Figure 8.34b shows that the electromagnetic torque reaches a very low value (negative value for a motor operation) after the startup period; the absolute value of this torque is suddenly increased due to the external torque variation, and comes back to the steady-state value once this torque is nullified.
- Figure 8.34c shows the current through one inverter semiconductor. The pattern is similar to that exhibited by the source and motor currents; however, this current is not continuous and oscillates with a frequency above 1 kHz.
- A very interesting aspect is the frequency of the motor currents, a value that is controlled from the control unit of the inverter. The system model includes a frequency sensor (see Figure 8.32) that can estimate the frequency of these currents. Figure 8.34d shows the variation of this frequency: it reaches a value very close to 25 Hz (the desired angular speed is

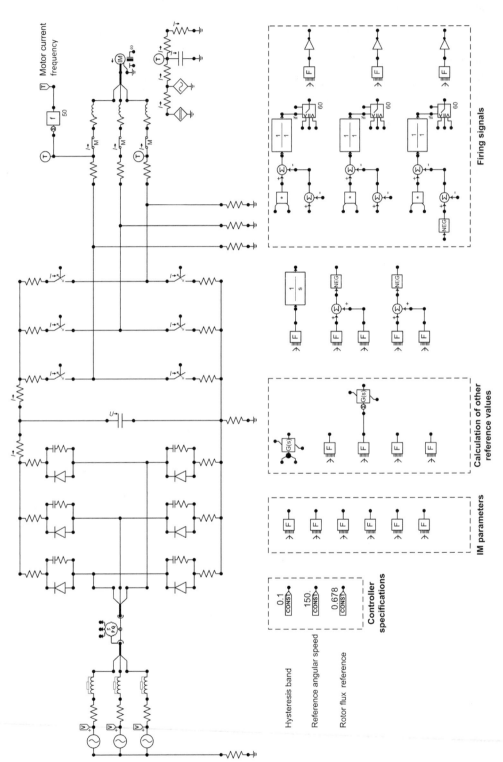

Figure 8.32 Case Study 2: ATPDraw implementation of the test system.

Table 8.2 Case Study 2: parameters of system components.

Component	Parameters
Voltage source	400 V (rms phase-to-phase value), 50 Hz, $R_{source} = 1\,\Omega$, $L_{source} = 1\,mH$
Rectifier-inverter set	DC link capacitor = 800 µF
	Snubber circuits: $R_{snubber} = 100\,\Omega$, $C_{snubber} = 1\,µF$
Motor circuit	$R_{motor} = 2.4\,\Omega$, $L_{motor} = 1.2\,mH$
Induction motor	Squirrel-cage, Stator connection = Y, 2 pole pairs, No saturable
	d-axis magnetization inductance/q-axis magnetization inductance = 0.1198 H
	Stator: d-axis impedance/q-axis impedance $R = 1.2420\,\Omega$, $L = 0.0012\,H$
	Stator: d-axis impedance/q-axis impedance $R = 1.4520\,\Omega$, $L = 0.0012\,H$
	Moment of inertia = 0.05 kg m^2, Damping factor = 1/20 Nms/rad

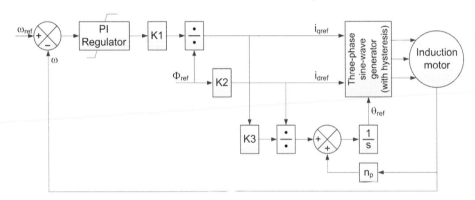

Figure 8.33 Case Study 2: scheme of the vector-controlled strategy.

150 rad/s and the motor has two pole pairs), and decreases during the transient caused by the external torque variation.

- The pattern of both active and reactive powers follow those of the currents mentioned above: once the angular speed of the motor has reached the specified value, the variation of both powers is very similar to that of the source current peaks; see Figure 8.34e.
- The last plot shows the variation of the dc link voltage (i.e. voltage between capacitor terminals in Figure 8.35). A fast initial transient elevates this voltage to a value of about 550 V. The variation caused by the external torque is very small: the voltage value ranges within a narrow band once the motor has reached the specified angular speed; see Figure 8.34f.

8.8.4 Case Study 3: Digitally-controlled Static VAR Compensator

SVCs can be used in medium voltage (MV) distribution systems to control voltage and reactive power [49, 50]. This case study presents a control strategy aimed at achieving optimal load compensation by means of an open-loop control. The strategy is based on the idea that optimizing the power factor is equivalent to minimize the current drawn from the supply source. In this study, the strategy is applied to a thyristor-based controlled-reactor SVC paralleled by a fixed capacitor bank. The control strategy was originally proposed in [55] and later implemented in ATP for single-phase and three-phase SVCs [56, 57]. This chapter details the implementation

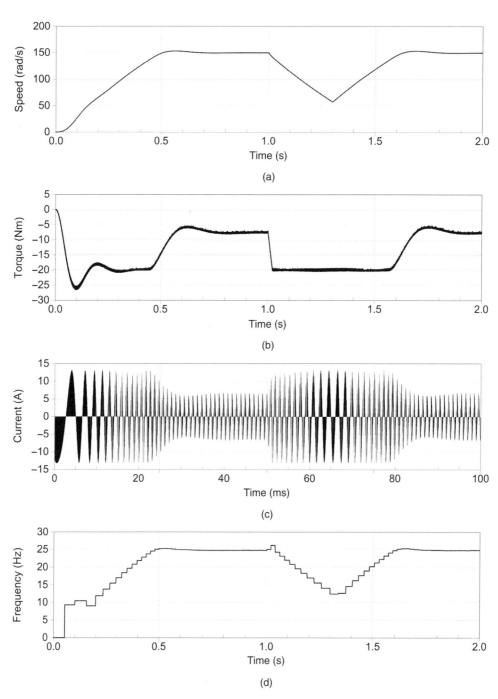

Figure 8.34 Case Study 2: simulation results – motor startup and mechanical torque change (continued). (a) Angular speed of the motor rotor, (b) Electromagnetic torque, (c) IGBT current, (d) Frequency of the motor currents, (e) Active and reactive powers from the grid, (f) DC link voltage.

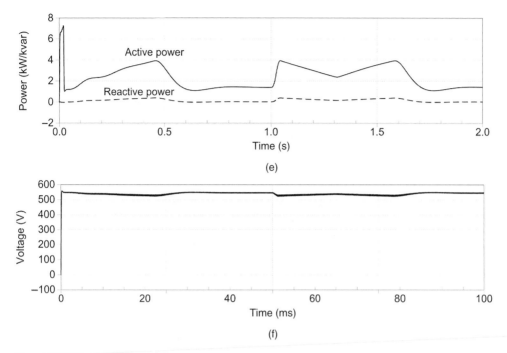

(e)

(f)

Figure 8.34 (Continued)

Figure 8.35 Case Study 3: test system configuration.

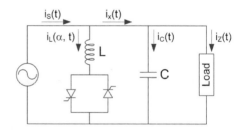

and simulation of this strategy with the model of a small single-phase test system. The main goal is to illustrate how some ATP capabilities, namely MODELS language, can be used to duplicate the performance of a digital control.

8.8.4.1 Test System
Figure 8.35 shows the configuration of the test system. A load, either linear or nonlinear, is connected to a node where the voltage is supposed to be sinusoidal and constant. The SVC is controlled to optimize the power factor of the source current. The capacitor is constant.

8.8.4.2 Control Strategy
The control problem is reduced to find the firing angle, α, that minimizes the rms value of the source current, $i_S(t)$. From the measured current, $i_x(t)$ (see Figure 8.35), the optimal firing angle is the one that satisfies the following equation [55]:

$$\frac{4V}{\omega^2 L}\left[\left(\frac{\pi}{2} - \alpha\right)\sin\alpha - \cos\alpha\right] = A(i_x, \alpha) \qquad (\omega = 2\pi f) \tag{8.1}$$

where

$$A(i_x, \alpha) = \int_{\alpha/\omega}^{(\pi-\alpha)/\omega} i_x dt - \int_{(\pi+\alpha)/\omega}^{(2\pi-\alpha)/\omega} i_x dt \tag{8.2}$$

V is the supply voltage peak, L is the reactor inductance, and f is the frequency of the source. Note that

- the left side of Eq. (8.1) is fixed for a given value of α as long as the value of V does not change;
- for $0 \leq \alpha \leq \pi/2$ the left side of (8.1) varies between $-4\,V/\omega^2 L$ and 0;
- $\alpha = 0$ is always a solution.

If no other value of α satisfies (8.1), the firing angle is chosen as follows:

- $\alpha = \pi/2$ if $A(i_x, 0) > -4\,V/\omega^2 L$, and consequently the capacitor should be increased;
- otherwise $\alpha = 0$, and the capacitor should be decreased.

Practical implementation

Several modifications to this strategy were suggested in [56]. Details of the hardware and software implementation for both single- and three-phase versions were also presented in [56, 57]. A solution of Eq. (8.1) was originally programmed in assembler using the following procedure [55]:

1. The left side of (8.1), which depends on the system configuration, is computed off-line and saved in a table with 256 32-bit integer words.
2. The measured current, i_x in Figure 8.35, is sampled 128 times every cycle and stored in the processor memory. This task is synchronized every time the value of the supply voltage passes from negative to positive.
3. The array of measured currents is compressed into a shorter 32-word array, taking advantage of the symmetry of the function (8.2).
4. Once a complete cycle is sampled, the processor computes the firing angle. This must be done before the next peak voltage.

MODELS implementation

A very similar procedure to that summarized above can be used to implement this open-loop control algorithm in MODELS.

1. The left side of Eq. (8.1) is computed and saved in a 32-element array before beginning the transient loop. This task is carried out in the INIT section.
2. Since ATP cannot perform the steady-state initialization of an SVC configuration, the simulation is started with the SVC deactivated; only when a complete cycle of the voltage waveform has been simulated, the firing angle of the SVC is computed. The simulation can start with any initial value of the voltage waveform, so this calculation is not performed before two zero-crossings are detected. This can assure that one complete cycle has been already simulated. The control algorithm checks whether the first zero-crossing of the voltage waveform has been detected; only then the measured current is saved.
3. The load current is sampled 128 times every cycle. This means that the time-step size has to be calculated from the source frequency and taking into account this sampling frequency.
4. After each zero crossing from negative to positive values of the voltage waveform, a 32-element array that contains the right side of Eq. (8.1) is computed.

5. This array is subtracted to that computed within the INIT section (before starting the transient loop). The firing angle is obtained using a linear interpolation between the last two elements once a change of sign is detected. If this change of sign is not detected, a firing angle of value 0 or $\pi/2$ must be used. This value is then selected following the logic described above.

ATP implementation

Figure 8.36 depicts the circuit implemented in ATPDraw to simulate the test system shown in Figure 8.35. The circuit includes the possibility of changing the load. The implementation of the control strategy in MODELS is shown in Figure 8.37.

Simulation results

As with previous case studies, thyristors have been represented as ideal switches and paralleled by a snubber circuit (see Figure 8.36) to avoid numerical oscillations. The main parameters of the test system are provided below:

- Source: Voltage (peak value) $= 326.6 \, (= 400/\sqrt{(3/2)})$ V; Frequency $= 50$ Hz
- Reactor: $L = 3$ mH, $R = 0.02 \, \Omega$
- Snubber: $R = 1000 \, \Omega$, $C = 0.2 \, \mu\text{F}$
- Capacitor: $C = 1200 \, \mu\text{F}$
- Load: Variable

Figure 8.38 shows some simulation results that correspond to a case in which the load is a resistance of $5 \, \Omega$ and the load variation is made by connecting in parallel a pure inductance of 10 mH; the inductance is connected at $t = 0.1$ second and disconnected at $t = 0.2$ second. The time-step size used in this simulation is $(1/50)/128 = 156.25\text{E-}6$ seconds.

The main conclusions from these results can be summarized as follows:

- The load-side current is initially leading the source voltage due to the effect of the capacitor current; after the connection of the reactive load it is almost in phase with the source voltage. The currents lead again once the reactance is disconnected; see Figure 8.38a.
- The reactor current is activated after the *initialization period* mentioned above. After this period and before the connection of the parallel load inductance, this current reaches a rather

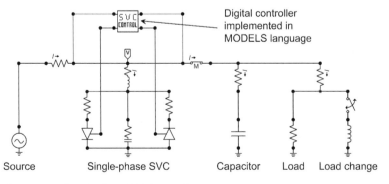

Figure 8.36 Case Study 3: ATPDraw implementation.

high value, which becomes almost zero during the connection period, when the shunt load reactance compensates the effect of the capacitor current. The SVC reactor current increases again once the load reactance is disconnected; see Figure 8.38b.

- After the initialization period, the source current is the sum of the resistive load current, the capacitor current and the SVC (reactor) current. Due to the highly distorted reactor current and to the fact that the test system does not include any current filter, this current is also distorted when the load is purely resistive, although it exhibits the effect of the reactor current and the compensation of the capacitive current; see Figure 8.38c.
- The comparison of the resistive load current, the capacitor current and the (SVC) reactor current, shown in Figure 8.38d, is useful to better analyse the impact that the SVC has on the load current in this particular case: the reactor current peak varies between a value close to that of the load current and an almost-zero value.

Figure 8.39 shows some simulation results corresponding to similar operating conditions but with a different sequence of events: initially, the shunt reactive load is connected so the total load current (that resulting from adding capacitor current and *RL* load current) is almost in phase with the source voltage, and the initial SVC current (after the *initialization period*) is very small; however, this current significantly increases once the load becomes purely resistive and the reactor current is required to compensate that of the capacitor. The conclusions about the various system currents are obviously the same that were derived from the previous cases, although the sequence of events is different.

Figure 8.37 Case Study 3: MODELS implementation of the SVC controller. (a) Main window, (b) MODELS code.

```
MODEL contrsvc
  INPUT    ix                    -- Load current                 [A]
           vt                    -- Source voltage               [V}
  OUTPUT   flag1                 -- Thyristor control signal 1
           flag2                 -- Thyristor control signal 2
  DATA     voltage               -- Source voltage peak          [V]
           frequency             -- Source frequency             [Hz]
           inductance            -- Reactor inductance           [H]
  VAR      omega     cycle    factor    da
           control1 control2 sig1      sig2
           ss1       ss2      ss3       n
           alfa      flag1    flag2
           t1        t11      t2        t22
           il[0..128]         s[0..128]         it[0..32]
  INIT
           omega:=2*pi*frequency    da:=pi/64
           control1:=0              control2:=0
           t1:=10                   t11:=10
           t2:=10                   t22:=10
           n:=0                     cycle:=1/frequency
           flag1:=0                 flag2:=0
           FOR  k:=0  TO  128  DO  s[k]:=0  ENDFOR
           factor:=4*voltage/(inductance*omega*omega)  -- Compute left-side array
           FOR k:=0 TO 32 DO
             il[k]:=factor*((pi/2-k*da)*sin(k*da)-cos(k*da))
             ENDFOR
  ENDINIT
  EXEC
    IF   (t=0)   THEN   sig1:=sign(vt)   ELSE   sig1:=sign(prevval(vt))   ENDIF
    sig2:=sign(vt)
    IF (sig1 < sig2) THEN                        -- Detect zero-crossing
        IF (control1=1) THEN                     -- Detect first complete cycle
           it[32]:=0
           FOR  k:=1  TO  32  DO                 -- Compute right-side array
             ss1:=s[64-k+1]+s[64+k-1]-s[128-k+1]-s[k-1]
             ss2:=s[64-k]+s[64+k]-s[128-k]-s[k]
             it[32-k]:=it[32-k+1]+(ss1+ss2)*timestep/2
           ENDFOR
           ss2:=it[0]-il[0]
           FOR  k:=1  TO  31  DO                 -- Compute the firing angle
             ss1:=ss2
             ss2:=it[k]-il[k]
             IF (sign(ss1)<>sign(ss2)) THEN
               alfa:=((k-1) + ss1/(ss1-ss2))*pi/64
               control2:=1
             ENDIF
             IF (control2=0) THEN                -- No solution
               IF  (it[0]<factor)  THEN  alfa:=pi/2  ELSE  alfa:=0  ENDIF
             ENDIF
           ENDFOR
           control2:=0                           -- Compute the triggering time
           t1:=t +(cycle/2)*(alfa/pi+0.5)-timestep/2    t11:=t1+10*timestep
           t2:=t1+cycle/2                               t22:=t2+10*timestep
           -- Restart sampling and saving the load current
           n:=1              s[n]:=ix            s[0]:=s[128]
        ELSE
           -- Zero-crossing - Start sampling and saving the load current
           control1:=1    n:=1      s[1]:=ix     s[0]:=s[128]
        ENDIF
      ELSE
        -- Sample and save the load current
        IF  (t>0)  THEN   n:=n+1    s[n]:=ix   ENDIF
    ENDIF
    flag1:=(t>=t1) AND (t<=t11)                  -- Control signal of valve 1
    flag2:=(t>=t2) AND (t<=t22)                  -- Control signal of valve 2
  ENDEXEC
ENDMODEL
```

Figure 8.37 (*Continued*)

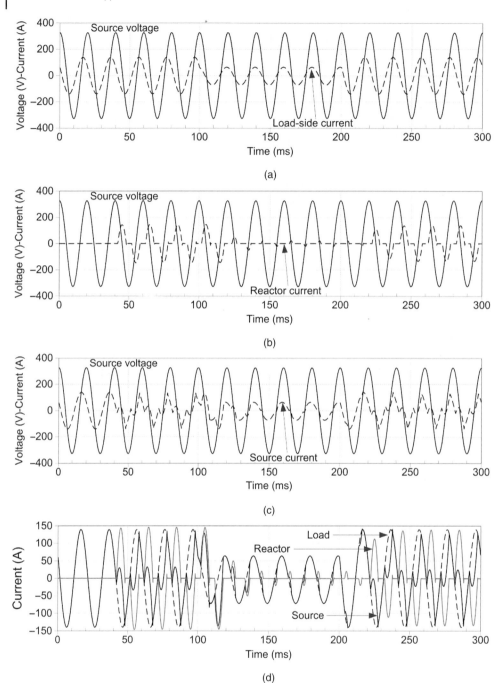

Figure 8.38 Case Study 3: simulation results with initial resistive load. (a) Source voltage and load (plus capacitor) current, (b) Source voltage and reactor current, (c) Source voltage and current, (d) Source, reactor and load current.

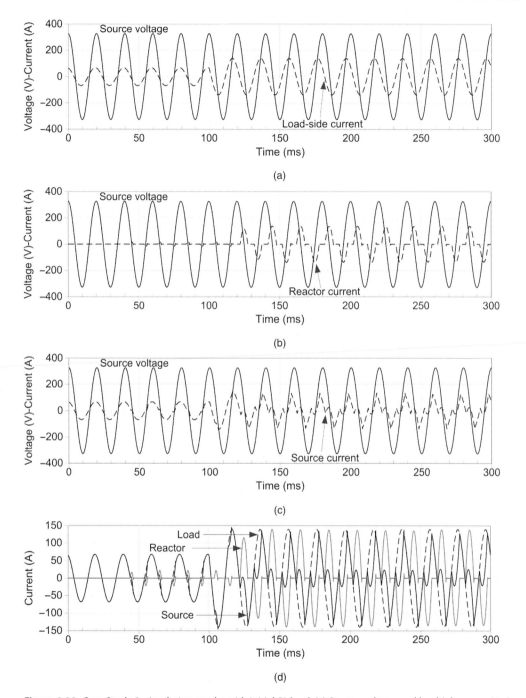

Figure 8.39 Case Study 3: simulation results with initial *RL* load. (a) Source voltage and load (plus capacitor) current, (b) Source voltage and reactor current, (c) Source voltage and current, (d) Source, reactor and load current.

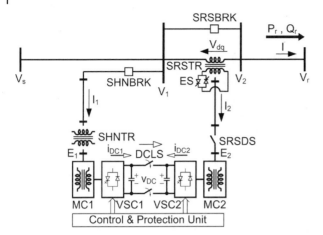

Figure 8.40 Case Study 4: diagram of the unified power flow controller.

8.8.5 Case Study 4: Unified Power Flow Controller

8.8.5.1 Configuration

Figure 8.40 shows the schematic diagram of the UPFC simulated in this case study; it consists of [46, 58, 59]: two voltage source converters, VSC1 and VSC2; two magnetic circuits, MC1 and MC2; a shunt-coupling transformer, SHNTR; a series-coupling transformer, SRSTR; a shunt breaker, SHNBRK; a series-disconnect switch, SRSDS; an electronic-bypass switch, ES; a dc link switch, DCLS; a series-bypass breaker, SRSBRK; current and voltage sensors; a control and protection unit.

The theory on the UPFC operation and control, as well as the model developed to represent this FACTS device, come from the work by K. Sen [46, 58]. The UPFC model implemented here is an ATPDraw version of a model whose input data file was implemented in plain text; this section shows how to implement an ATP code in ATPDraw by using a rather simple approach. The UPFC model was originally presented in [58]; readers interested in this and other FACTS devices can consult references [46, 59].

8.8.5.2 Control

Each UPFC VSC is coupled with a transformer at its output. Both VSCs can generate almost sinusoidal voltages with an acceptable amount of harmonic components. The two VSCs are identical: one operates as a static synchronous compensator (STATCOM) and the other operates as SSSC [46]. The control of the UPFC has, therefore, two parts dedicated respectively to control the STATCOM and the SSSC; their operation will depend on the DCLS status:

- When the DCLS is closed, the two VSCs share the dc link. When both VSCs operate together, the series-connected compensating voltage can be at any phase angle with respect to the prevailing line current. Therefore, the exchanged power at the terminal of each VSC can be reactive as well as real. The exchanged real power at the terminal of one VSC with the line flows to the terminal of the other VSC through the shared dc link. Both VSCs can also provide independent reactive power compensation at their respective AC terminals.
- When the DCLS is open, the two VSCs can be operated as stand-alone independent compensators (i.e. a shunt-connected compensator, STATCOM, and a series-connected compensator, SSSC) that will exchange almost exclusively reactive power at their respective

terminals. That is, during stand-alone operations, the SSSC injects a voltage in quadrature with the line current, thereby emulating an inductive or a capacitive reactance at the point of compensation in series with the line, and the STATCOM injects a reactive current, thereby also emulating a reactance at the point of compensation in shunt with the line.

The controller of the STATCOM is used to operate the VSC in such a way that the phase angle between the VSC output voltage and the line voltage is dynamically adjusted, so that the STATCOM generates or absorbs the desired reactive power at the point of compensation. Figure 8.41a shows the reactive current control block diagram. An instantaneous three-phase set of line voltages (v_1) is used to calculate the PLL angle (θ), which is phase-locked to the phase a of the line voltage (v_{1a}). An instantaneous three-phase set of measured currents (i_1) through the VSC is decomposed into its real or direct component (I_{1d}) and reactive or quadrature component (I_{1q}). The quadrature component is compared with the desired reference value (I_{1q}^*) and the error is passed through an error amplifier, which produces a relative phase angle (α) of the compensating voltage (E_1) with respect to the line voltage (V_1). The absolute phase angle (θ_1) of the compensating voltage is calculated by adding the relative phase angle of the compensating voltage with respect to the PLL angle θ. The reference quadrature component (I_{1q}^*) of the current is defined to be either positive if the STATCOM is emulating a capacitive reactance or negative if it is emulating an inductive reactance. The dc link capacitor voltage (v_{DC}) is dynamically adjusted in relationship with the compensating voltage.

The control scheme described above shows the implementation of the inner current control loop, which regulates the reactive current flow through the VSC regardless of the line voltage. However, if regulation of the line voltage is desired, an outer voltage control loop must be implemented. The outer voltage control loop automatically determines the reference reactive current for the inner current control loop, which, in turn, regulates the line voltage. Figure 8.41b shows the voltage control block diagram of the STATCOM. Using the reference PLL angle θ, the instantaneous three-phase set of measured line voltage v_1 is decomposed into its real or direct component, V_{1d}, and reactive or quadrature component, V_{1q}. The magnitude of V_{1dq} is calculated and compared with the desired reference value, V_1^* (adjusted by the droop factor, K_{drop}); the error is passed through an error amplifier that produces the reference current (I_{1q}^*) for the inner current control loop. The droop factor is the allowable voltage error at the rated reactive current flow through the STATCOM.

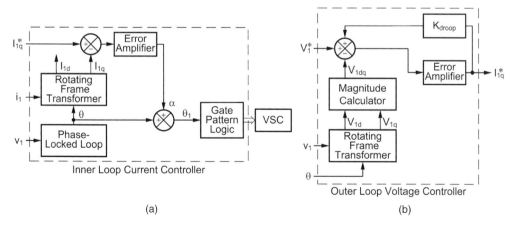

(a)   (b)

Figure 8.41 Case Study 4: (a) Reactive current control block diagram of the STATCOM. (b) Voltage control block diagram of the STATCOM.

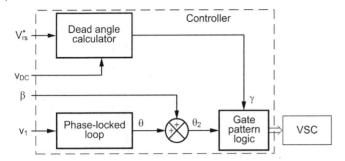

Figure 8.42 Case Study 4: control block diagram of the SSSC in an open loop voltage injection mode.

The SSSC can be operated in many different modes. In each mode of operation, the final outcome is such that the SSSC injects a voltage in series with the transmission line. As an example, Figure 8.42 shows the control block diagram in an open loop voltage injection mode. The desired peak fundamental voltage (V_{rs}^*) at the output of the VSC and its relative phase angle (β) with respect to the reference PLL angle (θ) are specified. The absolute phase angle (θ_2) of the VSC output voltage is calculated by adding the relative phase angle (β) of the VSC output voltage and the PLL angle (θ). The dead angle (γ) of each pole is calculated.

8.8.5.3 Modelling
An actual UPFC for high-voltage and high-power ratings contains many tens of series-connected GTO devices in one converter leg. If every individual switching device is represented, the model would be extremely complex and difficult for implementation. Except for some failure mode analyses and for the purposes of most application simulations, it is not necessary to represent all individual devices. Usually, the main concern is the terminal characteristics of a power electronics subsystem and how it interfaces with the connected system. Thus, some reduction of the modelling complexity is advisable. The guidelines followed to represent this test system are those recommended in [6, 7]. With these general guidelines, a VSC model for a system dynamic evaluation can be based on the approach shown in Figure 8.43. Irrespective of how many series and parallel GTO devices are used in an actual application, only two GTO devices are used in each phase of the model to form a converter leg. The UPFC model has been built with *harmonic neutralized* VSCs (HN-VSCs). A HN-VSC consists of a set of poles, a dc capacitor, a magnetic circuit, and a control and protection unit. Three stages are required to produce a sine wave-like voltage by means of a HN-VSC: (i) the control unit must produce a set of gate pulses; (ii) the gate pulses operate a set of poles, which produce a set of square wave voltages; (iii) the square wave voltages are combined with a magnetic circuit to produce a sinusoidal voltage. A HN-VSC with a finite pulse number produces a high-quality sinusoidal voltage with acceptable harmonic content. If a VSC produces a pure sine voltage of fundamental frequency, it can be thought of as a VSC of infinite pulse number. These topologies can be modelled step-by-step as detailed in [46, 59].

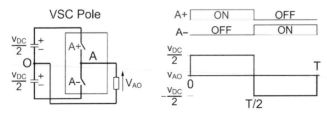

Figure 8.43 Case Study 4: VSC pole and its output voltage.

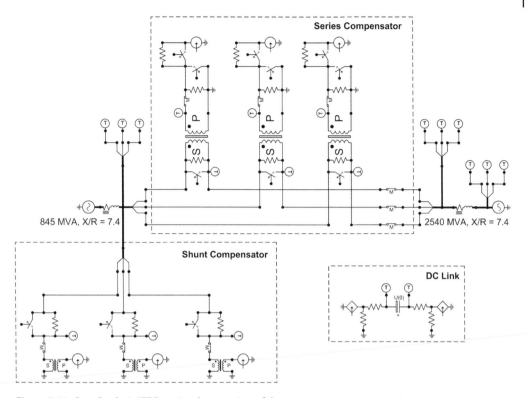

Figure 8.44 Case Study 4: ATPDraw implementation of the test system – power circuit.

8.8.5.4 ATPDraw Implementation

Figures 8.44 and 8.45 show the ATPDraw implementation of the test system under study. Figure 8.44 depicts the power system with the models of the series and shunt compensators. Note that both compensators are modelled by means of controlled switches and sources. The control section takes advantage of the already available UPFC model to achieve an easy implementation of the model: except for a few specifications (i.e. rated values), the entire control section is basically that presented in [46]. Figure 8.45 lists the various files that make the controller model and its functions. Some of the main parameters of the 60 Hz, 138 kV test transmission system are listed in Figure 8.44.

8.8.5.5 Simulation Results

Figure 8.46 shows some simulation results when the SSSC is operating in open-loop voltage-injection mode while the STATCOM is operated to deliver no reactive current. The sequence of events is as follows [46]:

- At the beginning of the operation, the series bypass breaker SRSBRK and the series disconnect switch SRSDS are open. The VSC2 generates no compensating voltage. The voltage v_{12a} at the terminal of the series coupling transformer SRSTR is the voltage across its leakage reactance. The real and reactive compensating powers exchanged at the terminal of the series coupling transformer SRSTR are mostly reactive due to the high-quality factor of the leakage reactance. The shunt breaker SHNBRK is open. The dc link capacitor is pre-charged.
- At 50 ms, the SHNBRK closes and the quadrature current demand of the VSC1 is set to zero.
- At 100 ms, a series compensating voltage at the VSC2 side is set at 0.2 pu with a relative phase angle $\beta = 300°$ leading the reference PLL angle θ. The series VSC2 output voltage e_{2q} leads the line current i by a phase angle φ; since $\varphi > 90°$, the SSSC emulates a negative resistance

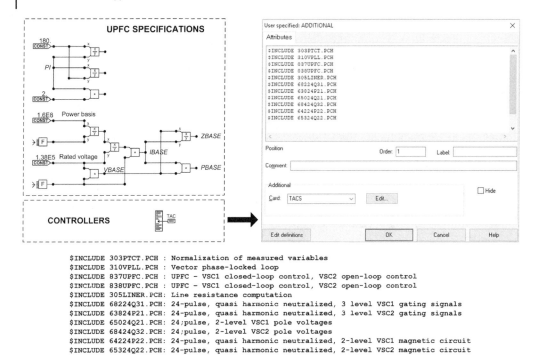

```
$INCLUDE 303PTCT.PCH : Normalization of measured variables
$INCLUDE 310VPLL.PCH : Vector phase-locked loop
$INCLUDE 837UPFC.PCH : UPFC - VSC1 closed-loop control, VSC2 open-loop control
$INCLUDE 838UPFC.PCH : UPFC - VSC1 closed-loop control, VSC2 open-loop control
$INCLUDE 305LINER.PCH: Line resistance computation
$INCLUDE 68224Q31.PCH: 24-pulse, quasi harmonic neutralized, 3 level VSC1 gating signals
$INCLUDE 63824P21.PCH: 24-pulse, quasi harmonic neutralized, 3 level VSC2 gating signals
$INCLUDE 65024Q21.PCH: 24;pulse, 2-level VSC1 pole voltages
$INCLUDE 68424Q32.PCH: 24;pulse, 2-level VSC2 pole voltages
$INCLUDE 64224P22.PCH: 24-pulse, quasi harmonic neutralized, 2-level VSC1 magnetic circuit
$INCLUDE 65324Q22.PCH: 24-pulse, quasi harmonic neutralized, 2-level VSC2 magnetic circuit
```

Figure 8.45 Case Study 4: ATPDraw implementation of the test system – control section.

in addition to an inductive reactance in series with the transmission line. The real power that is delivered to the line by the series VSC2 flows from V1 voltage bus through the STATCOM. The shunt VSC1 output voltage e_1 is almost in phase with the current i_1 flowing through it. The voltage v_2 at V2 voltage bus leads the voltage v_r at the receiving end. The real power P_r delivered at the receiving end decreases; the reactive power Q_r delivered at the receiving end becomes inductive.

- At 200 ms, the series-connected compensating voltage is maintained at 0.2 pu while the relative phase angle β is changed to 240°. The real power that is absorbed from the line by the series VSC2 flows to V_1 voltage bus through the STATCOM. The shunt VSC1 output voltage e_1 is almost 180° out of phase with the current i_1 flowing through it. The reactive power Q_r delivered at the receiving end becomes capacitive.
- At 300 ms, the series-connected compensating voltage is increased to 0.4 pu while the relative phase angle β stays at 240°. The voltage v_2 at V_2 voltage bus lags the voltage v_r and the real power flow P_r at the receiving end reverses. The series VSC2 output voltage e_2 lags the line current i by a phase angle φ. The reactive compensating power exchanged at the terminal of the series coupling transformer SRSTR becomes capacitive.

8.8.6 Case Study 5: Solid State Transformer

8.8.6.1 Introduction
The SST is foreseen as a fundamental component of the future smart grid since it offers several benefits [60]: enhanced power-quality performance, fast-voltage control, reactive-power compensation or reactive-power control at both primary and secondary sides, dc and high-frequency ac power supply. The SST can also provide some operational benefits, such as

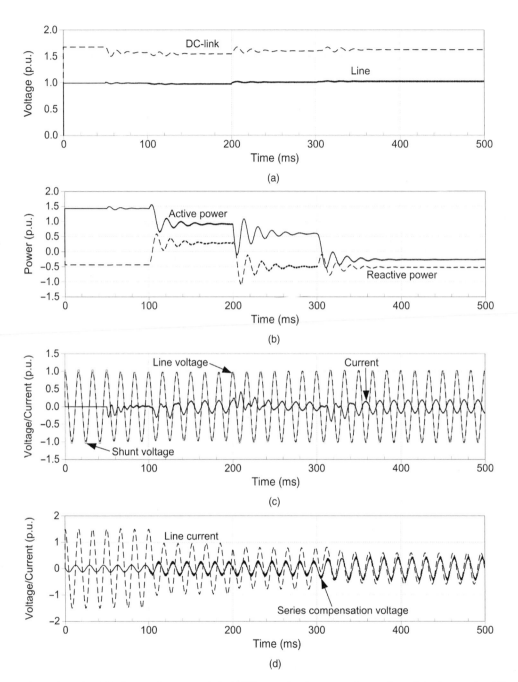

Figure 8.46 Case Study 4: performance of the UPFC operating in an open loop voltage injection mode while regulating a zero reactive current. (a) DC-link and line voltages, (b) Active and reactive power at the receiving end, (c) Shunt output and sending-end line voltages, shunt current, (d) Series compensation voltage and line current.

an efficient management of distribution resources, and can be used as a link between standard ac power-frequency systems and systems operating with either dc or ac at any power frequency [61, 62]. Since standardized voltages used by most utilities for MV distribution grids are usually equal or higher than 10 kV [63], multilevel topologies must be considered for the MV side of the SST if conventional Si-based semiconductors are used; see [64–66]. Actually, more than ten levels can be required if Si-based semiconductors with a blocking voltage of less than 2 kV are used for rated line voltages above 10 kVrms [67]. The section describes the topology and control strategies selected for a bidirectional multilevel SST design, details the distribution system used for testing the behaviour of the implemented SST model, and presents some simulation results that confirm the enhanced performance of the SST in comparison to the conventional transformer [68].

8.8.6.2 SST Configuration

The proposed bidirectional SST design is based on the commonly accepted three-stage design [69, 70]: (i) the input voltage at power frequency is first converted into dc voltage by the MV-side three-phase converter working as rectifier; (ii) the isolation stage is implemented by means of a dual active bridge (DAB) dc-dc converter, with an intermediate high-frequency transformer that reduces the MV square waveform into a LV square waveform; (iii) an LV three-phase dc/ac converter working as inverter provides the output power-frequency ac voltage to LV loads. Since generation can be connected to the secondary side of the transformer, the LV side should also be able to operate in generation mode to allow the power to flow from the secondary LV side to the primary MV side. Therefore, the converters and their respective switching strategies must be properly designed to work under bidirectional power flow conditions.

The implemented topology for the MV side is based on the cascaded connection of single-phase cells proposed or studied in several papers; see, for instance, [71–75]. The configuration selected for the basic cell is the single-phase DAB presented in [71, 72], although the present model incorporates some changes in the control strategies and the MV-side design is three-phase, while the configuration used in [71, 72] was single-phase. The LV side is based on the three-phase four-wire configuration and control strategy presented in [69, 70].

Figure 8.47a shows the configuration of the converter cell that will be the base of the modular design implemented here. Note that the cell consists of an input single-phase H bridge and an isolation stage made by a DAB dc-dc converter. Each single phase section of the SST is made out of a number of basic cells that will depend on the voltage level, the configuration being the result of series cascading in the primary MV side of the converter cell and connecting in parallel the secondary LV side, see Figure 8.47b. The proposed design consists of an ungrounded star connection of the three single-phase SST primary sides, see Figure 8.47c. The outputs of all DABs are connected in parallel to the LV-side dc bus. The LV converter feeds a three-phase four-wire system to which both load and generation can be connected, see Figure 8.47c.

8.8.6.3 Control Strategies

The configuration shown in Figure 8.47 implies that there will be as many input and isolation stages as single-phase cells are needed to build the final design (see Figure 8.47a). In addition, since it is assumed that all basic single cells are equally rated, the controllers have to achieve a homogeneously distributed voltage across all of them (i.e. voltages across each C_{dc1} in Figure 8.47a must be as equal as possible). Note that, given the series connection at the MV side, the current across any basic single cell is the same that in the corresponding phase of the MV level grid. The control strategies used for each stage as well as the controller required to voltage balance are summarized below.

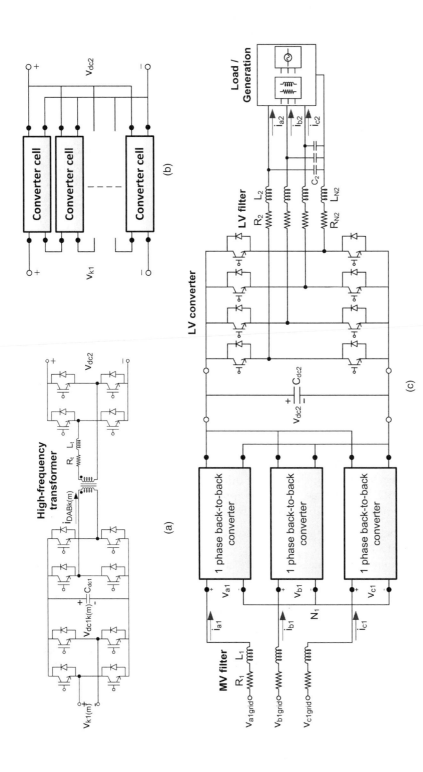

Figure 8.47 Case Study 5: cascaded modular bidirectional SST implementation: (a) Converter cell (with input H-bridge converter and dc-dc converter); (b) Modular design of a single-phase cell (back-to-back converter); (c) Overall SST configuration.

Input Stage

The control strategy for any single cell follows the principles used in [69, 70]. The input stage is implemented by means of a multilevel converter [71, 73]. The *abc*-frame model for the converter is as follows (see Figure 8.47):

$$v_{k1grid} = R_1 i_{k1} + L_1 \frac{d}{dt} i_{k1} + v_{k1} + v_{N1} \tag{8.3a}$$

$$v_{k1} = \sum_{m=1}^{M} v_{k1(m)} \tag{8.3b}$$

where $k \in a, b, c$, v_{k1grid} are the MV-side *abc* distribution system voltages at the point of coupling, i_{k1} are the MV-side *abc* distribution system currents, v_{k1} are the MV-side *abc* single-phase converter voltages, $v_{k1(m)}$ are the MV-side *abc* cell voltages, M is the number of cells that compose each single-phase MV-side converter, v_{N1} is the MV distribution system neutral, R_1 and L_1 are respectively the MV-side filter resistance and inductance.

The three-phase PWM converter obtained by applying the Park transform to Eq. (8.3a) may be represented by the following model:

$$\frac{d}{dt} i_{d1} = \omega_1 i_{q1} - \frac{R_1}{L_1} i_{d1} + \frac{1}{L_1} v_{d1conv} - \frac{1}{L_1} v_{d1grid} \tag{8.4a}$$

$$\frac{d}{dt} i_{q1} = -\omega_1 i_{d1} - \frac{R_1}{L_1} i_{q1} + \frac{1}{L_1} v_{q1conv} - \frac{1}{L_1} v_{q1grid} \tag{8.4b}$$

where i_{d1}, i_{q1} are the MV-side grid currents in the rotating *dq*-frame, v_{d1conv}, v_{q1conv} are the MV-side converter voltages in the rotating *dq*-frame, v_{d1grid}, v_{q1grid} are the MV-side grid voltages also in the rotating *dq*-frame, ω_1 is the MV-side grid angular frequency, R_1 and L_1 are respectively the resistance and inductance of the MV-side filter.

To achieve high-dynamic response and stability, the controller has an outer dc-link voltage control loop and an inner grid current control loop; see Figure 8.48. The reference for the grid current loop is given by the output of the outer dc-link voltage control loop. Figure 8.48 shows how the reference for the positive-sequence current value is obtained from the pu deviation of the dc link voltage average value with respect to the desired voltage.

A simple and effective strategy, the voltage oriented control (VOC), with feedforward of the negative-sequence grid voltage, has been applied in this work to control the MV-side converter [76]. A sequence separation method is applied to the grid voltages:

- The positive-sequence grid voltage is used to obtain the grid angle for synchronization purposes by means of a PLL [77].
- The negative-sequence grid voltage is fed-forwarded to the switching strategy that has to be generated at the converter terminals; therefore, no negative sequence voltage is seen by the inductive filter and only positive-sequence currents flow between the grid and the converter, even in presence of asymmetrical grid disturbances; see Figure 8.48.

This scheme ensures constant dc bus voltage, unity power factor condition at the input terminal in an average sense, and no ripple in the input active power. All the controllers have been implemented with conventional PI regulators. Because of the coupling between the *d*- and *q*- components of the grid currents, a conventional solution of adding two decoupling feed-forward inputs to each current control loop has been considered.

Since a single-phase section of the three-phase MV side is made by cascading single-phase cells, the voltage has to be uniformly distributed among the dc buses which are part of the topology implemented for the MV converter. A simple balance controller for the parallel operated DABs is used [71, 72]. The MV-side dc-link voltage of each module is selected as the feedback signal for the controller, being the reference V_{dc1}^* a predefined value. By default, it is assumed that the total voltage is equally shared among the modules, although the controller assumes that

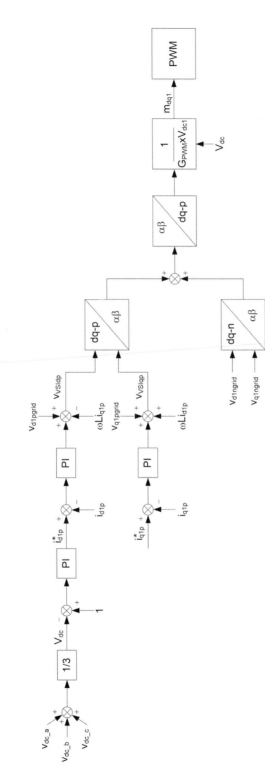

Figure 8.48 Case Study 5: medium-voltage side control.

there can be parameter mismatch (see below). In order to achieve voltage balancing among the dc links of each phase, a modification of the duty cycle, Δd, is added to the common duty cycle for each H-bridge (see Figure 8.49)

$$\Delta d_{1(m)} = \left(K_{p1} + \frac{K_{i1}}{s} \right) (V_{dc1}^* - v_{dc1(m)}) \tag{8.5}$$

A voltage balance among dc buses with the new duty cycle generated for each single-phase cell will also guarantee a power balance among the DABs: due to the series connection of the cells of each phase, the current that flows through each H-bridge cell is the same, therefore the power among the DABs will be balanced if the voltage balance is achieved. Note that the strategy depicted in Figure 8.49 guarantees that the new duty cycles will add to the same quantity that was previously used and, therefore, assures the same value of the total DC link voltage.

Isolation Stage
The amount and direction of the active power flow between primary and secondary of each high-frequency transformer is according to the following expression [78]

$$P_{k(m)} = \frac{v_{dc1k(m)}v_{dc2}}{2\pi f \cdot r \cdot L_t} \phi_{k(m)} \left(1 - \frac{|\phi_{k(m)}|}{\pi} \right) \tag{8.6}$$

where $v_{dc1k(m)}$ and v_{dc2} are the MV- and LV-side voltages of the DAB, r is the turns ratio of the transformers, f is the switching frequency of the converters, L_t is the transformer impedance value seen from the secondary side, and $\phi_{k(m)}$ is the phase between primary and secondary side voltages of the dual active H-bridge, being $k \in a, b, c$, and $1 \leq m \leq M$.

When more than a single DAB is used in parallel, as in Figure 8.47b, it has been observed that if the same phase shift is adopted for all DABs and the voltages in the MV-side dc links are regulated to the same value, a mismatch of parameters (e.g. transformer leakage inductances) can cause an unbalanced power share among them. Therefore, different phase shift may be needed for each DAB to ensure the power balance. The power balance controller is shown in Figure 8.50. The average power of each DAB is calculated and compared to the overall average, so a different phase shift can be generated for each of them if required. The dc-dc converter makes the power to flow towards the end-user side when voltage at the primary side of the transformer leads voltage at the secondary side. Power flows towards the MV distribution network when the

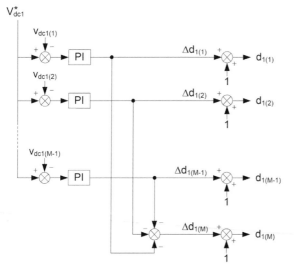

Figure 8.49 Case Study 5: DC voltage balance controller.

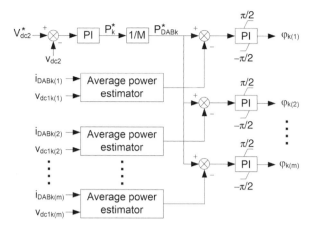

Figure 8.50 Case Study 5: isolation stage control.

voltage at the secondary side leads the voltage at the primary side. The absolute value of this reference phase-shift angle ϕ is limited to $\pi/2$, irrespective of the sense in which power flows.

Output Stage
The LV-side front end converter includes a four-leg converter for neutral currents, an inductor for filtering currents and a capacitor bank for filtering voltages. According to the configuration shown in Figure 8.47c, the LV-side converter may be connected to load and/or generation, and it is responsible for controlling the voltage (waveform and value) seen by load/generation. The three-phase four-wire LV converter allows connecting loads and/or generators with either one, two, or three phases. An SVM (Space Vector Modulation) PWM switching strategy has been used with this converter, see Figure 8.51 and references [69, 70, 79]. The main task of the LV-side converter controllers is to achieve positive-sequence capacitor voltages (i.e. to have balanced voltages at capacitor terminals) with stable frequency and voltage, independently of the power direction (i.e. load or generation) and the current balance. Each positive-, negative-, and zero-sequence has its respective controller. Negative- and zero-sequence capacitor voltage references are set to zero to cancel these components at the filter capacitor terminals at all time, even in presence of unbalanced load/generation currents. The positive sequence voltage controller regulates the filter capacitor voltages. For more details about the control strategy see [69, 70]. The controller shown in Figure 8.51 includes a current limiter that prevents the current peak caused during either an overload or a short-circuit from exceeding a specified value. The limiting effect is performed in the $\alpha\beta$ synchronous reference frame and does not affect the zero-sequence component of the current.

The limit of the peak currents specified in the simulation model, i_{lim}, is compared to the following value (see Figure 8.51):

$$i_{peak} = \sqrt{i_\alpha^2 + i_\beta^2} \tag{8.7}$$

If i_{peak} exceeds the specified i_{lim} then the dq values are obtained from the values i_α and i_β that result from changing i_{peak} to i_{lim}.

For more details on the four-leg converters and their control strategies, see [80–82].

8.8.6.4 Test System and Modelling Guidelines
The test system analysed in this section is a 50 Hz overhead distribution system based on previous models developed by the authors [68]. Figure 8.52 shows the configuration of the network,

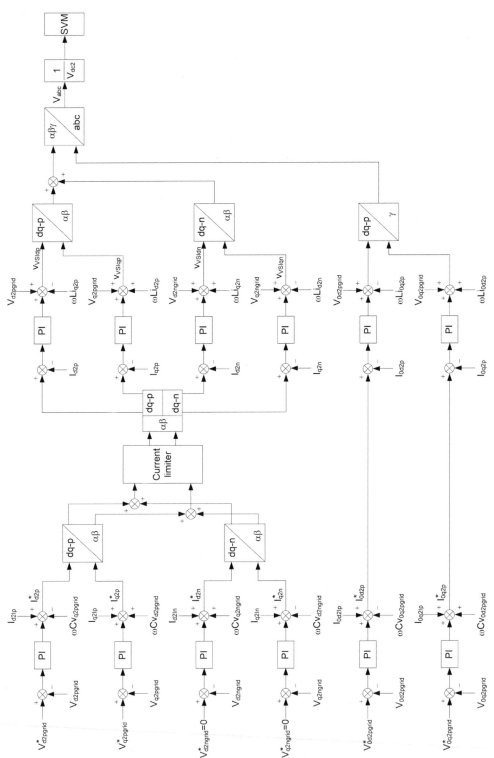

Figure 8.51 Case Study 5: low-voltage side control.

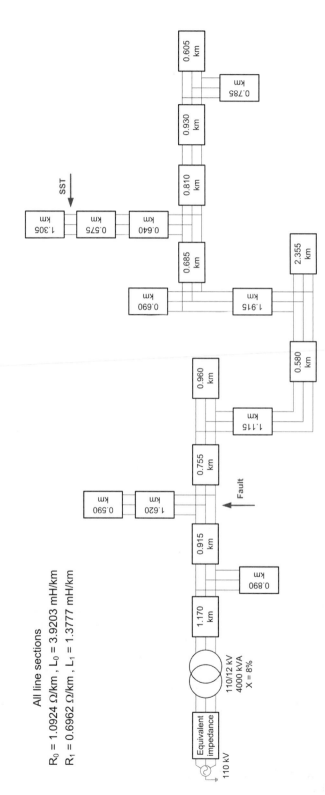

Figure 8.52 Case Study 5: diagram of the test system.

Table 8.3 Case Study 5: SST main parameters.

Parameter	Value
Primary side filter resistance (R_1)	0.5 Ω
Primary side filter inductance (L_1)	10 mH
MV DC link capacitance (C_{dc1})	500 μF
LV DC link capacitance (C_{dc2})	3000 μF
Secondary side filter resistance (R_2)	0.1 Ω
Secondary side filter inductance (L_2)	2 mH
Secondary side filter capacitance (C_2)	470 μF
Neutral resistance (R_n)	0.1 Ω
Neutral inductance (L_n)	1 mH
Rectifier/Inverter switching frequency	10 kHz
Transformer operating frequency	2 kHz
Transformer short-circuit resistance (R_t)	0.1 Ω
Transformer leakage inductance (L_t)	1 mH

pu length parameters and lengths of all line sections. Note that all line sections of the test system have the same pu length parameters. The delta-connected MV side of the substation transformer is grounded by means of a zig-zag reactor. The figure also indicates the node to which the SST is connected and the fault location considered in the first scenario, but it does not show the system loads.

The model has been implemented following the recommended guidelines: the semiconductors are modelled as ideal controlled switches and the high-frequency transformer is represented as an ideal transformer in series with its short-circuit impedance. The controllers have been implemented using TACS capabilities. Each single-phase section of the MV side is made up of six basic cells (see Figure 8.47). Taking into account the current semiconductor technology, this number is probably low for a realistic design of a multilevel converter that should be connected to a 12 kV grid. The number of basic cells has been reduced in this work to six in order to decrease the computing time, and because such quantity suffices to check the validity of the voltage balancing control strategy. Rated values of primary and secondary SST voltages are respectively 12 kV and 400 V.

The case studies presented below were carried out assuming that the rated power of the SST is 100 kVA [68]. Table 8.3 presents the main parameters used in this study; the parameter values of the high-frequency transformer are referred to its secondary side. The system loading is such that the steady-state voltage at the node to which the SST is connected is below 0.95 pu in all case studies. The controller of the MV-side converters will compensate the reactive power at the MV terminals of the SST; that is, reactive power will be close to zero under steady-state operating conditions and deviate from the zero value during transients. The implemented model is intended for distribution system studies; it can be used in power-quality studies (e.g. voltage dip propagation, current unbalance, overcurrents) and to test distribution system performance in front of many other steady-state and transient operating conditions.

8.8.6.5 Case Studies

Two scenarios have been simulated to illustrate the performance of the new SST model [68]. Figures 8.53 and 8.54 depict the main results of each case study. All figures show the same plots:

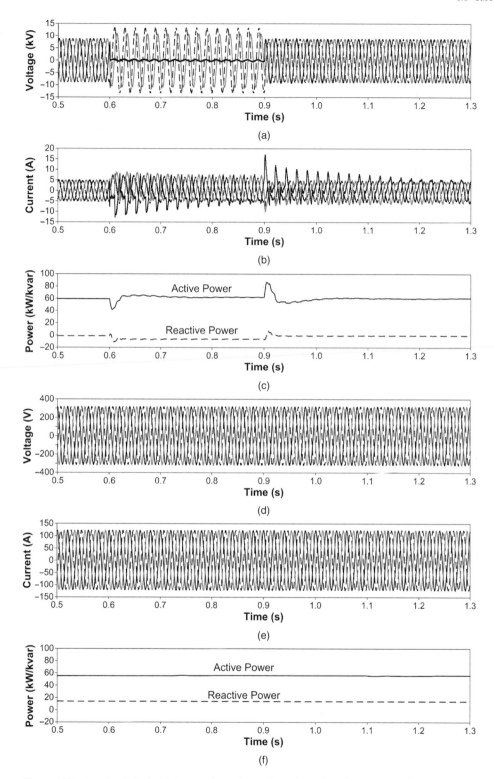

Figure 8.53 Case Study 5: simulation results: voltage dip and swells at the primary side. (a) Primary SST terminal voltages – MV level, (b) Primary SST terminal currents – MV level, (c) Primary SST terminal power – MV level, (d) Secondary SST terminal voltages – LV level, (e) Secondary SST terminal currents – LV level, (f) Secondary SST terminal power – LV level.

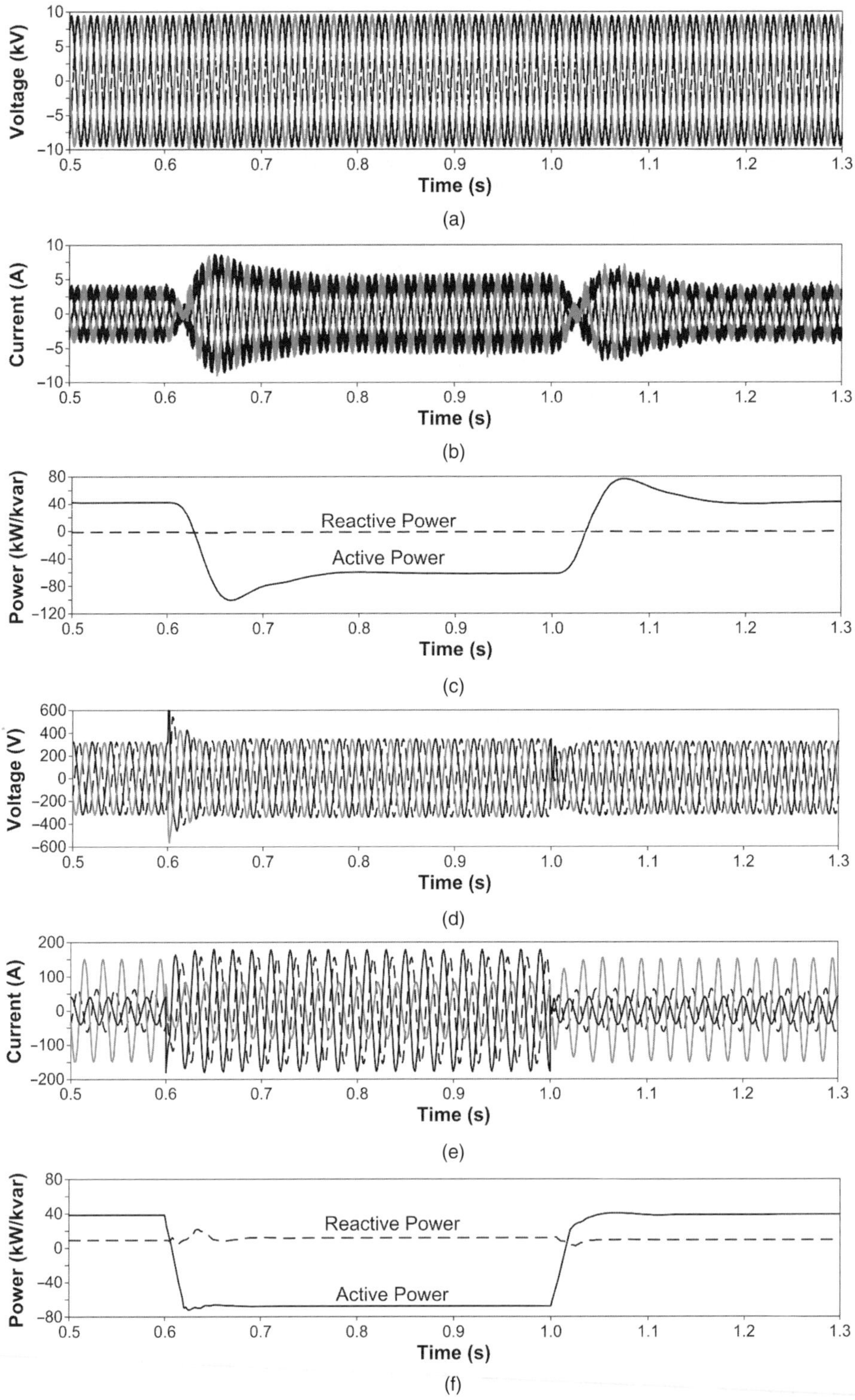

Figure 8.54 Case Study 5: simulation results: power flow reversal. (a) Primary SST terminal voltages – MV level, (b) Primary SST terminal currents – MV level, (c) Primary SST terminal power – MV level, (d) Secondary SST terminal voltages – LV level, (e) Secondary SST terminal currents – LV level, (f) Secondary SST terminal power – LV level.

voltages, currents, and powers measured at both MV and LV SST terminals. A short analysis of each case study is provided below.

1. *Voltage dips and swells at MV terminals.* Due to the grounding system selected for the test system, a single-phase-to-ground fault in the MV network (see fault location in Figure 8.52) will cause a voltage dip at the faulted phase MV terminals of the SST, and two voltage swells at the unfaulted phase terminals. The simulation results related to this case (see Figure 8.53) prove clearly that the voltage imbalance that occurs in the primary side of the SST is not propagated to the secondary side, where the phase voltages and currents remain constant and balanced. Note that the peak values of the currents during the fault condition are lower in the SST phases in which the fault causes swells and that the SST recovers current balance once the fault condition disappears.

2. *Power flow reversal.* A power flow reversal is caused by the presence of both load and generation at the secondary side. As deduced from Figure 8.54, the load initially exceeds the generation, but during a short period this situation is reversed. Note that, as with the previous case study, the currents at the LV terminals are always unbalanced but the currents measured at the MV terminals are always balanced. This case simultaneously illustrates two of the main advantages of the SST, its capabilities to quickly control a power flow reversal between its terminals and to balance MV currents, irrespective of the situation at the LV terminals.

These simulation results are those expected from the three-stage SST design simulated here; they confirm that intermediate capacitors provide stage decoupling and prevent disturbance at one side from propagating to the other side (e.g. secondary load immunity is achieved in front of dynamic unbalanced situations at the input side). The results have also shown that the new SST configuration incorporates the same advanced capabilities (e.g. fast voltage and power flow control, reactive power compensation, current balance) that other simplified models; see, for instance, [69, 70]; this supports its feasibility as a fundamental component of the future smart grid. However, it is important to remember that some aspects (e.g. semiconductor losses representation or the high-frequency transformer model) must be improved since they can have a significant influence in other studies. For more details and simulation results, see [68].

Acknowledgement

All case studies included in this chapter were developed and implemented by the authors. However, some case studies were based on examples previously developed by other authors. The authors are grateful to all those whose work inspired the examples. Thanks are also due to K. Sen; Case Study 4 is, as already mentioned, a new version of the UPFC model presented in his book on FACTS controllers (see reference [46]). Finally, the first author wants to thank some of his colleagues for the work made together in some IEEE Working Groups. The papers produced by those WGs were another source of inspiration for this chapter.

References

1 Mohan, N., Undeland, T.M., and Robbins, W.P. (2002). *Power Electronics: Converters, Applications, and Design*, 3e. Wiley.

2 Hingorani, N.G. and Gyugyi, L. (2000). *Understanding FACTS: Concepts and Technology of Flexible AC Transmission Systems*. New York: IEEE Press.

3 Kim, C.K., Sood, V.K., Jang, G.S. et al. (2009). *HVDC Transmission: Power Conversion Applications in Power Systems*. Wiley.

4 Yazdani, A. and Iravani, R. (2010). *Voltage-Sourced Converters: Modeling, Control, and Applications*. Hoboken (NJ, USA): Wiley.

5 Mohan, N., Robbins, W.P., Undeland, T.M. et al. (1994). Simulation of power electronics and motion control systems – an overview. *Proceedings of the IEEE* 82 (8): 1287–1302.

6 IEEE TF on Power Electronics (1997). Guidelines for modeling power electronics in electric power engineering applications. *IEEE Transactions on Power Delivery* 12 (1): 505–514.

7 Sen, K.K., Tang, L., Dommel, H.W. et al. (1999). Guidelines for modeling power electronics in electric power engineering applications. In: *Modeling and Analysis of System Transients Using Digital Programs* (eds. A. Gole, J.A. Martinez-Velasco and A. Keri). IEEE.

8 Johnson, B., Hess, H., and Martínez, J.A. (2005). Parameter determination for modeling systems transients. Part VII: semiconductors. *IEEE Transactions on Power Delivery* 20 (3): 2086–2094.

9 Chiniforoosh, S., Jatskevich, J., Yazdani, A. et al. (2010). Definitions and applications of dynamic average models for analysis of power systems. *IEEE Transactions on Power Delivery* 25 (4): 2655–2669.

10 Chiniforoosh, S., Jatskevich, J., Atighechi, H., and Martinez-Velasco, J.A. (2015). Dynamic average modeling of rectifier loads and AC-DC converters for power system applications. In: *Transient Analysis of Power Systems. Solution Techniques, Tools and Applications* (ed. J.A. Martinez-Velasco). Chichester (United Kingdom): Wiley/IEEE Press.

11 Kerkman, R.J., Leggate, D., and Skibinski, G.L. (1997). Interaction of drive modulation and cable parameters on AC motor transients. *IEEE Transactions on Industry Applications* 33 (3): 722–731.

12 Melfi, M., Sung, A.M.J., Bell, S., and Skibinski, G.L. (1998). Effect of surge voltage risetime on the insulation of low-voltage machines fed by PWM converters. *IEEE Transactions on Industry Applications* 34 (4): 766–775.

13 Wong, C. (1997). EMTP modeling of IGBT dynamic performance for power dissipation estimation. *IEEE Transactions on Industry Applications* 33 (1): 64–71.

14 Kraus, R. and Mattausch, H.J. (1998). Status and trends of power semiconductors device models for circuit simulation. *IEEE Transactions on Power Electronics* 13 (3): 452–465.

15 Berning, D.W. and Heffner, A.R. (1998). IGBT model validation. *IEEE Industry Applications Magazine* 4 (6): 23–34.

16 Heffner, A.R. and Diebolt, D.M. (1994). An experimentally verified IGBT model implemented in the saber circuit simulator. *IEEE Transactions on Power Electronics* 9 (5): 532–542.

17 Heffner, A.R. (1995). Modeling buffer layer IGBT's for circuit simulation. *IEEE Transactions on Power Electronics* 10 (2): 111–123.

18 Rashid, M. (1999). *Microelectronic Circuits: Analysis and Design*. Boston (MA, USA): PWS Publishers.

19 Dommel, H.W. (1986). *Electromagnetic Transients Program Reference Manual (EMTP Theory Book)*. Portland (OR, USA): Bonneville Power Administration.

20 Kulicke, B. (1981). Simulation program NETOMAC: difference conductance method for continuous and discontinuous systems. *Siemens Research and Development Reports* 10 (5): 299–302.

21 Kuffel, P., Kent, K., and Irwin, G. (1997). The implementation and effectiveness of linear interpolation within digital simulation. *Electrical Power and Energy Systems* 19 (4): 221–228.

22 Zou, M., Mahseredjian, J., Joos, G. et al. (2006). Interpolation and reinitialization in time-domain simulation of power electronic circuits. *Electric Power Systems Research* 76: 688–694.

23 Marti, J.R. and Lin, J. (1989). Suppression of numerical oscillations in the EMTP. *IEEE Transactions on Power Systems* 4 (2): 739–747.

24 Lin, J. and Marti, J.R. (1990). Implementation of the CDA procedure in the EMTP. *IEEE Transactions on Power Systems* 5 (2): 394–402.

25 Capolino, G.A. and Hénao, H. (1988). ATP simulation for power electronics and AC drives. Presented at the 15th European EMTP Users Group Meeting in Leuven, Belgium (17–18 October 1988).

26 Capolino, G.A. and Hénao, H. (1990). Simulation of electrical machine drives with EMTP. Presented at the 18th European EMTP Users Group Meeting in Marseille, France (May 1990).

27 Martinez, J.A. and Capolino, G.A. (1991). TACS and MODELS: Drive simulation languages in a general purpose program. Proceedings of MCED in Marseille, France.

28 Can/Am Users Group (2000). *ATP Rule Book*. Can/Am Users Group.

29 Krein, P.T. and Bass, R.M. (1992). Autonomous control technique for high performance switches. *IEEE Transactions on Industrial Electronics* 39 (3): 215–222.

30 Capolino, G.A., Henao, H. and Toury, O. (1989). Macro modelling of power electronics switches using ATP formulation. Presented at the 17th European EMTP User Group Meeting in Leuven, Belgium (November 1989).

31 Nagaoka, N. (1988). Large-signal transistor modeling using the ATP version of EMTP. *EMTP News* 1 (3): 14–24.

32 Capolino, G.A., Hefner, A.R., Henao, H. and Samihoeto, M. (1992). Modèle dynamique de transistor à grille isolée pour logiciel à usage général, (in French). Colloque SEE: Electronique de Puissance du Futur in Marseille, France (November 1992).

33 Dubé, L. (1996). How to use MODELS-based user-defined network components in ATP. European EMTP/ATP Users Group Meeting in Budapest, Hungary (November 1996).

34 Lauritzen, P. and Ma, C.L. (1991). Simple diode model with reverse recovery. *IEEE Transactions on Power Electronics* 2 (2): 188–191.

35 Ma, C.L. and Lauritzen, P. (1993). A simple power diode model with forward and reverse recovery. *IEEE Transactions on Power Electronics* 8 (4): 342–346.

36 Martinez-Velasco, J.A., Abdo, R. and Capolino, G. (1995). Advanced representation of power semiconductors using the EMTP. Proceedings of IPST in Lisbon, Portugal.

37 Badarou, R., Boussak, M. and Capolino, G.A. (1990). ATP simulation of transient behaviour of single phase capacitor run induction motor fed by a directional converter. Presented at the18th European EMTP Users Group Meeting in Marseille, France (May 1990).

38 Capolino, G.A. (1991). DC machines models: from UM to block scheme representation. *EMTP News* 4 (4): 4–13.

39 Martinez, J.A. (1993). Lectures Notes on Rotating Machines. EMTP Summer Course in Leuven, Belgium.

40 Domijan, A. and Yin, Y. (1994). Single phase induction machine simulation using the electromagnetic transients program: theory and test cases. *IEEE Transactions on Energy Conversion* 9 (3): 535–542.

41 Carrara, G., Casini, D., Cesario, P. and Taponecco, L. (1990). Modelling trapezoidal brushless motor drive using TACS. Presented at the 18th European EMTP Users Group Meeting in Marseille, France (May 1990).

42 Martinez, J.A. (1993). Educational use of EMTP MODELS for the study of rotating machine transients. *IEEE Transactions on Power Systems* 8 (4): 1392–1399.

43 Capolino, G.A. and Hénao, H. (1993). ATP advanced usage for electrical drives. EMTP Summer Course in Leuven, Belgium.

44 Knudsen, H. (1995). Extended Park's transformation for 2x3 phase synchronous machine and converter phasor model with representation of harmonics. *IEEE Transactions on Energy Conversion* 10 (1): 126–132.

45 Filizadeh, S., Sen, K.K., Jatskevich, J. et al. (2012). Power electronics in transmission and distribution systems. In: *The Encyclopedia of Life Support Systems (EOLSS)*. UNESCO.

46 Sen, K.K. and Sen, M.L. (2009). *Introduction to FACTS Controller: Theory, Modeling, and Applications*. IEEE Press/Wiley.

47 Jovcic, D. and Ahmed, K. (2015). *High Voltage Direct Current Transmission: Converters, Systems and DC Grids*. Wiley.

48 Sharifabadi, K., Harnefors, L., Nee, H.P. et al. (2016). *Design, Control, and Application of Modular Multilevel Converters for HVDC Transmission Systems*. Wiley.

49 Dugan, R.C., McGranaghan, M., Santoso, S., and Wayne Beaty, H. (2012). *Electrical Power Systems Quality*, 3e. New York (NY, USA): McGraw-Hill.

50 Ghosh, A. and Ledwich, G. (2002). *Power Quality Enhancement Using Custom Power Devices*. Kluwer Academic Publishers.

51 Gosh, A. and Shahnia, F. (2015). Applications of power electronic devices in distribution systems. In: *Transient Analysis of Power Systems. Solution Techniques, Tools and Applications* (ed. J.A. Martinez-Velasco). Chichester (United Kingdom): Wiley/IEEE Press.

52 Yazdani, A. and Alizadeh, O. (2015). Modelling of electronically interfaced DER systems for transient analysis. In: *Transient Analysis of Power Systems. Solution Techniques, Tools and Applications* (ed. J.A. Martinez-Velasco). Chichester (United Kingdom): Wiley/IEEE Press.

53 Funaki, T., Arita, H., Kan, M. and Kashiwagi, Y. (2001). Handling of switching transient and GIFU switch ability on the IEE-J's power electronics benchmark circuits. Proceedings of IPST in Rio de Janeiro, Brazil (June 2001).

54 Scott Meyer, W. and Liu, T.H. (1996). *Can/Am EMTP News* 96: 1–2.

55 Gómez Expósito, A., Gonzalez Vasquez, F., Izquierdo Mitchell, C. et al. (1992). Microprocessor-based control of an SVC for optimal load compensation. *IEEE Transactions on Power Delivery* 7 (2): 706–712.

56 Martinez, J.A. (1992). Simulation of a microprocessor-controlled SVC. Presented at the 21st European EMTP Users Group Meeting in Crete, Greece.

57 Martinez, J.A. (1995). EMTP simulation of a digitally controlled static var system for optimal load compensation. *IEEE Transactions on Power Delivery* 10 (3): 1408–1415.

58 Sen, K.K. and Stacey, E.J. (1998). UPFC-unified power flow controller: theory, modeling, and applications. *IEEE Transactions on Power Delivery* 13 (4): 1453–1460.

59 Sen, K.K. (2015). Analysis of FACTS controllers and their transient modelling techniques. In: *Transient Analysis of Power Systems. Solution Techniques, Tools and Applications* (ed. J.A. Martinez-Velasco). Chichester (United Kingdom): Wiley/IEEE Press.

60 Adabi, M.E. and Martinez-Velasco, J.A. (2018). Solid state transformer technologies and applications: a bibliographical survey. *AIMS Energy* 6 (2): 291–238.

61 Huber, J.E. and Kolar, J.W. (2016). Solid-state transformers: on the origins and evolution of key concepts. *IEEE Industrial Electronics Magazine* 10 (3): 19–28.

62 Huang, A.Q. (2016). Medium-voltage solid-state transformer. *IEEE Industrial Electronics Magazine* 10 (3): 29–42.

63 IEC Std 60038 IEC standard voltages; Edition 7.0, 2009.

64 Kouro, S., Malinowski, M., Gopakumar, K. et al. (2010). Recent advances and industrial applications of multilevel converters. *IEEE Transactions on Industrial Electronics* 57 (8): 2553–2580.

65 Rodríguez, J., Bernet, S., Wu, B. et al. (2007). Multilevel voltage-source-converter topologies for industrial medium-voltage drives. *IEEE Transactions on Industrial Electronics* 54 (6): 2930–2945.

66 Abu-Rub, H., Holtz, J., Rodriguez, J., and Baoming, G. (2010). Medium-voltage multilevel converters-state of the art, challenges, and requirements in industrial applications. *IEEE Transactions on Industrial Electronics* 57 (8): 2581–2596.

67 Backlund, B. and Carroll, E. (2006). *Voltage Ratings of High Power Semiconductors*. ABB Semiconductors.

68 Martin-Arnedo, J., González-Molina, F., Martinez-Velasco, J.A., and Adabi, M.E. (2017). EMTP model of a bidirectional cascaded multilevel solid state transformer for distribution system studies. *Energies* 10: 521.

69 Alepuz, S., González-Molina, F., Martin-Arnedo, J., and Martinez-Velasco, J.A. (2014). Development and testing of a bidirectional distribution electronic power transformer model. *Electric Power Systems Research* 107: 230–239.

70 Martinez-Velasco, J.A., Alepuz, S., González-Molina, F., and Martín-Arnedo, J. (2014). Dynamic average modeling of a bidirectional solid state transformer for feasibility studies and real-time implementation. *Electric Power Systems Research* 117: 143–153.

71 Zhao, T., Wang, G., Bhattacharya, S., and Huang, A.Q. (2013). Voltage and power balance control for a cascaded H-bridge converter-based solid-state transformer. *IEEE Transactions on Power Electronics* 28 (4): 1523–1532.

72 She, X., Huang, A.Q. and Ni, X. (2013). A cost effective power sharing strategy for a cascaded multilevel converter based solid state transformer. IEEE Energy Conversion Congress and Expo (ECCE) in Denver, USA (September 2013).

73 Huber, J.E. and Kolar, J.W. (2014). Common-mode currents in multi-cell solid-state transformers. International Power Electronics Conference in Hiroshima, Japan (May 2014).

74 Huber, J.E. and Kolar, J.W. (2014). Volume/weight/cost comparison of a 1 MVA 10 kV/400 V solid-state against a conventional low-frequency distribution transformer. IEEE Energy Conversion Congress and Expo (ECCE) in Pittsburgh, USA (November 2014).

75 Garcia Montoya, R.J., Mallela, A. and Balda, J.C. (2015). An evaluation of selected solid-state transformer topologies for electric distribution systems. IEEE Applied Power Electronics Conference Expo (APEC) in Charlotte, USA (March 2015).

76 Kim, D. and Lee, D. (2007). Inverter output voltage control of three-phase UPS systems using feedback linearization. Presented at the 33rd Annual Conference of the IEEE Industrial Electronics Society (IECON) in Taipei, Taiwan (November 2007).

77 Dai, J., Xu, D., and Wu, B. (2009). A novel control scheme for current-source-converter-based PMSG wind energy conversion systems. *IEEE Transactions on Power Electronics* 24 (4): 963–972.

78 De Doncker, R.W., Divan, D.M., and Kheraluwala, M.H. (1991). A three-phase soft-switched high-power-density dc-dc converter for high power applications. *IEEE Transactions on Industry Applications* 27 (1): 63–73.

79 Perales, M.A., Prats, M.M., Portillo, R. et al. (2003). Three-dimensional space vector modulation in abc coordinates for four-leg voltage source converters. *IEEE Power Electronics Letters* 99 (4): 104–109.

80 Zhang, R., Prasad, V.H., Boroyevich, D., and Lee, F.C. (2002). Three-dimensional space vector modulation for four-leg voltage-source converters. *IEEE Transactions on Power Electronics* 17 (3): 314–326.

81 Vechiu, I., Curea, O., and Camblong, H. (2010). Transient operation of a four-leg inverter for autonomous applications with unbalanced load. *IEEE Transactions on Power Electronics* 25 (2): 399–407.

82 Ebrahimzadeh, E., Farhangi, S., Iman-Eini, H., and Blaabjerg, F. (2016). Modulation technique for four-leg voltage source inverter without a look-up table. *IET Power Electronics* 9 (4): 648–656.

To Probe Further

The simulation results presented in the chapter are a very small sample of the results that can be derived from the case studies. ATP data files for most case studies are available in the companion website. Readers are encouraged to run the cases, check all results, and explore the performance of the test systems under different operating conditions.

9

Creation of Libraries

Juan A. Martinez Velasco and Jacinto Martin-Arnedo

9.1 Introduction

The Alternative Transients Program (ATP) is a circuit-oriented tool based on a time-domain solution method: users supply parameters of components and their interconnection, the program itself builds the system equations. A simulation is carried out with a fixed time-step, selected by the user, using the trapezoidal rule of integration [1]. The trapezoidal rule converts the differential equations of the network components into algebraic equations involving voltages, currents, and past values. These algebraic equations are assembled using a nodal approach. The resulting conductance matrix is symmetrical and remains unchanged as the integration is performed with a fixed time-step size. As this scheme can only be used to solve linear networks, some approaches to cope with nonlinear and time-varying elements have been implemented; see Chapter 3.

A steady-state phasor solution can be carried out to establish initial conditions for the transient solution, to analyse harmonic propagation (Harmonic Frequency Scan (HFS) [2]) or to obtain system impedance as a function of frequency (Frequency Scan (FS)). The steady-state solution of linear networks at a single frequency is a simple task, and can be obtained using nodal admittance equations. However, this task can be very complex in the presence of nonlinearities, which can produce steady-state harmonics. No solution method is available in ATP to obtain the initial state in systems with switching devices and nonlinear components.

ATP capabilities can be used to represent the most important power components, calculate control system dynamics, incorporate user-defined models, and develop module libraries [3]. The application of custom-made modules for representing new components has been presented in previous chapters (e.g. several case studies used modules for calculating active and reactive powers). This chapter shows how to create a library of custom-made modules for a specific type of studies, namely power quality studies.

Section 9.2 details some of the most common approaches for creating custom-made modules; two examples are used to illustrate how to create the modules and their implementation as ATPDraw components. Section 9.3 discusses the application of ATP in power quality studies, while Section 9.4 lists the modules created for analysing power quality problems. Three issues are covered in Sections 9.5: generation and propagation of harmonics, voltage dip studies in distribution networks, and application of mitigation techniques for reducing harmonic currents and voltage dips.

Transient Analysis of Power Systems: A Practical Approach, First Edition. Edited by Juan A. Martinez-Velasco.
© 2020 John Wiley & Sons Ltd. Published 2020 by John Wiley & Sons Ltd.
Companion Website: www.wiley.com/go/martinez/power_systems

9.2 Creation of Custom-Made Modules

9.2.1 Introduction

ATP built-in models cannot cover all models that are required in transient studies. However, ATP capabilities allow users to implement custom-made modules that can be useful for either facilitation of the simulation tasks (mostly the creation and edition of the input models) or representation of components not available in ATP (e.g. protective relays). Chapter 4 provided an introduction to those capabilities; case studies presented in other chapters have shown some potential applications.

The continuous improvements introduced in ATPDraw have facilitated these tasks to users: the current version of this graphical interface (GUI) allows users to create custom-made models by taking advantage of the routine DATA BASE MODULE and MODELS language, as well as to compress several components into a single group which can be nested as part of a higher-level group. These three options allow users to customize edition and simulation tasks. More sophisticated capabilities (e.g. compiled TACS 69, foreign models in MODELS sections, Type 94 branch component) can be used to implement advanced models. Although some of these capabilities are also used in case studies detailed in this book, this chapter shows how to use only the three options mentioned above. The rules to select one or other option might be as follows:

- If the model can be based on the conventional ATP models (i.e. branches, switches, sources, and TACS-based control sections) then use the supporting routine DATA BASE MODULE. If the features of the new model require a high-level language, cannot be based on TACS capabilities, and MODELS capabilities are enough to implement the desired model, then use MODELS.
- If the new model has to be based on the combination of MODELS sections and other conventional ATP capabilities (e.g. branches, switches, or sources), then use the option Group (available in ATPDraw).

These rules do not cover all scenarios. In fact, there can be applications for which a simple icon could be used to represent several components of the system model for which all parameters have been already specified; the option Group might be then the best one.

9.2.2 Application of DATA BASE MODULE

The application of this routine to the creation of modules is usually based on two ATP features [4, 5]:

1) *Data sorting by class.* An ATP data file has to be structured taking into account some rules that make both editing and correcting the file difficult. Data must be grouped by classes, such as control devices, branches, switches, and sources. Thanks to the use of '/' cards, it is possible to mix different data classes, and the user can group components of different classes which are geographically close without considering the traditional file structure. Whenever data class changes, it is necessary to mark a discontinuity point with a '/' card that indicates the subsequent data class (e.g. /BRANCH, /SWITCH, /SOURCE).
2) *Data modularization.* A component made by many network elements, either of the same or different classes, can be seen as a module. By means of the supporting routine DATA BASE MODULE the user can create a module for a complex component; this module will be stored as an ordinary file and used every time the component has to be simulated by means of a $INCLUDE request, followed by the list of arguments that describe the local component.

The input format for DATA BASE MODULE has two different sections:

1) An argument declaration with the list of parameter names. Three types of parameters can be specified in such a list: node names, numerical values, and dummy arguments.
2) A template that represents the ATP input of the module or network section under study.

The general form of a $INCLUDE statement is:

```
$INCLUDE, file-name, arg-1, ......., arg-n
```

where *file-name* is the name given to the file created by DATA BASE MODULE and *arg-1*, ..., *arg-n* are the arguments that describe the component. There is no restriction for a file name, although it is advisable to use a name related to the type of component that is being simulated.

The following example shows how this option can be used to create a very simple data module. A capacitor is a lumped component which is completely defined by a single parameter, its capacitance. Consider a 30 μF capacitor connecting BUSA and BUSB; the data card – a Branch card according to the ATP terminology – for this component would be

```
C 3456789012345678901234567890123456789012345678901234567890123456789012345678901234567890
  BUSA  BUSB                         30.0
```

The template for a capacitor could be

```
C 3456789012345678901234567890123456789012345678901234567890123456789012345678901234567890
ARG, NODE1, NODE2, _____C
NUM, _____C
/BRANCH
  NODE1 NODE2                              _____C
```

If the file created by DATA BASE MODULE is named CAPACITOR.LIB, the branch card can be replaced by the following $INCLUDE statement

```
$INCLUDE, capacitor.lib, BUSA#, BUSB#, 30.0
```

Some aspects are to be taken into account when a module like this one is used:

- It is compulsory to specify an argument in each field; otherwise, an error message will be generated and the program will stop the execution. Three arguments must be specified in the Capacitor module, the two terminal node names and the capacitance.
- The length that every field has in the template must be respected in the $INCLUDE statement; however, there is a difference between arguments specifying node names or signals and numerical arguments.

When the above capacitor file is used, node names must have an *exact length* of five characters (# signs may be added to match this length). It can be observed that terminal node names of that capacitor have four characters; a # sign is added to match the compulsory length.

As for the numerical argument of the module, it can have a *maximum length* of 6 digits; the underline sign '_' has been used in the template to match the desired length of this parameter.

This option has some interesting advantages:

- it is easy to understand what type of component is being simulated (a capacitor in this case);
- the card to be punched has an almost free-format;
- thanks to the/BRANCH card inserted at the beginning of template, the $INCLUDE statement aimed at simulating a capacitor card may be punched without taking into account the data class which is being simulated;
- the CAPACITOR.LIB file can be used as many times as necessary in the same file.

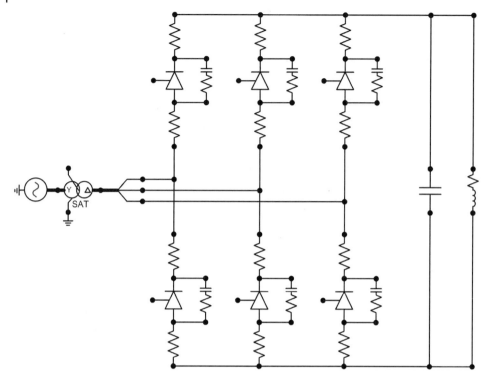

Figure 9.1 Three-phase controlled rectifier.

This section shows how DATA BASE MODULE can be used to create a custom-made module that can be implemented as an ATPDraw built-in model. Data modularization is especially useful for components or modules with repetitive connection of network elements. Figure 9.1 shows the schematic diagram of a three-phase controlled rectifier bridge.

The control strategy for this circuit might be summarized as follows:

```
INPUT            Va                                      {Phase A node voltage
                 Vb                                      {Phase B node voltage
                 Vc                                      {Phase C node voltage
OUTPUT           FPa , FNa                               {Phase A control signals
                 FPb , FNb                               {Phase B control signals
                 FPc , FNc                               {Phase C control signals
PARAMETERS       Alfa                                    {Delay angle
                 Freq                                    {Voltage frequency
INITIALIZATION   Delta = Alfa/(360 * Freq)
PROCEDURE        Vab = Va - Vb
                 Vbc = Vb - Vc
                 Vca = Vc - Va
                 D1ab = Vab[t - Delta]
                 D1bc = Vbc[t - Delta]
                 D1ca = Vca[t - Delta]
                 D2ab = D1ab[t - 0.0001]
                 D2bc = D2bc[t - 0.0001]
                 D2ca = D2ca[t - 0.0001]
                 FPa = (not . D1ca . and . D2ca) . or . (not . D2bc . and . D1bc)
                 FPb = (not . D1ab . and . D2ab) . or . (not . D2ca . and . D1ca)
                 FPc = (not . D1bc . and . D2bc) . or . (not . D2ab . and . D1ab)
                 FNa = (not . D2ca . and . D1ca) . or . (not . D1bc . and . D2bc)
                 FNb = (not . D2ab . and . D1ab) . or . (not . D1ca . and . D2ca)
                 FNc = (not . D2bc . and . D1bc) . or . (not . D1ab . and . D2ab)
```

Two modules can be built to respectively represent the power circuit and the control system. The template for the controlled rectifier using *data sorting by class* might be as follows:

```
BEGIN NEW DATA CASE  -NOSORT-
DATA BASE MODULE
C -------------------------------------------------------------------------------------------
C              FILE: THYBRIDGE.DBM - Thyristor Bridge Rectifier
C                 This file contains $PARAMETER expressions
C -------------------------------------------------------------------------------------------
$ERASE
ARG, VGEN_ , POS___, NEG___, GATE_ , KVOLTS, KVA___
NUM, KVOLTS, KVA___
DUM, INT___
DUM, VGENIA, VGENIB, VGENIC, VGENOA, VGENOB, VGENOC,
DUM, POSI_A, POSI_B, POSI_C, NEGI_A, NEGI_B, NEGI_C
/REQUEST
C ----- Calculation of snubber circuit parameters
$PARAMETER
MVA___ = KVA___ / 1000. $$
INT___R_SNUB = ( KVOLTS**2. / MVA___) * 100.
INT___C_SNUB = 1. / ( KVOLTS**2. / MVA___)
BLANK
C        1         2         3         4         5         6         7         8
C 345678901234567890123456789012345678901234567890123456789012345678901234567890
C ----- Thyristors
/SWITCH
11VGENIAPOSI_A                                           GATE_1          0
11VGENIBPOSI_B                                           GATE_3          0
11VGENICPOSI_C                                           GATE_5          0
11NEGI_AVGENOA                                           GATE_4          0
11NEGI_BVGENOB                                           GATE_6          0
11NEGI_CVGENOC                                           GATE_2          0
C ----- Connectivity resistors
/BRANCH
  VGEN_AVGENIA          0.0001
  VGEN_BVGENIB          0.0001
  VGEN_CVGENIC          0.0001
  VGENOAVGEN_A          0.0001
  VGENOBVGEN_B          0.0001
  VGENOCVGEN_C          0.0001
  POSI_APOS___          0.0001
  POSI_BPOS___          0.0001
  POSI_CPOS___          0.0001
  NEG___NEGI_A          0.0001
  NEG___NEGI_B          0.0001
  NEG___NEGI_C          0.0001
  POS___                1.0E+6
  NEG___                1.0E+6
C ----- Snubber circuits
$VINTAGE,1
  POSI_AVGENIA          INT___R_SNUB              INT___C_SNUB
  POSI_BVGENIBPOSI_AVGENIA
  POSI_CVGENICPOSI_AVGENIA
  VGENOANEGI_APOSI_AVGENIA
  VGENOBNEGI_BPOSI_AVGENIA
  VGENOCNEGI_CPOSI_AVGENIA
$VINTAGE,0
BEGIN NEW DATA CASE
$PUNCH
BEGIN NEW DATA CASE
BLANK
```

The template includes a section for calculating the parameters of the snubber circuits whose main goal is to avoid numerical oscillations. The values specified for calculating those parameters are the rated voltage and power of the controlled rectifier. Note also that the dc terminals of the rectifier are isolated from ground: a high-value resistance has been connected between the bridge terminals and ground.

The file created by DATA BASE MODULE would be the following one:

```
KARD  4  5  5  6  6  9  9  9 10 10 10 11 11 11 12 12 12 13 13 13 14 14 14 16 16
     17 17 18 18 19 19 20 20 21 21 22 22 23 23 24 24 25 25 26 26 27 27 28 29 31
     31 31 31 32 32 32 32 33 33 33 33 34 34 34 34 35 35 35 35 36 36 36 36
KARG  6  5 -1  5 -1  4 -2 -8  4 -3 -9  4 -4-10  4 -5-11  4 -6-12  4 -7-13  1 -2
      1 -3  1 -4  1 -5  1 -6  1 -7  2 -8  2 -9 2-10 3-11 3-12 3-13  2  3 -1
     -1 -2 -8 -2 -3 -8 -9 -2 -4 -8-10 -2 -5 -8-11 -2 -6 -8-12 -2 -7 -8-13
KBEG 10 18  1 23  1 65  3  9 65  3  9 65  3  9 65  9  3 65  9  3 65  9  3  3  9
      3  9  3  9  9  3  9  3  9  3  9  3  9  3  9  3  3  9  3  9  3  9  3  3 27
     59  9  3 21  9 15  3 21  9 15  3 21  3 15  9 21  3 15  9 21  3 15  9
KEND 15 23  6 28  6 69  8 14 69  8 14 69  8 14 69 14  8 69 14  8 69 14  8  7 14
      7 14  7 14 13  8 13  8 14  8 14  8 14  8 14  8 14  8 14  8  8 32
     64 14  8 26 14 20  8 26 14 20  8 26  8 20 14 26  8 20 14 26  8 20 14
KTEX  0  0  1  0  1  1  1  1  1  1  1  1  1  1  1  1  1  1  1  1  1  1  1  1  1
      1  1  1  1  1  1  1  1  1  1  1  1  1  1  1  1  1  1  1  1  1  1  1  1  1
      1  1  1  1  1  1  1  1  1  1  1  1  1  1  1  1  1  1  1  1  1  1  1
C -----------------------------------------------------------------------------
C               FILE: THYBRIDGE.DBM - Thyristor Bridge Rectifier
C                 This file contains $PARAMETER expressions
C -----------------------------------------------------------------------------
$ERASE
/REQUEST
C ----- Calculation of snubber circuit parameters
$PARAMETER
MVA___ = KVA___ / 1000. $$
INT___R_SNUB = ( KVOLTS**2. / MVA___ ) * 100.
INT___C_SNUB = 1. / ( KVOLTS**2. / MVA___ )
BLANK
C       1         2         3         4         5         6         7         8
C 345678901234567890123456789012345678901234567890123456789012345678901234567890
C ----- Thyristors
/SWITCH
11VGENIAPOSI_A                                                GATE_1          0
11VGENIBPOSI_B                                                GATE_3          0
11VGENICPOSI_C                                                GATE_5          0
11NEGI_AVGENOA                                                GATE_4          0
11NEGI_BVGENOB                                                GATE_6          0
11NEGI_CVGENOC                                                GATE_2          0
C ----- Connectivity resistors
/BRANCH
  VGEN_AVGENIA          0.0001
  VGEN_BVGENIB          0.0001
  VGEN_CVGENIC          0.0001
  VGENOAVGEN_A          0.0001
  VGENOBVGEN_B          0.0001
  VGENOCVGEN_C          0.0001
  POSI_APOS___          0.0001
  POSI_BPOS___          0.0001
  POSI_CPOS___          0.0001
  NEG___NEGI_A          0.0001
  NEG___NEGI_B          0.0001
  NEG___NEGI_C          0.0001
  POS___                1.0E+6
  NEG___                1.0E+6
C ----- Snubber circuits
$VINTAGE,1
  POSI_AVGENIA              INT___R_SNUB                  INT___C_SNUB
```

```
POSI_BVGENIBPOSI_AVGENIA
POSI_CVGENICPOSI_AVGENIA
VGENOANEGI_APOSI_AVGENIA
VGENOBNEGI_BPOSI_AVGENIA
VGENOCNEGI_CPOSI_AVGENIA
$VINTAGE,0
$EOF   User-supplied header cards follow.        06-Feb-19  19:40:07
ARG, VGEN_ , POS___, NEG___, GATE_ , KVOLTS, KVA___
NUM, KVOLTS, KVA___
DUM, INT___
DUM, VGENIA, VGENIB, VGENIC, VGENOA, VGENOB, VGENOC,
DUM, POSI_A, POSI_B, POSI_C, NEGI_A, NEGI_B, NEGI_C
```

This file is then inserted in an ATPDraw 'User Specified' window in which 'Data' and 'Nodes' have to be defined. The icon and the main window for this module could be those shown in Figure 9.2.

Every 'User Defined' model has a main window (i.e. that shown in Figure 9.2) and two additional windows in which the model is defined (see Figure 9.3):

- a 'Data' window with the template of the data (i.e. the parameters that define the implemented model) that have to be specified;
- a 'Nodes' window in which the icon connectivity is defined (note that the locations of the nodes that connect the implemented model to the rest of the circuit must be selected by the user).

The controlled rectifier model is based on the ideal thryristor model (i.e. the controlled Type 11 switch) available in ATP. As shown in Figure 9.2, the values to be specified are the rated voltage (phase-to-phase rms voltage value) and the rated power; they are internally used to estimate the snubber parameters. As for the connectivity, note that there are three nodes for connecting the model to the circuit and one node for connecting the thyristor gates to the control section.

Figure 9.4 shows the final circuit in which the power and the control parts of the controlled rectifier have been replaced by their corresponding *user-defined* modules.

Figure 9.5 shows some simulation results that have been obtained with the following values:

- High voltage power system: Rated voltage = 25 kV; Frequency = 50 Hz
- Step-down transformer: Rated voltages = 25/0.4 kV; Rated power = 1 MVA; YnD
- Controlled rectifier: Rated voltage = 0.4 kV; Rated power = 5 kVA
- Rectifier load: $R = 4\,\Omega, L = 1\,\text{mH}, C = 200\,\mu\text{F}$
- Control phase angle: 50°.

For more details about DATA BASE MODULE see Section XIX-F of the Rule Book [4].

9.2.3 Application of MODELS

A very similar procedure may be followed to implement a module that can perform some task for which capabilities of MODELS language are useful. One of these tasks might be the measurement of a system variable. Devices to perform a Fourier analysis of highly distorted variables can be very useful since they are frequently needed.

This section shows how to implement a single-phase device for measuring the rms value of a system variable (e.g. current, voltage). Remember that there is a TACS device that can fulfil this task (see Chapter 4), so the new module is not strictly necessary.

Figure 9.6 displays the diagram that could be needed for measuring currents and voltages with the new (MODELS-based) device when simulating the controlled rectifier implemented in the previous section. Just compare this figure with Figure 9.4.

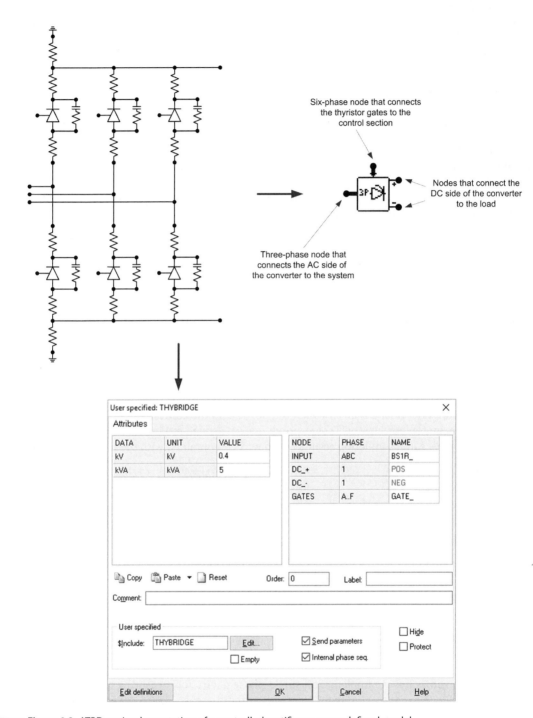

Figure 9.2 ATPDraw implementation of a controlled rectifier as a user defined model.

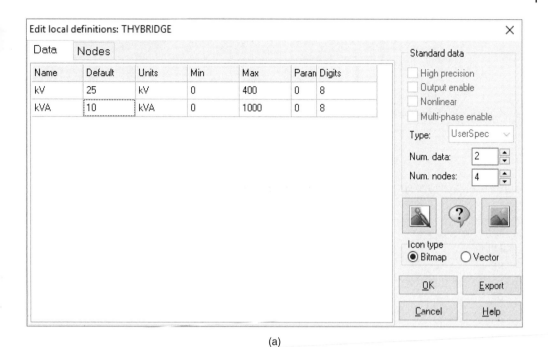

(a)

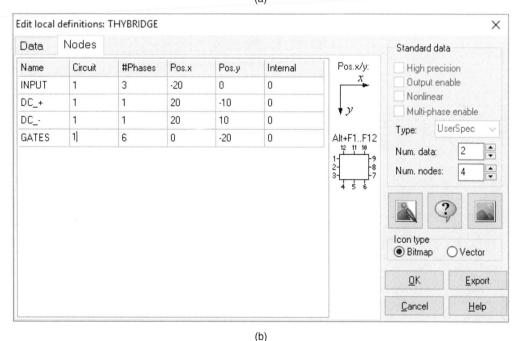

(b)

Figure 9.3 Menu windows of the controlled rectifier model. (a) 'Data' window, (b) 'Nodes' window.

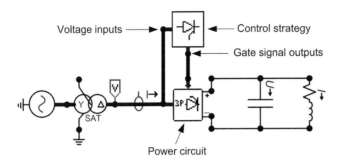

Figure 9.4 ATPDraw implementation of the test system with custom-made models.

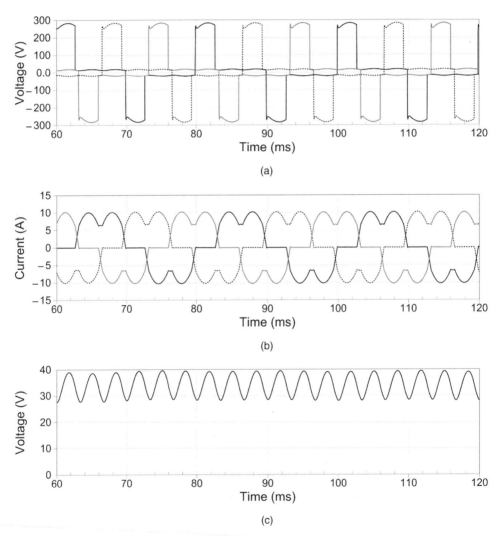

Figure 9.5 Test of the user-defined controlled rectifier – simulation results. (a) Voltages at the rectifier terminals – AC side, (b) Rectifier currents – AC side, (c) DC voltage.

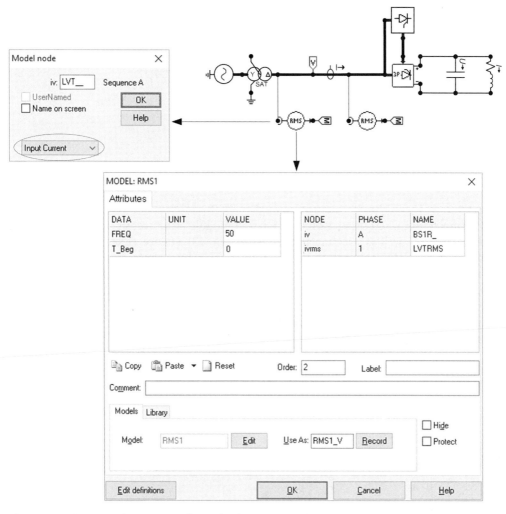

Figure 9.6 ATPDraw implementation of a single-phase MODELS-based measurement device.

The implemented MODELS section is as follows:

```
MODEL RMS1
  INPUT    iv          -- Input variable (current,
                          voltages, TACS variable)
  OUTPUT   ivrms        -- RMS value of the input variable
  DATA     Freq         -- Power system frequency (Hz)
           T_Beg        -- Time at which the device is
                          activated (s)
  VAR      iv2, ivrms, NSCICLO, SPER, aux, TIT
  HISTORY  ivrms {DFLT:0}
           INTEGRAL(iv2) {DFLT:0}
  INIT
     SPER:=RECIP(Freq)/2  NSCICLO:=1  TIT:=OFF
  ENDINIT
```

```
        EXEC
          IF t < T_Beg THEN iv2:=0 ELSE
            iv2:=iv**2
            aux:=INTEGRAL(iv2)
            IF (t-T_Beg) >= NSCICLO*SPER THEN
              ivrms:=sqrt(aux/SPER)
              INTEGRAL(iv2):=0
              NSCICLO:=NSCICLO+1
            ENDIF
          ENDIF
        ENDEXEC
      ENDMODEL
```

Figure 9.6 also shows the window used to implement this model. As a user-specified module, the implementation when using a MODELS-based module also involves defining input data and connectivity, as well as selecting an icon. As shown in the figure, when implementing the device, it is important to select the type of input variable whose rms value is to be measured. The two devices are used to measure the rms values of the ac-side current and voltage of the controlled rectifier. Since the operation of the system is balanced, measuring rms values of one phase will suffice.

Figure 9.7 shows the results that correspond to the same case simulated in the previous section. The oscillograms for any other system variable would be those shown in Figure 9.5. Since the two devices are activated at $t = 0$, one can note that the algorithm implemented in MODELS for measuring rms values needs half a cycle (i.e. 10 ms for a frequency of 50 Hz) to

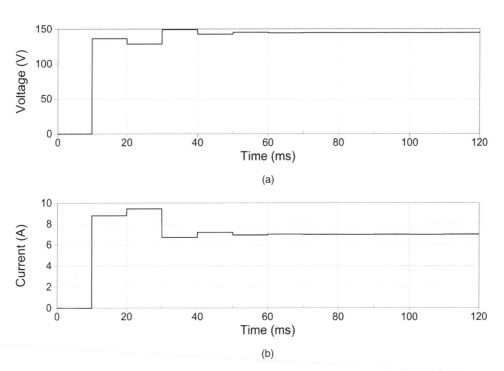

(a)

(b)

Figure 9.7 Test of the single-phase MODELS-based measurement device – simulation results. (a) Rectifier voltage – AC side, (b) Rectifier current – AC side.

estimate the rms value. This means that the measured value is updated every 10 ms, as shown in Figure 9.7.

9.2.4 The Group Option

The application of the Group option is illustrated with the same test system used in the previous section but replacing the controlled rectifier by an uncontrolled rectifier. Figure 9.8 shows the configuration of the ATPDraw-implemented converter that will be substituted by a compressed group. The task to be carried out with this example consists of substituting the rectifier bridge shown in Figure 9.8 by an icon in which the entire rectifier will be compressed: the icon implemented here will have some connection nodes (e.g. those shown in Figure 9.8) but no parameter/data to be specified.

The procedure that can lead to a grouped rectifier might be the following one:

1) Name all those elements one of whose terminals will be a connection node. The upper part of Figure 9.9 shows the names given to these elements (i.e. $R1$, $R2$, $R3$, RU, RD).
2) Apply the option Compress under the menu Edit. A window such as that shown in the middle part of Figure 9.9 will show up. Since no parameter is to be specified once the group has been created, only connection nodes are to be defined. The window shows the names given to the connection nodes of the new group. Note that at the right bottom corner of the window there is a figure that can be used to locate all external nodes.
3) Name the group. The selected name in this case is 'BRIDGE', as shown in the figure.
4) Once the group has been created, users can design a specific icon. The icon designed for this group is shown at the bottom of Figure 9.9.

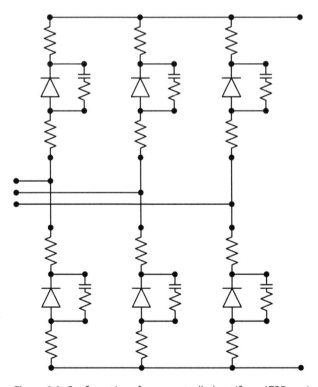

Figure 9.8 Configuration of an uncontrolled rectifier – ATPDraw implementation.

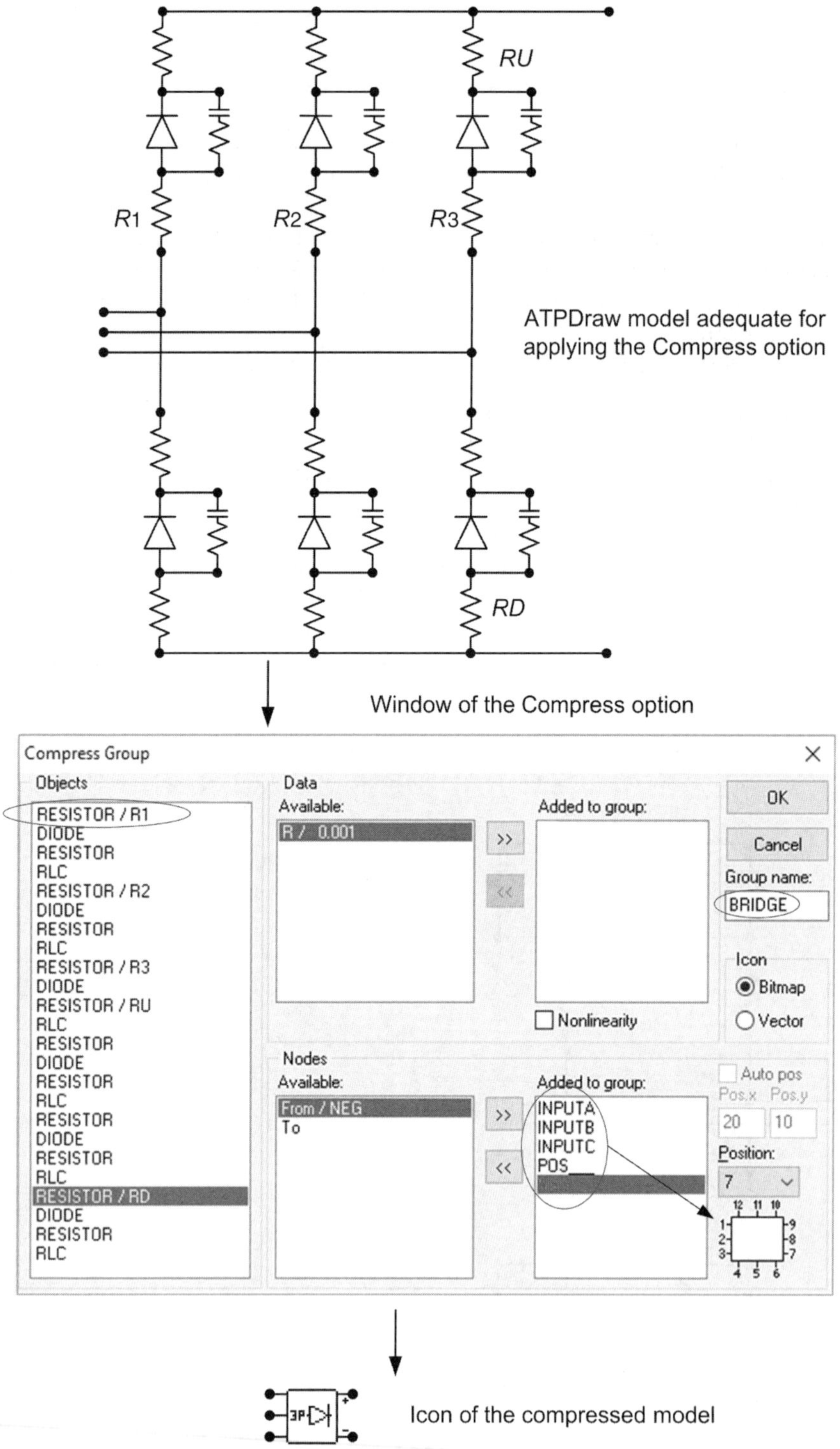

ATPDraw model adequate for applying the Compress option

Window of the Compress option

Icon of the compressed model

Figure 9.9 ATDraw implementation of an uncontrolled rectifier using the Group/Compress option.

When this group is opened either to change the name, the icon or the location of external nodes, the user will see the window shown in Figure 9.10. The figure includes the window for local definitions. The diagram implemented in ATPDraw for testing the new custom-made module is similar to the diagram of the previous test system (see Figure 9.11); however, the rectifier is now uncontrolled and no TACS section is needed. Figure 9.12 shows some results that correspond to those obtained with the system tested in previous sections and a phase angle equal to 0.

9.3 Application of the ATP to Power Quality Studies

9.3.1 Introduction

Disturbances have been present in power systems since their inception. Although most power equipment can operate with relatively wide variations in voltage, current, and frequency, the proliferation of electronic equipment sensitive to these variations has increased the concern of utilities and their customers. The term *power quality* has been frequently used to describe disturbances which can lead to equipment misoperation [6–9]. Power quality can have a different definition to different people. To users with sensitive electronic equipment the main concern is the voltage waveform distortion. To industrial users with insensitive equipment, the main concern is the continuity of service. A broad definition of power quality would also include voltage unbalance in multiphase networks, and dielectric selection in equipment [7].

Power disturbances can have a very adverse effect on costumer equipment; many of these effects are well known and documented [10–12]. Standards are being continuously updated and developed to define, characterize and evaluate power quality disturbances, and to provide measures for mitigating their effects [13]. Although the causes can be originated in either the power system or customer locations, users with sensitive equipment can install their own uninterruptible power supply (UPS), which provide ride-trough capabilities, or custom power devices, which provide utility solutions to power quality problems [14–16].

Digital simulation is a very effective way of performing power quality studies. Several techniques have been developed over the last decades to simulate and analyse power quality problems using a digital computer; these techniques are based on transformed methods, frequency-domain methods [17–20], and time-domain methods [21]. A time-domain tool like ATP can be used to predict the waveform and magnitude of power disturbances, analyse the influence of system and component parameters on these disturbances, validate modelling of power system components and equipment, or test and design mitigation techniques.

9.3.2 Power Quality Issues

Phenomena that can cause power-quality disturbances are many, and they can be originated from several sources. An important effort has been made by the power industry and international associations to classify and characterize these phenomena. A first classification might distinguish between steady-state and transient problems. Variations of the first group are characterized by a waveform distortion (i.e. a deviation from the ideal waveform at power frequency); the most common power quality problems included in this group are harmonics, interharmonics, notches, and noise. The term transient is generally used to denote a transition from one steady state to another. Transient power quality problems are characterized by a frequency spectrum and duration. They can be classified into two subcategories: impulsive and oscillatory. There are, however, other power quality variations, such as dips and swells, in which transient events are involved, although they are usually characterized as steady-state

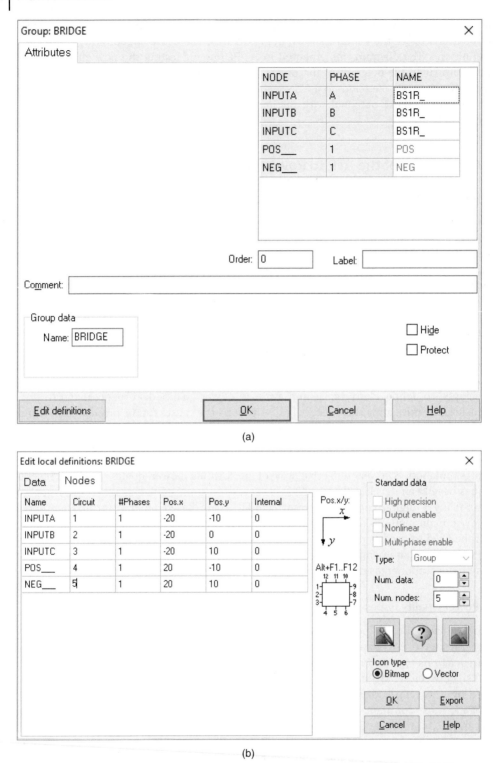

Figure 9.10 Custom-made uncontrolled rectifier model – group option. (a) Main window of a compressed uncontrolled rectifier, (b) 'Nodes' window of a compressed uncontrolled rectifier.

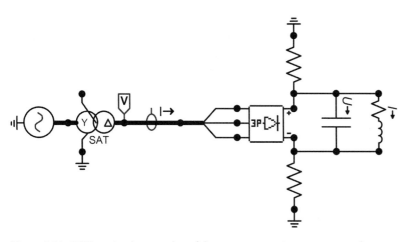

Figure 9.11 ATPDraw implementation of the test system using a compressed custom-made model.

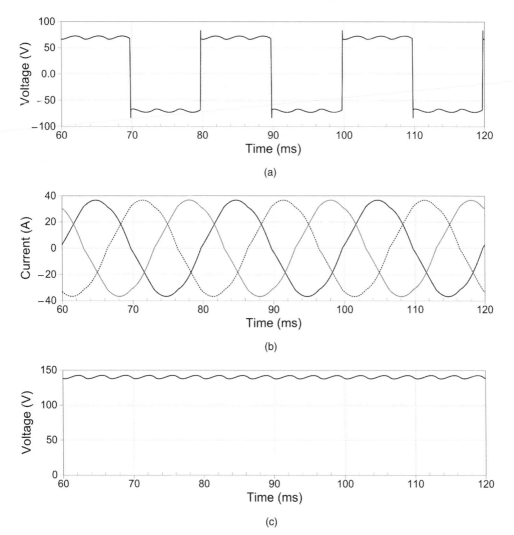

Figure 9.12 Test of the compressed uncontrolled rectifier model – simulation results. (a) Voltages at the rectifier terminal – phase a, (b) Rectifier currents – AC side, (c) DC voltage.

Table 9.1 Power quality categories and characterization.

Type of distortion	Duration	Method of characterizing
Harmonics	Steady state	Harmonic spectrum
		Harmonic distortion
Phase-unbalance	Steady state	Unbalance factor
Interruptions	------------	Duration
Notches	Steady state	Duration
		Magnitude
Voltage flicker	Steady state	Variation magnitude
		Frequency of occurrence
		Modulation frequency
Dips/swells	Transient	Magnitude
		Duration
		Rms vs. time
Oscillatory transients	Transient	Waveform
		Peak magnitude
		Frequency range
Impulsive transients	Transient	Rise time
		Peak magnitude
		Duration
Noise	Steady state/transient	Magnitude
		Frequency spectrum

variations. A very complete set of definitions and descriptions of power quality variations can be found in [7], which also includes a classification of power quality disturbances taking into account frequency spectrum, duration, and voltage magnitude. Table 9.1 shows a summary of some categories and their characterization.

9.3.3 Simulation of Power Quality Problems

Digital simulation of power quality issues can be useful to understand how disturbances propagate into the network, determine waveform distortion caused by different sources, quantify the impact of some disturbances, test mitigation techniques, or design power conditioning solutions. Another important possibility is the application with educational purposes.

Taking into account the nature of power quality disturbances, their sources and the behaviour of the power network, the list of basic capabilities that are required in simulation tools may include:

1) *Frequency-domain simulation* to analyse harmonic propagation, identify resonance conditions, and design filter banks.
2) *Time-domain simulation* to simulate those variations of transient nature and determine their impact.
3) *Accurate modelling* to represent the power network and disturbance sources, with capabilities to model nonlinear and frequency-dependent behaviour.
4) *Multi-level modelling* to allow users the representation of different parts of a system using different approaches. For instance, a detailed model is usually needed for static converters and rotating machines in transient analysis, but a less complex representation can be used in frequency-domain simulations.

5) *Post-processing capabilities* to display simulation results and quantify the impact of power quality variations. Several indices are currently in use to quantify in a single number a waveform distortion.

6) *Interface to external programs* to take advantage of capabilities available in other software tools.

9.3.4 Power Quality Studies

The applications of the ATP package to power quality studies are many. Five major areas are analysed in this section [22–24]:

a) Modelling of power system components and sources of power quality problems.
b) Simulation of the effects of a power quality disturbance, using either a frequency- or a time-domain simulation.
c) Analysis and design of mitigation techniques.
d) Postprocessing of simulated results using built-in options or an interface to an external tool.
e) Development of new simulation tools by integrating capabilities from other several tools to obtain a custom-made package.

Table 9.2 shows a list of tasks in power quality studies which can be accomplished by the ATP package, and some of the most usual applications.

Table 9.2 ATP capabilities for power quality studies.

Task	Capability	Applications
Pre-processing	ATPDraw	Input file edition
Frequency-domain simulation	HARMONIC FREQUENCY SCAN	Harmonic analysis
	FREQUENCY SCAN	Resonance analysis
		Filter bank design
Time-domain simulation	TPBIG	Steady-state initialization
		Transient analysis (switching, dips, and swells, flicker)
		Harmonic reduction (active filter, wave-shaping techniques)
Statistical simulation	Statistical switches	Monte Carlo analysis
	MODELS	
Development of user-defined models	MODELS	Semiconductor model
	DATA BASE MODULE	Arc furnace model
	ATPDraw capabilities	Harmonic source model
Development of library modules	DATA BASE MODULE	Creation of library modules
Development of simulation tools	ATPDRAW	User-designed software tool
	DATA BASE MODULE	
Plotting	Calcomp plot	Visual analysis
	Plotting tools	Validation
Postprocessing	Postprocess plot file	Index calculation
	TACS/MODELS	Harmonic source model

A) *Modelling.* Two tasks related to this topic can be usually performed: the choice of the most adequate representation for every component and the development of custom-made models, using capabilities available in ATP.

   1) *Modelling guidelines.* They were discussed in Chapter 2. Several documents are currently available to consult modelling guidelines of power components for digital simulations [25–27]. The guidelines are valid for both frequency-domain and time-domain simulations.

   2) *Development of custom-made models.* Many component models needed in power quality studies are not available in ATP (e.g. frequency-dependent models of loads and rotating machines for harmonics studies, an arc furnace model for flicker studies). As shown in the previous section, ATP capabilities can be used to create and implement custom-made models.

B) *Harmonics analysis.* Harmonics studies are aimed at investigating the generation and propagation of harmonic components generated by nonlinear loads, switched loads and other harmonic sources. They are used to quantify the distortion in voltage and current waveforms, determine resonant conditions, and analyse mitigation techniques [19, 24]. Several methodologies have been proposed to perform this type of analyses; they are based on power-flow studies, current injection analysis, transform methods, and probabilistic methods. Several approaches using ATP capabilities can be used to perform harmonics analysis. Two main aspects are to be considered: the representation of the harmonic source and the calculation method.

   1) *Representation of harmonic sources.* There are at least three approaches for representing harmonic sources using a transients program [28]: (i) a voltage/current injection through an equivalent source composed of the harmonic spectrum of the load, (ii) a switching function, used to represent the terminal characteristics of a converter, which appears as a current source seen from the power network; (iii) a detailed representation of the static converter and the associated control strategy.

   2) *Calculation method.* Harmonic distortion is a steady-state power quality variation. Two methods can be used to quantify this distortion by means of ATP:

      – *Steady-state solution.* ATP can obtain the initial steady state of a linear system excited by sinusoidal sources only; a true nonlinear steady state in systems with complex harmonic sources is not available in ATP (i.e. steady-state initialization cannot be applied to systems with power electronics converters for which a detailed representation of the power circuit and the control strategy is used). ATP can also perform a Frequency Scan to obtain the driving point impedance at a particular node versus frequency; it is used to detect resonance conditions and design filter banks. Frequency Scan can only be used with linear systems.

      – *Transient solution.* Harmonic distortion can also be calculated during the ATP transient loop. The actual steady state in a power system can be obtained by starting the system from standstill or by obtaining the steady state of a linear system, where nonlinear loads are ignored or linearized. Using both approaches, the simulation has to be carried out to let the program calculate the actual steady state.

   Table 9.3 shows a summary of the approaches that can be used in harmonics analysis.

C) *Transient analysis.* Several transient categories can affect power quality; they are classified into two groups according to the current and voltage waveshapes: impulsive and oscillatory. Transient events are usually associated to lightning and power equipment switching; however, other power quality categories, such as dips or flicker, are also associated to transient events. A short summary on the most common power quality categories for which a tool such as ATP can be used is presented below.

Table 9.3 Harmonic analysis using ATP.

Calculation method		Harmonic source representation	Limitations
Steady-state solution	True-nonlinear initialization	Nonlinear components	Not yet implemented
	Frequency scan	Sinusoidal source of variable frequency	Only one source Only linear systems
	Harmonic frequency scan	Sinusoidal source of variable frequency	Only one source per harmonic Only linear systems
Transient solution	From a standstill system	All approaches can be used	Slow convergence in systems with light damping
	From a linearized steady state	All approaches can be used	Some initialization could be needed for static converters

1) The calculation and mitigation of overvoltages were the primary goal of ATP; see Chapter 6. An important aspect in these studies is the choice of the most adequate model for each network component taking into account the frequency spectrum of the transient disturbance [25, 26]. The simulation of transients caused by capacitor switching is one of the most typical applications [29–33].

2) Voltage dips are usually originated by faults and motor starting [34–36]. ATP can be used to quantify dip magnitudes and evaluate ride-through capabilities [37]. A detailed representation can be used for all components involved in a voltage dip study.

3) Voltage fluctuations known as voltage flicker are associated to lighting effects [8]. They can be classified into two categories: cyclic (produced by arc furnaces) and noncyclic (produced by starting of large motors) [38]. Transients programs have been extensively used to simulate voltage flicker sources, mainly arc furnaces, and quantify their impact [39–43].

4) Due to the random nature of many transient events, a statistical approach can be useful to evaluate their impact and design mitigation techniques. The statistical evaluation of voltage dips in distribution systems using ATP was presented in [44]. See also [45–47].

5) A high percentage of power quality problems are due to improper grounding [8, 11]. A grounding system must provide a low-impedance path between the neutral of an electric system and the earth. This low impedance path can be needed for many reasons [48]. Transients originated in the power system or at the customer facility will travel through the grounding system and reach sensitive equipment. An ATP model can be very useful to optimize the design of a grounding system and analyse its behaviour; see [49, 50].

D) *Mitigation techniques.* The analysis and design of devices for power quality enhancement using ATP is one of the most interesting areas. Techniques for mitigating power quality problems in distribution networks use passive and active filters [51–53], UPS [8], static var compensation (SVC) [16], superconducting magnetic energy storage (SMES) [54], or custom power devices [14, 15]. Most of these techniques are based on power electronics devices. Guidelines for modelling power electronics systems using ATP were discussed in Chapter 8. Electromagnetic Transients Program (EMTP)-like programs have been routinely used to simulate and design passive filters [55, 56], active filters [57], and control strategies based on current waveshaping [58, 59]; to analyse the performance of micro-SMES systems for protection of customer facilities [60]; or to simulate and design custom power devices

[61–64]. The solid state-transformer (SST) offers enhanced power quality performance, fast voltage control, reactive power compensation or reactive power control at both primary and secondary sides, dc and high-frequency ac power supply [65]; see also Chapter 8.

E) *Postprocessing*. Some ATP features can be applied to analyse simulation results, calculate power quality indexes, and test algorithms aimed at identifying disturbances. Postprocess Plot File is a capability that allows users to process simulation results from a previous run inside a control section, using either TACS or MODELS [66]. A foreign routine edited inside a MODELS section using a high-level language (e.g. C) can be used to establish a link between ATP and MATLAB during run time; processing and graphical capabilities available in MATLAB can be then used for analysing power quality disturbances [67, 68].

F) *Development of simulation tools*. Capabilities available in other software packages can also be used to develop custom-made simulation tools. Reference [69] describes the design of a MATLAB-EMTP based tool for power quality analysis in distribution systems; the new tool can perform input file assembly, simulation running, and postprocessing of simulation results. A similar approach can be used with ATP to develop new custom-made tools for power quality analysis.

9.4   Custom-Made Modules for Power Quality Studies

A library of modules has been developed to facilitate the usage of the ATP package in power-quality studies; namely, for carrying out harmonic analyses and voltage dip studies. The library allows users to represent power-quality disturbances (harmonics, dips), measure variables affected by the disturbances, or analyse the performance of traditional (e.g. passive filters) and advanced mitigation techniques (e.g. active filters, custom power devices).

Table 9.4 shows a selected list of the implemented modules. Although the library is mainly an educational tool, many modules could be used in advanced studies. Since it is intended for power-quality studies, the modules can cover most topics related to this area; that is, the representation of power quality disturbances, or measuring and mitigation devices. Some modules have been added to facilitate some studies; for instance, Frequency Scan is a task that can be easily performed by using ATPDraw capabilities, the implemented module is mostly aimed at facilitating the task.

9.5   Case Studies

9.5.1   Overview

The main goal of this section is to present the ATP as an adequate tool for power quality studies. The section covers the application of ATP to the analysis of two power quality categories: harmonics and voltage dips. The cases study the causes, propagation and effects of harmonics and voltage dips, and how to mitigate both types of power quality categories. Although the case studies do not cover all scenarios related to harmonics and voltage dips, they will serve to illustrate how the creation of custom-made modules can help ATP users to simulate and analyse these and other power quality problems.

9.5.2   Harmonics Analysis

The case studies presented in this section illustrate how to use the ATP capabilities in harmonic analysis (i.e. generation and propagation of harmonics, detection of resonance problems, application of filters for compensation of harmonics). The cases show the application of

Table 9.4 Custom-made modules for power quality studies.

Type of applications	Module	
Source of power quality disturbances	3 PHASE HARMONIC SOURCE	Harmonic current injection
	3P DIODE RECTIFIER	Uncontrolled rectifier
	3P	Three-phase controlled bridge rectifier
Measuring devices	S P Q	Measurement of active and reactive powers
	SMD	Measurement of voltage dip magnitude and phase jump
Harmonic analysis	FOURIER	Fourier analysis
	F.S.	Frequency Scan
Voltage dip studies	V-DEP	Voltage-dependent load with sensitivity curve
		Overhead line model
Mitigation devices	3P f	Three-phase passive filter
	ACTIVE FILTER	Active filter converter
	DVR	Dynamic voltage restorer (DVR)

custom-made modules for generating harmonic waveforms, carrying out the Fourier analysis of test system variables (i.e. node voltages, branch currents, and voltages), or compensating harmonic currents by means of both passive and active filters.

9.5.2.1 Case Study 1: Generation of Harmonic Waveforms

Several custom-made modules based on MODELS language have been created to generate harmonic waveforms. The modules are three-phase and can be used as either voltage or current source. A second module is attached to define the type of source (voltage or current).

Figure 9.13 shows the icon, the main window, and the MODELS code implemented in one of these modules. The user has to specify the fundamental frequency, and the magnitude and phase for each harmonic. The waveform can include up to 15 components. It is possible to parallel more than one module if more than 15 harmonic components are needed to fully define the (voltage or current) source; the option 'Order' has been implemented in each module to increase the harmonic components as much as required.

Figure 9.14 shows two small test systems in which this module has been used to represent a three-phase harmonic current source. The load is a simple three-phase 1-Ω resistor. The figure includes an additional custom-made module for Fourier analysis; this module asks for a harmonic analysis of the specified variable using one of the following options: '4' for node voltages; '8' for branch voltages; '9' for branch currents. Table 9.5 lists the harmonic contents. Figure 9.14 also displays some simulation results corresponding to two different scenarios. In the first scenario, a single module is used to inject a three-phase current, 13 being the highest harmonic number; in the second scenario the additional module is used to add the 17th and 19th harmonics.

The results provided by the ATP Fourier analysis of the second scenario are shown below.

Harmonic number	Cosine coefficient	Sine coefficient	Complex amplitude	Fraction of fundamental
0	-1.05943323847E-01	0.00000000000E+00	1.05943323847E-01	0.001060689
1	9.98813114146E+01	2.27854857324E-01	9.98815713119E+01	1.000000000
2	-5.00113749394E-02	1.69138908105E-02	5.27941031336E-02	0.000528567
3	-8.93069321532E-02	2.97470503560E-02	9.41308405120E-02	0.000942425
4	-1.35036357135E-01	5.73947321520E-02	1.46727546926E-01	0.001469015
5	2.17093886261E+01	-1.22296951695E+01	2.49171225959E+01	0.249466666
6	-3.36879255637E-02	2.26756769509E-02	4.06086524520E-02	0.000406568
7	9.78908101842E+00	-1.13426748590E+01	1.49827360700E+01	0.150005010
8	-1.85123744872E-03	-3.70524809875E-02	3.70986984599E-02	0.000371427
9	-3.79344395721E-02	6.14274899877E-03	3.84285710235E-02	0.000384741
10	-6.78882648139E-02	5.46784923764E-02	8.71696852581E-02	0.000872730
11	3.58772205028E+00	-9.30733226298E+00	9.97487761146E+00	0.099867047
12	-6.00701867084E-02	-3.85433427962E-02	7.13723798474E-02	0.000714570
13	3.57808596270E+00	-3.45840278142E+00	4.97626857746E+00	0.049821689
14	-5.23322426454E-02	-3.08641300823E-02	6.07557252120E-02	0.000608278
15	-1.07810526216E-01	9.54916976242E-03	1.08232602325E-01	0.001083609
16	-2.20811923785E-01	7.39831336455E-02	2.32876382979E-01	0.002331525
17	1.76281612911E+01	-9.35755672947E+00	1.99578540542E+01	0.199815179
18	7.84556935333E-02	-8.61298814632E-02	1.16506018423E-01	0.001166442
19	9.32512259705E+00	-3.92311156333E+00	1.01167542121E+01	0.101287496
20	1.72575402719E-01	-1.07046198120E-01	2.03079191833E-01	0.002033200
21	1.01302765073E-01	-7.40180829005E-02	1.25462850310E-01	0.001256116
22	7.27632927097E-02	-5.97166167254E-02	9.41306064948E-02	0.000942422
23	5.66698990649E-02	-5.11254980302E-02	7.63236137042E-02	0.000764141
24	4.61647170942E-02	-4.52015430607E-02	6.46092918972E-02	0.000646859
25	3.87151266798E-02	-4.07874134724E-02	5.62358793974E-02	0.000563026
26	3.31391473341E-02	-3.73298421342E-02	4.99171333291E-02	0.000499763
27	2.88036561299E-02	-3.45243785265E-02	4.49620208520E-02	0.000450153
28	2.53348182719E-02	-3.21884489644E-02	4.09627790025E-02	0.000410113
29	2.24973409559E-02	-3.02032317954E-02	3.76611943646E-02	0.000377058

Derived from table: 1) RMS value = 7.56532960E+01   2) THD = 3.83924579E+01 %

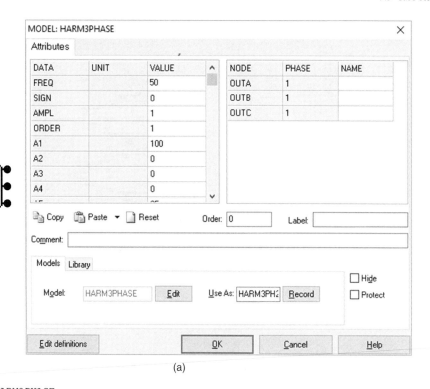

(a)

```
MODEL HARM3PHASE
DATA    FREQ, SIGN, AMPL, ORDER
        A1, A2, A3, A4, A5, A6, A7, A8, A9, A10, A11, A12, A13, A14, A15
        P1, P2, P3, P4, P5, P6, P7, P8, P9, P10, P11, P12, P13, P14, P15
OUTPUT OUTA, OUTB, OUTC
VAR     OUTA, OUTB, OUTC, WTA, WTB, WTC, A[1..15], P[1..15]
INIT    OUTA:=0  OUTB:=0  OUTC:=0
        A[1..15]:=[A1,A2,A3,A4,A5,A6,A7,A8,A9,A10,A11,A12,A13,A14,A15]
        P[1..15]:=[P1,P2,P3,P4,P5,P6,P7,P8,P9,P10,P11,P12,P13,P14,P15]
ENDINIT
EXEC
 WTA:=2*PI*FREQ*T
 WTB:=WTA-PI*(120/180)
 WTC:=WTB-PI*(120/180)
 OUTA:=0  OUTB:=0  OUTC:=0
 FOR k:=1 TO 15 DO
      OUTA:=OUTA + A[k]*COS((15*(ORDER-1)+k)*WTA + P[k]*PI/180)
      OUTB:=OUTB + A[k]*COS((15*(ORDER-1)+k)*WTB + P[k]*PI/180)
      OUTC:=OUTC + A[k]*COS((15*(ORDER-1)+k)*WTC + P[k]*PI/180)
 ENDFOR
 IF SIGN = 0 THEN
   OUTA:=+OUTA*AMPL    OUTB:=+OUTB*AMPL    OUTC:=+OUTC*AMPL
   ELSE
   OUTA:=-OUTA*AMPL    OUTB:=-OUTB*AMPL    OUTC:=-OUTC*AMPL
 ENDIF
ENDEXEC
ENDMODEL
```

(b)

Figure 9.13 Generation of harmonic waveforms. (a) Icon and main window, (b) MODELS code for generation of harmonic waveforms.

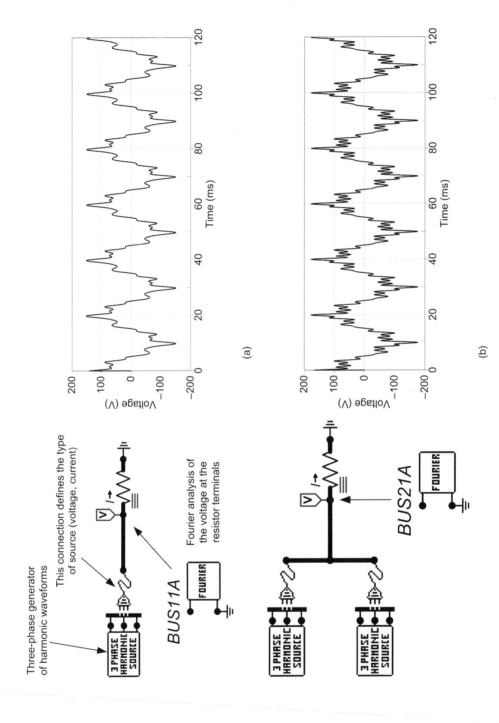

Figure 9.14 Case Study 1: generation of harmonic waveforms – test system implementation and simulation results. (a) First scenario, (b) Second scenario.

Table 9.5 Case Study 1: harmonic content of the nonlinear load.

Order	Magnitude (A)	Phase (°)
1	100	0
5	25	30
7	15	50
11	10	70
13	5	45
17	20	30
19	10	25

Figure 9.15 Case Study 2: diagram of the test system.

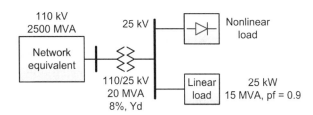

These results can be used to check the accuracy of the Fourier routine implemented in ATP: they should be the same that were specified in the input window of the harmonic waveform generator (see Table 9.5).

9.5.2.2 Case Study 2: Harmonic Resonance

Common means for preventing resonance are the use of filters, or the application of technologies that limit the harmonics at the source. Figure 9.15 shows the diagram of the 60-Hz test system: a linear load, paralleled by a nonlinear load represented by a source of harmonic currents, is being fed from the lowest voltage side of a step-down transformer. To improve the load power factor, a capacitor bank will be installed at the point of common coupling (PCC). After installing the capacitor bank, a resonance problem may occur due to the presence of harmonic currents injected by the nonlinear load. This problem can be predicted by performing a frequency scan of the system once the capacitor bank has been placed. Figure 9.16 shows the diagram of the original test system as implemented in ATPDraw. Note that a new MODELS-based module for generating harmonic waveforms is used now; this module can specify voltages and currents

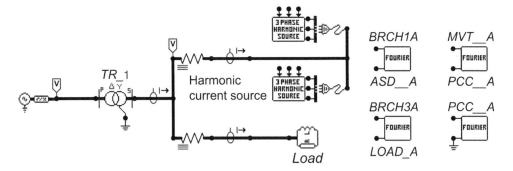

Figure 9.16 Case Study 2: ATPDraw implementation of the test system.

taking a system node as reference. Figure 9.17 shows the icon, main window, and MODELS code of the new module. Table 9.6 lists the harmonic content of the current source that represents the nonlinear load: highly distorted currents with a total harmonic distortion (THD) above 100%. Figure 9.18 depicts some simulation results: it is evident that the currents injected by the nonlinear load have a significant effect on the system voltages and currents.

Figures 9.19 and 9.20 show the diagram of the test system after installing a 3 Mvar capacitor bank and some simulation results corresponding to the new system configuration. It is evident the impact that the installation of the capacitor bank has on the capacitor current: once the nonlinear load currents have been activated the capacitor current increases beyond expected.

A frequency scan of the system seen from the PCC after installing the capacitor bank can detect resonance problems. The diagram of the test system is now that depicted in Figure 9.21. The plot shown in Figure 9.22 confirms that a resonance can occur at a frequency close to the 5th harmonic, which is present in the current injected by the nonlinear load. The nonlinear currents will flow to the HV system through the transformer and divide into other components connected to the PCC, depending on the harmonic impedances.

This problem can be solved by installing a filter tuned to the 5th harmonic. The new frequency scan is also presented in Figure 9.22, and compared to the frequency response obtained with the capacitor bank. Note that the application of a band-pass filter has not eliminated resonances, but now they occur at frequencies below the tuned frequency and correspond to points away from the load-generated harmonics. For a practical filter design, it is important to account for some variations in the tuned and resonant frequency with varied system switching conditions.

Figure 9.23 displays the diagram of the test system for time-domain simulations after replacing the capacitor with the passive filter. A custom-made module for representing three-phase passive filters is now used: the parameters to be specified are the power frequency, the rated voltage, the rated power and the harmonic order to which the filter is tuned. The new simulation

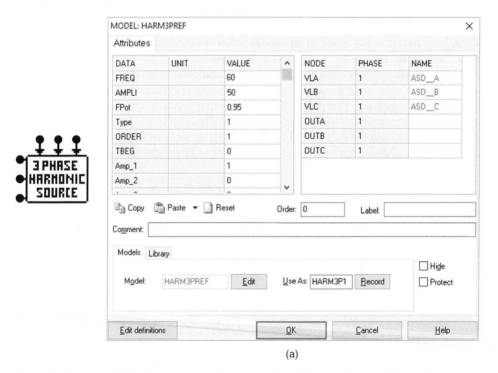

(a)

Figure 9.17 New module for harmonic waveform generation. (a) Icon and main window, (b) MODELS code.

```
MODEL HARM3PREF
INPUT  VLA,  VLB,  VLC
DATA   FREQ,  AMPLI,  FPot,  Type,  ORDER, TBEG
         Amp_1,  Amp_2,  Amp_3,  Amp_4,  Amp_5,  Amp_6,  Amp_7,  Amp_8,  Amp_9,
         Amp_10,Amp_11,Amp_12,Amp_13,Amp_14,Amp_15
          Pha_1,  Pha_2,  Pha_3,  Pha_4,  Pha_5,  Pha_6,  Pha_7,  Pha_8,  Pha_9,
          Pha_10,Pha_11,Pha_12,Pha_13,Pha_14,Pha_15
VAR    OUTA, OUTB, OUTC, WT, NWIN, OK, AMORT
         A[1..15], P[1..15], CoefA[1..3], CoefB[1..3], VLPh[1..3]
OUTPUT OUTA, OUTB, OUTC
INIT
  OUTA:=0   OUTB:=0   OUTC:=0   NWIN:=1   OK:=0
  CoefA[1..3]:=[0 0 0]   CoefB[1..3]:=[0 0 0]   VLPh[1..3] :=[0 0 0]
  A[1..15]:=[Amp_1, Amp_2, Amp_3, Amp_4, Amp_5, Amp_6, Amp_7, Amp_8, Amp_9,
             Amp_10,Amp_11,Amp_12,Amp_13,Amp_14,Amp_15]
  P[1..15]:=[Pha_1, Pha_2, Pha_3, Pha_4, Pha_5, Pha_6, Pha_7, Pha_8, Pha_9,
             Pha_10,Pha_11,Pha_12,Pha_13,Pha_14,Pha_15]
ENDINIT
EXEC
  IF T >= TBEG THEN
        IF (T-TBEG) >= (NWIN-1)*(1/(FREQ*256)) AND NWIN <= 256 AND OK = 0 THEN
            CoefA[1]:=CoefA[1]+VLA*COS((NWIN-1)*2*PI/256)/(256/2)
            CoefB[1]:=CoefB[1]-VLA*SIN((NWIN-1)*2*PI/256)/(256/2)
            CoefA[2]:=CoefA[2]+VLB*COS((NWIN-1)*2*PI/256)/(256/2)
            CoefB[2]:=CoefB[2]-VLB*SIN((NWIN-1)*2*PI/256)/(256/2)
            CoefA[3]:=CoefA[3]+VLC*COS((NWIN-1)*2*PI/256)/(256/2)
            CoefB[3]:=CoefB[3]-VLC*SIN((NWIN-1)*2*PI/256)/(256/2)
            NWIN:=NWIN+1
        ENDIF
        IF NWIN > 256 THEN
            VLPh[1..3]:=[atan2(CoefB[1],CoefA[1])
                 atan2(CoefB[2],CoefA[2])   atan2(CoefB[3],CoefA[3])]
            FOR K:=2 TO 15 DO
                P[K]:=(P[K]-P[1]*K)*PI/180-K*ACOS(FPot)
            ENDFOR
            P[1]:=-ACOS(FPot)
            OK:=1
        ENDIF
    IF OK = 1 THEN
        WT:=2*PI*FREQ*T
        AMORT:=AMPLI*(1-EXP(-(T-TBEG-1/FREQ)/0.01))
        OUTA:=0   OUTB:=0   OUTC:=0
        FOR k:=1 TO 15 DO
            OUTA:=OUTA+A[k]*COS((15*(ORDER-1)+k)*WT+P[k]+VLPh[1]*K)
            OUTB:=OUTB+A[k]*COS((15*(ORDER-1)+k)*WT+P[k]+VLPh[2]*K)
            OUTC:=OUTC+A[k]*COS((15*(ORDER-1)+k)*WT+P[k]+VLPh[3]*K)
        ENDFOR
        IF Type = 0 THEN
            OUTA:=+OUTA*AMORT   OUTB:=+OUTB*AMORT   OUTC:=+OUTC*AMORT
         ELSE
            OUTA:=-OUTA*AMORT   OUTB:=-OUTB*AMORT   OUTC:=-OUTC*AMORT
        ENDIF
     ENDIF
  ENDIF
 ENDIF
ENDEXEC
ENDMODEL
```

(b)

Figure 9.17 *(Continued)*

Table 9.6 Harmonic content of the nonlinear load current – Frequency = 60 Hz; Base current = 50 A.

Order	Magnitude (A)	Phase (°)
1	1	−13
5	0.765	118
7	0.627	−90
11	0.248	31
13	0.127	169
17	0.071	−163
19	0.084	−34
23	0.044	60
25	0.033	176

results are shown in Figure 9.24: the passive filter currents are much lower than those with the capacitor bank, the voltage at PCC and the linear load current are less distorted.

Table 9.7 provides a summary of results; namely, the THD of the main system variables in the three scenarios: (i) without capacitor bank, (ii) with capacitor bank, (iii) with passive filter.

9.5.2.3 Case Study 3: Harmonic Frequency Scan

There are some important differences between the ATP options for harmonic analysis:

- the FS option solves the network for the specified sources, incrementing, in each subsequent step, the frequency of the sources but not their amplitudes;
- the HFS performs harmonic analysis by executing a string of phasor solutions determined by a list of sinusoidal sources entered by the user [2].

The main advantage of HFS compared with the time-domain harmonic analysis is a reduction in runtime, and avoidance of accuracy problems with Fourier analysis.

Several models have been developed for HFS analysis; they are frequency-dependent RLC elements, frequency-dependent load based on the CIGRE load model [70], and harmonic current/voltage sources with frequency-dependent amplitude and phase.

A very simple test system is used to introduce this option. Consider the configuration shown in Figure 9.25: an ideal voltage source supplies a linear impedance. Assume the source voltage has a fundamental component (50 Hz) and three harmonics (100, 150, and 200 Hz). To obtain the current that will flow through the load impedance, HFS can be applied. The HFS option is selected in the Settings window (see Figure 9.26a) while the harmonic components of the system sources are specified in the HFS Source option (see Figure 9.26b). Note that the user has to select between current or voltage source and specify: (i) the actual frequency or the harmonic number (F/n is treated as actual frequency if $F/n >=$ power frequency; if not, F/n is the multiple of the power frequency); (ii) the amplitude of each harmonic, in A or V; (iii) the angle of each harmonic, in degrees. The results are plotted in Figure 9.27: the figure shows the rms current flowing through the load impedance versus the frequency. Since the load inductance increases with the frequency, and the resistance is assumed constant, the load current decreases as the source frequency increases.

All harmonic analyses presented in this section were carried out by assuming that impedances were frequency-independent. The next study is based on a case included in the ATP package to illustrate the application of the HFS option and define some branches as a function of frequency.

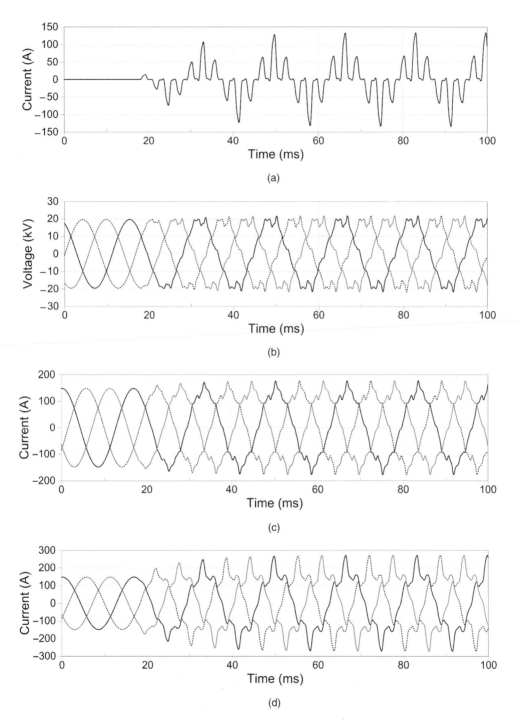

Figure 9.18 Case Study 2: simulation results – original test system configuration. (a) Nonlinear load current, (b) Voltages at the PCC, (c) Linear load currents, (d) MV-side substation transformer currents.

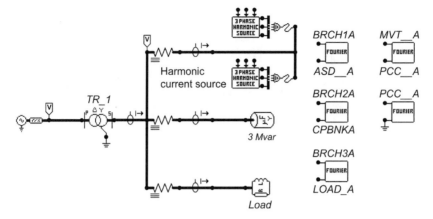

Figure 9.19 Case Study 2: ATPDraw implementation of the test system after installing the capacitor bank.

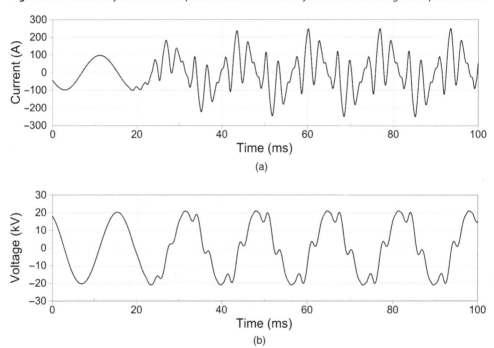

Figure 9.20 Case Study 2: simulation results with capacitor bank. (a) Capacitor bank current, (b) Voltage at the PCC.

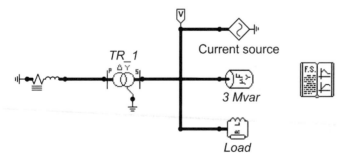

Figure 9.21 Case Study 2: ATPDraw diagram for frequency scan of the test system.

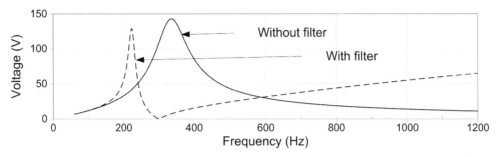

Figure 9.22 Case Study 2: frequency scan with capacitor bank and with passive filter.

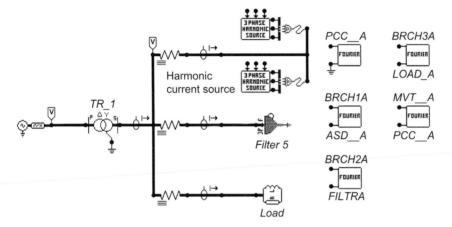

Figure 9.23 Case Study 2: ATPDraw diagram of the test system with a passive filter.

Table 9.7 Case Study 2: total harmonic distortion.

Nonlinear load current		103.47%
Load current	Without capacitor bank	8.96%
	With capacitor bank	18.71%
	With passive filter	3.38%
Voltage at PCC	Without capacitor bank	11.14%
	With capacitor bank	23.16%
	With passive filter	4.21%
MV-side substation current	Without capacitor bank	23.51%
	With capacitor bank	71.91%
	With passive filter	7.63%
Capacitor bank current		129.35%
Passive filter current		43.01%

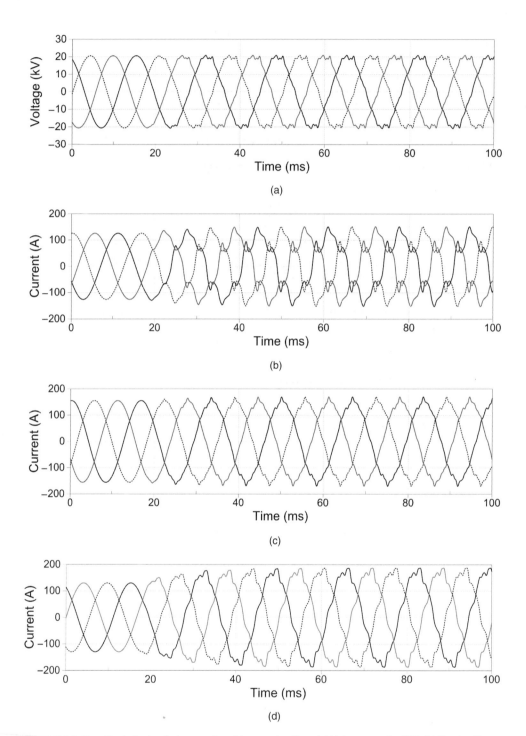

Figure 9.24 Case Study 2: simulation results with a passive filter. (a) Voltages at the PCC, (b) Passive filter currents, (c) Linear load currents, (d) MV-side substation transformer currents.

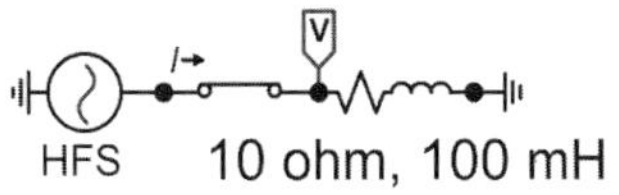

Figure 9.25 Case Study 3 – HFS application: ATPDraw implementation of the test system.

HFS   10 ohm, 100 mH

Consider the test system depicted in Figure 9.28 : a 10-kV 50-Hz network supplies some medium and low-voltage loads, and includes two filters tuned to the 5th and the 7th harmonic, respectively. An LV load has been represented with the CIGRE load model: it is a frequency-dependent model proposed by CIGRE for representing composite distribution loads and based on tests carried out by Electricité de France (EdF) [70]; the model consists of two parallel branches and is available in ATP and ATPDraw as a built-in model. The harmonic currents injected by the converter are modelled with the HFS source (see Table 9.8); since the converter is three-phase, three current sources are now needed to represent the full converter. This case is based on the DCN21 case. A list of the main system component parameters follows:

- Source   10 kV, 50 Hz
- Source impedance $Z_0 = 0.300 + j3.100$; $Z_+ = 0.011 + j1.053\,\Omega$
- Cable C0 $R_0 = R_+ = 0.38\,\Omega\,\mathrm{km^{-1}}$; $X_0 = X_+ = 0.41\,\Omega\,\mathrm{km^{-1}}$
      $C_0 = C_+ = 0.30\,\mu\mathrm{F\,km^{-1}}$; Length $= 2$ km
- Cable C1, C2, C3 $R_0 = 1.280$, $R_+ = 0.164\,\Omega\,\mathrm{km^{-1}}$; $X_0 = 0.152$, $X_+ = 0.0987\,\Omega\,\mathrm{km^{-1}}$
      $C_0 = C_+ = 0.408\,\mu\mathrm{F\,km^{-1}}$; Length $= 0.1$ km

- Filter – 5th harmonic $X = 11.47\,\Omega$; $C = 11.10\,\mu\mathrm{F}$
- Filter – 7th harmonic $X = 11.66\,\Omega$; $C = 5.57\,\mu\mathrm{F}$
- Load 1   10 kV, 5.5 MVA, 0.9 p.f.
- Load 2 (CIGRE model)   0.380 kV, 400 kW, 199.5 kvar; A = 0.073; B = 2; C = 0.74
- Transformer TR1   10/0.38 kV, 1 MVA, $x = 6.5\%$, X/R ≈ 10
- Transformer TR2   10/0.60 kV, 3.75 MVA, $x = 6.5\%$, X/R ≈ 10

The variation of transformer impedance parameters as a function of frequency is according to the following forms:

$$R_h = R_{50} \cdot (1 + 0.2(h - 1)^{1.5}) \qquad\qquad L_h = L_{50} \cdot h^{-0.03} \tag{9.1}$$

where h is the harmonic order. These variations have been implemented in the $PARAMETER settings. The model includes capacitances to tank (0.0033 μF per phase) and between phases (0.0015 μF). All values are referred to the MV side.

- Converter: See Table 9.8 and Figure 9.28.

Figure 9.29 shows some simulation results:

- the voltages developed at the grid terminal and the load L1 are both measured at the MV level and exhibit very similar behaviour as a function of frequency;
- the voltages measured at the terminal of Load 2 (CIGRE model) and the converter exhibit some differences, being voltages at any harmonic higher at the converter terminals;
- the currents through the two MV-level filters show a different pattern that allows distinction of the current through each one (e.g. the highest current to the filter tuned at the 5th harmonic is that corresponding to this harmonic).

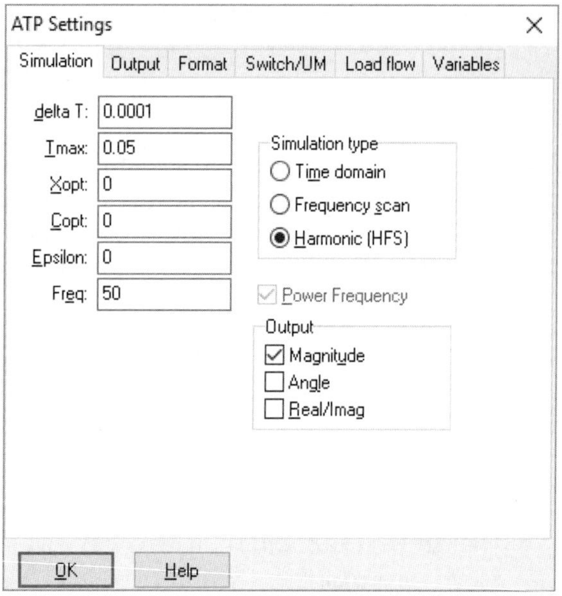

(a)

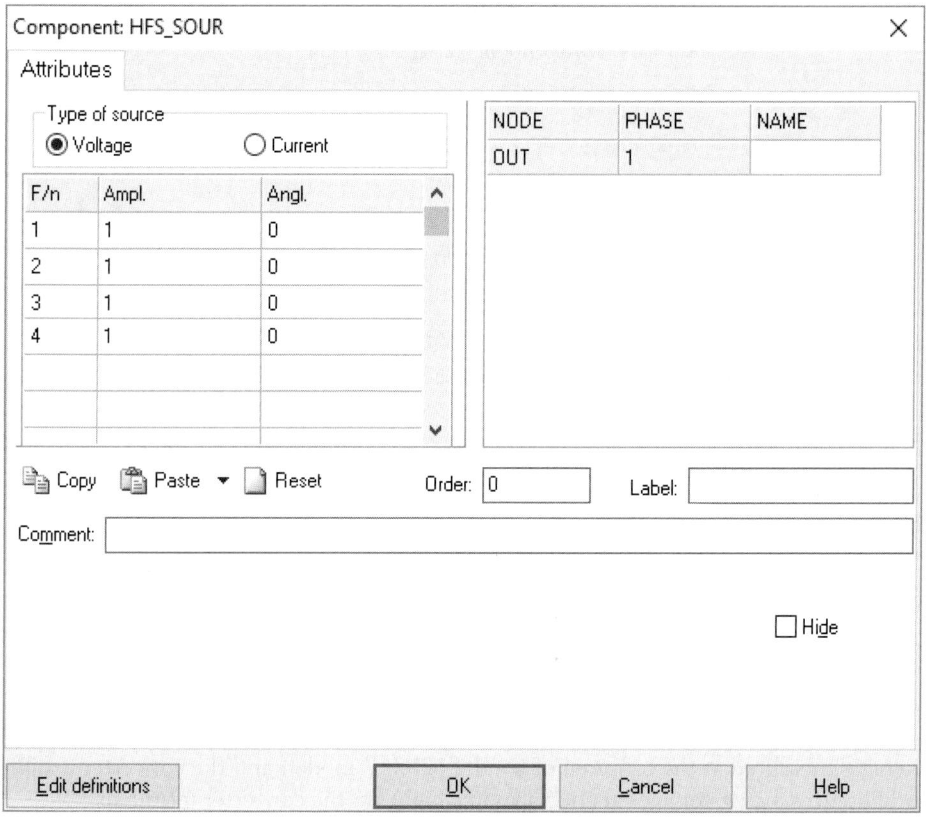

(b)

Figure 9.26 Case Study 3 – HFS application: selection and specification of the HFS source. (a) HFS selection – ATP settings window. (b) HFS source (Type 14) window.

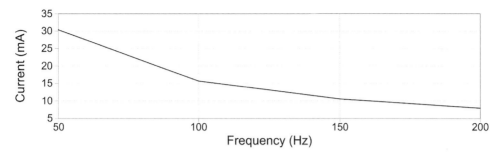

Figure 9.27 Case Study 3 – HFS application: simulation results – load current.

Table 9.8 Case Study 3: harmonic content of the current injection.

Harmonic order	Magnitude (A)	Phase (°)
0.33	727.321	−310
0.5	727.321	−310
1	4075	−170
5	727.321	−310
7	463.262	−110
11	206.488	−250
13	137.259	−50
17	62.675	−190
19	48.096	−350

9.5.2.4   Case Study 4: Compensation of Harmonic Currents

Problems associated with the propagation of harmonic currents produced by variable topology converters can be mitigated by using active filters. Figure 9.30 shows the scheme of a low-voltage network that feeds a diode rectifier. An active filter is placed between the source and the rectifier to compensate for the harmonic currents produced by the latter. Figure 9.31 depicts the configuration of the active filter converter, as implemented in ATPDraw [3]. The full active filter model includes a control section, here implemented in TACS, and an external passive circuit for filtering the currents injected by the active filter at the PCC, see Figure 9.30.

This case study analyses the performance of an LV active filter and shows how to use the ATP capabilities for implementing custom-made modules that could represent the active filter. Two modules (and two icons) have been developed for representing the power and the control sections: the Group option is used for implementing the power part (see Figure 9.31); the model of the control section is based on a user-defined module and DATA BASE MODULE.

Modelling guidelines are basically those applied to previous cases: the basic cell of the active filter inverter is an IGBT with an antiparallel diode and a parallel snubber circuit; very small resistances are used to separate semiconductors of different cells; see Figure 9.31.

The implementation of the two modules that will represent the active filter (without the external passive filter) must account for both parameters and connectivity. The following information is to be specified to define the inverter depicted in Figure 9.31:

- Parameters: system frequency, ratios of current and voltage transformers, dc link voltage, hysteresis band, initial time for activating active filter, dc link capacitances.

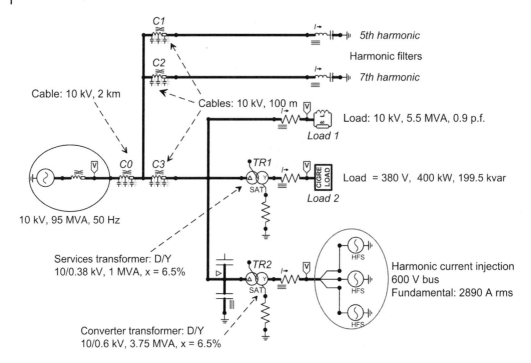

Figure 9.28 Case Study 3: HFS application – ATPDraw implementation of the test system.

- Connection nodes (i.e. inputs from the network, outputs to the network): voltages at the PCC, voltages at the dc link terminals of the active filter, nonlinear load currents, inverter output currents, IGBT switch status.

All parameters, except the dc link capacitances are specified in the control section. The common connection nodes between the icons of the two sections are the IGBT-gating signals: these signals are outputs for the control module and inputs for the power module.

Figure 9.32 shows the application of the Compress option to group the inverter model in a single module. Note that those elements (i.e. IGBTs, resistors, dc capacitors) that define the connectivity of the inverter have been named in order to facilitate the task. This has also been used with the dc-side capacitors whose parameters must be specified. Figure 9.32b shows the icon and the main window of the new grouped model. Although this is a rather simple case, a sophisticated control strategy is required for the filter inverter. In order to understand the main features of the implemented strategy, a short summary is provided below.

1) Calculation of the reactive reference current
   - Read and normalize nonlinear load currents.
   - Extract fundamental components of nonlinear load current (e.g. c_1 and s_1).
   - Delay terms and calculate tracked voltages.
   - Obtain the imaginary component of the compensation current.
2) Calculation of the active reference current.
   - Read dc voltages, calculate and normalize voltage between dc link terminals.
   - Input this voltage to a PI controller.
   - Read phase-to-ground voltages at the PCC.
   - Extract fundamental components of PCC voltage (e.g. c_2 and s_2).
   - Delay terms and calculate tracked voltages.

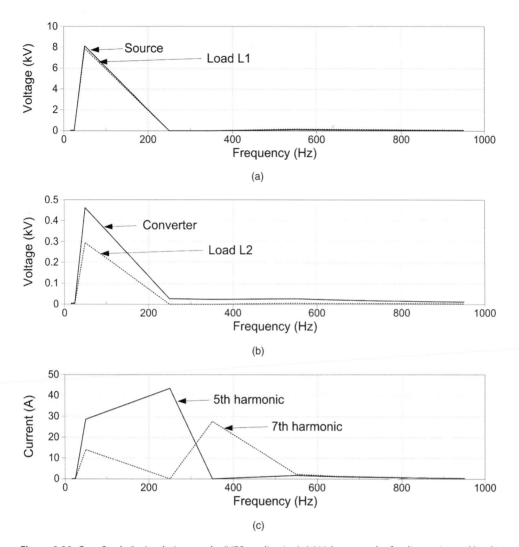

Figure 9.29 Case Study 3: simulation results (HFS application). (a) Voltages at the feeding point and load L1, (b) Voltages at load L2 and the converter, (c) Currents through filters.

Figure 9.30 Case Study 4 – configuration of the test system.

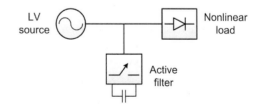

- Obtain the active component of compensation current for dc voltage regulation.
3) Set up IGBT gating signals
  - Obtain compensation current reference.
  - Read and normalize active filter currents.
  - Obtain current errors.

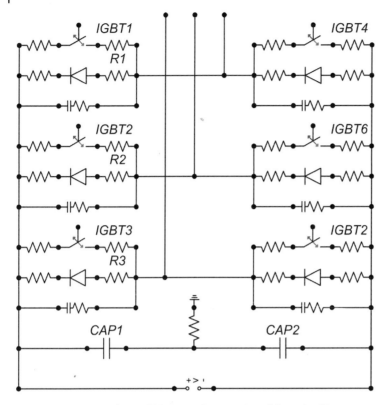

Figure 9.31 Case Study 4 – ATPDraw implementation of the active filter converter.

- Generate gating signals according to the following procedure

$$g_1 = [s_{S1} - (e_{ia} < -h) + (e_{ia} > h)] \cdot [t \geq t_{beg}]$$
$$g_4 = [not.g_1] \cdot [t \geq t_{beg}]$$
$$g_3 = [s_{S2} - (e_{ib} < -h) + (e_{ib} > h)] \cdot [t \geq t_{beg}]$$
$$g_6 = [not.g_3] \cdot [t \geq t_{beg}]$$
$$g_5 = [s_{S3} - (e_{ic} < -h) + (e_{ic} > h)] \cdot [t \geq t_{beg}]$$
$$g_2 = [not.g_5] \cdot [t \geq t_{beg}] \tag{9.2}$$

where h is the hysteresis band for the error currents, s_{S1}, s_{S3}, s_{S5} are the statuses of IGBTs 1, 3, and 5, respectively (see Figure 9.31), and e_{ia}, e_{ib}, e_{ic} are the currents errors for phases a, b, and c, respectively.

The strategy has been implemented in TACS code and embedded in a user-defined module developed with DATA BASE MODULE. Figure 9.33 shows the selected icon and the main window. The meaning of the parameters is straightforward (see above).

Figure 9.34 shows the circuit implemented in ATPDraw for representing the full test system. This circuit includes an uncontrolled three-phase diode rectifier implemented in

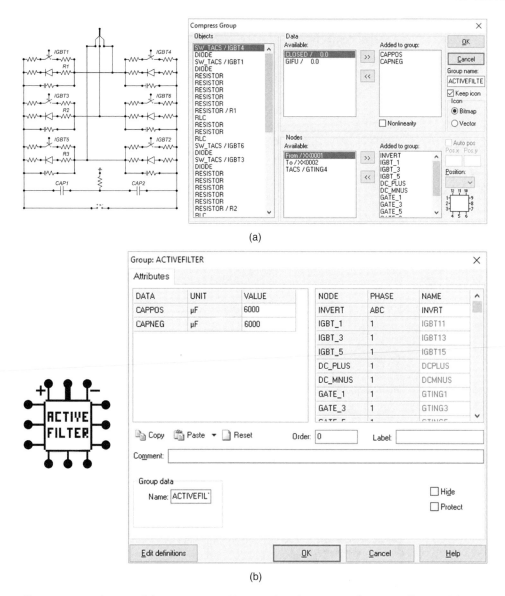

(a)

(b)

Figure 9.32 Application of the group option (Compress) to the creation of an active filter model. (a) Application of the group option (Compress), (b) Icon and main window of the implemented model.

a custom-made module; the dc-side configuration of this rectifier (the parallel connection of a capacitor and the RL load branch); the impedance that connects this nonlinear load to the PCC; a T-shaped circuit for filtering the inverter currents to be injected at the PCC; the equivalent impedance of the LV network seen from the PCC; and several modules for Fourier analysis (see Table 9.7). The circuit includes two very small resistors for connectivity and measuring purposes.

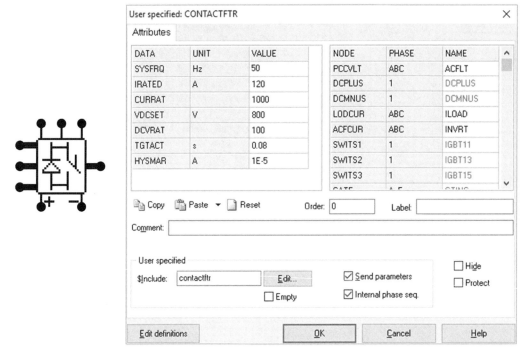

Figure 9.33 Case Study 4 – main window of the active filter controller.

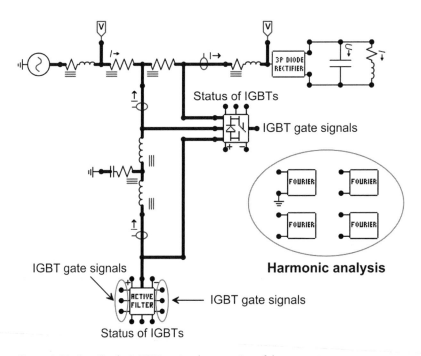

Figure 9.34 Case Study 4: ATPDraw implementation of the test system.

The main parameters of this circuit are listed below.

- Source: 400 V, 50 Hz
- Source impedance: $R = 0.023\,\Omega$, $L = 0.061\,\text{mH}$
- Rectifier impedance: $L = 1.808\,\text{mH}$
- Current filter: $L = 0.75\,\text{mH}$ (inverter side), $L = 0.16\,\text{mH}$ (grid side), $C = 30\,\mu\text{F}$
- Filter capacitors: $6000\,\mu\text{F}$
- Rectifier capacitor: $1000\,\mu\text{F}$
- Rectifier load: $R = 10\,\Omega$, $L = 1\,\text{mH}$
- Rectifier snubbers: $R = 100\,\Omega$, $C = 1\,\mu\text{F}$
- Inverter controller: Frequency = 50 Hz, CT ratio = 1000, DC voltage = 800 V, PT ratio = 100, Activation time = 8 ms, Hysteresis band = 1E-5 A.

The simulation results displayed in Figure 9.35 confirm the task of the active filter for compensating the harmonic currents of the rectifier load. It is evident that the highly distorted rectifier currents become almost purely sinusoidal currents at the source terminals. Table 9.9 shows the harmonic distortions calculated from the application of the ATP capability for Fourier analysis: the active filter significantly reduces the THD of the rectifier currents (i.e. the value of 26.18% is decreased to a 1.6%). Interestingly, to obtain such a low distortion the active filter has to inject highly distorted currents (i.e. with a THD close to 250%).

9.5.3 Voltage Dip Studies in Distribution Systems

9.5.3.1 Overview

An interruption of the power service can originate significant economic losses to affected customers. However, the main concern for many industrial and commercial users is the mal-operation originated by voltage dips. A voltage dip is a sudden reduction of the voltage below a specified dip threshold followed by its recovery after a brief interval (e.g. within one minute) [34, 71]. The interest in voltage dips is due to the problems they cause on several types of equipment: adjustable-speed drives, process-control equipment and computers are notorious for their sensitivity. A voltage dip is not as damaging as an interruption but, as there are far more voltage dips than interruptions, the total damage due to dips can be larger: although voltage dips are less severe than interruptions, they are more frequent, and the consequences for sensitive equipment can be as important as those by an interruption. Given the diversity of causes and the difficulty of preventing all these causes, voltage dips are one of the most frequent power quality disturbances. These reasons have increased the interest on mitigation equipment aimed at preventing or reducing voltage dip effects [72–75].

Voltage dips are, in general, a three-phase phenomenon in which all three-phase voltages are involved, and sometimes even the neutral-to-ground voltage. A common way of presenting a voltage dip is through the rms voltages as a function of time: a voltage dip is usually characterized by the remaining (retained) voltage, its duration and the phase jump, that is the difference between the phase voltage before and after the dip.

The main cause of voltage dips is a short-duration increase in current. The most severe voltage dips are originated by short-circuits, although they can also be originated by transformer energizing, motor starting, or other sudden load changes [76]. These causes produce a different pattern of the voltage profile; for instance, voltage dips due to motor starting show a rather small sudden voltage drop followed by a gradual recovery similar for all phases, while a temporary

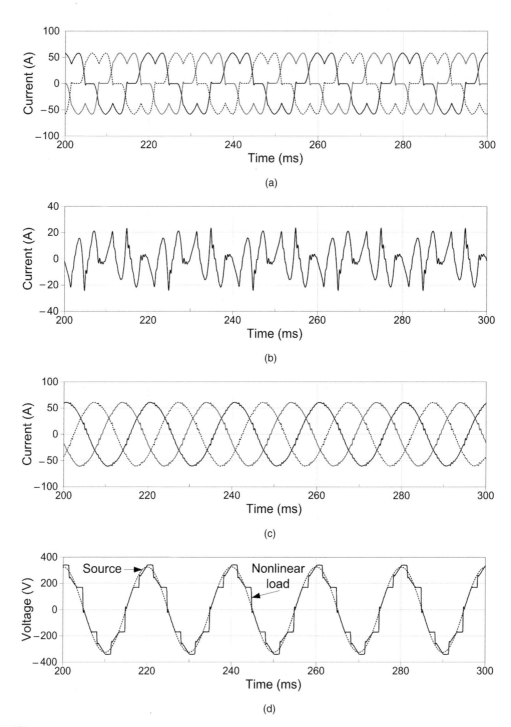

Figure 9.35 Case Study 4: simulation results. (a) Nonlinear load currents, (b) Active filter current – phase a, (c) Source currents, (d) Voltages at the supply system and nonlinear load terminals.

Table 9.9 Case Study 4: total harmonic distortion.

Nonlinear load currents	26.18%
Active filter currents	249.5%
Source currents	1.566%
System terminal (PCC) voltages	0.168%
Nonlinear load (AC side rectifier) terminal voltages	15.50%

single-phase-to-ground fault will produce a sudden voltage drop of shorter duration followed by a sharp recovery.

The reduction in voltage caused during a voltage dip leads to a reduction in energy-transfer capability of the system, limits the fault-clearing time in transmission systems, and rules the connection of wind farms to the grid.

Processing of the sampled voltage waveforms is needed to characterize voltage dip events. This is detailed in IEC 61000-4-30 [76], where the remaining voltage and the duration are defined as the two main characteristics to quantify a voltage dip. Both are obtained from the rms voltage as a function of time. For processing of event indices into site and system indices, some approaches have been proposed; for instance, the voltage dip performance is summarized in a table with the number of events per year for different remaining voltages and durations. Some examples are the Unipede Disdip table [73], IEC 61000-2-8 [71], IEEE Std.493 [74], and IEEE Std.1346 [75]. SARFI (System Average RMS Variation Frequency) indices belong to this group [72, 77].

Several alternative methods can be considered to reduce the tripping of equipment due to voltage dips. A list of the most effective methods would include reducing the number of faults, improving the performance of the protection system to obtain faster fault clearing, improving the network design and operation, installing mitigation equipment at the interface, or improving end-user equipment.

This section presents some cases that will illustrate how to measure the characteristics of voltage dips and mitigate them. All case studies deal with voltage dips caused by faults (short-circuits). The library designed for voltage dip analysis is basically a revised version of that used in some previous works [78]. As for harmonic analysis, the new library is mainly an educational tool.

9.5.3.2 Case Study 5: Voltage Dip Measurement

Figure 9.36 shows the diagram of the test system as implemented in ATPDraw: it is a two-feeder distribution system supplied from a 110/25 kV substation transformer. A fault occurs in the middle of the line that supplies the upper load. The goal is to measure the voltage waveshapes at each load terminal. The module implemented for representing faults allows users to simulate each type of fault.

The main parameters of the system are as follows:

- High voltage system: 110 kV, 50 Hz, 1500 MVA, $R_+ = 0.1$, $X_+ = 1.0$, $R_0 = 0.11$, $X_0 = 1.1$ p.u.
- Transformer: 110/25 kV, 5 MVA, Ynd, 8%, $X/R = 10$, Zig-zag grounding impedance
- Overhead lines: $R_+ = 0.614$, $X_+ = 0.391$, $R_0 = 0.764$, $X_0 = 1.564 \, \Omega \, km^{-1}$
- Distribution transformers: 25/0.4 kV, 2 MVA, Dyn, 6%, $X/R = 10$
- Loads: 0.4 kV, 1 MVA, pf = 0.95 (lg).

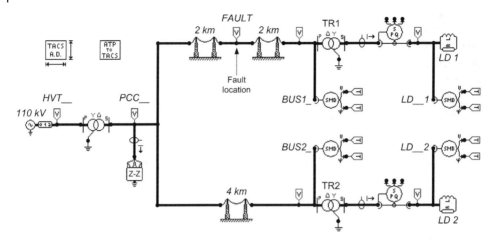

Figure 9.36 Case Study 5: voltage dip measurement.

The system model implemented in ATPDraw incorporates custom-made modules for representing the HV system equivalent, the substation and distribution transformers, the zig-zag grounding impedance, overhead lines, loads, and voltage dip monitors.

Figure 9.37 shows the simulation results derived from a phase-to-phase-to-ground fault. The simulated fault is located at the middle of the upper overhead line, occurs at 0.06 seconds and last 0.04 seconds. This can easily be deduced from the oscillograms of the fault currents shown in Figure 9.37a. and the rms voltages at the MV-side terminals of the upper distribution transformer (see Figure 9.37b). Although the short-circuit is between phases 'a' and 'b', only the phase 'a' voltage at the terminal of Load 1 reaches a zero value during the fault. This is due to the connection of the distribution transformer.

An important aspect to account for when characterizing the voltage dip is that the duration of the fault is not that specified in the switches that represent the fault: as mentioned above, the fault duration to be simulated is 0.04 seconds; however, from the current oscillograms shown in Figure 9.37a one can easily observe that the simulated duration is longer than 0.04 seconds. This is due to the fact that the instant at which a switch is fully open is the instant at which the switch current is zero, and this instant is different from the instant at which the order to open the switch arrives.

One can also observe that, as expected, the retained voltage at each phase of both loads is different. As for the phase shift caused by the fault at each load phase it is phase 'c' the value that exhibits the highest decrement. Note also that since not all phase voltages are touching ground, the fault causes a dip in the faulted phases and a swell in the healthy phase.

Another important aspect to account for is the waveshape exhibited by all rms voltages during the fault. Although the fault can cause an almost instantaneous variation of the voltages (e.g. an instantaneous reduction to a zero voltage), the rms value needs a power frequency period to reach the zero value due to the manner in which this rms value is determined. It is a factor to be accounted for when estimating the dip duration.

Figures 9.38 and 9.39 show some simulation results obtained with phase-to-ground and three-phase-to-ground faults, respectively. In both figures, the results correspond to the transformer and the load located at the faulted feeder. Note how the voltage dip caused by the first fault is almost negligble due to the effect that the winding connections of the distribution transformer TR1 has on the load terminal voltages. Obviously, the three-phase fault is much more severe and its effect more important (see next case study).

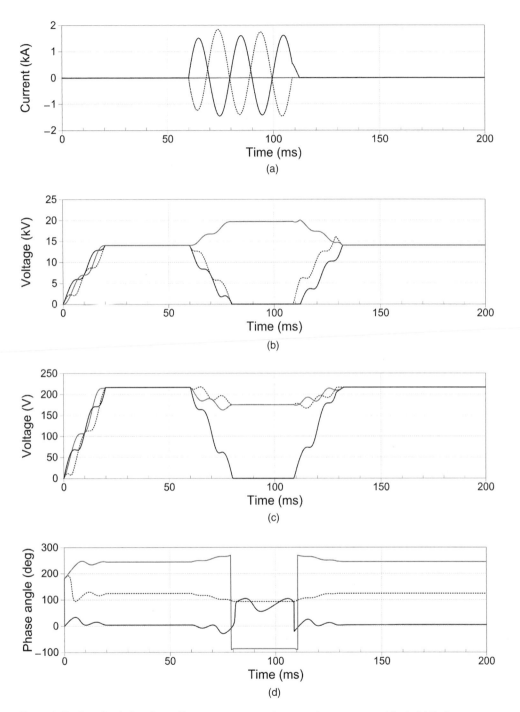

Figure 9.37 Case Study 5: voltage dip measurement – phase-to-phase-to-ground fault. (a) Fault currents, (b) MV-side distribution transformer TR1 terminal voltages, (c) Load 1 terminal voltages, (d) Phase angle shift of load 1 terminal voltages. (e) MV-side distribution transformer TR2 terminal voltages, (f) Load 2 terminal voltages, (g) Phase angle shift of load 2 terminal voltages, (h) Active powers at the load terminals.

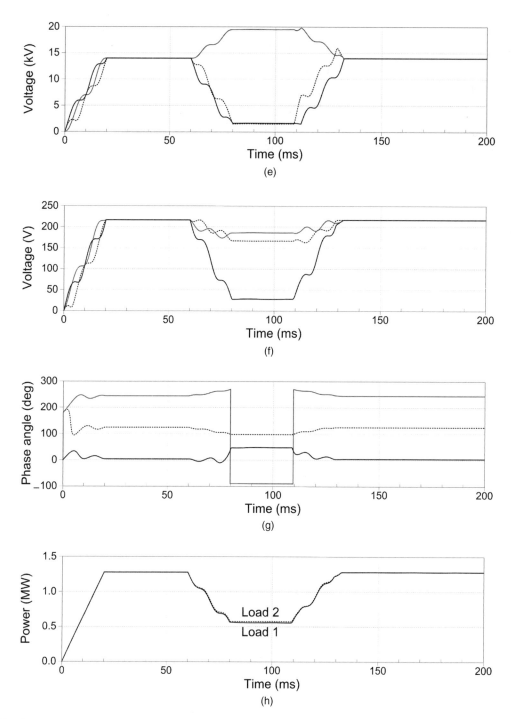

Figure 9.37 (*Continued*)

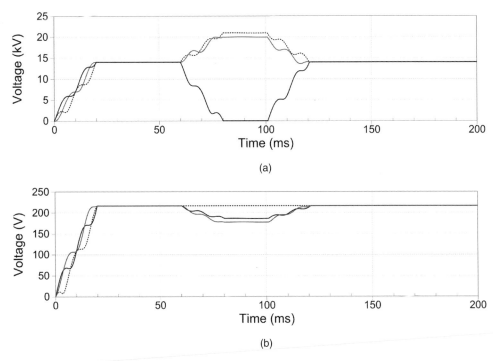

Figure 9.38 Case Study 5: voltage dip analysis and characterization – phase-to-ground fault. (a) MV-side distribution transformer TR1 terminal voltages, (b) Load 1 terminal voltages.

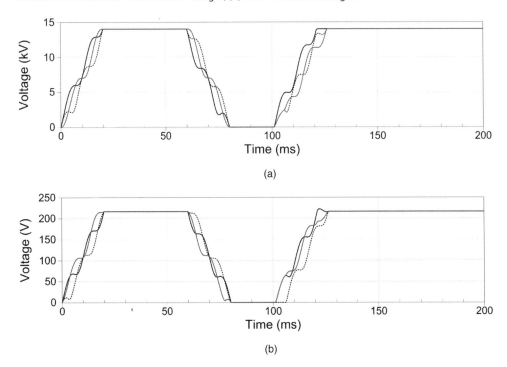

Figure 9.39 Case Study 5: voltage dip analysis and characterization – three-phase fault. (a) MV-side distribution transformer TR1 terminal voltages, (b) Load 1 terminal voltages.

9.5.3.3 Case Study 6: Voltage Dip Characterization

This case study illustrates how to use ATP capabilities for characterizing a voltage dip and represent a load according to a given voltage-tolerance curve. A summary of these concepts is presented prior to their implementation and application.

Voltage dip characterization

Magnitude and duration are the main characteristics of a voltage dip (see previous case study). However, other characteristics, such as phase-angle jump, the point on wave at both the initiation and the ending of the dip have to be considered. Three-phase equipment will typically experience three different magnitudes, as the majority of dips are due to single-phase or phase-to-phase faults. The characterization is generally based on the lowest of the three voltages and the longest duration. Sensitivity of three-phase equipment will be assessed against single-phase dips or symmetrical three-phase dips and interruptions. This section is dedicated to define magnitude, duration, and phase-angle jump of single-phase dips. For more details on voltage dip characterization see [34, 79, 80].

Voltage dip magnitude The magnitude of a voltage dip is usually determined from the rms voltage, which can be obtained by means of the following expression:

$$V_{rms}(k) = \sqrt{\frac{1}{N} \sum_{i=k-N+1}^{i=k} v(i)^2} \qquad (9.3)$$

where N is the number of samples per cycle, and $v(i)$ is the recorded voltage waveform.

Figure 9.40 shows an example. The rms voltage has been calculated over a window of one cycle: each point in the right plot is the rms voltage. The figure shows how the rms value drops from its pre-event value to a value of about 30% where it stays for almost 1.5 cycles after which it recovers to a value similar to its pre-event value. The voltage dip magnitude is usually characterized through the remaining voltage and given as a percentage of the nominal voltage. In this chapter, the term *magnitude* denotes the remaining voltage during the fault.

Voltage dip duration The duration of a voltage dip caused by a fault is determined by the fault-clearing time. However, the duration of a dip is normally longer than the fault-clearing

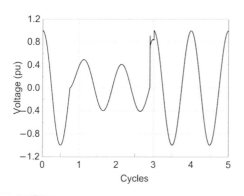
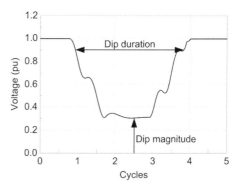

Figure 9.40 Example of voltage dip.

time. In a magnitude versus duration plot, a number of areas can be now distinguished; this is shown in Figure 9.40.

Standards define duration as the time difference between the moment at which the voltages fall below the corresponding dip threshold (usually 90%), and the moment at which the dipped rms voltage rises above it. However, these two instants (initiation and ending of disturbance) are related to the instantaneous voltage waveform, not to the rms voltage. The use of the instantaneous voltage waveforms for determination of the during-disturbance magnitudes and durations can exactly correlate the instants of initiation and ending with the corresponding points on wave of disturbance initiation and ending. The one cycle or one half-cycle error in dip duration will introduce some error that will only be significant for short-duration dips; for very long dips, it does not really matter. For an illustrative example of the consequences that the use of rms voltages can cause, see [80]. The so-called post-fault dip can give a more serious error in dip duration for long dips. When the fault is cleared the voltage does not always recover immediately. The effect can be especially severe for dips due to three-phase faults. The post-fault dip can last several seconds, it is much longer than the actual dip and can cause a serious error on the dip duration as obtained by a power quality monitor. Another problem is that different monitors can give different results; see [80].

Voltage angle jump Since voltage is a complex quantity, a change in the system, like a short-circuit, causes a change in phase angle. The latter will be referred to as the phase-angle jump associated with the voltage dip. The phase-angle jump manifests itself as a shift in voltage zero crossing compared to a synchronous voltage (e.g. as obtained by using a phase-locked loop). Phase-angle jumps are not of concern for most equipment, although power electronics converters using phase-angle information for their firing instants could get disturbed. Phase-angle jumps during three-phase faults are due to the difference in X/R ratio between the source and the feeder; a second cause is the transformation of dips to lower voltage levels. To obtain the phase-angle jump, the phase angle of the voltage during the dip must be compared with the phase angle of the voltage before the dip. The phase angle of a waveform can be obtained from the voltage zero-crossing or from the phase of the fundamental component of the voltage. There is no unique value for the phase-angle jump due to a dip since an average value or the largest value during a dip could be used. The oscillation of the phase angle around dip initiation and voltage recovery are due to the shift of the window in and out of the dip. It takes about one cycle before the phase-angle jump reaches a reliable value. This could lead to erroneous values.

Since most power systems are three-phase, and a large fraction of equipment tripping due to voltage dips concerns three-phase loads, additional characterization is needed for three-phase unbalanced dips (i.e. voltage dips due to non-symmetrical faults); see [80]. Transformers change the character of three-phase unbalanced voltage dips. Different winding connections affect the voltages in different ways. Symmetrical components can be used to study the transfer of three-phase unbalanced voltage dips through transformers in a systematic way. The transfer of the zero-sequence voltage depends on the neutral grounding of the transformer. When both sides are solidly grounded, the zero-sequence impedance is not affected, unless there is a tertiary delta winding present. In the latter case, the zero-sequence voltage is damped. However, the zero-sequence voltage is not considered in the classification of three-phase unbalanced voltage dips. The absolute value (in pu) of positive and negative-sequence voltages remain the same but are rotated, being the rotation of the negative-sequence voltage opposite to the rotation of the positive-sequence voltage.

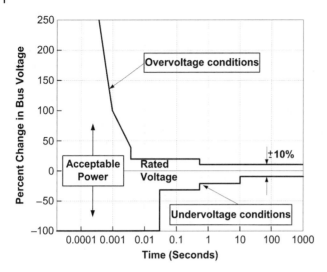

Figure 9.41 The ITI curve.

Voltage-tolerance curves

Electrical equipment prefers a constant rms voltage. The other extreme is no voltage for a longer period of time; in such cases, the equipment will simply stop operating. Some equipment will stop within one seconds, other equipment can withstand a supply interruption for much longer. For each piece of equipment, it is possible to determine how long it will continue to operate after the supply becomes interrupted. A rather simple test would give the answer.

Connecting the points obtained by performing these tests results in the so-called *voltage-tolerance curves* (also known as acceptability curves). This concept was introduced by T. Key, who developed a voltage-tolerance curve known as the CBEMA (Computer Business Equipment Manufacturers Association) curve [81]. This curve has two loci: the overvoltage locus above the $\Delta|V| = 0$ axis and the undervoltage locus below this axis. The ordinates of these loci represent the intensity of a bus voltage amplitude disturbance, which is measured as a deviation in voltage amplitude from the rated value. For the CBEMA curve, the ordinate is shown in percent deviation from rated voltage, while the abscissa represents the duration of the event being studied. The disturbance time duration is usually expressed in either cycles or seconds. Overvoltages of extremely short duration are usually tolerable if the event occurs below the upper limb of the power acceptability curve, that is, in the 'acceptable power' region. Although the CBEMA curve was originally designed to identify computer vulnerability to power supply disturbances, it has been applied to adjustable speed drives, fluorescent lighting, general (unspecified) loads, and modern solid-state (and microprocessor-based) computer loads. The CBEMA curve was redesigned in 1996 and renamed for its supporting organization, the Information Technology Industry Council (ITI) [82]. The newer curve describes an acceptable operating region in steps. The ITI curve, depicted in Figure 9.41, is recommended as a design target for manufacturers of computer equipment [8]. IEC 61000-4-11 describes how to obtain voltage tolerance of equipment [83].

Test systems

Figure 9.42 shows the system used to test a voltage-sensitive load model. The system is the same as that tested in the previous case study. The new circuit model incorporates new user-defined modules for representing sensitive loads, estimating voltage dip characteristics, and reporting

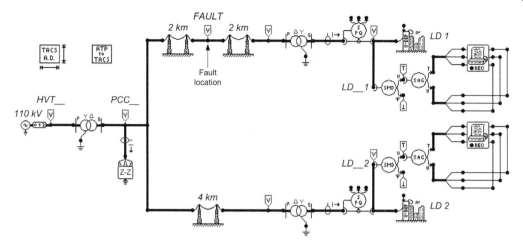

Figure 9.42 Case Study 6A: voltage dip analysis and characterization – sensitive load model.

Figure 9.43 Modules for voltage dip
measurement and characterization.

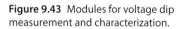
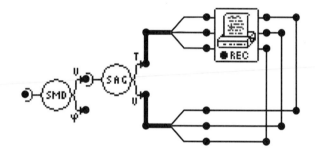

these characteristics in an output text file. Figure 9.43 displays the modules used to obtain and report voltage dip characterization.

The sensitive load model is a static voltage-dependent load model that behaves according to the following equation:

$$S = P_0 \sum_{k=0}^{n_p} a_k V^k + jQ_0 \sum_{k=0}^{n_q} b_k V^k \quad \left(\sum_{k=0}^{n_p} a_k = 1 \quad ; \quad \sum_{k=0}^{n_q} b_k = 1 \right) \tag{9.4}$$

where P_0 and Q_0 are respectively the rated active and reactive power at nominal voltage, and V is the pu voltage.

This equation assumes that there could be a part of a power demand that is voltage-independent, which is the approach implemented in some load flow programs. By using this approach, the power demand remains the same, irrespective of the values of bus voltages. This is not a realistic model for voltage dip calculation, as it would mean that even for very low retained voltages, the demand will be the same as that prior to the dip. A V^1 dependence means that the load behaves as a constant current source, while a V^2 dependence means that a load behaves as a constant impedance.

The rated powers are the same that were used in the previous case study. The voltage tolerance is in both load models the ITI curve and the percentage of load sensitive to voltage dip (i.e. the percentage that behaves according to the ITI curve) is 40%.

Figure 9.44 shows some simulation results that correspond to the same fault that was simulated in Case Study 5. From the comparison between these results and those depicted in Figure 9.37 one can deduce that the fault currents are the same, the rms voltages at the load

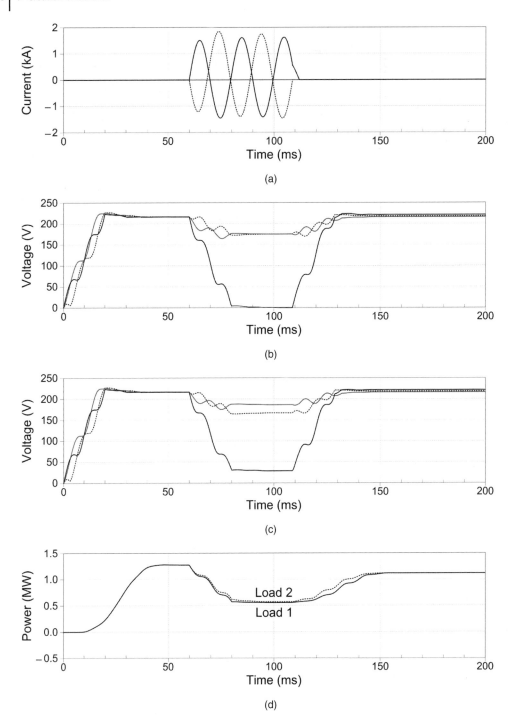

Figure 9.44 Case Study 6A: voltage dip analysis and characterization (sensitive load model – phase-to-phase-ground fault. (a) Fault currents, (b) Load 1 terminal voltages, (c) Load 2 terminal voltages, (d) Active powers.

terminals exhibit some differences during the initial transient and fault periods. Finally, the active power curves show what happens in front of this fault: since the voltage-tolerance curve applies to each load phase, it is the load of phases affected by the fault (namely only phases 'a') that will be reduced, which in this case is about 13.3% of the total load. The percentage is the same for both loads; although the retained voltage is not the same at the terminals of the two loads, the voltage dip affects both loads in the same quantity. Just consider the 'undervoltage condition' curve in Figure 9.41, the voltage reduction is more than 50% for both loads and the voltage dip duration is longer than 0.01 seconds. In other words, during the voltage dip, the operation point for both loads is below the ITI tolerance curve. The characteristics of the voltage dips measured at the two load terminals are reported by a MODELS-based module (see Figure 9.43). The output of this module provides some information with the following format:

Run	PCT1A	PCT1B	PCT1C	PCV2A	PCV2B	PCV2C	NBUS
0.0	.066315	.050575	.059135	.614748E-3	169.973758	164.221959	1.0
0.0	.066135	.051145	.051405	29.1052786	164.508012	174.999481	2.0

where PCT1A, PCT1B, and PCT1C are the voltage dip duration (in seconds) for phases 'a', 'b' and 'c', respectively, while PCV2A, PCV2B, and PCV2C are the retained voltages, in volts. Note that the voltages of the healthy phases are very similar for the two loads, but the phase that experiences the voltage reduction is very different for each load. As for the voltage dip durations, they are very similar for both loads, but exhibit some differences between phases of the same load. The run number is also printed; it can be important in case of simulating multiple runs.

The new load model generates a message indicating the sensitive percentage of the phase 'a' load has been disconnected in both loads at 0.08 seconds. One can check that this is the moment at which the rms voltage curve of the phase that experiences the highest voltage reduction in the two loads reaches its minimum value (see Figures 9.44b,c).

A second larger system model has been created to test the library. Figure 9.45 shows an MV distribution network with 14 loads and 26 overhead line sections. All LV loads have the same ratings. A new load model that combines the distribution transformer and the load has been created; see figure. The main parameters of the system are as follows:

- High voltage system: 110 kV, 50 Hz, 1500 MVA, $R_+ = 0.1$, $X_+ = 1.0$, $R_0 = 0.11$, $X_0 = 1.1$ p.u.
- Transformer: 110/25 kV, 20 MVA, Yd, 8%, $X/R = 10$, Zig-zag grounding impedance
- Overhead lines: $R_+ = 0.614$, $X_+ = 0.391$, $R_0 = 0.764$, $X_0 = 1.564\,\Omega\,\mathrm{km}^{-1}$
- Loads: Distribution transformer 25/0.4 kV, 1 MVA, Dyn, 6%, $X/R = 10$, Power 750 kW, pf = 0.8 (lg).

The number attached to each overhead line is the line length.

Measurement and report of voltage dips for all loads is made with the same modules depicted in Figure 9.43, although in this case the measurements are made at the MV level since the LV side is not accessible with the new load model.

The new distribution system model is tested with a three-phase fault at node BUS21; see Figure 9.45. The fault occurs at 0.1 seconds and lasts 0.1 seconds. Figure 9.46 provides the results corresponding to some load nodes.

As for the previous case study, the simulation of this system will provide the voltage dip characteristics at the MV-side terminals of each load. The simulation of this case study creates the following file:

Run	PCT1A	PCT1B	PCT1C	PCV2A	PCV2B	PCV2C	NBUS
0.0	.10634	.08692	.078245	10056.1068	11173.5656	11436.8184	1.0
0.0	.10638	.08699	.078255	9932.52242	11054.2146	11318.8619	2.0
0.0	.10638	.08699	.078255	9932.52242	11054.2146	11318.8619	3.0

0.0	.10658	.08698	.078275	9726.72715	10855.7685	11122.4009	4.0
0.0	.10669	.08717	.078285	9465.68726	10603.6725	10873.2777	5.0
0.0	.10669	.08717	.078285	9465.68726	10603.6725	10873.2777	6.0
0.0	.10665	.08707	.078285	9584.42567	10718.3808	10986.5909	7.0
0.0	.122915	.1133	.109745	5341.49945	5906.98812	6040.14321	8.0
0.0	.122955	.11333	.109715	5264.0086	5841.65395	5977.83142	9.0
0.0	.122955	.11333	.109715	5264.0086	5841.65395	5977.83142	10.
0.0	.124795	.11736	.11233	18.8826988	23.1169799	33.4131608	11.
0.0	.125195	.11742	.1125	2.88966104	9.94042602	32.6305692	12.
0.0	.125195	.11742	.1125	2.88966104	9.94042602	32.6305692	13.
0.0	.124965	.11739	.11242	.38035958	15.8252774	32.9865838	14.

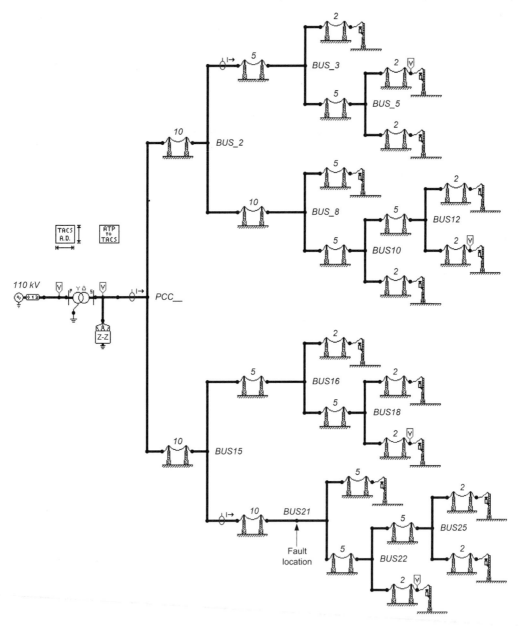

Figure 9.45 Case study 6B: voltage dip analysis and characterization.

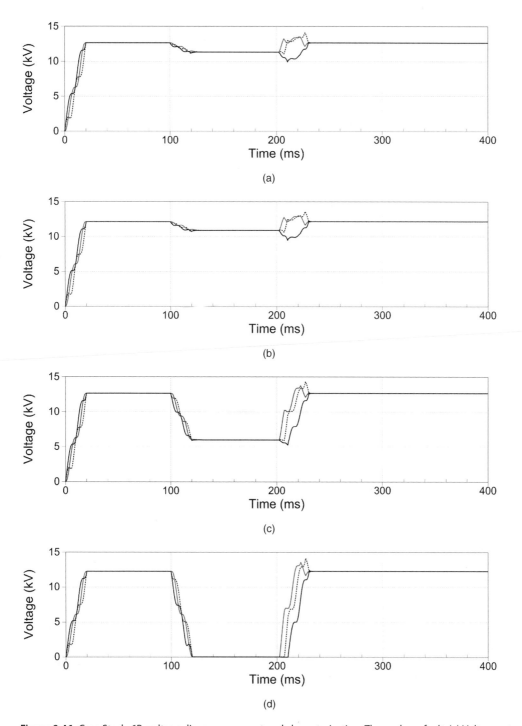

Figure 9.46 Case Study 6B: voltage dip measurement and characterization. Three-phase fault. (a) Voltages at load 2 MV-side terminals, (b) Voltages at load 6 MV-side terminals, (c) Voltages at load 10 MV-side terminals, (d) Voltages at load 14 MV-side terminals.

It is evident from these results that the retained voltage decreases at the distance between the load node and the fault location decreases, and the voltage dip duration decreases as this distance increases. This can also be checked from the results shown in Figure 9.46. As for the previous case study, the report also provides the number of the run; this is again meaningless but could be very useful in a stochastic prediction, for which a high number of runs should be carried out [44–47].

9.5.3.4 Case Study 7: Voltage Dip Mitigation

The list of methods for mitigating voltage dips includes: (i) reducing the number of short-circuit faults; (ii) reducing the fault-clearing time; (iii) changing the system such that short-circuit faults result in less severe events at the equipment terminals or at the customer interface; (iv) improving the immunity of the equipment; (v) connecting mitigation equipment between the sensitive equipment and the supply [34]. The method selected in this case study is related to the last item.

For industrial customers, who do not normally have access to system or equipment improvement, the installation of additional mitigation equipment is often the only option left to achieve the desired quality of supply at the system-load interface. Traditional devices include motor-generator sets, and constant-voltage or ferroresonant transformers. Modern custom power devices based on power electronics are one of the best options [14–16].

A static series compensator, also known as dynamic voltage restorer (DVR), is a voltage source converter (VSC) connected in series on the distribution feeder to provide a controllable source, whose voltage adds to the source voltage to obtain the desired load voltage [84]. Depending on the control strategy, it is possible to correct supply voltage imbalance, perform load voltage regulation, compensate for voltage dips, and cancel low-order supply voltage harmonics. This solution is very attractive for large sensitive industrial customers, as it allows for protection of the entire plant through the installation of a single device; however, this compensator cannot protect the load against interruptions.

Figure 9.47 shows a configuration that performs the series-injection principle. The converter configuration is similar to that used with the active filter (see Figure 9.31). The voltage rating of the converter dictates the maximum injected voltage, which is thus the maximum (three-phase) dip magnitude that can be compensated for. The controller can be designed to compensate for voltage dips by only providing reactive power (i.e. by injecting a voltage in quadrature with the load current). However, the compensation capability of the device is very limited in these conditions. Therefore, an energy storage device is normally connected to the dc bus of the converter to provide the energy necessary for the compensation. The capacity of the energy storage device has a big impact on the compensation capability of the system, as it ultimately determines the ride-through time for the load. An alternative is to connect the dc link of the VSC to the grid by means of a converter that could control the dc voltage within reasonable limits. See references [14–16] for more details on custom power devices.

Figure 9.48 displays the circuit implemented in ATPDraw to test the performance of the DVR model. The test system is the same that has been simulated in previous case studies. The only difference is the incorporation of a DVR for mitigating voltage dips that could affect sensitive Load 2. The installation of a DVR includes the series three-phase transformer (that will inject the series compensating voltages in case of dip), the VSC plus its controller, the supply system of the converter dc link, here represented as double constant dc voltage source, a filter between the VSC and the series three-phase transformer. The filter is a T-shaped circuit (a series *RL* branch plus a shunt *C* branch).

The VSC and its controller have been implemented in a single custom-made module. As mentioned above, the VSC configuration is basically that used with the active filter; however,

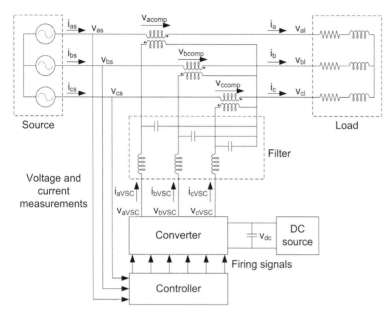

Figure 9.47 Scheme of a series compensator for voltage dip mitigation.

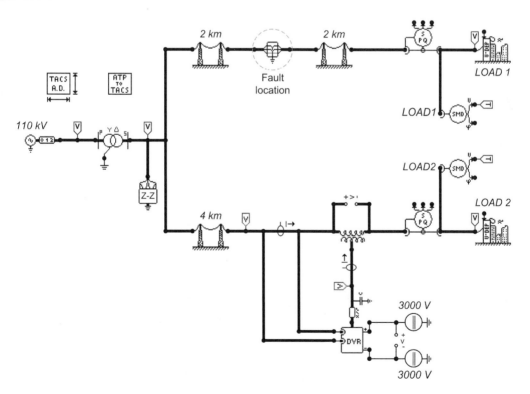

Figure 9.48 Case Study 7: voltage dip mitigation – dynamic voltage restorer.

the control strategy is very different. The main aspects of this strategy can be summarized as follows:

1) Read voltages and currents at the source side of the series transformer (see Figure 9.48)
2) Calculate α and β components of voltages and currents.
3) Obtain positive- and negative-sequence of the previously calculated α and β components.
4) Obtain homopolar components of voltages.
5) Obtain positive- and negative-sequence dq components of voltages and currents.
6) Compensate positive- and negative-sequence dq components of voltages for filter (located between the DVR and the series transformer) correction.
7) Obtain dq components of the voltages to be injected in series for voltage dip compensation.
8) Antitransform positive- and negative-sequence dq components of series voltages: first, obtain $\alpha\beta$ components; next, obtain phase-sequence series voltages.
9) Remove homopolar voltage components previously calculated and normalize the resulting voltages using the DC voltage as reference.
10) Generate a triangular-shaped waveform (with a frequency specified by the user) and compare it to the calculated voltages. The pulse width modulation (PWM) control strategy is based on this comparison. If the IGBTs and their gating signals are numbered as in Figure 9.31, then the signals in this case would be obtained as follows:

$$g_1 = [v_{aref} - v_{triang}] \qquad g_4 = [not.g_1]$$
$$g_3 = [v_{bref} - v_{triang}] \qquad g_6 = [not.g_3]$$
$$g_5 = [v_{cref} - v_{triang}] \qquad g_2 = [not.g_5] \tag{9.5}$$

where v_{aref}, v_{bref}, v_{cref} are the voltages obtained in the previous step, and v_{triang} is the triangular-shaped waveform.

The user-defined model of the three-phase transformer that will inject the series compensating voltages in case of dip in the path of Load 2 is based on a conventional transformer model with the following ratings and parameters: 4/25 kV, 4 MVA, $x = 6\%$, $X/R = 10$. To define the three-phase T-shaped filter installed between the DVR converter and this transformer, the user has to specify the value of the capacitors to grounds, C, the order of the resonant frequency, n, and the ratio X/R.

The formulas implemented within the user-defined filter model to obtain the values of R and L are as follows:

$$L = \frac{1}{(\omega n)^2 C} \quad \Rightarrow \quad R = \frac{\omega L}{r} \tag{9.6}$$

where f is the power frequency.

The parameters used in this case study are $C = 200\,\mu F$, $n = 11$, $r = 1$.

Figure 9.49 shows some simulation results that support the application of a DVR for mitigating voltage dips. These results correspond to the transient response of the test system for a two-phase-to-ground fault. The fault occurs at 0.2 seconds and clears at 0.4 seconds.

- Since the DVR is installed to compensate voltage dips that could affect only Load 2, the voltage dip will affect to Load 1 terminal voltages, but not to Load 2; the voltages at the Load 2 terminal exhibit some deviation from the normal operation at the instants the fault occurs and clears.
- The series voltages injected by the DVR are different in magnitude and phase angle for each phase since the voltage dip at the input terminals of the DVR has different voltages (see Figure 9.49c).
- The terminal voltages at Load 1 node cause some disconnection of sensitive power, but the installation of the DVR avoids any disconnection of sensitive power in Load 2 node.

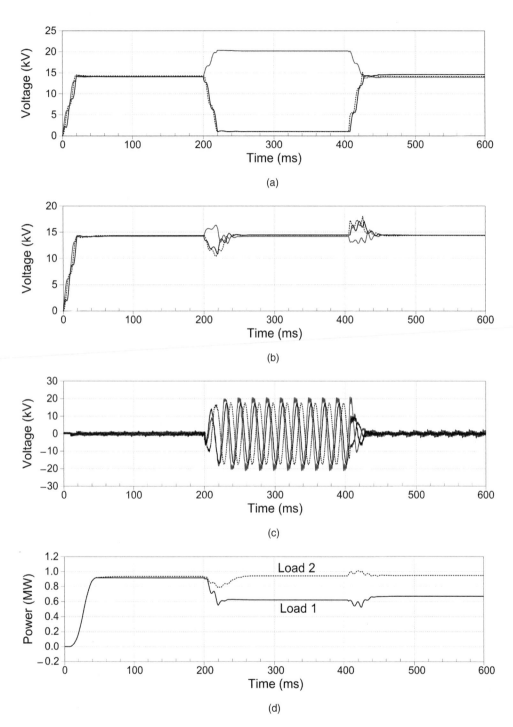

Figure 9.49 Case Study 7: voltage dip mitigation. (a) Load 1 terminal voltages, (b) Load 2 terminal voltages, (c) Series voltages injected by the DVR, (d) Active powers.

References

1 Dommel, H.W. (1986). *Electromagnetic Transients Program Reference Manual (EMTP Theory Book)*. Portland (OR, USA): Bonneville Power Administration.

2 Furst, G. (1998). Harmonic analysis with the new HARMONIC FREQUENCY SCAN (HFS) in ATP. *EEUG News* 4 (1): 18–25.

3 Martinez-Velasco, J.A. and Martin-Arnedo, J. (1999). EMTP modular library for power quality analysis. IEEE Budapest Power Tech in Budapest, Hungary (29 August–2 September 1999).

4 Can/Am Users Group (2000). *ATP Rule Book*. Can/Am Users Group.

5 Prikler, L. and Høidalen, H.K. (2002). ATPDraw for Windows User's Manual, SEFAS TR F5680.

6 Bollen, M.H.J. (2003). What is power quality. *Electric Power Systems Research* 66 (1): 6–14.

7 Heydt, G.T. (1998). Electric power quality: a tutorial introduction. *IEEE Computer Applications on Power* 1 (1): 15–19.

8 Dugan, R.C., McGranaghan, M.F., Santoso, S., and Wayne Beaty, H. (2012). *Electrical Power Systems Quality*, 3e. New York: McGraw-Hill.

9 Heydt, G.T. (1991). *Electric Power Quality*. Stars in a Circle Publications.

10 IEEE WG on Power System Harmonics (1983). Power system harmonics: an overview. *IEEE Transactions on Power Apparatus Systems* 102 (8): 2455–2460.

11 Domijan, A., Heydt, G.T., Meliopoulos, A.P.S. et al. (1993). Directions of research on electric power quality. *IEEE Transactions on Power Delivery* 8 (1): 429–436.

12 IEEE TF on the Effects of Harmonics on Equipment (V.E. Wagner, Chairman) (1993). Effects of harmonics on equipment. *IEEE Transactions on Power Delivery* 8 (2): 672–680.

13 IEEE Std 519-2014. (2014). IEEE Recommended Practice and Requirements for Harmonic Control in Electric Power Systems.

14 Hingorani, N.G. (1995). Introducing custom power. *IEEE Spectrum* 32 (6): 41–48.

15 Ghosh, A. and Ledwich, G. (2002). *Power Quality Enhancement Using Custom Power Devices*. Norwell (MA, USA): Kluwer Academic Publishers.

16 Ghosh, A. and Shahnia, F. (2015). Applications of power electronic devices in distribution systems. In: *Transient Analysis of Power Systems. Solution Techniques, Tools and Applications* (ed. J.A. Martinez-Velasco)), 177–249. Wiley–IEEE Press.

17 Arrillaga, J., Smith, B.C., Watson, N.R., and Wood, A.R. (1997). *Power System Harmonic Analysis*. Wiley.

18 Task Force on Harmonics Modeling and Simulation, Harmonics Working Group (1998). *Tutorial on Harmonics Modeling and Simulation*. IEEE Special Publication.

19 TF on Harmonics Modeling and Simulation (1996). Modeling and simulation of the propagation of harmonics in electric power networks. Part I: concepts, models, and simulation techniques. *IEEE Transactions on Power Delivery*: 452–465.

20 TF on Harmonics Modeling and Simulation (1996). Modeling and simulation of the propagation of harmonics in electric power networks. Part II: sample systems and examples. *IEEE Transactions on Power Delivery* 11 (1): 466–474.

21 Dommel, H.W. (1997). Techniques for analyzing electromagnetic transients. *IEEE Computer Applications on Power* 10 (3): 18–21.

22 Martinez-Velasco, J.A. (ed.) (1997). *Computer Analysis of Electric Power System Transients*. IEEE Press.

23 Martinez-Velasco, J.A. (1998). Power quality analysis using electromagnetic transients programs. Presented at the 8th International Conference on Harmonics and Quality of Power in Athens, Greece (October 1998).

24 Martinez, J.A. (2000). Power quality studies using electromagnetic transients programs. *IEEE Computer Applications on Power* 13 (3): 14–19.

25 CIGRE WG 33.02. (1990). Guidelines for Representation of Network Elements when Calculating Transients CIGRE Brochure 39.

26 Gole, A., Martinez-Velasco, J.A., and Keri, A. (eds.) (1998). *Modeling and Analysis of Power System Transients Using Digital Programs*. IEEE Special Publication.

27 IEC TR 60071-4. (2004). Insulation Co-ordination–Part 4: Computational Guide to Insulation Co-ordination and Modeling of Electrical Networks.

28 Hatziadoniu, C.J. (1998). Time domain methods for the calculation of harmonic propagation and distortion. In: *Harmonics Modeling and Simulation*. IEEE.

29 Grebe, T.E. (1996). Application of distribution system capacitor banks and their impact on power quality. *IEEE Transactions on Industry Applications* 32 (3): 714–719.

30 McGranaghan, M.F., Zavadil, R.M., Hensley, G. et al. (1992). Impact of utility switched capacitors on customer systems – magnification at low voltage capacitors. *IEEE Transactions on Power Delivery* 7 (2): 862–868.

31 McGranaghan, M.F., Grebe, T.E., Hensley, G. et al. (1991). Impact of utility switched capacitors on customer systems – part II: adjustable-speed drive concerns. *IEEE Transactions on Power Delivery* 6 (4): 1623–1628.

32 Bellei, T.A., O'Leary, R.P., and Camm, E.H. (1996). Evaluating capacitor-switching for preventing nuisance tripping of adjustable-speed drives due to voltage magnification. *IEEE Transactions on Power Delivery* 11 (3): 1373–1378.

33 Adams, R.A., Middlekauff, S.W., Camm, E.H. and McGee, J.A. (1998). Solving customer power quality due to voltage magnification. Paper PE-384-PWRD-0-11-1997. IEEE PES Winter Meeting in Tampa, USA (February 1998).

34 Bollen, M.H.J. (2000). *Understanding Power Quality Problems. Voltage Dips and Interruptions*. New York: IEEE Press.

35 McGranaghan, M.G., Mueller, D.R., and Samotyj, M.J. (1993). Voltage sags in industrial systems. *IEEE Transactions on Industry Applications* 29 (2): 397–403.

36 Lamoree, J., Mueller, D.R., Vinett, P. et al. (1994). Voltage sag analysis case studies. *IEEE Transactions on Industry Applications* 30 (4): 1083–1089.

37 Key, T.S. (1995). Predicting behaviour of induction motors during service faults and interruptions. *IEEE Industry Applications Magazine* 1 (1): 6–11.

38 Bishop, M.T., Do, A.V., and Mendis, S.R. (1994). Voltage flicker measurement and analysis system. *IEEE Computer Applications in Power* 7 (2): 34–38.

39 Manchur, G. and Erven, C.C. (1992). Development of a model for predicting flicker from electric arc furnaces. *IEEE Transactions on Power Delivery* 7 (1): 416–426.

40 Tang, L., Mueller, D., Hall, D. et al. (1994). Analysis of DC arc furnace operation and flicker caused by 187 Hz voltage distortion. *IEEE Transactions on Power Delivery* 9 (2): 1098–1107.

41 Montanari, G.C., Logingi, M., Cavallini, A. et al. (1994). Arc-furnace model for the study of flicker compensation in electrical networks. *IEEE Transactions on Power Delivery* 9 (4): 2026–2036.

42 Tang, L., Kolluri, S., and McGranaghan, M.F. (1997). Voltage flicker prediction for two simultaneously operated ac arc furnaces. *IEEE Transactions on Power Delivery* 12 (2): 985–992.

43 Varadan, S., Makram, E.B., and Girgis, A.A. (1996). A new time domain voltage source model for an arc furnace using EMTP. *IEEE Transactions on Power Delivery* 11 (3): 1685–1691.

44 Martinez, J.A. and Martin-Arnedo, J. (2004). Voltage sag stochastic prediction using an electromagnetic transients program. *IEEE Transactions on Power Delivery* 19 (4): 1975–1982.

45 Martinez, J.A. and Martin-Arnedo, J. (2006). Voltage sag studies in distribution networks. Part I: system modelling. *IEEE Transactions on Power Delivery* 21 (3): 1670–1678.

46 Martinez, J.A. and Martin-Arnedo, J. (2006). Voltage sag studies in distribution networks. Part II: voltage sag assessment. *IEEE Transactions on Power Delivery* 21 (3): 1679–1688.

47 Martinez, J.A. and Martin-Arnedo, J. (2006). Voltage sag studies in distribution networks. Part III: voltage sag index calculation. *IEEE Transactions on Power Delivery* 21 (3): 1689–1697.

48 Sakis Meliopoulos, A.P. (1988). *Power System Grounding and Transients: An Introduction*, 17–135. New York: Marcel Dekker.

49 Menter, F.E. and Grcev, L. (1994). EMTP-based model for grounding system analysis. *IEEE Transactions on Power Delivery* 9 (4): 1838–1849.

50 Heimbach, M. and Grcev, L. (1997). Grounding system analysis in transients programs applying electromagnetic field approach. *IEEE Transactions on Power Delivery* 12 (1): 186–193.

51 Grady, W.M., Samotyj, M.J., and Noyola, A.H. (1990). Survey of active power line conditioning methodologies. *IEEE Transactions on Power Delivery* 5 (3): 1536–1542.

52 Rastogi, M., Naik, R., and Mohan, N. (1994). A comparative evaluation of harmonic reduction techniques in three-phase utility interface of power electronic loads. *IEEE Transactions on Industry Applications* 30 (5): 1149–1155.

53 Akagi, H. (1996). New trends in active filters for power conditioning. *IEEE Transactions on Industry Applications* 32 (6): 1312–1322.

54 DeWinkel, C. and Lamoree, J.D. (1993). Storing power for critical loads. *IEEE Spectrum* 30 (6): 38–42.

55 Girgis, A.A., Quaintance, W.H., Qiu, J., and Makram, E.B. (1993). A time-domain three-phase power system impedance modeling approach for harmonic filter analysis. *IEEE Transactions on Power Delivery* 8 (2): 504–510.

56 Makram, E.B., Subramanium, E.V., Girgis, K.A., and Catoe, R.C. (1993). Harmonic filter design using actual recorded data. *IEEE Transactions on Industry Applications* 29 (6): 1176–1183.

57 Aredes, M. and Watanabe, E.H. (1995). New control algorithms for series and shunt three-phase four-wire active power filters. *IEEE Transactions on Power Delivery* 10 (3): 1649–1656.

58 Mohan, N., Rastogi, M., and Naik, R. (1993). Analysis of a new power electronics interface with approximately sinusoidal 3-phase utility currents and a regulated DC output. *IEEE Transactions on Power Delivery* 8 (2): 540–546.

59 Yin, Y. and Wu, A.Y. (1998). A low-harmonic electric drive system based on current-source inverter. *IEEE Transactions on Industry Applications* 34 (1): 227–235.

60 Lamoree, J., Tang, L., DeWinkel, C., and Vinett, P. (1994). Description of a micro-SMES system for protection of critical customer facilities. *IEEE Transactions on Power Delivery* 9 (2): 984–991.

61 Venkataramanan, G., Johnson, B.K., and Sundaram, A. (1996). An AC-AC power converter for custom power applications. *IEEE Transactions on Power Delivery* 11 (3): 1666–1671.

62 Venkataramanan, G. and Johnson, B.K. (1997). A pulse width modulated power line conditioner for sensitive load centers. *IEEE Transactions on Power Delivery* 12 (2): 844–849.

63 Hochgraf, C. and Lasseter, R.H. (1997). A transformer-less static synchronous compensator employing a multi-level inverter. *IEEE Transactions on Power Delivery* 12 (2): 881–887.

64 Halpin, S.M., Spyker, R.L., Nelms, R.M. and Burch, R.F. (1996). Application of double-layer capacitor technology to static condensers for distribution system voltage control. 1996 IEEE/PES Winter Meeting in Baltimore, USA (21–25 January, 1996).

65 Martin-Arnedo, J., González-Molina, F., Martinez-Velasco, J.A., and Adabi, M.E. (2017). EMTP model of a bidirectional cascaded multilevel solid state transformer for distribution system studies. *Energies* 10: 521.

66 Kizilcay, M. and Dubé, L. (1997). Post processing measured transient data using MODELS in the ATP-EMTP. *EEUG News* 3 (2): 24–36.

67 Gole, A. and Daneshpooy, A. (1997). *Towards Open Systems: A PSCAD/EMTDC to MAT-LAB Interface*, 446–451. Seattle: IPST.

68 Wehrend, H. (1997). Calling MATLAB out of an ATP/MODELS simulation. European EMTP/ATP Users Group Meeting in Barcelona, Spain (November, 1997).

69 Vu, K., Timko, K., Novosel, T., and Hakola, T. (1997). *Simulation Tool for Distribution-System Modeling, Analysis and Algorithm Testing*. Seattle (USA): IPST.

70 CIGRE Working Group 36-05 (1981). Harmonics, characteristic parameters, methods of study, estimates of existing values in the network. *Electra* 77: 35–54.

71 IEC 61000-2-8. (2002). Electromagnetic compatibility (EMC) – Part 2-8: Environment - Voltage dips and short interruptions on public electric power supply systems with statistical measurement results.

72 IEEE Std. 1564-2014. (2014). IEEE Guide for Voltage Sag Indices.

73 DISDIP Group. (1990). Voltage Dips and Short Interruptions in Medium Voltage Public Electricity Supply Systems. Report from the International Union of Producers and Distributors of Electrical Energy (UNIPEDE).

74 IEEE Std. 493-2007 (2007). IEEE Recommended Practice For Design of Reliable Industrial and Commercial Power Systems.

75 IEEE Std. 1346-1998. (1998). IEEE Recommended Practice for Evaluating Electric Power System Compatibility with Electronic Process Equipment.

76 IEC 61000-4-30. (2015). Electromagnetic compatibility (EMC) – Part 4-30: Testing and measurement techniques–Power quality measurement methods.

77 Brooks, D.L., Dugan, R.C., Waclawiak, M., and Sundaram, A. (1998). Indices for assessing utility distribution system RMS variation performance. *IEEE Transactions on Power Delivery* 13 (1): 254–259.

78 IEEE Std. 1159-2009 (2009). *IEEE Recommended Practice for Monitoring Electric Power Quality*. IEEE.

79 Bollen, M.H.J. and Zhang, L.D. (2003). Different methods for classification of three-phase unbalanced voltage dips due to faults. *Electric Power Systems Research* 66 (1): 59–69.

80 CIGRE WG C4.102 (J.A. Martinez-Velasco, Convenor). (2009).Voltage Dip Evaluation and Prediction Tools. CIGRE Brochure 372.

81 Key, T.S. (1979). Diagnosing power quality related computer problems. *IEEE Transactions on Industry Applications* 15 (4): 381–393.

82 Information Technology Industry Council (ITI). www.itic.org (accessed 03 June 2019).

83 IEC 61000-4-11. (2004). Electromagnetic compatibility (EMC) – Part 4-11: Voltage dips, short interruptions and voltage variations immunity tests.

84 Woodley, N., Morgan, L., and Sundaram, A. (1999). Experience with an inverter based dynamic voltage restorer. *IEEE Transactions on Power Delivery* 14 (3): 1181–1186.

To Probe Further

The simulation results presented in the chapter are a very small sample of the results that can be derived from the case studies. ATP data files for all case studies are available in the companion website. Readers are encouraged to run the cases, check all results, and explore the performance of the test systems under different operating conditions. An important aspect of this chapter is the creation of new custom-made models; readers are also encouraged to take advantage of the models implemented in this chapter to create new versions by changing or improving every part of the models (i.e. the icon, the inside code, the menus).

10

Protection Systems

Juan A. Martinez-Velasco and Jacinto Martin-Arnedo

10.1 Introduction

Figure 10.1 shows the scheme of a protection system that consists of three major parts: instrument transformers, protective relays, and circuit breakers [1–5]. The instrument transformers lower the power system voltages and currents to safe working levels. The protective relays receive information about the operating conditions of the high-voltage power system via the instrument transformers. The relays check whether the power system is in a healthy or an unhealthy state. In the latter case, the relays send trip signals to circuit breakers, which will interrupt the large currents.

Actual power systems and their protection systems operate in a closed-loop manner: information about the status of the power system reaches protective relays from instrument transformers, relays analyse this information, and take a decision that can modify the topology of the power system by opening circuit breakers. Since faults can be temporary, many protection schemes include reclosing. Modern software tools allow users to represent a closed-loop interaction, which is the most realistic approach for analysing the performance of a combination of power and protection systems under transient conditions [6–9].

The application of time-domain simulation tools on the modelling and simulation of protection systems have been discussed in several tutorial publications [7–10]. Relay models are now being used in the design stage to analyse system performance or to study the impact of relay operations on system security. These models may reproduce relay characteristics, relay input pre-processing, measurement processes and the outputs provided by relays [9, 10].

To ensure that a protection system model will perform as expected, it must be tested under realistic power system conditions. The power system model that generates the data must be represented with enough detail and accuracy. Therefore, it is desirable that the simulation environment be capable of simulating any fault scenario and system configuration.

The main goal of this chapter is to show how the Alternative Transients Program (ATP) capabilities can be used for modelling and simulating protection systems. The chapter provides modelling guidelines and summarizes the work done to date on the simulation of protection systems using digital simulation tools; lists some of the uses, advantages, disadvantages, and limitations of protection system models, and discusses the modelling of instrument transformers, relay technologies, and other protective devices (e.g. fuses).

Section 10.2 reviews guidelines for representing power system components. Modelling guidelines for representing protection systems using ATP are discussed in three sections dedicated respectively to instrument transformers, protective relays at transmission levels, and protection systems at distribution levels. Section 10.3 discusses instrument transformer models and presents two illustrative case studies. Section 10.4 analyses the implementation of relay models

Transient Analysis of Power Systems: A Practical Approach, First Edition. Edited by Juan A. Martinez-Velasco.
© 2020 John Wiley & Sons Ltd. Published 2020 by John Wiley & Sons Ltd.
Companion Website: www.wiley.com/go/martinez/power_systems

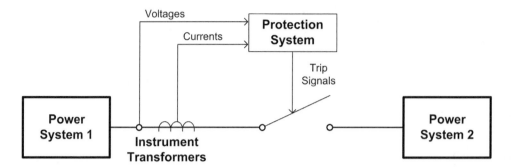

Figure 10.1 Scheme of a protection system for implementation in a digital simulation tool.

and their applications; the section includes two case studies with different relay technologies. The design of the protection systems for distribution systems uses a different methodology from that followed at transmission levels; the modelling of protective devices at distribution levels and their coordination are discussed in Section 10.5, which also covers the study of distribution systems with distributed energy resources (DERs). The section presents several cases studies aimed at testing the custom-made models implemented for representing distribution-level protective devices and their coordination.

The cases studied in this chapter are just a small sample of the ATP applications in this field. Section 10.6 provides a very short summary of the main conclusions.

The principles on which the devices installed to protect power systems are based and coordinated are beyond the goals of this chapter. Readers interested in these aspects should consult the specialized literature [1–5, 11–20].

10.2 Modelling Guidelines for Protection Studies

The representation of power system components for protection studies is usually made taking into account guidelines recommended for low-frequency and switching transients [21]. This section provides a summary of modelling guidelines for those power systems components that can be affected by a fault and whose behaviour is critical for the behaviour of the protection system [7, 8, 22].

10.2.1 Line and Cable Models

Transmission line parameters are evenly distributed along the line length, and some of them are functions of frequency. For steady-state studies, such as short-circuit calculations, positive- and zero-sequence parameters calculated at the power frequency from tables and simple handbook formulas may suffice. For electromagnetic transient studies, line parameters are generally computed using auxiliary subroutines available in ATP [23]. This tool has two major categories of transmission line models: constant- and frequency-dependent models. In both cases, the models can be based either on a lumped- or a distributed-parameter representation. In the constant-parameter category, there is a variety of options such as positive- and zero-sequence lumped-parameter representation, PI-section representation, or distributed-parameter transposed and untransposed line representation. In the frequency-dependent category, the distributed-parameter approach may be considered for either transposed or untransposed lines.

10.2.1.1 Models for Steady-State Studies

Some protection studies can be carried out with the system operating under steady state, so modelling transmission lines at only one particular frequency may suffice. ATP has a number of models that could be used for this purpose.

- *Exact-PI circuit model.* This lumped-parameter model can represent the line accurately at one specific frequency. This model is based on the hyperbolic equations, may take into account skin effect and ground return corrections, and may be multiphase in the phase domain with constant parameters. The model is not adequate for transient studies.
- *Nominal-PI circuit model.* This model is derived from the exact-PI model and used when the frequency of transients is low. The line is generally assumed to be untransposed, so the approach can be used to model particular transposition schemes in great detail by cascaded connection of PI sections. This model can only be used for representing short lines and cannot represent frequency-dependent parameter lines.

10.2.1.2 Models for Transient Studies

Distributed- and frequency-dependent parameter models are the most adequate for transient studies. They use travelling wave solutions and can be valid over a very wide range of frequencies.

- *Constant and distributed-parameter line model.* This model assumes that the line parameters are constant. The line inductance and capacitance are distributed, and losses are lumped. The above conditions are met for positive-sequence parameters to approximately 1–2 kHz, but not for zero-sequence parameters, so the model is good only where the zero-sequence currents are very small, or oscillate with a frequency close to that at which the parameters were calculated. This frequency should not be very high to meet the condition $R_l << Z_{surge}$, otherwise the line must be split into smaller sections.
- *Frequency-dependent and distributed-parameter line model.* The line parameters are a function of frequency. If frequency-dependent models are based on the modal theory, the transformation matrices for untransposed lines are complex and frequency-dependent; however, it is also possible to obtain a good accuracy by using real and constant transformation matrices. Frequency-dependent line models can also be based on a phase-domain model, which does not use the modal transformation matrix.

In a few words, the exact-PI model can be used when a steady-state analysis suffices; for transient analysis, the frequency-dependent and distributed-parameter model must be used for the lines of main interest, while the constant and distributed-parameter model can be used for lines of secondary interest.

The application of insulated cable models for protection studies follow the same guidelines that for overhead lines. Cable models may also use either a lumped- or a distributed-parameter representation, and their parameters can also be computed using auxiliary routines available in ATP [24]; see Chapter 2.

10.2.2 Transformer Models

Transformer modelling over a wide frequency range still presents substantial difficulties: the transformer inductances are nonlinear and frequency dependent, the distributed capacitances between turns, between winding segments, and between windings and ground, produce resonances that can affect the terminal and internal transformer voltages. Models of varying complexity can be developed for power transformers using supporting routines and models available in ATP. Although none of the existing models can portray the physical layout of the transformer or the high-frequency characteristics introduced by inter-winding capacitance effects, ATP capabilities can be used to model any transformer type at a particular frequency range. A summary of built-in capabilities available in ATP for representing power transformers is provided below [21, 22].

- *Ideal transformer* This model neglects magnetization currents. This approach can be used together with other linear and nonlinear components to represent more complex transformer models not available in ATP.
- *Saturable transformer model.* This model uses a star-circuit representation for single-phase transformers with multiple windings and considers that around each individual coil, there are a magnetic leakage path and a magnetic reluctance path. It is good for low-frequency transients. Since the winding resistances are frequency dependent, they need to be modified to reflect proper behaviour at high frequencies.
- *Matrix model.* In a transformer bank of single-phase units, the individual phases are not magnetically coupled, and their modelling is balanced assuming all three phases have equal parameters. In three-phase core transformers, there is magnetic coupling between windings. In addition, they may have asymmetry of magnetic path lengths, which results in asymmetrical flux densities in the individual legs of the transformer core. The core asymmetry effects are more noticeable under unbalanced operation. An accurate representation requires the use of a model that takes into consideration coupling of every phase winding with all other phase windings. Models based on matrices of mutually coupled coils can represent quite complex coil arrangements. The matrix elements for transformers with any number of windings can be derived from the short-circuit impedances between pairs of windings. The BCTRAN routine can generate the branch matrices from the positive- and zero-sequence short-circuit and excitation test data. The resulting models are good for less than 2 kHz. They can take into account excitation losses, but nonlinear behaviour is not represented and must be added externally. Two matrix representations are possible for transformer modelling: they are based on either admittance or impedance matrix. The impedance matrix representation is only possible if the exciting current is nonzero; otherwise the matrix is singular.

Transformer saturation should be modelled if the flux will exceed the linear region. The auxiliary routine SATURA can be used to calculate the magnetization branch saturation parameters: this supporting routine generates the data for the piecewise linear inductance by converting the rms voltage – current data into peak flux-peak current data. The resulting curve is single valued (without hysteresis). Although the nonlinear inductor model works well for a number of cases, it also exhibits several limitations: it is frequency independent and does not represent hysteresis effects, which means that remanent flux in the core cannot be represented. ATP provides also a pseudo-nonlinear hysteretic reactor, which can overcome some of these limitations. See Chapter 4 for more details on ATP capabilities for representing hysteretic reactances.

10.2.2.1 Low-frequency Transformer Models

They can be based on the built-in capabilities summarized above. Supporting routines may be used for modelling transformer windings as mutually coupled branches. When matrix models are used, the magnetic core of the transformer is typically represented with a nonlinear or a hysteretic reactance branch connected externally to the terminals of the windings. The built-in saturable transformer component is simpler to use than the matrix models; however, if zero-sequence behaviour of three-phase core-type transformers must be modelled, then the matrix approach should be used. The transformer models discussed here are valid only at moderate frequencies, but accurate enough for overcurrent protection studies. Eddy currents in the transformer core introduce losses and delay flux penetration into the core. Modelling of eddy currents is not an easy task since data are not usually available. No-load losses include hysteresis and eddy current losses, and can be represented by a (nonlinear) resistor in parallel with the magnetizing inductance branch.

10.2.2.2 High-frequency Transformer Models

At frequencies above 2 kHz, capacitances and capacitive coupling between windings can be important or very important. For frequencies of up to 30 kHz, the simple addition of total capacitances of windings to ground and between windings is sufficient for many purposes. For frequencies above 30 kHz, a more detailed representation of the internal winding arrangement is required and capacitances between winding and among winding segments must be modelled. The values of terminal-to-ground capacitances, including bushing capacitances, vary considerably, with typical values in the range of 1–10 nF. This is due mainly to the physical arrangement of the transformer windings and the overall transformer design. High-frequency transformer models can be derived from its terminal behaviour: if explicit representation of a transformer is not required, the user can model transformer effects without modelling the transformer itself (e.g. a black-box representation can be used in these situations); however, in relaying studies, the interest may be in internal faults [25]. Several high-frequency transformer models have been presented in the literature [26, 27].

10.2.3 Source Models

Source models used in protection studies are represented by means of detailed machine models or as ideal sinusoidal sources behind subtransient reactances or the equivalent Thevenin impedances of the system [22]. The choice of a specific model depends on system configuration, the location of the fault, and the objectives of the study.

- A detailed model of the machines involved in a disturbance is mostly used for representing small generating stations where the system disturbance is likely to cause change in frequency. The model requires complete machine data, including the mechanical part and the control systems, depending upon the timeframe of the study and their response time.
- A representation based on an ideal source with subtransient reactance is used for modelling large generating stations. The assumption is that the system inertia is infinite and the disturbance under study does not cause system frequency variation. The timeframe of interest is small (approximately 10 cycles) and the machine controls such as excitation system and governor will not respond to the disturbance. Large systems can be divided into subsystems, and each subsystem can then be reduced to an ideal three-phase source in series with equivalent positive- and zero-sequence Thevenin impedances. The main advantage of this model is that the computation requirements are significantly reduced; its main disadvantage is that the Thevenin impedance represents the system equivalence at power-frequency only, so the transient response is not as accurate as when the complete system is represented. A solution to this limitation is to implement a dynamic equivalent model for some parts of the test system [28].

10.2.4 Circuit Breaker Models

Circuit breakers are usually represented as ideal switches (i.e. the switch opens at a current zero and there is no representation of arc dynamics and losses), although custom-made circuit breaker models can be implemented if a detailed arc modelling is required [29, 30]. Two types of switches are applicable for protection studies [22, 23]:

- *Controlled switch.* This type of switch can be controlled by either using a closed-loop strategy or specifying the times at which it has to close and/or open. The actual opening time will occur at the next current zero after the time at which it is required to open. To simulate current chopping, a current margin is also specified so the switch actually opens at the instant the current magnitude falls below the current margin and the time is greater than the time at which it has to open.
- *Statistical/systematic switch.* This type of switch is used to open or close the circuit breaker either randomly (with a predetermined distribution function) or systematically. These switches can be employed to determine the maximum peak currents that can flow through a relay when closing into a fault. Usually, a few hundred simulations are run to obtain the information of interest (e.g. the maximum relay currents).

10.3   Models of Instrument Transformers

10.3.1   Introduction

To achieve the proper performance of the protection system model during transient phenomena, an accurate representation of instrument transformers is needed. Since instrument transformer transients considerably influence the performance of the relays, the models must be as realistic as possible [7, 8, 31]. The transient performance of current transformers (CTs) is influenced by a number of factors; for instance, the build-up of the core flux is likely to cause saturation and subsequently substantial errors in the magnitude and phase angle of the generated signals. The CT core may also retain an unknown amount of flux because of the ferromagnetic character, which can contribute to saturation in fault conditions. The transient response of magnetic voltage transformers (VTs) and coupling capacitor voltage transformers (CCVTs) depends on several phenomena taking place in the primary network, such as sudden decrease of voltage at the transformer terminals due to a fault or sudden overvoltages on the sound phases caused by the line-to-ground faults in the network. These phenomena can generate high-frequency (VT) or low-frequency aperiodic (CCVT) oscillations on the secondary side and saturation of the magnetic core. This section summarizes the models proposed to represent instrument transformers in power system transient analysis, and presents two case studies with instrument transformer models implemented in ATP. For more details see [31–40].

10.3.2   Current Transformers

Figure 10.2 shows the conventional equivalent circuit of a CT [7, 8, 31]. One of the main concerns in this model is the representation of the nonlinear magnetization characteristic. A saturated CT core will usually result in distorted and reduced secondary current that may cause relay misoperation. An accurate representation of the magnetization characteristic should include the hysteresis loop and simulate minor loops. Inter-winding capacitances can generally be neglected at the frequencies of interest in protection studies.

To study the impact of saturation, the CT equivalent circuit may be represented as shown in Figure 10.3. The CT primary wire resistance and inductance can usually be neglected. The CT secondary wire inductance can also be neglected although, in some cases, it can be taken into consideration. The magnetizing branch L_m can be modelled on either the primary or the secondary side. Modelling on the secondary is preferred because V-I curve measurements are regularly performed from the CT secondary. The magnetizing branch is represented by a nonlinear inductor whose characteristic is specified in piecewise linear form by the user. Since the

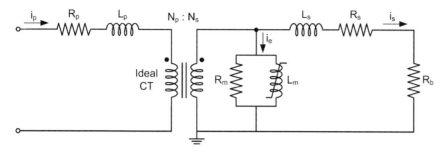

i_p Primary current
i_s Secondary current
i_e Exciting current
R_p, L_p Primary winding resistance and leakage inductance
R_s, L_s Secondary winding resistance and leakage inductance
L_m Magnetizing inductance
R_m Core loss resistance
R_b CT burden resistance, including lead resistance

Figure 10.2 Equivalent circuit of a current transformer.

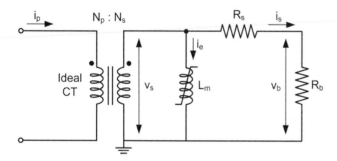

Figure 10.3 Current transformer model for relay protection applications.

flux-current data points are not readily available, ATP users will usually apply the SATURA routine to convert the V_{RMS}-I_{RMS} characteristic into an equivalent flux-current set.

Hysteresis representation is important if the study is intended to include the effects of remanence on CT performance. If the model does not represent the hysteresis, it may still allow the specification of a steady-state flux level at the beginning of a study. Specification of an initial value of flux will simulate the presence of remanent flux as if the model had included hysteresis. Models that do not allow specification of initial value of flux and do not represent hysteresis are valid and produce satisfactory results for studies where remanent flux is not a concern.

Modern protection devices have very low impedance that may be neglected when considering the saturation. The impedance of the CT wire and the leads that interconnect CTs and protective relays should be represented when studying the CT saturation. The R_b value can represent a combination of CT secondary winding resistance, lead resistance, and CT burden.

Figure 10.4 shows the impact of remanence on a CT without an air gap for symmetrical and asymmetrical 60 Hz fault current 20-times the CT-rated current and at rated resistive load. The secondary current is displayed for a remanent flux of +80%, 0%, and −80% of the saturation flux. This figure shows how a +80% positive remanence reduces time-to-saturation to about

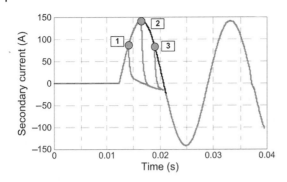

Figure 10.4 Impact of remanence for symmetrical fault currents: (1) 80% remanent flux; (2) 0% remanent flux; (3) −80% remanent flux.

2 ms, while time-to-saturation for 0% remanence is about 5 ms. For −80% remanence almost no saturation occurs in the first half-cycle.

10.3.3 Coupling Capacitor Voltage Transformers

CCVTs are widely used in high-voltage power systems to obtain standard low-voltage signals for protective relaying. A CCVT transforms the line voltage to low voltage through a sequence of capacitive potential dividers and an electromagnetic voltage transformer (VT). A typical circuit connection is shown in Figure 10.5 [7, 8, 31]. A relatively heavy current may be drawn from the protective device when the burden is an electromechanical relay; in such a situation a large error can result. To avoid this problem, the loading effect on the capacitive divider is tuned by a compensating reactor on the primary side of the VT. An additional circuit, designed to suppress ferroresonance, is added at the secondary side. All these components make circuitry quite complex and influence its transient response.

Ferroresonance may occur in a circuit containing capacitors and iron-core inductors. It is usually characterized by overvoltages and distorted waveforms of currents and voltages. Ferroresonance suppression circuits (FSCs) are incorporated in CCVTs [40] and designed to attenuate ferroresonance quickly after it occurs. A typical common FSC design includes capacitors and iron-core inductors connected in parallel, tuned to the fundamental frequency, and permanently connected to the secondary side of the CCVT (Figure 10.6). Capacitor C_f is connected in parallel with an iron-core inductor L_f tuned to the fundamental frequency;

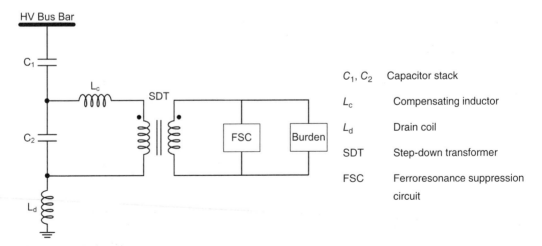

C_1, C_2	Capacitor stack
L_c	Compensating inductor
L_d	Drain coil
SDT	Step-down transformer
FSC	Ferroresonance suppression circuit

Figure 10.5 A CCVT circuit connection.

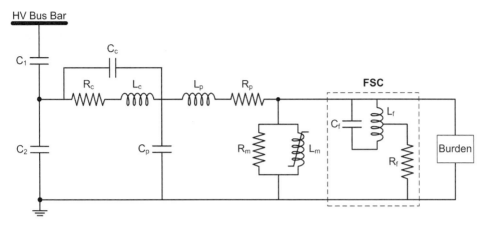

Figure 10.6 Reduced CCVT model with ferroresonant suppression circuit.

resistor R_f is a damping resistor designed to damp ferroresonance oscillations within one cycle. The circuit is tuned with a high Q factor in order to attenuate ferroresonance oscillations at any harmonic except the fundamental.

FSCs do not affect transient response unless an overvoltage occurs. The FSC can be modelled using the non-saturable transformer. Primary and secondary windings are connected in such a way that parallel resonance occurs only at the fundamental frequency. At other frequencies, only the leakage inductance is involved, so the damping resistor is the one that attenuates ferroresonance oscillations.

Other designs may include a resistor connected to the step-down transformer secondary side, or even a series gap, which is activated whenever an overvoltage occurs.

Validation tests performed without stray capacitances have shown that the CCVT model presented here can reproduce the subsidence transients and the ferroresonance behaviour; see, for instance, [35]. Simulation results have also shown that the magnetization inductance has a negligible effect on the subsidence transients, the magnitude of the subsidence transient increases with an increased burden and when the values of the divider capacitances are decreased, and the subsidence transient has the largest magnitude and lasts the longest time, when the fault occurs at a primary voltage zero.

10.3.4 Voltage Transformers

Modelling of magnetic voltage transformers (VTs) is similar to modelling any other power transformer. Figure 10.7 shows one model that can be used to accurately simulate the transient response of VTs [7, 8, 31].

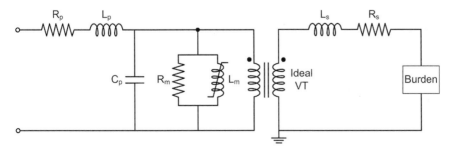

Figure 10.7 Voltage transformer model.

10.3.5 Case Studies

This section presents the test of two instrument transformer models that will later be used in another case study. An accurate simulation aimed at testing a relay performance must be based on an accurate representation of the instrument transformers. The introduction of this section has listed some concerns that are to be accounted for when simulating or testing a protection scheme model. CT saturation, CCVT ferroresonance, a subsidence transient on the CCVT secondary, or the relaxation current in a CT secondary are some aspects that can lead to misoperation of the protection system and have to be predicted before designing the protection scheme.

10.3.5.1 Case Study 1: Current Transformer Test

Figure 10.8 shows the diagram of the circuit implemented in ATPDraw for testing a CT model: a 60-Hz 34.6 kV (phase-to-phase rms value) source supplies a linear *RL* branch divided into several sections that will allow moving the fault location and check the transient response of the CT model. The transformer core is modelled using the Type-96 pseudo-nonlinear hysteretic inductor. Figure 10.9 depicts the current-flux curve specified in the nonlinear branch. The resistance of the burden is 3 Ω.

Figure 10.10 shows simulation results derived from applying a permanent fault and moving the fault location, as indicated in Figure 10.8. The phase angle of the source voltage remains

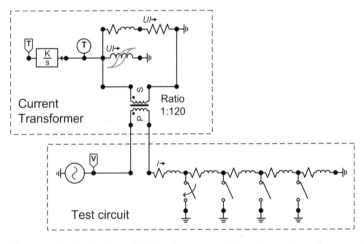

Figure 10.8 Case Study 1: ATPDraw implementation – Current transformer test.

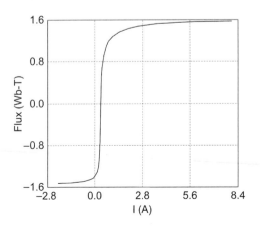

Figure 10.9 Case Study 1: Current-flux characteristic curve of the current transformer core.

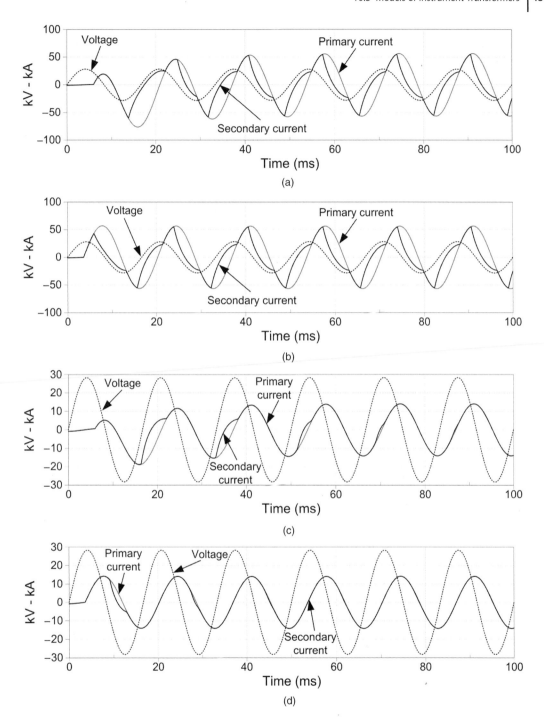

Figure 10.10 Case Study 1: Current transformer test (secondary currents are scaled). (a) Fault location: Node 1 – Tclose = 6 ms, (b) Fault location: Node 1 – Tclose = 3.6 ms, (c) Fault location: Node 3 – Tclose = 6 ms, (d) Fault location: Node 3 – Tclose = 3.6 ms.

the same in all simulations, although the instant at which the fault occurs is varied. It is evident that, as the fault location approximates the source, the fault current increases and the CT core reaches saturation more quickly. The figure also shows that the distortion initially exhibited by the secondary current of the transformer depends on the moment at which the fault occurs, although the final waveform is the same irrespective of the fault time.

10.3.5.2 Case Study 2: Coupling Capacitor Voltage Transformer Test

Figure 10.11 shows the scheme of a coupling-capacitor voltage transformer implemented in ATPDraw. The circuit includes an FSC. The main parameters of this CCVT model are listed below:

- Divider capacitances: $C_1 = 5$ nF; $C_2 = 82$ nF.
- Compensating inductor: $R_c = 228\,\Omega$, $L_c = 58$ H, $C_c = 100$ nF.
- Step-down transformer: Ratio $= 6584/115$ V, $R_p = 400\,\Omega$, $L_p = 2997$ mH .
- FSC: $C_f = 9.6\,\mu$F, $R_f = 40\,\Omega$.

The step-down transformer is modelled by means of the saturable transformer model available in ATP and the current-flux curve shown in Figure 10.12.

Figure 10.13 shows simulation results corresponding to a burden resistance of $200\,\Omega$ with and without FCS. The fault occurs once the initial transient in the implemented circuit settled down. The effect of the FSC is evident from these results: with a high-resistance burden, ferroresonance can occur and the voltage between the burden terminals can increase. The addition of the FSC can avoid this phenomenon and limit the burden voltage. The results when the burden resistance decreases are basically the same with and without FCS.

Figure 10.14 illustrates another concern. The plot displays some difference between the voltage measured in the high-voltage network and the voltage measured between the burden terminals. If this *subsidence voltage* reached a large value, it could cause relay misoperation.

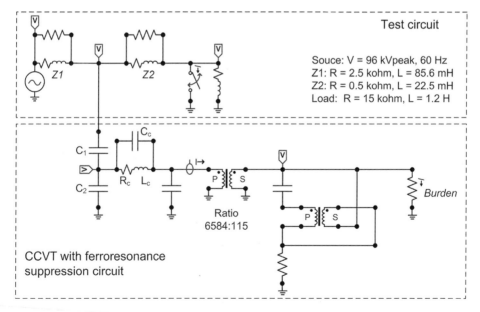

Figure 10.11 Case Study 2: ATPDraw implementation – Coupling capacitor voltage transformer test.

Figure 10.12 Case Study 2: Current-flux characteristic curve of the transformer core.

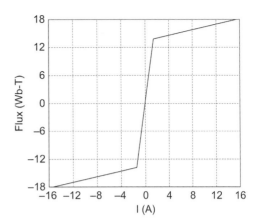

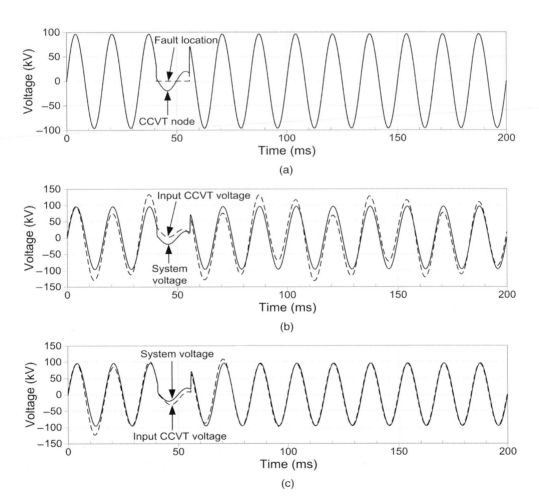

Figure 10.13 Case Study 2: Coupling capacitor voltage transformer test. (a) Source and system node voltages, (b) Bus 1 and Node 1 voltages – Burden = 200 Ω – Without FSC, (c) Bus 1 and Node 1 voltages – Burden = 200 Ω – With FSC.

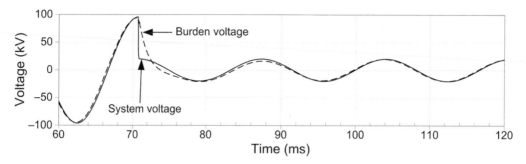

Figure 10.14 Case Study 2: CCVT test. System and burden voltages with FSC (Burden = 40 Ω).

10.3.6 Discussion

The simulation results presented above have shown some areas of concern related to the performance of instrument transformers under transient conditions [7, 8, 31]:

- CT saturation reduces the magnitude of the secondary current. This will reduce the operating force or torque of electromechanical relays; a reduced torque increases the operation time and reduces the reach of the relay.
- CT saturation affects the zero-crossings of the current wave; this will affect schemes that depend on the zero crossings, such as phase comparators.
- The relaxation current in the CT secondary is the current that flows when the primary circuit is de-energized. This current is more pronounced in the case of CTs with an anti-remanent air gap. The relaxation current can delay the resetting of low-set overcurrent relays and also cause the false operation of breaker failure relays.
- The reduction of the primary voltage to zero creates a subsidence transient on the CCVT secondary, because of the stored energy in the capacitive and inductive elements. This transient can affect the speed of the protection scheme and cause relay misoperation.

10.4 Relay Modelling

10.4.1 Introduction

The development of a relay model for use in transients simulations will usually follow an iterative process. The first step is to identify and implement the components of the relay; next, the relay model must be validated to confirm that it can represent the behaviour of the original relay. If the developed model does not meet the requirements, the components that need to be remodelled or additional components that need to be added should be identified, and a revised model should be developed. This revised model should again be validated. The process is continued until a suitable model is achieved.

This section provides some insight about the type of relay models that can be implemented in a time-domain simulation tool and the options that can be considered for their simulation, details the sources from which this information can be obtained, and discusses the accuracy and limitations of those models. The section includes two simple case studies that illustrate how ATP capabilities can be used to implement and simulate electromechanical and numerical relays.

10.4.2 Classification of Relay Models

Different types of relay models have been developed and used. The classification may be based on the source of input waveforms, relaying system structure, modelling details, and relay technology [9].

- *Input waveforms.* Two different types of relay models can be defined based on the source of the input waveforms [41]. *Steady-state models* represent a relay characteristic that is valid for stationary voltages and currents without any consideration of transients; the models reflect the relations between phasors in the original relays (e.g. relations between voltage and current phasors). These models can be used to: (i) evaluate the relay design concept and its main characteristics; (ii) make general decisions regarding the selection of the protection type, (iii) calculate the relay parameters (or settings), (iv) evaluate the relay settings which determine the relay actions (e.g. tripping signals) initiated by network calculation results, or (v) establish relay coordination margins. These models cannot provide all the information that is required in transient studies. *Transient models* use instantaneous samples obtained from a transients simulation tool, from data recorded by digital fault recorders (DFRs), or from numerical relays during actual power system disturbances. This type of model can be used to investigate the behaviour of the relay models in front of critical transients or during network disturbances (e.g. time sequence of tripping commands or pick up signals), or analyse the influence of settings on the protection behaviour.
- *Relaying system structure.* Relay models can be divided into two categories. *Structural models* mimic the actual relays by adequately representing their internal structure; they are applicable when a detailed relay description is available, and imitate relay performance on the basis of mathematical models of the physical phenomena. *Black-box models* reflect the relay performance by mapping input signals to the output, without taking into consideration the internal logical or physical structure of the relay.
- *Model details.* The models can also be divided into two categories: detailed and generic models [41]. A *detailed model* reproduces all the characteristics, algorithms, and behaviour of the actual relay; it is usually very complex and can be developed only when there is a full knowledge of the relay structure and parameters. Practically, only manufacturers can develop such a type of model. A *generic model* is a simplified model based on the relay performance and characteristics; it behaves in a similar manner to the actual relay within specified and clearly understood bounds. This type of model can be very useful as, in many situations, there is no reason to use a detailed model, and the required information may not be available. Generic models may not be adequate for some marginal cases and for precise timing.
- *Relay technology.* The relay models reflect the technology used in relay construction. The following three groups of the relays should be considered [7–10]:
  - *Electromechanical relays.* They are difficult to model because they involve a combination of different physical phenomena [42–44]. References [43, 44] present the modelling of a Type 50 overcurrent relay as three subsystems: electrical, magnetic, and mechanical. This approach offers some advantages since the three subsystems can be developed separately and combined as desired.
  - *Static relays.* Static electronic relays use solid-state components such as transistors, diodes, gates, flip-flops, comparators, counters, level detectors, or integrators. They were mainly developed from the late 1960s until early 1990s. Not much work has been carried out to implement these types of relays using a time-domain simulation tool, and they are no longer manufactured. An early static model of a distance relay was presented in [45].

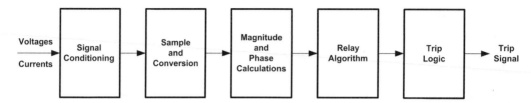

Figure 10.15 Schematic diagram of a numerical relay.

The model of a static mho distance relay intended for the protection of transmission lines was implemented into an EMTP-like tool, and presented in references [46, 47].

– *Numerical relays.* Detailed models of these relays represent measurement and decision-making algorithms in the same or similar form as in the physical relay [16, 17, 48–50]. Numerical relays include frequency tracking algorithms to make relays insensitive to frequency excursions, and additional protection functions such as directional elements, detection of power-swing, computation of sequence quantities, and phase selection logic. Some protection algorithms are based on voltage and current phasors. Since a time-domain simulation tool calculates voltage and current waveforms such as functions of time, it may be necessary to convert the sequences of voltage and current values to their equivalent phasors as functions of time. This conversion can be done by using signal-processing techniques such as the *Discrete Fourier Transform algorithm* [5] and the *Least Squares algorithm* [51]. Currents and voltages during faults contain high-frequency components. To obtain phasor values of voltages and currents, many digital relays first sample (digitize) the input voltages and currents, then recover the estimates of the phasor values. The general organization of a numerical relay can be that displayed in Figure 10.15: (i) the first section may include input auxiliary transformers and anti-aliasing low-pass filters; (ii) the second section may be the analogue-to-digital converter (ADC); (iii) the third section is the detector that estimates fundamental frequency information; (iv) the relay measuring principles are next; (v) finally, a section that represents the trip logic.

The main limitation of relay models may be the level of approximation used to represent an actual relay. The existing practice in the industry is to make available only limited information regarding relay design and its behaviour. Most of the information provided in relay manuals is given in the form of operating characteristics and related phasor equations. In general, this is not sufficient for a full description of the relay behaviour under transient conditions, although the limitations of relay models are relative depending on their application. If the purpose is to provide general education about relay designs and behaviour, phasor-based models can suffice. If the purpose is to determine actual behaviour of a relay under fault condition, phasor-based models can be quite limited and very often inaccurate.

10.4.3 Implementation of Relay Models

A relay model is composed of modules representing its various parts and can be implemented either using a single tool or a combination of different software and hardware platforms [52, 53]. The behaviour of a relay model can be simulated and checked by interfacing the relay model to the power system model. The interface between relay models and the power system model can be realized by means of either open- or closed-loop interaction [9]: see Figure 10.16. Each method has its advantages and disadvantages and has to be selected depending on the type of study and its purpose.

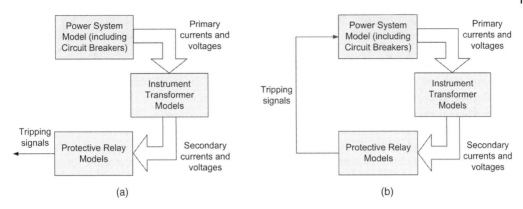

Figure 10.16 Interaction between power system and protection system: (a) Open-loop interaction; (b) Closed-loop interaction.

- *Open-loop method.* The principle is illustrated in Figure 10.16a. Primary current and voltage waveforms are obtained from the transient response of the power system under fault conditions. CT and CCVT models use the simulated waveforms to calculate the secondary voltages and currents, which may exhibit distortion introduced by the transducers. Finally, the computer model processes secondary values and produces the response. In some cases, the tripping response of the relay provides sufficient information, but in many complex scenarios it is impossible to obtain the correct answer.
- *Closed-loop method.* It is particularly suitable for protection schemes involving tripping activity, see Figure 10.16b. It is the most challenging option since it requires the interaction between the power system and the protection model on a step-by-step basis. Reference [54] gives an example using MATLAB and an EMTP-like tool.

To develop a reliable relay model, it is necessary to know the details of the design and the operational characteristics of the relay. This knowledge can be obtained from technical design characteristics and operational data of all relay elements (transducers, filters, A/D converters, etc.), test results carried out on the relays, and information on the algorithms that govern the relay operation. The development of a model may be relatively simple for numerical relays, provided sufficient information about the algorithms is available.

- *Manufacturers.* Manufacturers usually release a certain amount of information that may include service, instruction, or technical manuals, application or relay setting notes, testing and approval documentation, and technical description of the components. With some exceptions, relay circuit diagrams and algorithm source codes are not usually available for confidentiality reasons. If the available data is not sufficient for building a complete model, at least it can be used to calibrate generic simulation models. The information provided by the manufacturer enables some appreciation of the operating principle; the most fundamental source of information is always the instruction manual, but very rarely is the full documentation made available. It is important to perform an extensive validation procedure in order to assess the accuracy of the simulation results under varying power system fault conditions and, if possible, introduce additional corrections to the relay model.
- *Published literature.* This includes books, papers, special reports, courses, and patents. Most relaying principles belong in the public domain and in many industrial products the proprietary information is usually minimal. Patents taken out by the manufacturer are very often a source of detailed information.

- *Relay test data information.* Test data can be used to evaluate relay characteristics and support model development. Manufacturers carry out tests during product development using digital and analogue power systems simulators. Several open- and closed-loop real-time simulators have been utilized for relay testing by providing an interface between simulation computers and relays to be tested. References [55–63] discuss some developments in digital simulators and relay testing methodologies. Many utilities have DFRs distributed across the whole power system, so waveforms of actual faults recorded in DFR files are becoming a valuable source of information to achieve more realistic protection system models. However, the dynamic range of the DFR measurements can limit accuracy of the information. Proper settings of protection relays are a prerequisite for a reliable operation, and for this purpose, it is necessary to determine the fault values at various places at the time of fault occurrence. Also, it is necessary to consider the coordination task for the adjustment of multiple protective relays and different types of protective relays [64–66]. When the manufacturer's data is not available, empirical data can be used to develop a black-box model.

10.4.4 Applications of Relay Models

An accurate relay model can be used to investigate different relay measuring principles for a particular protection application and to optimize relay settings. Since simulations are performed without using physical equipment, a significant potential cost reduction in testing can be achieved. An important application of computer models is the analysis of protection system misoperations and their causes. This can help to gain insight into the relaying process and for the analysis of power system disturbances and blackouts. In addition, computer modelling can automate the calculation of the response of every relay measuring element to a large number of simulated fault conditions. Usefulness and applications of relay models are analysed below [9].

- *Manufacturers.* Several benefits can be derived from the use of relay models; for instance, careful investigation of the protection software (i.e. filter designs, sampling rates, relay algorithms, communication interfaces), fast testing of software changes, or testing the ability of a protection device to cooperate with other devices (e.g. remote control systems). Although these possibilities cannot completely replace tests with actual protection devices, they are rather fast and can be carried out without using a complete set of lab equipment.
- *Utilities.* Relay models can be used to evaluate and select the proper protection according to network conditions. In this evaluation, one can include the power system model and the complete protection system. These models can be used to select a relay measuring principle and to evaluate how different measuring principles perform in a specific power system protection application. Relay models can also be used to optimize relay settings, determine the security margins, predict protection actions under different system conditions, and help identifying power system constraints.
- *Training and education.* Relay models can effectively support the education of students, engineers, and researchers since they can easily demonstrate the general working principles of protection systems. The interaction between the power system and the protection system can be shown in an easy manner. It is also possible to show the protection coordination process and how it actually works. Students can learn how to improve relay settings in order to avoid protection failures and maloperations.

10.4.5 Testing and Validation of Relay Models

Testing procedures depend on the type of model and relay design. A general principle is that tests should include a range of realistic conditions under which the device is expected to operate. Due to the limited number and range of recorded network faults, only a restricted test set can generally be formed from fault records. Simulation then becomes an effective way of creating a suitably extensive test set. The tests to be conducted for model validation should contain, depending on the type of relay and the type of model, some or all of the following test categories [9]:

- *Operational tests* (static conditions). They are performed to ensure proper steady-state operation. Simulations include marginal conditions for relay operation such as faults on the verge of a zone in the case of distance relays, or high-resistance faults and external worst case scenario faults for unit protection relays. Operational tests may be sufficient for static relay models, for which input signals must be in the form of phasors.
- *Timing tests* (dynamic conditions). They are designed to verify the transient response of the relay. A number of simulations are performed and the tripping times are recorded under various fault conditions. Input signals are sampled current and voltage traces. Variable fault parameters may include: (i) the type of protected circuit (overhead line, cable, transformer, etc.), (ii) the length of the protected circuit, (iii) the position of the fault along the protected circuit, (iv) power system fault level, (v) fault connection and resistance, (vi) type of earthing in the grid. This category of tests is necessary for transient models.
- *Transient traces* (dynamic conditions). If possible, it is recommended to utilize the recording of transient signals from inside the relay. In the case of transient models, this helps to verify individual sections of the model by comparing model and relay transient responses. This is the most reliable method of validation of the dynamic protection models.

A model is an approximation of the actual device and requires careful validation to ensure sufficient accuracy. The general principle of validation is based on the comparison of the results produced by the model and those from the actual device under various system fault conditions. The following validation procedures have been used [9]:

1. *Modular validation.* It may concern either individual relay components or the whole relay. Validation of the individual components is feasible when each relay module or component can be monitored separately. The range of input/output signals should be sufficient to capture the required operational characteristics of the modelled component. Validated model components may then be utilized to build models of larger relay structures. The validation of the components, however, does not eliminate the need for validation of the whole relay.
2. *Use of secondary injection.* Relays can be tested by using specialized injection equipment that reproduce current and voltage waveforms from digital recorder data. Records can be obtained from an actual network event or by means of simulation. A common practice is to inject only fundamental frequency components of waveforms, without adding frequency components that may exist in the actual recordings. This type of test is based on the assumption that the relay measures only fundamental frequency components. The injection of both the real relay and the model with the same signal is an effective method as both the relay and the model are then subjected to the same input signals and both should exhibit the same response. When using a phasor-based test method, differences between the model and the real relay may be observed due to the non-realistic nature of this type of test.
3. *Use of published test results and relay technical specification.* It is also possible to validate a relay model using public domain data or the data from manufacturer testing procedures. This approach is particularly practical in the case of phasor-based models, which do not require detailed dynamic behaviour to be taken into account. In such cases, theoretical characteristics may suffice to validate the required scope of operation of the model. When an accurate dynamic behaviour is necessary, it is important to perform the validation procedure under varying system fault conditions.

The application of a given validation method depends on some factors [9]:

- *Relay type*. In the case of electromechanical relays, validation primarily involves the calibration of model parameters such as resistances and magnetic properties of the relay mechanism. Because of the strong interdependence between all the parts, it is usually difficult to apply the modular approach. However, electronic and numerical relays are suited to a modular validation due to their internal modular architecture.
- *Modelling extent*. The validation of static relay models can be constrained to the comparison of the theoretical (or measured) static tripping characteristics with the model simulation results. The time factor is not taken into account since the relay model does not have enough detail to respond dynamically to the input signals, although a constant time delay can still be incorporated. Transient models require more extensive measurement and validation procedures, since not only the static tripping characteristic must be consistent, but also tripping times under various system conditions have to be tested.
- *Data availability*. Different relay designs require different levels of information.

10.4.6 Accuracy and Limitations of Relay Models

The proper assessment of a model accuracy is essential for further interpretation of relay simulation results and diagnoses. The ways of expressing an error depend on the type of model and the extent of validation [9]. For transient relay models, two general categories can be distinguished in terms of time response: for fast operating relays (tripping within one or two cycles), the error may be characterized as the absolute time difference between the model and the relay (in ms); for slower operating schemes (operation from a few hundred milliseconds to seconds), the error may be expressed as the percentage of the recorded actual tripping time.

Several factors must be considered before a relay model can be widely and safely used. It is important to first verify the different modules, and then verify the accuracy of the assembled set. Moreover, the usage of a simplified model requires a full understanding of the relay model limitations. However, the main limitation may be the level of approximation used to represent an actual relay. Relay models can be inadequate in evaluating the performance of protection systems because of several limitations:

- *Valid operating range*. Studies conducted with relay models can be used to identify critical cases for which relay systems should be tested. Relay models can be used for studies in which the range of system operation is limited. In addition, several types of studies cannot be performed using relay models; for instance, the impact of radio-frequency radiation on relays, seismic disturbances, contact bouncing, and environmental factors.
- *Modelling of critical components*. In general, not all relay components are modelled. Factors that affect the performance of numerical relays cannot be or are not modelled in sufficient detail; specific issues include modelling of A/D converters, deviation of frequency, and accumulation of errors in recursive computations and integrity of components.
- *Modelling of communication channels*. The performance of some relay systems is affected by the performance of the communication channels used for inter-trip or blocking. Although a communication channel can be modelled in ATP, it cannot be based on an accurate representation because of the limited capabilities of the tool for these purposes.

10.4.7 Case Studies

10.4.7.1 Overview
This section presents two examples that illustrate the scope of models and case studies that can be carried out with ATP in protection studies. The first example deals with an electromechanical

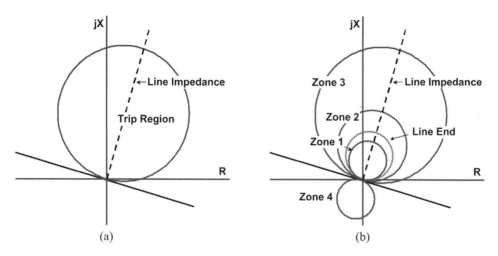

Figure 10.17 MHO characteristic: (a) Typical MHO characteristic; (b) MHO characteristic zones.

distance relay, while the second example is based on the model of a numerical distance relay. The study is carried out in both cases with the same test system model.

Distance relays compute the ratio voltage/current (i.e. the apparent impedance) on a transmission line and operate when the value is less than a preset value. The operation characteristic is often plotted on the impedance plane. Figure 10.17a shows the characteristic for phase and ground fault protection, which is considered the classic distance characteristic. Distance relays can have multiple protection zones (forward and reverse). Figure 10.17b shows a typical case with three forward and one reverse zone:

- **Zone 1**. This zone provides high-speed protection for any type of fault within most of the line segment. Experience suggests setting this zone to reach less than 100% of the line since it could risk over-reaching for close-in faults just beyond the remote end substation. The Zone 1 reach can be set to 85% of the line.
- **Zone 2**. This zone is set to protect the remainder of the line plus a safety margin, typically greater than 20%. Zone 2 delay time is coordinated with instantaneous fault clearing at the remote bus, typically 15–30 cycles.
- **Zone 3**. This section is set with progressively longer time delays and provides extended back-up protection.
- **Zone 4**. This may be set as reverse-looking section with the reach exceeding the remote end Zone 2 reach, plus a safety margin.

10.4.7.2 Case Study 3: Simulation of an Electromechanical Distance Relay

This case study presents the implementation and application of an electromechanical relay model, which is applied to the protection study of a transmission line. Although the test system is a real one, the study is theoretical, since no validation has been performed. The main goals of the case are to illustrate how to represent the protection system in a transients program, test the relay model performance, and show how to use ATP in protection studies [67, 68].

A dynamic state-space model of the electromechanical MHO distance relay model used in this study was presented in [69]. The ATP model of the relay using TACS capabilities was detailed in [42]. The approach used in this chapter to represent the relay uses MODELS code [70]. A diagram of a typical MHO distance relay is shown in Figure 10.18. The relay has two coils: a polarizing coil and an operating coil. The polarizing coil is connected to a voltage transformer through a memory circuit consisting of a variable inductor and a capacitor. The operating coil

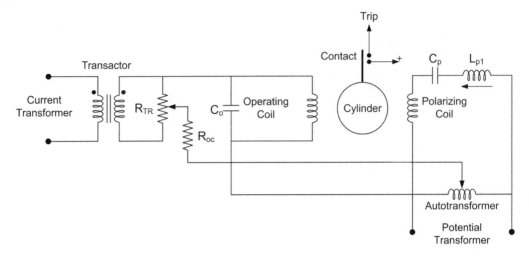

Figure 10.18 Case Study 3: Diagram of an MHO distance relay.

is connected to a CT through a transactor. The electromagnetic torque is developed by the interaction of currents through the two coils; if the torque is of adequate direction, magnitude, and duration, the relay will trip. The magnitude of the impedance setting can be adjusted by a potentiometer or transactor and autotransformer taps. The angle of maximum torque can be adjusted by the variable inductor.

From this diagram, a state-space model can be obtained. A ninth-order model was presented in [69] with the following form:

$$\frac{d}{dt}[X] = [A][X] + [B][Y] \tag{10.1}$$

where

$$[X] = [x_1 \quad x_2 \quad x_3 \quad x_4 \quad x_5 \quad x_6 \quad x_7 \quad x_8 \quad x_9]^T \tag{10.2a}$$

$$[Y] = [y_1 \quad y_2 \quad y_3]^T \tag{10.2b}$$

x_1 current through the transactor secondary
x_2 current through the operating coil
x_3 voltage across the operating coil
x_4 voltage across the memory circuit capacitor
x_5 current through the polarizing coil
x_6 maximum density of induced current by the polarizing coil
x_7 maximum density of induced current by the operating coil
x_8 angular displacement of the cylinder
x_9 angular velocity of the cylinder
y_1 input voltage from the VT
y_2 derivative of the input current from the CT
y_3 electromagnetic torque.

Details about matrices, the calculation of their elements, as well as the electromagnetic torque, were presented in [69]. It is important to note that the torque acts as an input, but its value is calculated from values of some state variables (i.e. x_2, x_5, x_6, x_7).

A hybrid approach is used here for simulating this relay model [67, 68]. The model is split up into two parts, and different ATP capabilities are used to represent each part in a data file:

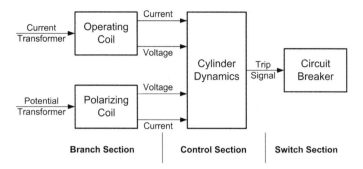

Figure 10.19 Case Study 3: Schematic representation of the electromechanical relay in ATP.

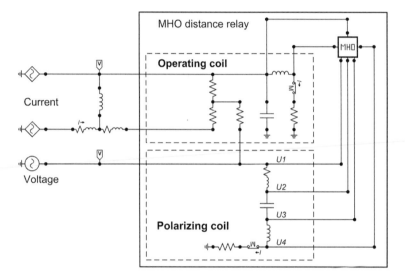

Figure 10.20 Case Study 3: ATPDraw implementation of the distance relay.

circuits of the operating and polarizing coils are represented in a BRANCH section, cylinder dynamics are represented by means of a MODELS section; see Figure 10.19. The circuit implemented in ATPDraw is shown in Figure 10.20.

The MODELS section implemented for representing the cylinder dynamics is the following one:

```
MODEL MHO
    CONST  mu {VAL:12.566371E-7}  - Permeability of the air
    DATA   npc  - Number of turns of the polarizing circuit
           noc  - Number of turns of the operating coil
           lh   - Length of the cylinder                              (mm)
           h    - Total length of the air gaps                        (mm)
           r    - Radius of the cylinder                              (mm)
           th   - Thickness of the cylinder wall                      (mm)
           rc   - Resistivity of the cylinder material               (ohm.m)
           ro   - Resistance of the operating coil                   (ohm)
           lo   - Inductance of the operating coil                    (H)
           rp   - Resistance of the polarizing coil                  (ohm)
           lp   - Inductance of the polarizing coil                   (H)
           rp1  - Resistance of the variable inductor                (ohm)
           lp1  - Inductance of the variable inductor-Polarizing circuit  (H)
```

```
              maxd - Maximum angular displacement                              (rad)
              ks   - Elastic coefficient of the spring                         (kg.m)
              j    - Moment of inertia of the cylinder                         (kg.m2)
      VAR     a64  a65 - Coefficients of the state-equations
              a66  a72 - Coefficients of the state-equations
              a73  a77 - Coefficients of the state-equations
              a89  a98 - Coefficients of the state-equations
              b61  b93 - Coefficients of the state-equations
              x6   y6  - Maximum density of induced current by the polarizing coil
              x7   y7  - Maximum density of induced current by the operating coil
              x8   y8  - Angular displacement of the cylinder
              x9   y9  - Angular velocity of the cylinder
              tor      - Auxiliary variable
              torque   - Electromagnetic torque
      INPUT   v1       - Voltage across the operating coil
              u1       - Voltage from the potential transformer
              u2       - First node voltage of capacitor in polarizing circuit
              u3       - Second node voltage of capacitor in polarizing circuit

              x2       - Current trough the operating coil
              x5       - Current trough the polarizing coil
      HISTORY x6 {DFLT:0}  x7{DFLT:0}  INTEGRAL(y8){DFLT:0} INTEGRAL(y9){DFLT:0}
      INIT
          a64:=-4.E6*npc/(lp+lp1)/pi/r/th        a65:=a64*(rp+rp1)
          a66:=-4.E6*rc*h*(lh+2*r)/mu/pi/(r*r)/th/lh
          a72:=-4.E6*noc*ro/pi/r/th/lo
          a73:=-a72/ro                           a77:=a66
          a89:=1
          a98:=-ks/j
          b61:=-a64
          b93:=1/j
          tor:=8*mu*r*r*th*lh/3/h/1.E9
          x6:=0    x7:=0    x8:=0    x9:=0
      ENDINIT
      EXEC
        y6:=a64*(u2 - u3) + a65*x5 + b61*u1
        LAPLACE(x6/y6):=1.0| / (-a66| + 1|s)
        y7:=a72*x2 + a73*v1
        LAPLACE(x7/y7):=1.0| / (-a77| + 1|s)
        torque:=tor*(noc*x2*x6 - npc*x5*x7)
        y8:= a89*x9
        x8:= INTEGRAL(y8) {DMIN:0   DMAX:maxd}
        y9:= a98*x8 + b93*torque
        x9:= INTEGRAL(y9) {DMIN:0}
      ENDEXEC
    ENDMODEL
```

The parameters are those presented in [42, 69]. This relay model does not incorporate the representation of protection zones; that is, the user cannot specify protection zones as depicted in Figure 10.17.

Figure 10.21 shows the results of a simple test: a current is injected into the operating coil and a voltage is applied between the terminals of the polarizing coil. The figure shows the plots of the angle, speed, and torque. Remember that these three values are calculated within the MODELS section. Since the maximum displacement of the relay cylinder has been fixed in 0.10 rad, the relay will trip with the simulated operating conditions.

The analysis of the relay model behaviour is an important aspect. The relay response can be crucial for selecting the instrument transformers and very useful to understand its behaviour when applied to protect transmission lines.

The circuit shown in Figure 10.20 has been grouped to create a single block that could represent the complete relay model. The circuit implemented in ATPDraw for testing the relay model and the menu window of this model are shown in Figure 10.22.

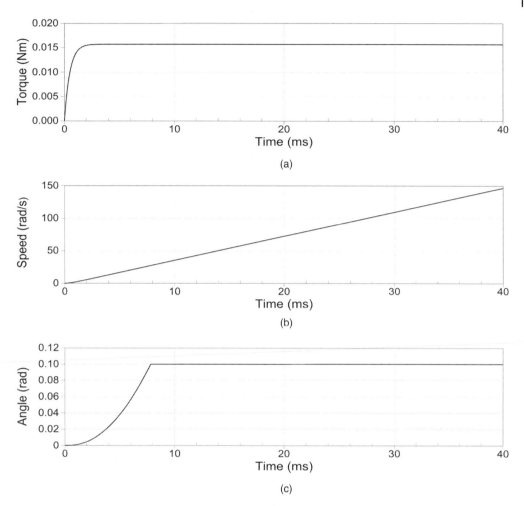

Figure 10.21 Case Study 3: Diagram of an MHO distance relay (Voltage = 30 V, $\phi = 0°$; Current = 1 A, $\phi = 30°$). (a) Electromagnetic torque, (b) Angular velocity, (c) Angular displacement.

The relay test is basically aimed at estimating the *R-X* values for which the relay trip. With the implemented model, a trip occurs when the relay angle reach its maximum value; therefore, the test can be based on a parametric analysis in which the values of *R* and *X* are varied within the specified limits. Simulation results will provide the combinations of *R* and *X* for which the relay cylinder reaches its maximum displacement. In this example, the values of *R* and *X* vary between 0 and 1.5 Ω in steps of 0.05 Ω. The relay model response will depend on the input levels; that is, if the source voltage remains constant, the relay response will depend on the ratios selected for the voltage and CTs. Figure 10.23 shows some results obtained with a 60 Hz voltage source whose peak value was 10 kV. The figure show some expected results: each plot corresponds to a different combination of voltage and current reductions (provided by the instrument transformers); the greater the reductions the smaller the impedance area (or distance) covered by the relay. It is evident that the distance covered when the instrument transformers decrease the voltage and the current 250 and 2000 times, respectively, is much shorter than that covered when the reductions are respectively 100 and 500 times.

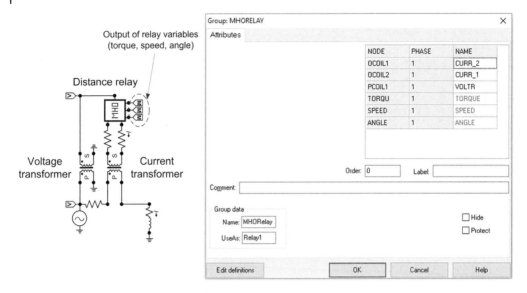

Figure 10.22 Case Study 3: Testing the relay model.

Figures 10.24 and 10.25 show respectively the geometry and the diagram of the 60-Hz 125.6 mi 115-kV (sub)transmission line simulated in this case study. The model implemented in ATP-Draw is depicted in Figure 10.26. Some important aspects of this model are listed below:

- the transmission line is represented by a frequency-dependent distributed-parameter model;
- network equivalents at both ends of the line are modelled as linear lumped-parameter impedances;
- the voltage level of the left-side system is 230 kV and the system includes a step-down 230/115 kV transformer represented by its short-circuit impedance: all parameters of the left-side equivalent are reduced to the 115 kV level (see Figure 10.26);
- the protection system model includes instruments transformer models, but only the models for phase *A* are included in the circuit implemented in ATPDraw.

Only phase *A* to ground faults with very low fault resistance are simulated.

The models used to represent instrument transformers are those analysed in the previous section. Note that a second CT has been added in the model to adapt the transmission current level to that required by the relay. To test the model under transient conditions, the following cases were run:

- Symmetrical single-phase-to-ground fault at location FAULT1, phase *A*.
- Asymmetrical single-phase-to-ground fault at location FAULT2, phase *A*.

An asymmetrical fault is obtained by closing the switch at voltage zero, while symmetrical faults were caused by closing the switch at peak voltage.

Figures 10.27 and 10.28 show results obtained from the simulation of the two scenarios. One can observe the distorted waveforms of currents and voltages caused by the fault located at FAULT1; this distortion can occur with either symmetrical or asymmetrical faults, depending of the fault location. Remember that voltages at the CCVT location, after the fault is caused, should be close to zero; however, one can observe that an oscillatory voltage, the subsidence transient, does appear at both sides of the voltage transformer. As expected, faster relay

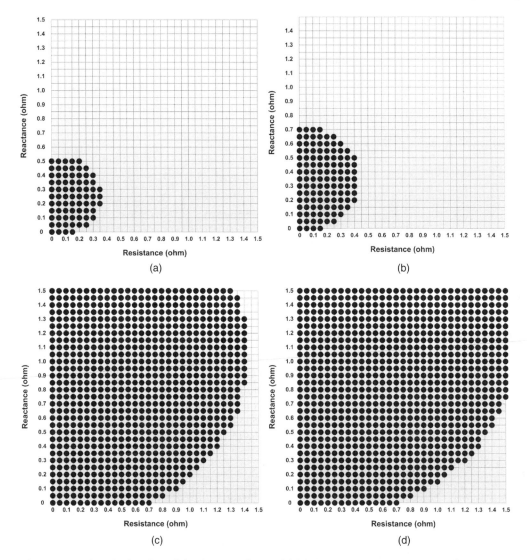

Figure 10.23 Case Study 3: Test of the distance relay model. (a) CT ratio = 2000; VT ratio = 250, (b) CT ratio = 2000; VT ratio = 100, (c) CT ratio = 500; VT ratio = 250, (d) CT ratio = 500; VT ratio = 100.

operations are caused by faults which are closer to the relay location. For faults caused at the same location, the relay response is faster with asymmetrical faults.

10.4.7.3 Case Study 4: Simulation of a Numerical Distance Relay

Numerical distance relays are digital realizations of proven electromechanical distance relay designs: voltage and current signals from the CCVT and CT secondary are input for a low-pass filter that removes frequency content above the sampling frequency; the signal is then sampled by the A/D converter; the sampled data is passed again through a low-pass filter that removes the frequency content above the fundamental frequency. Most digital relays estimate amplitudes and phase angles of phasors using digital filters. This information is then used to detect abnormal conditions.

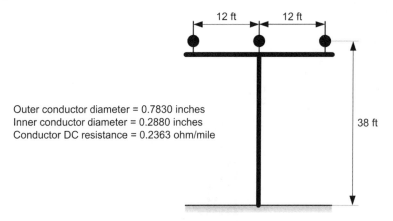

Figure 10.24 Case Study 3: Configuration of the test line.

This new case study uses the same transmission line model that was simulated in the previous case and is based on a distance relay model developed by B. Wilson [48]. The model simulated here is a simplified version of the original implementation.

Figure 10.29 shows a diagram of the model organization: it is divided into four sections or submodels whose roles within the model are analysed below.

1) *Low Pass Filter Model.* The first section is analogue anti-aliasing low-pass filter. The model uses a third-order Butterworth design and has a 3-db roll-off frequency of 235 Hz. The frequency, order, and design were listed in the manufacturer's instruction book for the actual physical relay. The Butterworth filter with a roll-off frequency of 1 rad/second has poles equally spaced along a unit circle in the complex plane and located at: (i) $p_1 = -1.0 + j0$; (ii) $p_2 = -0.5 + j0.866$; (iii) $p_3 = -0.5 - j0.866$. The transfer function is:

$$H(s) = \frac{1}{(s - p_1)(s - p_2)(s - p_3)} \tag{10.3}$$

For a filter with a roll-off frequency of 235 Hz, a simple frequency scaling results in: (i) $p_1 = -1476.5 + j0$; (ii) $p_2 = -738.3 + j1278$; (iii) $p_3 = -738.3 - j1278$. The denominator polynomial becomes: $s^3 + 2.953 * 10^3 s^2 + 4.358 * 10^6 s + 3.216 * 10^9$.

2) *Analogue-to-digital converter model.* An analogue-to-digital converter model samples the 60 Hz fundamental eight times per cycle (480 Hz). The model receives a vector of dimension four as input and produces four vector outputs of dimension nine. The first and the last entries in the output vector correspond to samples of the input signal one cycle apart.

3) *Detector model.* The relay uses a Fourier notch filter. The Fourier detector recovers fundamental frequency (phasor) information. The purpose of the detector is to estimate the real and imaginary components of the fundamental frequency phasor. The implemented detector is a recursive-realization of a discrete Fourier transform, it is a three-phase (plus zero sequence) Fourier detector that recovers Fourier coefficients from a single-phase input signal. Both inputs to and outputs from this model are vectorial. Four different phase-specific vectors are the inputs. For the inputs, the vector components correspond to different samples in time. There are two output vectors of the real and imaginary Fourier components.

4) *Distance measuring unit model.* The relay uses the mho relay measuring principle and makes decisions whether to open the circuit breaker. The circular mho relay characteristic was developed from comparisons made between two different inputs. The inputs are called S_1 and S_2. The equations for the two inputs to the comparator are all that must be known for steady-state modelling. In this study, the S_1 signal is called the operating signal and S_2 the

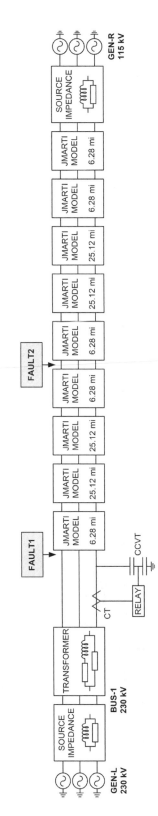

Figure 10.25 Case Study 3: ATPDraw implementation of the test system.

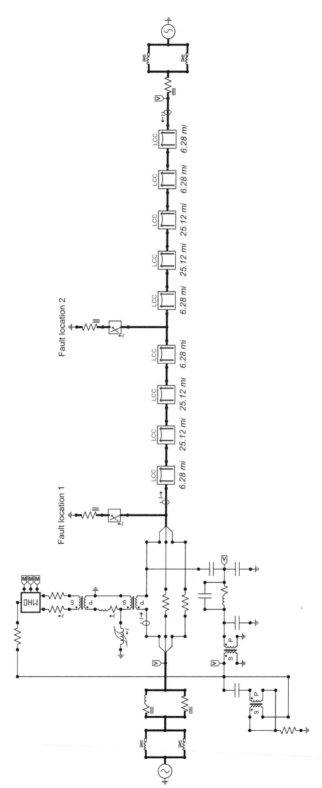

Figure 10.26 Case Study 3: ATPDraw implementation of the test system.

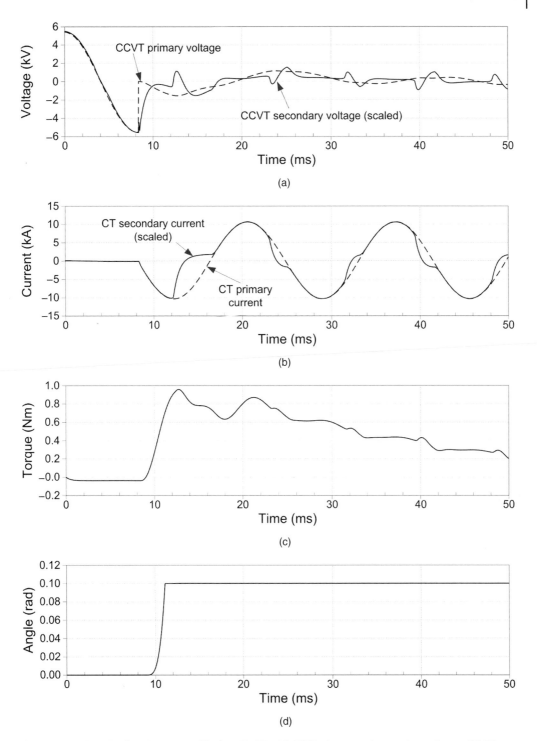

Figure 10.27 Case Study 3: Symmetrical fault at FAULT1: (a) CCVT primary and secondary voltages; (b) CT primary and secondary currents; (c) Relay electromagnetic torque; (d) Relay angular displacement.

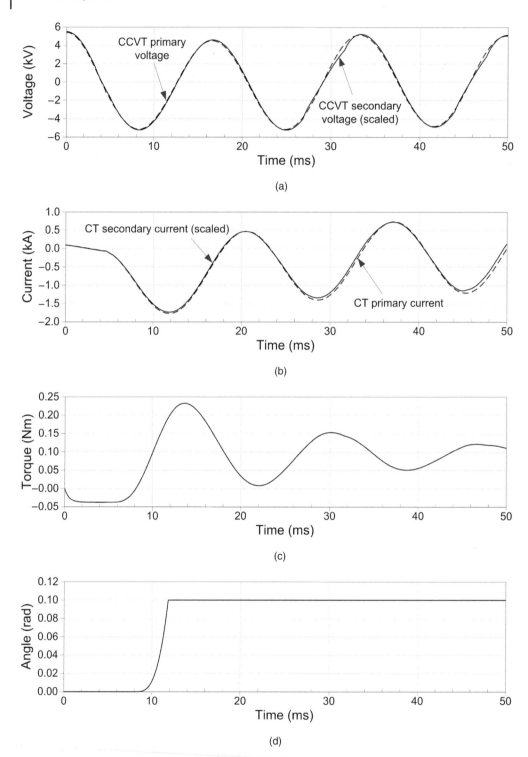

Figure 10.28 Case Study 3: Asymmetrical fault at FAULT2: (a) CCVT primary and secondary voltages; (b) CT primary and secondary currents; (c) Relay electromagnetic torque; (d) Relay angular displacement.

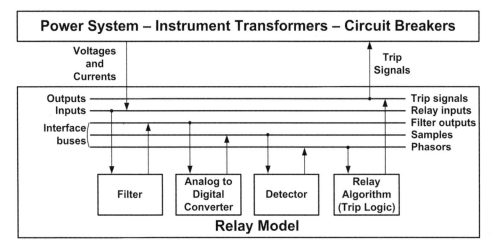

Figure 10.29 Case Study 4: Diagram of a distance relay model and interface to the power system model.

polarizing. The trip decision can be made with an amplitude or phase angle comparison criteria. Models of both the uncompensated self-polarized and the zero-sequence compensated quadrature-voltage polarized mho measuring units were implemented.

- For the phase A self-polarized mho characteristic, the two inputs are:

$$S_1 = I_a Z_{cg} - V_{fa} \qquad S_2 = V_{fa} \tag{10.4}$$

- For the phase A zero-sequence compensated quadrature-voltage polarized mho characteristic, the two inputs are:

$$S_1 = (I_a - kI_0)Z_{cg} - V_{fa} \qquad S_2 = V_{fbc} \tag{10.5}$$

where I is the secondary current, k is the zero-sequence compensation factor ($= (Z_0 - Z_1)/3Z_1$), Z_0 is the zero-sequence transmission line impedance, Z_1 is the positive-sequence line impedance, Z_r is the relay replica impedance, and V_f is voltage at the relay location.

The relay has been implemented in MODELS code. Figure 10.30 depicts the scheme of the test system: a relay model is located at each end of the transmission line. The selected settings are the same for both relays.

Figure 10.31 displays the new test system model, as implemented in ATPDraw. The relay model is a simplified version of the original model and can only be used for single-phase-to-ground faults. Note that the three circuit-breaker phases will simultaneously open; that is, each relay will create a single trip signal that will open (if required) the three phases of its corresponding circuit breaker. Instrument transformers are not modelled, so the study is limited since relay responses are idealized by assuming a perfect behaviour of the instrument transformers. There will not be reclosing operations.

The selected settings for both relays are as follows:

$$Z_{pos} = 125\angle 68° \qquad Z_{zero} = 425\angle 80° \quad \text{ohm} \tag{10.6}$$

where Z_{pos} and Z_{zero} are respectively the positive- and zero-sequence impedances of the protected zone.

Two fault locations have been selected to check the performance of the relay model; see Figure 10.30.

Figures 10.32 and 10.33 shows some of the results obtained with a single-phase-to-ground fault (phase A) at each location indicated in Figure 10.30. The fault characteristics are the same

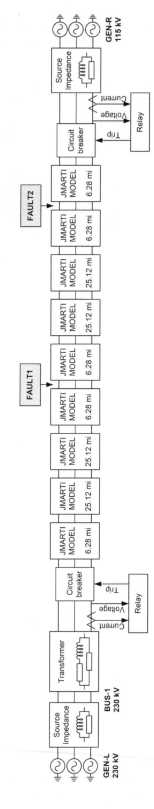

Figure 10.30 Case Study 4: Diagram of the test system.

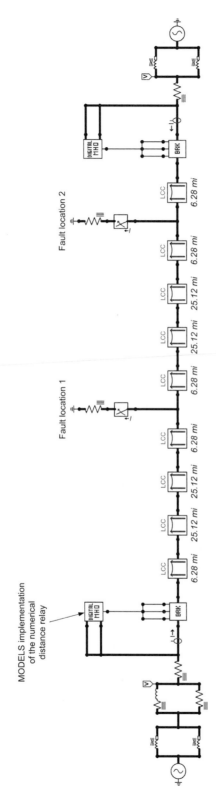

Figure 10.31 Case Study 4: ATPDraw implementation of the test system.

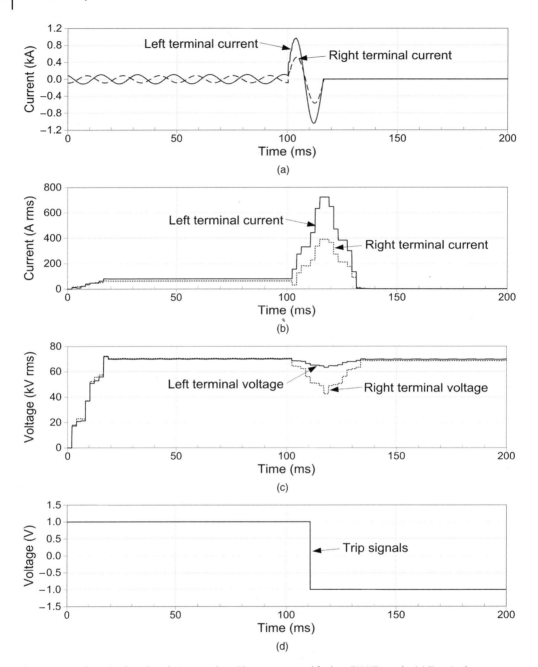

Figure 10.32 Case Study 4: Simulation results – Phase-to-ground fault at FAULT1 node. (a) Terminal currents, (b) Current magnitudes, (c) Voltage magnitudes, (d) Circuit breaker trip signals.

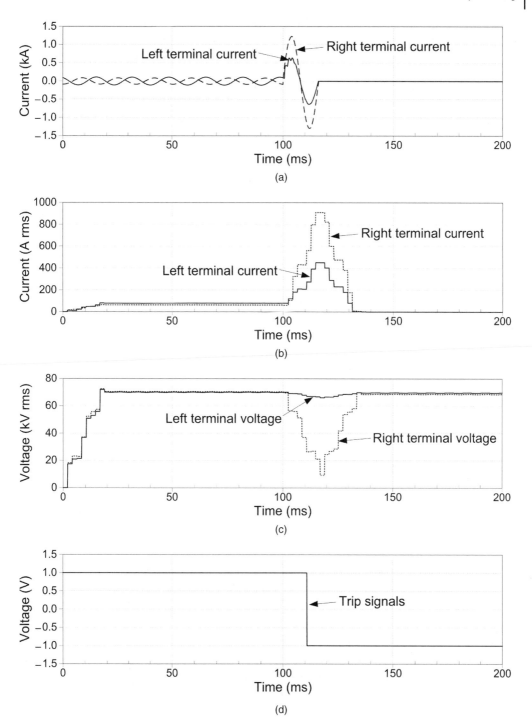

Figure 10.33 Case Study 4: Simulation results – Phase-to-ground fault at FAULT2 node. (a) Terminal currents, (b) Current magnitudes, (c) Voltage magnitudes, (d) Circuit breaker trip signals.

in both cases: the fault is permanent and occurs at 0.1 seconds, so after the two circuit breakers open they should remain opened. The two figures show the line currents measured at the two line terminals, the internal relay variables for terminal voltages and currents, and the relay signals sent to the circuit breakers.

The waveforms of the currents in both figures are as expected: the higher current is that measured at the line terminal that is closer to the fault location. Note, however, that the highest voltage drop occurs in both cases at the right terminal; this is due to the values of the equivalent impedances at the two-line terminals.

An interesting aspect of these results is that the two relays generate the trip signals simultaneously, the instant at which the signals are generated does not depend on the fault location. This is due to the algorithm implemented for this case study and to the fact that there is no communication between relays. The original model included a simple model of a communication channel between the relays located at each end of the line, so the model simulated the time interval required to transmit a digital signal between the two line ends. For more details, see references [48, 49].

10.5 Protection of Distribution Systems

10.5.1 Introduction

The main priority of the protection system is to prevent damage to utility equipment; secondary goals are reliability and power quality. In general, distribution protection is based on standardized settings, equipment, and procedures [71–73].

Although it is commonly accepted that the installation of small distributed generation (DG) units at distribution levels has many advantages, utilities are generally concerned that the installation of small DG units may result in damage to their equipment or to the equipment of their customers. The connection of DG units raises several challenges, such as islanding or the impact of DG on power quality [74–80].

This section summarizes the main features of the devices commonly used to protect radial distribution systems with or without DG, discusses the models implemented in ATP, and presents some case studies that illustrate the modelling of distribution-level protective devices and some of the common problems that can be encountered in distribution system protection.

10.5.2 Protection of Distribution Systems with Distributed Generation

10.5.2.1 Distribution Feeder Protection

The distribution system has, in general, a radial design and its protection is basically overcurrent protection. A highly reliable performance of a distribution network can be achieved by installing different types of protective devices. The coordination between protective devices is another important aspect, and this is particularly difficult in distribution networks given the number of different protective devices (breaker-relay sets, reclosers, fuses, sectionalizers). This coordination is not easy and not always possible. In addition, it can be altered by the presence of distributed generators [74, 75, 81].

10.5.2.2 Interconnection Protection

Small DG units that operate interconnected with the utility distribution system can be classified into three groups: synchronous generators (recuperating engines, combustion turbines, small hydro), induction generators (wind generators), and non-traditional asynchronous generators (fuel cells, photovoltaic) [77–79]. Generator protection is typically connected at

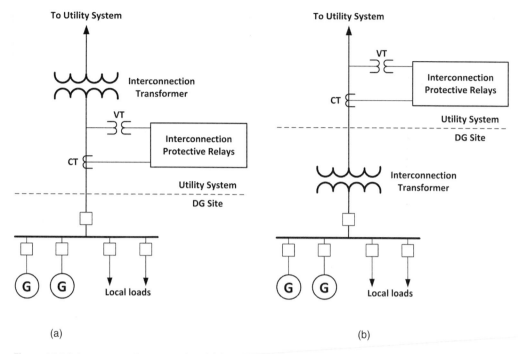

Figure 10.34 Interconnection protection: (a) Protection at the transformer primary; (b) Protection at the transformer secondary.

the terminals of the generator and provides detection of internal short circuits and abnormal operating conditions (loss-of-field, reverse power, overexcitation, and unbalanced currents). For smaller generating units, utilities usually leave to the DG owners the responsibility of selecting the protection level they believe is appropriate. Utilities, however, become involved in specifying interconnect protection, which provides the protection that allows DG units to operate in parallel with the utility grid. Interconnection protection for small generators is established at the point of common coupling (PCC) between the utility and the generator. This point can be at the primary of the interconnection transformer as illustrated in Figure 10.34a, or at the secondary of the transformer as illustrated in Figure 10.34b, depending on ownership and utility interconnect requirements. Interconnection protection satisfies the utility's requirements to allow the DG to be connected to the grid. Its function is threefold [77–79]:

1. Disconnects the DG when it is no longer operating in parallel with the utility system.
2. Protects the utility system from damage caused by connection of the DG, including the fault current supplied by the DG for utility system faults and transient overvoltages.
3. Protects the generator from damage from the utility system, especially through automatic reclosing.

10.5.3 Modelling of Distribution Feeder Protective Devices

This subsection presents a review of the protective devices used to protect radial distribution systems and a summary of the approaches followed here for representing each one. Readers are referred to textbooks and standards for more details [71–73, 82, 83].

10.5.3.1 Circuit Breakers – Overcurrent Relays
The performance of a breaker during an opening operation is governed by the characteristics of the overcurrent relay. There are two types of relays: instantaneous and time-delay. The

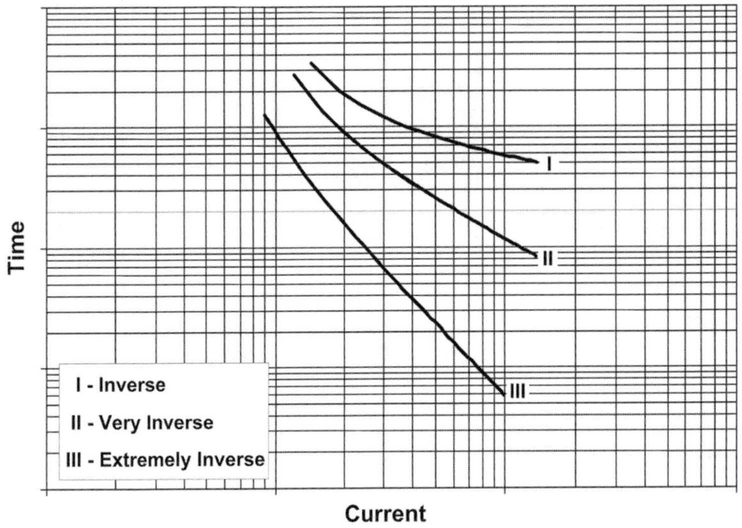

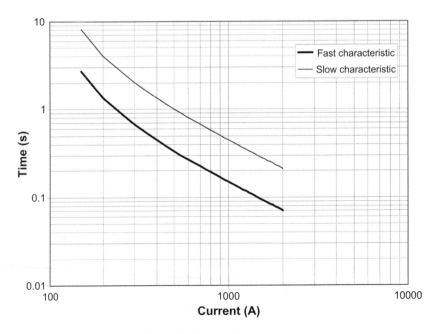

Figure 10.35 Time-current characteristics of overcurrent relays.

time-current characteristic of an overcurrent relay can consist of two sections, the first one is independent of the current, and the second one has an operating time that varies inversely with current. Depending on the rate with which the operating time and current are related, the time-overcurrent characteristic can be classified as inverse, very inverse and extremely inverse, see Figure 10.35. Typical values of these parameters can be consulted in the literature for each type of characteristic [83, 84]. Circuit-breaker models for protection studies do not usually include dynamic arc representation and, by default, all opening operations are successful. A breaker module is usually based on an ideal switch controlled by an overcurrent relay whose time-current curve and parameters have to be specified by users.

Figure 10.36 Time-current characteristics of a recloser.

10.5.3.2 Reclosers

A recloser is an overcurrent protective device that can sense and interrupt fault currents, and reclose a predefined number of times automatically. In general, reclosers have less interrupting capability than breakers [71–73]. Recloser operation uses two time-current curves, see Figure 10.36. The first one, known as fast or instantaneous, is mainly used to save lateral fuses under temporary fault conditions. The second curve is known as slow or time-delay, and its main purpose is to delay recloser tripping, and allow fuses to blow under permanent fault conditions. A recloser can be set for a number of different operations, although a very common reclosing sequence has two fast operations followed by two time-delay trips. The recloser model is based on an ideal controlled switch, and allows users to include two tripping curves (fast and slow), select the type of time-current characteristic, and specify the number of reclosing operations for each characteristic, as well as the duration of each reclosing interval.

10.5.3.3 Fuses

Current fuses can be classified into two groups: current limiting and expulsion fuses. Fuses of the first group limit the magnitude as well as the duration of the current; the second group limits only the duration of the fault, allowing the flow of overcurrents, which will be interrupted after one zero pass. An expulsion fuse interrupts a fault current at current zero; a current limiting fuse interrupts a fault current by forcing a current zero. These differences can significantly affect power quality.

A detailed fuse model should duplicate the following stages [82]: current sensing, arc initiation, arc interruption and, current interruption. The melting period, during which temperatures rise, begins with the fault and finishes when the fuse melts; during this stage, the current flows without limitation. The melting period of a fuse depends on the magnitude and the duration of the current, as well as on the electrical properties of the fuse. This characteristic is shown in the so-called time-current curve provided by manufacturers. The performance of a fuse is displayed by means of the minimum melting and the total clearing curves; see Figure 10.37.

Expulsion fuses typically interrupt fault currents at the first current zero. In some cases, expulsion fuses may not be successful in interrupting the current at the first current zero, but

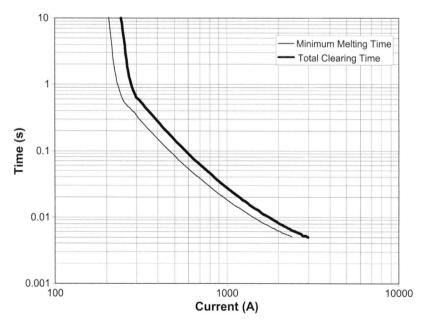

Figure 10.37 Time-current characteristics of a fuse.

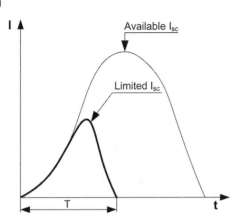

Figure 10.38 Effect of a current limiting fuse.

succeed in interrupting current at the second current zero; if a fuse fails to interrupt current at the second current zero, it will usually be destroyed and the current will be interrupted by a backup device. When the fault occurs, an expulsion fuse heats to its melting point; the current continues to flow in the form of an arc, at zero current the arc is extinguished, being the fuse subjected to a transient recovery voltage (TRV), whose magnitude depends on the operating conditions [85]. One or several arc reignitions can be caused by the TRV; the process only stops when the dielectric strength build-up is faster than that caused by the TRV.

Upon interruption, the operation of a current limiting fuse results in the insertion of additional impedance and the development of an arc voltage; when this voltage exceeds the system voltage, the arc is extinguished and the action is accomplished [82]. Figure 10.38 shows the operation of a current limiting fuse, from the instant at which the fault is caused until the instant at which the fuse melts and the current is interrupted. The operating time of a current limiting fuse can be shorter than half a cycle; however, this high speed has a cost, since an overvoltage will usually occur. The fuse limits the let-through energy to a fraction of the energy available from the system. The two main parameters in the current limiting fuse operation are the melt I^2t and the nonlinear resistance characteristic after melting.

The approach chosen for representing every fuse type has been different. Since the action of a current limiting fuse results in the sudden insertion of a high-value resistance, this type of fuse has been modelled as a variable resistance controlled from a TACS section, in which users can specify the parameters that govern the fuse resistance [86]. The input of the minimum melting time-current curve is made by means of a regression equation that matches the characteristic provided by the manufacturer; this means that different equation coefficients have to be considered for each manufacturer. An expulsion fuse can be represented as a switch that opens at the first zero-current; therefore, these fuses have been modelled as controlled switches. The module required for this type of fuse is basically the same as developed for representation of current limiting fuses, but without including any post-melting resistance. For more details, see references [10, 85–88].

10.5.3.4 Sectionalizers

A sectionalizer is a circuit-opening device that has no capability to break fault current, and it is usually installed downstream from a breaker or a recloser. After a circuit has been de-energized by a backup protective device (e.g. a recloser or a reclosing breaker), a sectionalizer isolates the faulted portion of a distribution network [71, 73], after which, the rest of the circuit is returned to service upon reclosure of the backup device. A sectionalizer counts the interruptions of the backup device; it can be set to open after one, two, or three counts have been registered within a predetermined timespan, see Figure 10.39. A sectionalizer opens during the open interval of the backup device. Although it cannot interrupt faults, it can be closed into them.

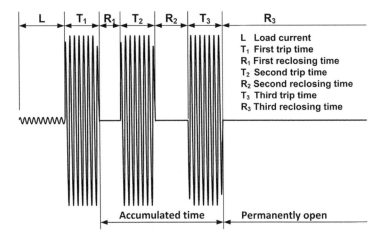

Figure 10.39 Sectionalizer performance.

Sectionalizers can be used in place of fuses or between a reclosing device and a fuse. They only detect current interruptions above a predetermined level and have no time-current characteristics, permitting easy coordination with other protective devices on the system. Their advantages over fuse cutouts are many: they offer safety and do not open under load; after a permanent fault, the fault-closing capability of the sectionalizer greatly simplifies circuit testing; if a fault is still present, interruption takes place safely at the backup recloser; the replacement of fuse links is not required; the possibility of error in the selection of the correct size is eliminated. The sectionalizer model implemented in ATP is based on an ideal controlled switch: the user has to specify the number of interruptions of the backup device and the minimum time during which the current is interrupted in each operation.

10.5.4 Protection of the Interconnection of Distributed Generators

Interconnection protection design depends on factors such as generator size, point of interconnection, type of generator, and interconnection transformer connection. The winding connection of the interconnection transformer plays an important role in how DG units will interact with the utility system under faulted conditions. All connections have advantages and disadvantages which need to be addressed by the utility in their guidelines, since the choice of a transformer connection has a deep impact on interconnection protection requirements [77–79]. Transformer connections used to interconnect dispersed generators to distribution systems can be classified into two groups, depending on the primary winding grounding.

- *Ungrounded-primary transformer winding.* The major concern with this connection is that after the substation breaker is tripped for a permanent ground fault, the system is ungrounded. This subjects line-to-neutral equipment on the unfaulted phases to an overvoltage that will approach line-to-line voltage. This occurs if the DG is near the capacity of the load on the feeder when the substation breaker trips. Some utilities use ungrounded interconnection transformers only if an overload (200% or more) on the DG occurs when the breaker trips. During ground faults, this overload level will not allow the voltage on the unfaulted phases to rise higher than the normal line-to-neutral voltage. Ungrounded primary windings should generally be reserved for smaller DGs where high overloads are expected on islanding.
- *Grounded-primary transformer winding.* This connection provides an unwanted ground fault current for supply circuit faults and reduces the current from substation breaker at the utility

Table 10.1 Interconnection protection objectives.

Objective	Protection function
Detection of loss of parallel operation	81O/U, 27, 59, 59I
Fault backfeed detection	Phase faults: 51V, 67, 21
	Ground faults: 51N, 67N, 59N, 27N
Detection of damaging system conditions	Negative sequence: 46, 47
	Loss of synchronism: 27
Abnormal power flow detection	32
Restoration	25

substation, which can result in a loss of relay coordination. When the DG is off-line (i.e. the generator breaker is open), the ground fault current will still be provided to the utility system if the interconnect transformer remains connected. This would be the case when interconnect protection trips the generator breaker. The transformer at the DG site acts as a grounding transformer with zero-sequence current circulating in the delta secondary windings. An interconnection transformer with grounded both primary and secondary windings also provides a source of unwanted ground current for utility feeder faults similar to that described above, as it also allows ground relays at the substation to respond to ground faults on the secondary of the DG transformer. This can require the utility to increase ground relay pickup and/or delay tripping to provide coordination, which reduces the sensitivity and speed of operation for feeder faults and can increase wire damage.

Table 10.1 lists some specific objectives of an interconnection protection system, as well as the relay requirements to accomplish each objective [76].

In general, all these types of relays can be based on modern numerical technologies. The models implemented in ATP to mimic these relays follow a different approach: they use a TACS section that duplicates the relay characteristic curves: users have to specify the rated values and the parameters that are needed to match the characteristics curves.

10.5.5 Case Studies

The main goal of this section is to present a scope of the distribution protective device models implemented in ATP. The models can represent circuit breakers, reclosers, fuses, sectionalizers, and several relay types. The case studies will show the performance of the protective relays, how they can be coordinated when more than one device is installed to protect a distribution system, and how to use the various relay models for simulating the interconnection protection of an embedded generator. The chapter does not cover any microgrid scenario, so by default, it must be assumed that an islanding operation is not accepted, and embedded generators must be disconnected in case of loss of parallel operation. The section also provides some details about the approaches used to implement protective devices and relays in ATP and ATPDraw.

10.5.5.1 Case Study 5: Testing the Models

The cases included in this subsection present several circuits implemented in ATPDraw for testing the custom-made models of the circuit breaker and overcurrent relays, recloser, fuse, and sectionalizer. Each case has been organized using a similar pattern: first, the test system implemented in ATPDraw is presented; next, the characteristics of the developed models are detailed; finally, some simulation results illustrate the behaviour of the models.

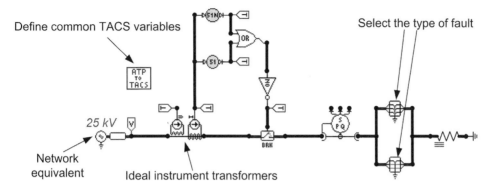

Figure 10.40 Case Study 5: ATPDraw implementation of the system for testing the circuit breaker and overcurrent relay models.

Test of the circuit breaker and overcurrent relay models

Figure 10.40 shows the circuit implemented in ATPDraw for testing the circuit breaker and the overcurrent relays models (i.e. 51 and 51N relay models). The supply source is represented by means of a simple system equivalent with the following parameters: Voltage $= 25\,\text{kV}$ (phase-to-phase rms value); Frequency $= 50\,\text{Hz}$; Short-circuit capacity $= 300\,\text{MVA}$; Ratio $X/R = 8$.

This system representation will be used with other protective device models tested in this section. It is a very simplified representation that does not account for frequency dependence and assumes that positive- and zero-sequence impedance values are the same.

A custom-made model representing ideal instrument transformers with unity ratios is used to pass the feeder currents to the overcurrent relays. All cases studied in this section are carried out without including instrument transformer models or using the ideal models shown in Figure 10.40. Therefore, simulation results should be analysed with some care since the cases studied in previous sections have shown how the instrument transformers can distort voltage and current waveforms, and then affect the relay responses.

Relay models are implemented in TACS code and mimic the characteristic curves, without any connection to the physical implementation of the actual relays. The time-current characteristics of the two overcurrent relays are represented by the following expression [83, 84]:

$$t(I) = \frac{K}{(I/I_a)^n - 1} \tag{10.7}$$

where n is a factor that characterizes each type of relay, K is a factor to distinguish each member of a family, and I_a is the pickup current (i.e. the smallest value of the current that will trigger the breaker to operate). Figure 10.41 depicts the characteristic curves and parameters selected for the two overcurrent relays, while Figures 10.42 and 10.43 show the main menu window of the circuit breaker and Relay 51 models.

Figures 10.44 and 10.45 display some results derived from the simulation of the response of the circuit breaker and relay models to a three-phase and a single-phase-to-ground fault, respectively. The comparison of these results provides some insight about the implemented models: the three-phase fault causes the opening of the circuit breaker due to an operation of the Relay 51, while the single-phase-to-ground fault causes the operation of the Relay 51N; although the current level of the three-phase fault is higher than that due to a single-phase-to-ground fault, the response of the Relay 51N is a little bit faster than that of the Relay 51 due to the

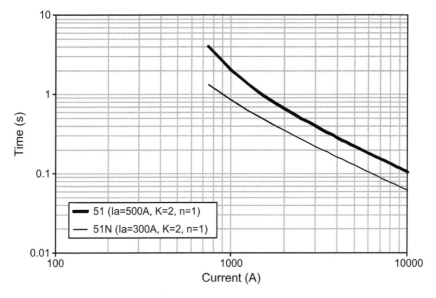

Figure 10.41 Case Study 5: Characteristic curves of the overcurrent relays.

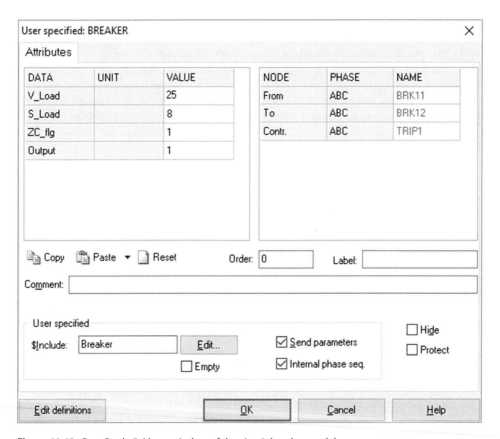

Figure 10.42 Case Study 5: Menu window of the circuit breaker model.

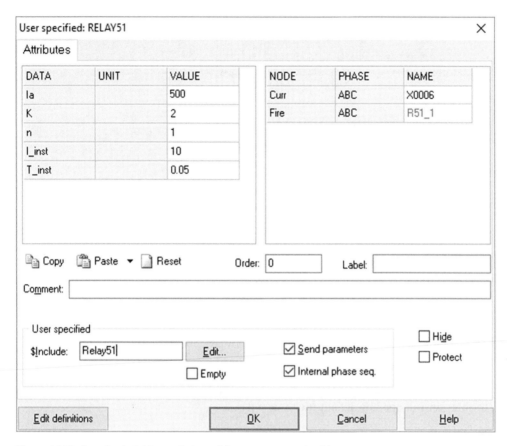

Figure 10.43 Case Study 5: Menu window of the overcurrent relay 51.

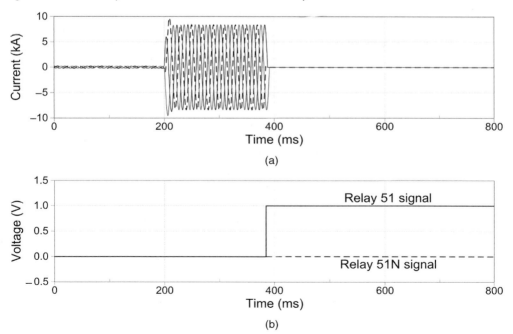

Figure 10.44 Case Study 5: Circuit breaker and overcurrent relay performance – Three-phase fault. (a) Circuit currents, (b) Trip signals.

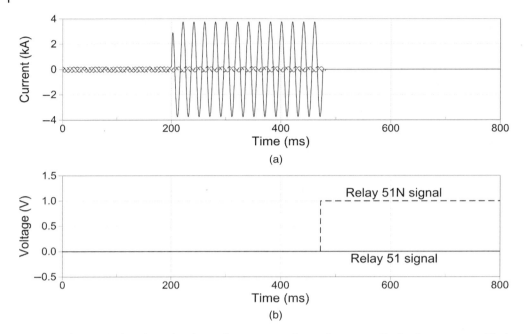

(a)

(b)

Figure 10.45 Case Study 5: Circuit breaker and overcurrent relay performance – Single-phase-to-ground fault. (a) Circuit currents, (b) Trip signals.

selected relay settings for this case study. Irrespective of the number of phases involved in the fault, the breaker will open by default the three poles.

Recloser test

The recloser model is based on a similar approach to that used with overcurrent relays and includes the switch model required for opening and closing operations. Figure 10.46 shows the circuit implemented in ATPDraw for testing the recloser model. The supply side is represented with the same equivalent system model that was used in the previous case (see Figure 10.40). The implemented recloser model is based on a controlled switch that allows users to represent two tripping curves (fast and slow), select the type of time-current characteristic, specify the number of reclosing operations for each characteristic and the duration of each reclosing interval [86, 89–91]. The model allows up to four operations. If less than four operations are to be simulated then a very long delay time (to indicate the recloser will remain opened) has to be specified for the last operation. This model does not include any instrument transformer.

Figure 10.47 displays the fast and slow characteristic curves selected for the recloser. Figure 10.48 shows the main window of the implemented recloser model.

The results derived from the simulation of a three-phase fault with different recloser settings are shown in Figure 10.49. The fault occurs at 0.2 seconds and clears at 2 seconds; see figure. The first plot corresponds to a recloser model with one fast and two slow operations; the second plot shows the behaviour of the recloser model with one fast and three slow operations.

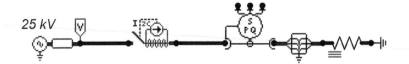

Figure 10.46 Case Study 5: ATPDraw implementation of the system for testing the recloser model.

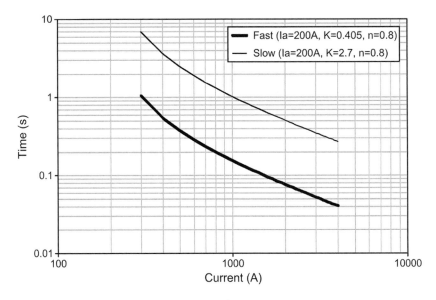

Figure 10.47 Case Study 5: Recloser tripping characteristics.

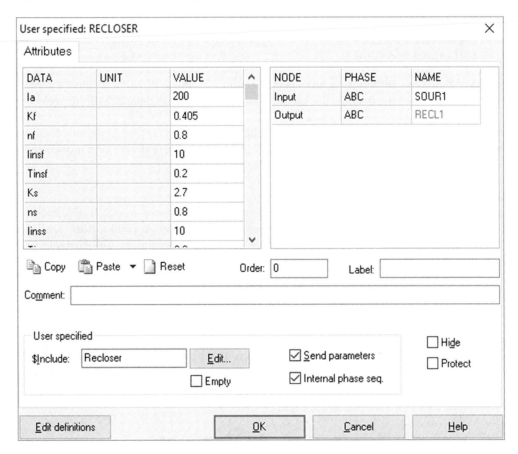

Figure 10.48 Case Study 5: Recloser menu window.

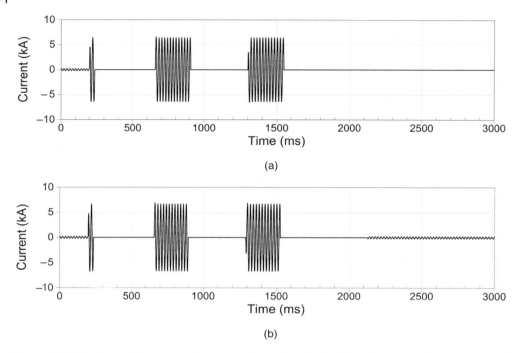

(a)

(b)

Figure 10.49 Case Study 5: Recloser performance – Three-phase fault. (a) One fast and two slow operations, (b) One fast and three slow operations.

Note that due to the fault duration, the selected settings for the first case will cause a permanent disconnection of the load. In the second case, the fault clears before the third reclosing operation.

The recloser settings might not be realistic enough; that is, actual reclosing periods could be longer than assumed in this study. Obviously, this is not a limitation of the recloser model, since, if required, the user can select longer reclosing periods.

Fuse test

The configuration of the system implemented for testing the fuse model is displayed in Figure 10.50; it is the same system that has been used to test the breaker and recloser models.

Figure 10.51 shows the main window of the implemented fuse model. The characteristic curves selected for this case study are basically those shown in Figure 10.37. The fuse model is three-phase but it can open one, two, or three phases, depending on the fault, and can be applied to reproduce the behaviour of either an expulsion or a current limiting fuse. Figure 10.52 shows the response of the two fuse models in front of a single-phase-to-ground fault.

Sectionalizer test

The configuration of the test system is shown in Figure 10.53: the series connection of a recloser and sectionalizer will protect the supply side from a three-phase fault at the load terminals. This

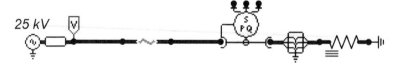

Figure 10.50 Case Study 5: ATPDraw implementation of the system for testing the fuse model.

User specified: FUSE ✕

Attributes

DATA	UNIT	VALUE	˄	NODE	PHASE	NAME
kVnom		10		Input	ABC	
Ia		220		Output	ABC	FUSA1
T_Blow		0.0001				
Rperm		0.02				
Kmult		100000				
Kexp		0.75				
Type		1				
PI		1	˅			

🗐 Copy   📋 Paste ▾ 📄 Reset   Order: 0   Label:

Comment:

User specified
$Include: Fuse   Edit...   ☑ Send parameters   ☐ Hide
             ☐ Empty   ☑ Internal phase seq.   ☐ Protect

Edit definitions   OK   Cancel   Help

Figure 10.51 Case Study 5: Fuse menu window.

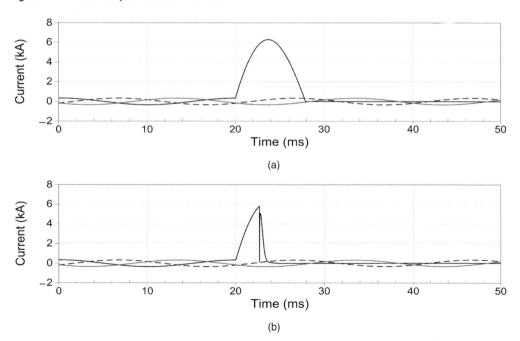

Figure 10.52 Case Study 5: Fuse model performance: Fault currents with a single-phase-to-ground fault.
(a) Expulsion fuse, (b) Current limiting fuse.

Figure 10.53 Case Study 5: ATPDraw implementation of the system for testing the sectionalizer model.

is not a realistic configuration since the sectionalizer will usually disconnect only a portion of a distribution network (i.e. the faulted section) and not the entire feeder, as in the case shown in this figure. Such a scenario will later be analysed; the goal now is to check the coordination of the sectionalizer model with the recloser model.

The sectionalizer model measures the current flowing through it and the line voltage. These values are used, according to its settings, to determine the instant at which, being the line de-energized, the downstream circuit must be opened. Figure 10.54 shows the main window of the implemented model: the user has to specify voltage and power rated values (they are only used to obtain the parameters of the internal snubber circuit), a current value needed to detect that a fault current is passing through the sectionalizer, the minimum delay time to start the operation, the number of recloser operations prior to the sectionalizer operation, a flag to select between single-phase or three-phase operation, and a flag for output purposes. Remember that the sectionalizer does not have any characteristic curve.

Figure 10.55 shows results derived from the simulation of a temporary three-phase fault that occurs at 0.2 seconds and clears at 2 seconds. The recloser model will use one fast and two slow

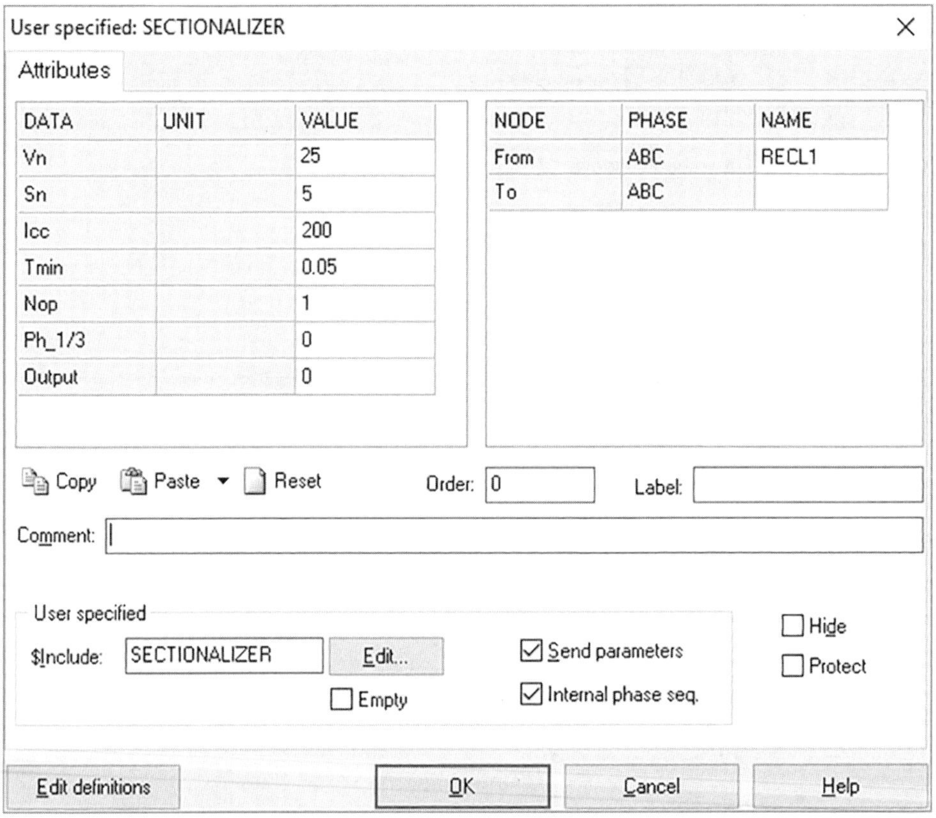

User specified: SECTIONALIZER   ✕

Attributes

DATA	UNIT	VALUE
Vn		25
Sn		5
Icc		200
Tmin		0.05
Nop		1
Ph_1/3		0
Output		0

NODE	PHASE	NAME
From	ABC	RECL1
To	ABC	

📋 Copy   📋 Paste ▾   🗋 Reset   Order: 0   Label:

Comment: |

User specified

$Include: SECTIONALIZER   Edit...   ☑ Send parameters   ☐ Hide

☐ Empty   ☑ Internal phase seq.   ☐ Protect

Edit definitions   OK   Cancel   Help

Figure 10.54 Case Study 5: Sectionalizer menu window.

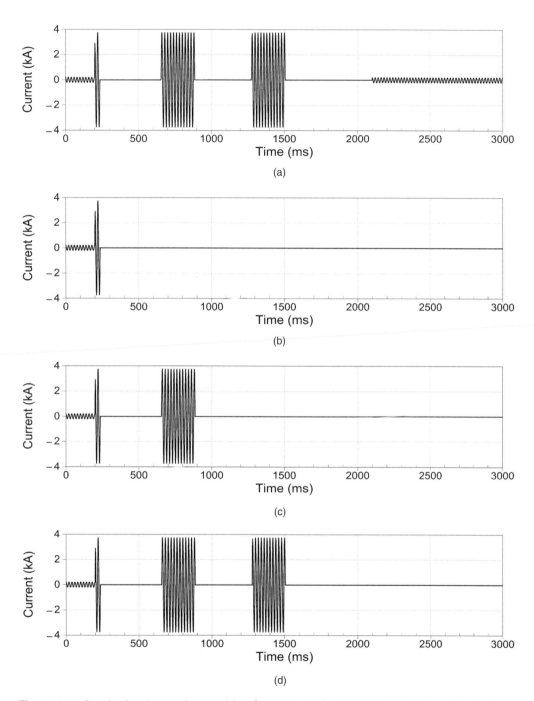

Figure 10.55 Case Study 5: Sectionalizer model performance – Fault currents with a three-phase fault. (a) Recloser without sectionalizer, (b) Recloser with sectionalizer – Open after 1 recloser operation, (c) Recloser with sectionalizer – Open after two recloser operations, (d) Recloser with sectionalizer – Open after three recloser operations.

operations. The figure plots show results corresponding to the following settings: (i) without sectionalizer; (ii) one sectionalizer operation; (iii) two operations; (iv) three operations. Obviously, other settings for both the recloser and the sectionalizer could produce different results.

10.5.5.2 Case Study 6: Coordination Between Protective Devices

Some selectivity between devices that protect radial distribution systems is usually required. In general, breaker and reclosers are set to trip for faults beyond a downstream fuse (i.e. before the fuse blows). This practice, known as *fuse saving*, can avoid fuse replacement and long outage times. A fuse saving scheme must use instantaneous overcurrent characteristics to trip the breaker/recloser before the fuse can blow. This will usually be followed by a fast reclosing operation, and if the fault persists, by the application of a slower inverse time characteristic to trip the breaker/recloser. As detailed above, the sectionalizer is an alternative to the fuse and its coordination with a recloser considering a realistic scenario will also be tested in this case study.

Figure 10.56 shows the system implemented in ATPDraw for testing the coordination between protective devices. It is a 25 kV, 50 Hz distribution network that supplies three LV loads. The lower voltage side of the substation transformer is grounded by means of a zig-zag reactor that limits the maximum ground current to 800 A. The parameters of the main components are provided in the figure.

Depending on the fault location, two scenarios can be simulated with the same file:

- In the first one, the fault is located at F1 and a fuse-saving strategy is used between the recloser and the FUSE 2, so in case of fault downstream of the fuse, the recloser will perform two fasts and one slow operation before allowing the fuse to blow.
- In the second one, the fault is located at F2, and the sectionalizer will open after one fast and one slow operation of the recloser.

The characteristic curves of the recloser and the fuses are depicted in Figure 10.57. All the fuses have been modelled using the same fuse type and ratings. The selected fuse model is of current limiting type and can operate the three poles independently.

Figure 10.58 shows the results obtained from the simulation of a fuse-saving strategy. A temporary three-phase fault downstream FUSE 2, at F1, occurs at 0.2 seconds and clears at 1.2

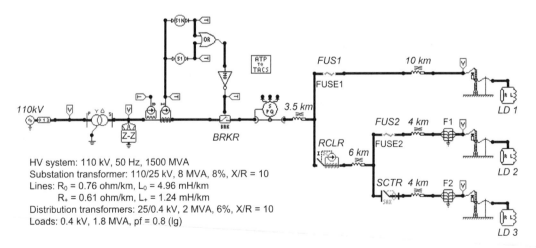

Figure 10.56 Case Study 6: Coordination between protective devices – Three-phase fault.

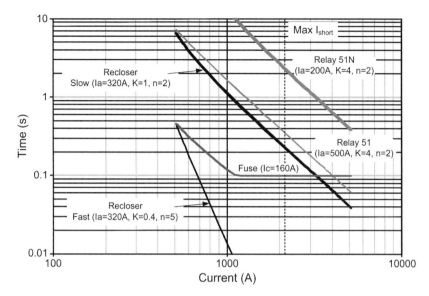

Figure 10.57 Case Study 6: Characteristic curves of protective devices.

seconds. It is evident from these results that the fast recloser characteristic avoids FUSE 2 to blow: during the first two operations, the recloser is faster than the fuse, but the fuse is faster during the third operation, and it opens before the fault is cleared. One can observe from the currents through the breaker, the recloser and the fuses, that only the Load 2 has been disconnected. Obviously, very different results could be obtained with different recloser settings. Note that the first two reclosing periods last for 0.4 seconds; if the first two reclosing periods were longer (i.e. 0.6 seconds), the fuse would not operate and no load would be disconnected.

Different results would also be derived with a different fault location if the sectionalizer has to operate. Assume that the location is moved to a position downstream of the sectionalizer, and the recloser settings are changed: one fast and two slow operations. Figure 10.59 shows the results obtained when the sectionalizer has to operate after two recloser operations. A temporary three-phase fault with the same initiation and clearing instants is again simulated. The plots of the currents prove that the sectionalizer will open after detecting two recloser operations, although the recloser settings were changed. As for the current plots, it is evident that only the current corresponding to Load 3 is now missed in the currents through the breaker and the recloser. The current through FUSE 1 becomes the same as that before the transient caused by the fault once the sectionalizer has operated.

Obviously, different results should be expected if the recloser settings were different; for instance, if the reclosing period after the first recloser operation was longer and the fault duration was shorter, the sectionalizer would not operate and the currents in the entire system would be the same that were prior to the fault occurrence.

10.5.5.3 Case Study 7: Protection of Distributed Generation

Figure 10.60 shows the ATPDraw implementation of the system used to test some interconnection protection features. The main goal is to check the transient response of the protection system in a distribution network with distributed generation. Given the system configuration, several fault locations can be used for testing the response of protective devices: (i) the fault

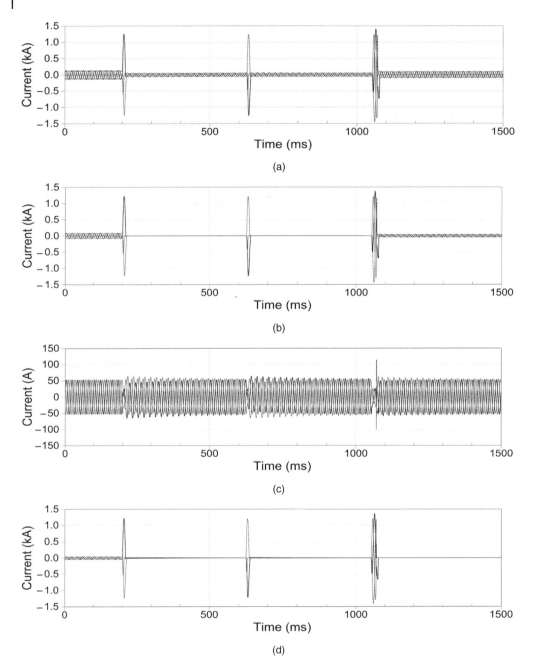

(a)

(b)

(c)

(d)

Figure 10.58 Case Study 6: Simulation results – Fuse saving – Three-phase fault at the MV terminals of Load 2 – Node F1. (a) Circuit breaker currents, (b) Recloser currents, (c) Fuse 1 currents, (d) Fuse 2 currents.

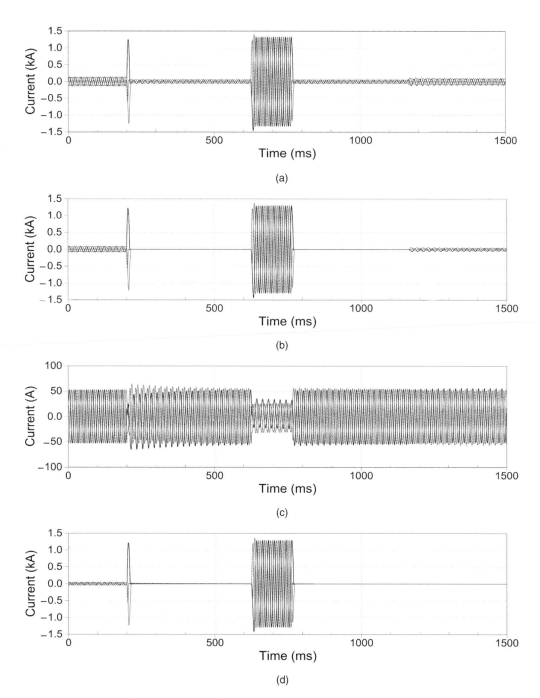

Figure 10.59 Case Study 6: Simulation results – Sectionalizer operation – Three-phase fault at the MV terminals of Load 3 – Node F2. (a) Circuit breaker currents, (b) Recloser currents, (c) Fuse 1 currents, (d) Sectionalizer currents.

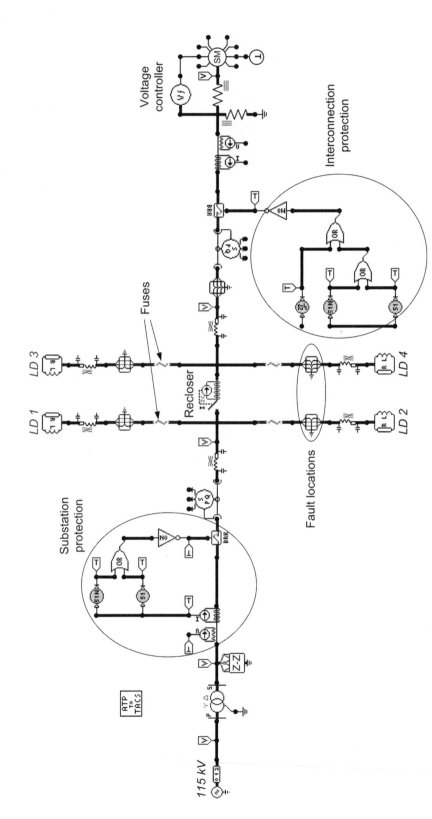

Figure 10.60 Case Study 7: ATPDraw implementation of the test system.

occurs on any lateral feeder; (ii) the fault location is on the path between the main power supply and the distributed generator.

If generation islands are not accepted and a fuse protection has been selected for all lateral feeders, a fuse-blowing strategy can be implemented: for any fault located on any lateral feeder, the corresponding fuse has to operate and isolate the faulted section before any other device (i.e. any protective device located between the main supply and the local generator) could operate. However, for a fault location between the main supply and the generator, either the interconnection protection or the substation protection should operate; such selective design would avoid the entire system load being disconnected.

The main parameters of the supply system and the local generator are listed below:

- Transmission system: 115 kV (phase-to-phase rms value), 50 Hz
- System equivalent: 1500 MVA, $Z_{1/2} = 0.1 + j1\,\Omega$, $Z_0 = 1.1 + j1.1\,\Omega$
- Substation transformer: 110/6 kV, 8 MVA, 8%, $X/R = 10$
- Generator: 6 kV, 2 MVA, 2 poles
- Loads: Loads 1–2 = 2 MVA, pf = 0.9 (lg); Loads 3–4 = 1 MVA, pf = 0.8 (lg).

The MV side of the substation transformer is grounded by means of a zig-zag reactor that limits the ground currents to 800 A. The diagram shows the fault locations that have been selected to analyse the performance of the protective device models and the coordination between them.

The characteristic curves of overcurrent relays that control substation circuit breakers are depicted in Figure 10.61. A Relay 27, which operates in case of undervoltages, is included in the interconnection protection of the generator. Its settings are as follows: it will open the generator circuit breaker if the terminal voltage decreases below 88% of the generator rated voltage for more than 0.5 seconds, or below 50% for more than 0.16 seconds.

Figure 10.62 shows some results obtained from the simulation of a three-phase fault located at the load-side of Fuse 4. The fault occurs at 0.2 seconds and clears at 2 seconds. For this scenario, a fuse-blowing strategy could be recommended to avoid the fault impact could be propagated beyond the faulted section. The plots show that the recloser and the fuse that protects the faulted section open simultaneously, so there will be an island formed by the generator and

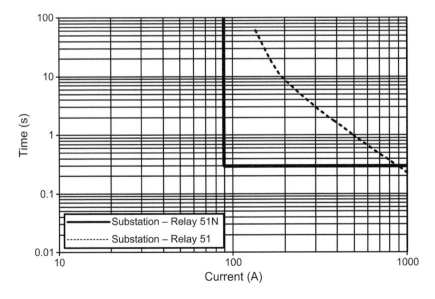

Figure 10.61 Case Study 7: Time-current characteristics of overcurrent relays.

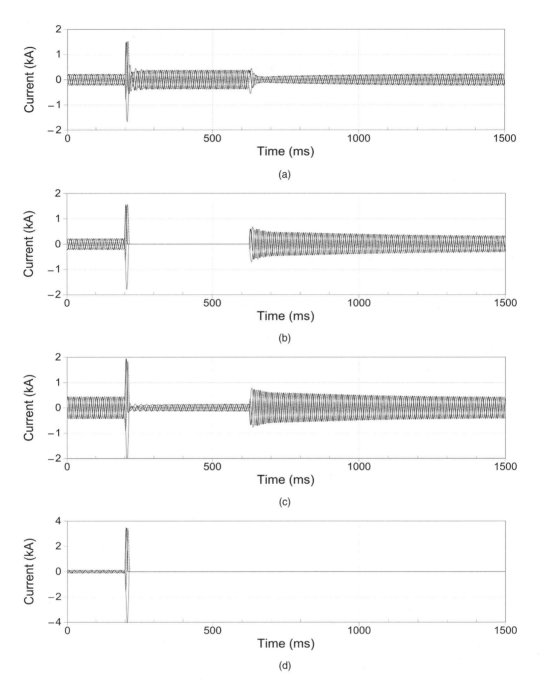

Figure 10.62 Case Study 7: Simulation results – Fuse blowing – Three-phase fault downstream Fuse 4. (a) Substation circuit breaker currents, (b) Recloser currents, (c) Generator currents, (d) Fuse 4 currents.

load LD 3 during the period needed by the recloser to close again. Since after the reclosing operation the fault has been isolated, the system will continue operating satisfactorily.

The currents through the breakers that protect both the substation transformer and the local generator exhibit some variation as a consequence of the fault occurrence, but the fast operation of the fuse avoids significant variations. Actually, the currents supplied from the substation and the generator remain almost the same after the reclosing operation. Although the final load is lower than at the beginning of the simulation, the final currents through the recloser and the two breakers are almost the same that before the fault; this is basically due to the changes in the active and reactive powers (not shown in the figure); for instance, the reactive power supplied from the substation decreases, while that supplied from the generator increases. An interesting conclusion from this case study is that the temporary island does not affect the behaviour of the system after the reclosing operation, and there is no need to disconnect the generator during the islanding period. A deeper analysis of the case would provide additional conclusions about the generator controllers; they are beyond the goal of the study.

The response of protective devices for a three-phase fault located at the generator interconnection terminals is summarized in Figure 10.63. As for the previous case, the three-phase fault occurs at 0.2 seconds and clears at 2 seconds. The plots show that both the recloser and the generator interconnection circuit breaker open; in the later case due to the operation of the Relay 27 that detects a voltage decrement below the 50% of the rated voltage for a period longer than acceptable. However, the recloser remains closed after a first fast operation because the fault clears before the second opening operation takes place. As expected, the currents through the substation breaker increase because all loads are now supplied from the substation; however, the recloser currents decrease although one would expect some increment since there is no contribution from the generator; this decrement is due to a significant voltage drop that takes place in the feeder from the substation terminals.

10.6 Discussion

During the first few cycles following a power system fault, the protection system is expected to make a correct decision to minimize the extent of equipment damage. Protective systems are a critical part of the power system, since an incorrect operation can have detrimental consequences to the power equipment and power system behaviour.

Although computer models of major power system components have been used in software packages for many years, models of relays and other protective devices have been applied to a lesser degree. Protection system models and their simulation with a time-domain tool are now used to: (i) investigate and improve protection designs and algorithms; (ii) select relay types suited for a particular application; (iii) verify, test, and properly adjust settings of relays and other protective devices; (iv) make relay design easier and less costly.

The case studies presented in this chapter have shown that ATP can represent the behaviour of physical relays, and it is possible to create relay models with different levels of sophistication. Simple models only need mathematical equations to describe the pick up and tripping characteristics of the relays, and they can be used to make general decisions for the selection of relay types or to derive relay settings. It is important to remember that the case studies presented here do not analyse real systems and the implemented models have not been validated; the main goal was to show how ATP capabilities can be applied to the simulation of protection systems.

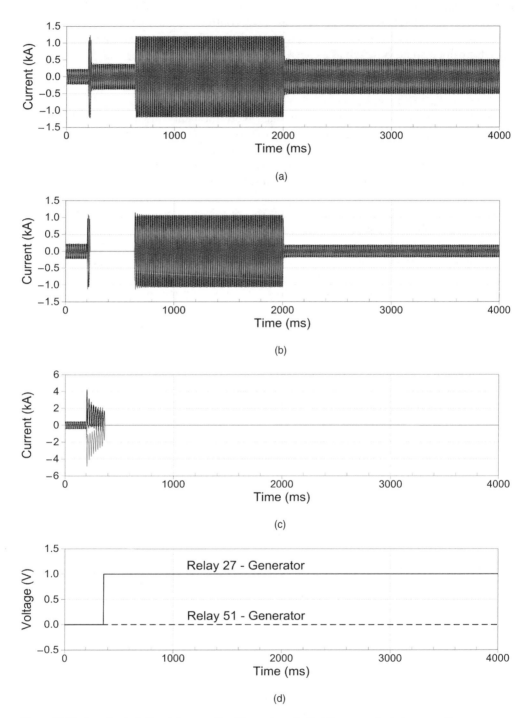

Figure 10.63 Case Study 7: Simulation results – Three-phase fault at the generator interconnection. (a) Substation circuit breaker currents, (b) Recloser currents, (c) Generator currents, (d) Interconnection relay signals.

Acknowledgement

All case studies included in this chapter were developed and implemented by the authors. However, some case studies were based on models previously developed by other authors. Thanks are due to R. Wilson, who developed the model used in Case Study 4, and L. Kojovic, who provided parameters used in some case studies.

References

1 Horowitz, S.H. and Phadke, A.G. (2008). *Power System Relaying*, 3e. Wiley.

2 Lewis Blackburn, J. and Domin, T.J. (eds.) (2007). *Protective Relaying. Principles and Applications*, 3e. CRC Press.

3 Elmore, W.A. (2003). *Protective Relaying: Theory and Applications*, 2e. CRC Press.

4 Anderson, P.M. (1998). *Power System Protection*. Mc-Graw Hill – IEEE Press.

5 Phadke, A.G. and Thorp, J.S. (2009). *Computer Relaying for Power Systems*, 2e. Wiley.

6 Garrett, B.W., Dommel, H.W. and Engelhardt, K.H. (1987). Digital simulation of protection systems under transient conditions. Ninth Power Systems Computation Conference in Cascais, Portugal (September 1987).

7 Chaudhary, A.K.S., Wilson, R.E., Glinkowski, M.T. et al. (1998). Modeling and analysis of transient performance of protection systems using digital programs. In: *Chapter 7 of Modeling and Analysis of System Transients using Digital Systems* (eds. A. Gole, J.A. Martinez-Velasco and A.J.F. Keri). IEEE Special Publication, TP-133-0.

8 IEEE Tutorial Course. (1999). Electromagnetic Transient Program Applications to Power Systems Protection, IEEE.

9 CIGRE Working Group B 5.17 (2006). Relay software models for use with electromagnetic transient analysis program. *CIGRE Brochure*: 295.

10 Martinez-Velasco, J.A. and Gonzalez-Molina, F. (2015). Calculation of power system overvoltages. In: *Chapter 5 of Transient Analysis of Power Systems. Solution Techniques, Tools and Applications* (ed. J.A. Martinez-Velasco). Wiley-IEEE Press.

11 Cook, V. (1985). *Analysis of Distance Protection*. Research Study Press Ltd., Wiley.

12 IEEE Tutorial Course. (1988). Microprocessor Relays and Protection Systems. Publication No 88EHO269-1-PWR.

13 IEEE Tutorial Course. (1997). Advancements in Microprocessor Based Protection and Communication. Publication No. 97TP120-0.

14 Horowitz, S.H. (1981). *Protective Relaying for Power Systems*. IEEE Press, Selected Reprint Series.

15 Horowitz, S.H. (ed.) (1992). *Protective Relaying for Power Systems II*. IEEE Press, Selected Reprint Series.

16 Ziegler, G. (2011). *Numerical Distance Protection. Principles and Applications*, 4e. Siemens.

17 Ziegler, G. (2012). *Numerical Differential Protection. Principles and Applications*, 2e. Siemens.

18 Paithankar, Y.G. (1997). *Transmission Network Protection: Theory and Practice*. Marcel Dekker.

19 Johns, A.T. and Salman, S.K. (1997). *Digital Protection for Power Systems*, vol. 15. Peter Peregrinus Ltd, IEE Series.

20 Reimert, D. (2005). *Protective Relaying for Power Generation Systems*. CRC Press.

21 Durbak, D.W., Gole, A.M., Camm, E.H. et al. (1998). Modeling guidelines for switching transients. In: *Chapter 4 of Modeling and Analysis of System Transients using Digital*

Systems (eds. A. Gole, J.A. Martinez-Velasco and A.J.F. Keri). IEEE Special Publication, TP-133-0.

22 IEEE Power System Relaying Committee. (2005). EMTP Reference Models for Transmission Line Relay Testing. http://www.pes-psrc.org (accessed 04 July 2019).

23 Dommel, H.W. (1986). *Electromagnetic Transients Program Reference Manual (EMTP Theory Book)*. Portland, OR, USA: Bonneville Power Administration.

24 Can/Am Users Group (2000). *ATP Rule Book*. Can/Am Users Group.

25 Bastard, P., Bertrand, P., and Meunier, M. (1994). A transformer model for winding fault studies. *IEEE Transactions on Power Delivery* 9 (2): 690–699.

26 Martinez-Velasco, J.A. (2012). Basic methods for analysis of high frequency transients in power apparatus windings. In: *Chapter 2 of Electromagnetic Transients in Transformer and Rotating Machine Windings* (ed. C.Q. Su). IGI Global.

27 Popov, M., Gustavsen, B., and Martinez-Velasco, J.A. (2012). Transformer Modelling for impulse voltage distribution and terminal transient analysis. In: *Chapter 6 of Electromagnetic Transients in Transformer and Rotating Machine Windings* (ed. C.Q. Su). IGI Global.

28 Annakkage, U.D., Nair, N.K.C., Liang, Y. et al. (2012). Dynamic system equivalents: A survey of available techniques. *IEEE Transactions on Power Delivery* 27 (1): 411–420.

29 CIGRE Working Group 13.01 (1993). Applications of black box modelling to circuit breakers. *Electra* 149: 40–71.

30 Martinez-Velasco, J.A. and Popov, M. (2009). Circuit Breakers. In: *Chapter 7 of Power System Transients. Parameter Determination* (ed. J.A. Martinez-Velasco). CRC Press.

31 Tziouvaras, D., McLaren, P., Alexander, C. et al. (2000). Mathematical models for current, voltage and coupling capacitor voltage transformers. *IEEE Transactions on Power Delivery* 15 (1): 62–72.

32 IEEE Special Publication, Power System Relaying Committee. (1976). Transient Response of Current Transformers. Report 76-Ch1130-4 PWR.

33 IEEE Power System Relaying Committee (1977). Transient response of current transformers. *IEEE Transactions on Power Apparatus Systems* 96 (6): 1809–1814.

34 IEEE Committee Report (1981). Transient response of coupling capacitor voltage transformer. *IEEE Transactions on Power Apparatus Systems* 100 (12): 4811–4814.

35 Lucas, J.R., McLaren, P.G., Keerthipala, W.W.L., and Jayasinghe, R.P. (1992). Improved simulation models for current and voltage transformers in relay studies. *IEEE Transactions on Power Delivery* 7 (1): 152–159.

36 Kezunovic, M., Kojovic, L., Skendzic, V. et al. (1992). Digital models of coupling capacitor voltage transformers for protective relay transient studies. *IEEE Transactions on Power Delivery* 7 (4): 1927–1935.

37 IEEE Committee Report (1993). Relay performance considerations with low ratio CTs and high fault currents. *IEEE Transactions on Power Delivery* 8 (3): 884–897.

38 Kezunovic, M., Kojovic, L., Abur, A. et al. (1994). Experimental evaluation of EMTP-based current transformer models for protective relay transient study. *IEEE Transactions on Power Delivery* 9 (1): 405–413.

39 Kojovic, L., Kezunovic, M., and Fromen, C.W. (1994). A new method for the CCVT performance analysis using field measurements, signal processing and EMTP modelling. *IEEE Transactions on Power Delivery* 9 (4): 1907–1915.

40 Iravani, M.R., Wang, X., Polishchuk, I. et al. (1998). Digital time-domain investigation of transient behavior of coupling capacitor voltage transformer. *IEEE Transactions on Power Delivery* 13 (2): 622–629.

41 McLaren, P.G., Benmouyal, G., Chano, S. et al. (2001). Software models for relays. *IEEE Transactions on Power Delivery* 16 (2): 238–245.

42 Domijan, A. and Emami, M.V. (1990). State space relay modeling and simulation using the electromagnetic transient program and its transient analysis of control system capability. *IEEE Transactions on Energy Conversions* 5 (4): 697–702.

43 Glinkowski, M.T. and Esztergalyos, J. (1996). Transient modeling of electromechanical relays. Part I: armature type relay. *IEEE Transactions on Power Delivery* 11 (2): 763–770.

44 Glinkowski, M.T. and Esztergalyos, J. (1996). Transient modeling of electromechanical relays, part II: plunger type 50 relays. *IEEE Transactions on Power Delivery* 11 (2): 771–782.

45 Garrett, B.W. (1987). Digital Simulation of Power System Protection under Transient Conditions. Ph.D. Thesis. University of British Columbia.

46 Chaudhary, A.K.S. (1991). Protection System Representation in the Electromagnetic Transients Program. Ph.D. Thesis. Virginia Tech.

47 Chaudhary, A.K.S., Tam, K.S., and Phadke, A.G. (1994). Protection system representation in the electromagnetic transients program. *IEEE Transactions on Power Delivery* 9 (2): 700–711.

48 Wilson, R.E. and Nordstrom, J.M. (1993). EMTP transient modeling of a distance relay and a comparison with EMTP laboratory testing. *IEEE Transactions on Power Delivery* 5 (3): 984–990.

49 Wilson, R.E. (1992). A new method using relay macromodels for simulation of the transient response of distance relays. Ph.D. Thesis. University of Idaho.

50 EPRI. (2004). Protective Relays. Numerical Protective Relays. EPRI Report 1009704.

51 Sachdev, M.S. and Baribeau, M.A. (1979). A new algorithm for digital impedance relays. *IEEE Transactions on Power Apparatus Systems* 98 (6): 2232–2240.

52 Dysko, A., McDonald, J.R., Burt, G.M. et al. (1999). Integrated modeling environment: A platform for dynamic protection modeling and advanced functionality. IEEE PES Transmission & Distribution Conference in Atlants, USA (April, 1999).

53 Peterson, J.N. and Wall, R.W. (1991). Interactive relay controlled power system modelling. *IEEE Transactions on Power Delivery* 6 (1): 96–102.

54 Li, H. (1999). A New Adaptive Distance Relay. Ph.D Thesis. University of Manitoba.

55 Nimmersjö, G., Hillström, B., Werner-Erichsen, O., and Rockefeller, G.D. (1988). A digitally-controlled, real-time, analog power system simulator for closed-loop protective relaying testing. *IEEE Transactions on Power Delivery* 3 (1): 138–115.

56 McLaren, P.G., Kuffel, R., Weirckx, R. et al. (1992). A real time digital simulator for testing relays. *IEEE Transactions on Power Delivery* 7 (1): 207–213.

57 Montmeat, A., Giard, A. and Roguin, J. (1995). MORGAT for testing MV and EHV protective relays. International Conference on Power Systems Transients in Lisbon, Portugal (September 1995).

58 Kezunovic, M., Domaszewicz, J., Skendzic, V. et al. (1996). Design, implementation and validation of a real time digital simulator for protection relay testing. *IEEE Transactions on Power Delivery* 11 (1): 158–164.

59 Friedland, R., Pannhorst, H.D., and Kulicke, B. (1996). Digital network model for tests of different equipment in electric power systems. *Electrical Power Systems Research* 36: 197–202.

60 Sidhu, T.S., Sachdev, M.S., and Das, R. (1997). Modern relays: research and teaching using PCs. *IEEE Computer Applications Power* 50–55.

61 IEEE Committee Report. (1996). Relay performance testing, IEEE Special Publication, No. 96 TP 115-0.

62 Kezunovic, M., Pickett, B.A., Adamiak, M.G. et al. (1998). Digital simulator performance requirements for relay testing. *IEEE Transactions on Power Delivery* 13 (1): 78–84.

63 CIGRE WG 34-10 (2000). Analysis and guidelines for testing numerical protection schemes. *CIGRE Brochure* 159.

64 Fugita, N., Ichie, R., Ogawa, S. et al. (1995). Computer-aided protective relay settings for large power networks. *Electrical Engineering in Japan* 115 (5): 22–37.

65 Baumann, U. and Wellssow, W.H. (1996). Computer aided analysis of protection relay settings with respect to starting conditions. Presented at the 12th Power System Computation Conference in Zurich, Switzerland (August 1996).

66 De Sa Pinto, J., Afonso, J., and Rodrigues, R. (1997). A probabilistic approach to setting distance relays in transmission networks. *IEEE Transactions on Power Delivery* 12 (2): 681–686.

67 Martinez-Velasco, J.A. and Kojovic, Lj.A. (1997). Modeling of electromechanical distance relays using the ATP. Presented at the 32nd Universities Power Engineering Conference in Manchester, UK (September 1997).

68 Martinez-Velasco, J.A. and Kojovic, Lj.A. (1997). ATP modeling of electromechanical distance relays. International Conference on Power Systems Transients in Seattle, USA (June 1997).

69 Peng, Z., Li, M.S., Wu, C.Y. et al. (1985). A dynamic state space model of a MHO distance relay. *IEEE Transactions on Power Apparatus Systems* 104 (12): 3558–3564.

70 Dubé, L. and Bonfanti, I. (1992). MODELS: a new simulation tool in the EMTP. *European Transactions on Electric Power Engineering* 2 (1): 45–50.

71 Gers, J.M. and Holmes, E.H. (2004). *Protection of Distribution Networks*, IEE Power and Energy Series, 2e.

72 Short, T.A. (2004). *Electric Power Distribution Handbook*. CRC Press.

73 Gönen, T. (2008). *Electric Power Distribution System Engineering*. CRC Press.

74 Walling, R.A., Saint, R., Dugan, R.C. et al. (2008). Summary of distributed resources impact on power delivery systems. *IEEE Transactions on Power Delivery* 23 (3): 1636–1644.

75 IEEE Power System Relay Committee. (2004). Impact of distributed resources on distribution relay protection.

76 IEEE Std. 1547-2003. (2003). IEEE Standard for Interconnecting Distributed Resources with Electric Power Systems.

77 Mozina, C.J. (2006). Distributed generator interconnect protection practices. IEEE PES T&D Conference in Dallas, USA (June 2006).

78 Mozina, C.J. (2001). Interconnect protection of dispersed generators. IEEE PES T&D Conference in Atlanta, USA (November 2001).

79 Mozina, C.J. (2001). Interconnection protection of IPP generators at commercial/industrial facilities. *IEEE Transactions on Industry Applications* 37 (3): 681–688.

80 IEEE IAS WG Report. (2006). Application of islanding protection for industrial and commercial generators. Presented at the 59th Annual Conference for Protective Relay Engineers in Texas, USA (April 2006).

81 Girgis, A. and Brahma, S. (2001). Effect of distributed generation on protective device coordination in distribution networks. Large Engineering Systems Conference on Power Engineering in Halifax, Canada (February 2001).

82 IEEE Std C37.48.1-2002. (2002). IEEE Guide for the Operation, Classification, Application and Coordination of Current-Limiting Fuses with rated Voltages 1-38 kV.

83 IEEE Std C37.112-1996. (1996). IEEE Standard Inverse-Time Characteristic Equations for Overcurrent Relays.

84 Computer Representation of Overcurrent Relays Characteristics Working Group of the PSRC (1989). Computer representation of overcurrent relay characteristics. *IEEE Transactions on Power Delivery* 4 (3): 1659–1667.

85 Leix, K.L., Kojovic, Lj.A., Marz, M. and Lampley, G.C. (1999). Applying current-limiting fuses to improve power quality and safety. IEEE T&D Conference in New Orleans (April 1999).

86 Martinez, J.A. and Martin-Arnedo, J. (2006). Voltage sag studies in distribution networks. Part I: system modeling. *IEEE Transactions on Power Delivery* 21 (3): 1670–1678.

87 Kojovic, L.A., Hassler, S.P., Leix, K.L. et al. (1998). Comparative analysis of expulsion and current-limiting fuse operation in distribution systems for improved power quality and protection. *IEEE Transactions on Power Delivery* 13 (3): 863–869.

88 Kojovic, L.A. and Hassler, S.P. (1997). Application of current-limiting fuses in distribution systems for improved power quality and protection. *IEEE Transactions on Power Delivery* 12 (2): 791–800.

89 Martinez, J.A. and Martin Arnedo, J. (2010). EMTP modeling of protective devices for distribution systems with distributed generation. IEEE PES General Meeting in Minneapolis, USA (July 2010).

90 Martinez, J.A. and Martin Arnedo, J. (2009). Impact of distributed generation on distribution protection and power quality. IEEE PES General Meeting in Calgary, Canada (July 2009).

91 Martínez-Velasco, J.A., Martin-Arnedo, J. and Castro-Aranda, F. (2010). Modeling protective devices for distribution systems with distributed generation using an EMTP-type tool. *Ingeniare* 18 (2): 258–273.

To Probe Further

The simulation results presented in the chapter are a very small sample of the results that can be derived from the case studies created for this chapter. Data files for all case studies are available on the website. Readers are encouraged to run the cases, check all results, and explore the performance of the test systems with different settings of protective devices or under different operating conditions. Several custom-made models have been created to represent protective devices; readers are encouraged to use these models and create new versions by changing or improving every aspect of the modules (i.e. icon, code, menu windows).

11

ATP Applications Using a Parallel Computing Environment

Javier A. Corea-Araujo, Gerardo Guerra and Juan A. Martinez-Velasco

11.1 Introduction

The previous chapters have shown how the Alternative Transients Program (ATP) package can be used for both analysis and design tasks: post-mortem studies can detect the causes of a dielectric failure, the simulation of a distribution system with some penetration of power electronics converters can be useful for estimating the harmonic distortion produced by those converters, an insulation coordination study can be used for selecting the insulation levels of an overhead transmission line or the ratings of surge arresters aimed at protecting substation equipment.

Although it has been shown how ATP capabilities can be used to carry out parametric and statistical studies, the fact is that previous chapters only presented case studies in which ATP was applied as a standalone simulation tool. ATP users can go a step further by developing simulation environments in which ATP capabilities are combined with capabilities from other simulation tools. Such combinations can create powerful tools that significantly expand ATP applications.

MATLAB/Simulink is a well-known simulation environment with features that can be used for prototyping new models and solving large-scale systems in combination with other tools. In addition, several MATLAB-based toolboxes can be used for very different applications. ATP offers capabilities for developing custom-made models that can easily be accommodated to studies for which a transient analysis is required.

This chapter presents the application of a MATLAB-ATP environment to some advanced applications in which time-consuming parametric and statistical studies are to be carried out. A common feature to the studies presented in this chapter is the usage of a multicore environment to reduce simulation times. The following three applications are detailed in this chapter:

1) Generation of 3D bifurcation diagrams for ferroresonance characterization.
2) Lightning performance analysis of overhead transmission lines.
3) Optimum selection of parameters for hybrid high-voltage direct current (HVDC) circuit breakers.

A dedicated procedure for each application using a MATLAB-ATP link and a multicore environment has been developed. This chapter includes a section for each application. Each section is organized in the same manner: a short introduction to the problem to be analysed, a summary of modelling guidelines to be considered for the study under consideration, a short description of the implemented procedure, and some illustrative examples.

Transient Analysis of Power Systems: A Practical Approach, First Edition. Edited by Juan A. Martinez-Velasco.
© 2020 John Wiley & Sons Ltd. Published 2020 by John Wiley & Sons Ltd.
Companion Website: www.wiley.com/go/martinez/power_systems

The procedures implemented for all the case studies presented in this chapter use the library of MATLAB modules developed by M. Buehren for allocating ATP tasks in a multicore environment; see [1]. A summary of the main characteristics of the multicore installation are listed in Appendix A of this chapter.

11.2 Bifurcation Diagrams for Ferroresonance Characterization

11.2.1 Introduction

Ferroresonance normally refers to a series resonance that typically involves the interaction of a saturable transformer and a distribution cable or transmission line [2]. Due to the involved nonlinearities and the various situations that can lead to ferroresonance, it is difficult to predict whether it will occur or not, and the severity with which it can occur [2]. Traditionally, the analysis of ferroresonance has been difficult due to its unpredictability and the lack of an overall understanding. See Chapters 5 and 6 for more details about this phenomenon.

This section proposes the application of bifurcation diagrams for ferroresonance characterization. Since the computing task can be very tedious even for very simple cases, a procedure based on a MATLAB-ATP link and using parallel computing has been implemented. The next two subsections provide a short introduction to ferroresonance characterization and the modelling guidelines to be followed when analysing ferroresonance phenomena using a time-domain simulation tool. The procedure developed for generating 3D bifurcation diagrams and its implementation in a multicore environment is summarized in the subsequent subsection. The last subsections present three ferroresonance case studies of increasing complexity and a discussion about the advantages of the implemented procedure.

11.2.2 Characterization of Ferroresonance

Recorded waveforms corresponding to actual events and results from numerical simulations have led to classify ferroresonance states into four different modes [3]:

- *Fundamental mode.* Voltages and currents are periodic with a period equal to the power frequency period.
- *Subharmonic mode.* The signals are periodic with a period that is a multiple of the source period.
- *Quasi-periodic mode.* This mode is not periodic and exhibits a discontinuous spectrum.
- *Chaotic mode.* The signals show an irregular and unpredictable behaviour.

Since it is difficult to distinguish normal transient states from ferroresonant transient states, this classification corresponds to the steady-state condition, once the initial transient state is over.

In many scenarios, more than one ferroresonant state is possible, and the operation may jump in and out of ferroresonance modes depending on switching angle or nonlinear circuit parameters. Ferroresonance can be characterized by using either a spectral study method based on Fourier's analysis or a stroboscopic analysis obtained by measuring current i and voltage v at a given point of the system and plotting in plane v-i the instantaneous values at instants separated by a system period. The stroboscopic method, also known as *Poincaré map* [4], can be used to differentiate between ferroresonance states (e.g. a quasi-periodic state from a chaotic state). A *bifurcation diagram* records the locations of all the abrupt changes in a system signal (e.g. voltage) when a system parameter is varied (either parametrically or quasi-statically).

The bifurcation diagram is an alternative to traditional parametric methods for understanding system behaviour in ferroresonance analysis. The application of this option has been the subject of many works [5–11]. Techniques applied to the calculation of a bifurcation diagram may be based on the principle of continuation [5, 9], experimentation, or time-domain simulation [6, 8]. Reference [12] proposed the application of 3D bifurcation diagrams for complementing ferroresonance characterization, mainly when more than one parameter can significantly affect the behaviour of the ferroresonant nonlinear circuit. That reference presented the MATLAB-ATP link used in this chapter to obtain bifurcation diagrams.

11.2.3 Modelling Guidelines for Ferroresonance Analysis

The transformer model is probably the most critical part of any ferroresonance study. Different models and different means of determining the parameters are required for each type of core. Several transformer models based on the principle of duality have been presented in the literature [13–17]. By default, the transformer model must have an accurate representation of all nonlinear inductances of the core, including hysteresis effects [18–20]. Internal capacitances of the transformer might also be considered since they could be crucial in the final development of a ferroresonant process. The application of some specialized transformer models is limited, since no standard test has been proposed for determining all of their parameters, and tests suggested in the literature are not always performed. Besides, many of these models have not been implemented in all simulation packages. Although much effort has been made on refining models and performing simulations using transient circuit analysis programs such as the Electromagnetic Transients Program (EMTP) and like [21], determining nonlinear parameters is still a challenge.

Other aspects to be considered, besides the transformer model, are the study zone that must be represented in the model, the source impedance, the transmission or distribution line(s)/cable(s), and any other capacitance not already included from the previous components. Source representation is not generally critical; if the source does not contain nonlinearities, it is sufficient to use the steady-state Thevenin impedance and the open-circuit voltage. Lines and cables may be represented as *RLC* coupled PI equivalents, cascaded for longer lines/cables. Shunt or series capacitors may be represented as a standard capacitance, paralleled with the appropriate resistance. Transformer stray capacitances may also be incorporated either at the corners of an open-circuited delta transformer winding or midway along each winding.

Since ferroresonance is a nonlinear phenomenon, the conditions with which it is initiated will have a significant impact; therefore, residual fluxes, initial capacitance voltages, shifting angles from both source and switch cannot be neglected; see [22, 23]. This aspect was analysed in [12] by varying the switch angle and the phase shift of the source. Both values play a very important role and can decide the final state of the ferroresonance and the length of the transient between the normal operation and the ferroresonant final state [24].

11.2.4 Generation of Bifurcation Diagrams

A roadmap for ferroresonance analysis, going from system modelling to ferroresonance behaviour analysis and prediction using bifurcation diagrams was presented in [12]. A bifurcation diagram records on a plane the locations of all abrupt changes experienced by the final operating state of the test system while one or more system parameters are quasi-statically varied [25]. The 2D bifurcation diagram is an alternative to traditional parametric and sensitivity methods for understanding system behaviour in ferroresonance analysis. A procedure based on a MATLAB-ATP link was proposed in [12]: (i) ATP is used to implement the test system

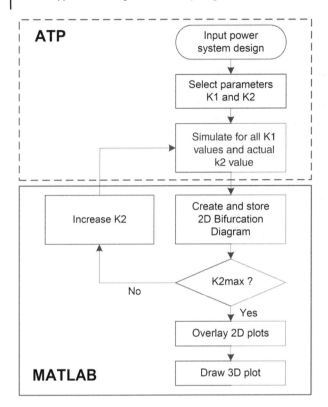

Figure 11.1 Flow chart for generating 3D bifurcation diagrams [12].

model and perform a parametric study by varying parameters of concern within selected limits; (ii) MATLAB reads output files, forms an array that links parameter values and voltage peaks, and creates the diagram.

A 3D bifurcation diagram is based on the variation of two system parameters, stacking as many 2D planes as values given to the second parameter range. The implementation of 3D bifurcation diagrams uses the experimentation method, which consists of repeating time-domain simulations followed by frequency-domain sampling of the same output to determine its periodicity. The procedure proposed in [12] for obtaining 3D bifurcation diagrams is shown in Figure 11.1. The first three blocks of the scheme (inside the dashed box) are entirely processed in ATP and the resulting outputs are exported to MATLAB (notice that there is one output for every value of parameter K1). Sampling is performed once every duty cycle in order to create the 2D plane. The stacking is developed using a 3D matrix; every single layer contains a 2D plane. In such manner, the stacking planes generate a volume, with K1 and K2 as x and y axes, respectively, and the ferroresonant voltage value expressed in pu as z axis. The base for pu calculations is always the source voltage. The resulting map is easy to read, presents straight lines when the output solution is a fundamental mode and broken lines when the output is in any period different than period-1 or in chaotic mode [3].

11.2.5 Parametric Analysis Using a Multicore Environment

The generation of bifurcation diagrams can be a very tedious task since a high number of simulations has to be run in order to obtain a rigorous representation of the whole phenomenon;

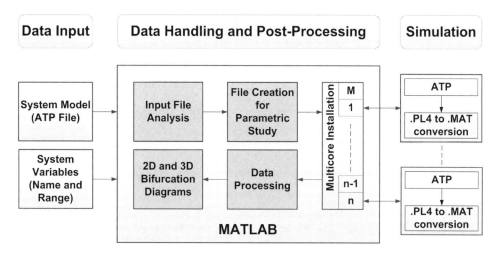

Figure 11.2 Procedure for generation of bifurcation diagrams using a multicore installation.

that is, a complete parametric study of a ferroresonant scenario can require thousands of runs. In addition, a rather short time step (i.e. less than 10 μs) has usually to be considered due to the nonlinearities involved in this phenomenon. A simple solution to this problem is to distribute the tasks among several cores; that is, the application of parallel computing.

A MATLAB procedure, based on the approach presented in [12], has been developed to drive ATP within a multicore installation and collect the information generated to obtain the bifurcation diagram that will characterize the behaviour of the test system. Figure 11.2 presents a schematic diagram of the procedure implemented in MATLAB to run ATP in a multicore environment [26].

A template of the input file corresponding to the test case is firstly elaborated. This template includes POCKET CALCULATOR VARIES PARAMETERS (PCVP), normally used to perform parametric studies. By means of the MATLAB procedure, the PCVP section is varied as many times as required to cover the whole range of the first parameter; PCVP is also used to vary the second parameter while the first one remains constant. The user has to specify the range of values and the number of runs for each parameter.

The diagram presented in Figure 11.2 shows a procedure covering three main parts:

1. Data input preparation (aimed at editing input files).
2. Data handling and generation of plots.
3. Conversion of simulation results produced by ATP (i.e. the so-called PL4 files) to MATLAB format (i.e. MAT files).

Note the directions in which the information is flowing: the MATLAB code generates the final ATP input files, distributes them between cores, controls the ATP runs and the data conversion, reads and manipulates simulation results once they have been translated into in MATLAB code, and finally takes care of the generation of bifurcation diagrams. Data conversion and further post-processing steps of this procedure take advantage of the previous developments, used for the case studies presented in reference [12]. For this expanded version, additional MATLAB code was needed to edit ATP input files and distribute them among cores.

Another important aspect was the size of the files generated by ATP when using PCVP. Since a PL4 file for each combination of the two parameters that are varied is needed, the size of the information that has to be manipulated can be huge (e.g. if every parameter is varied 200 times, up to 40 000 PL4 files will have to be read). To avoid the storage and manipulation

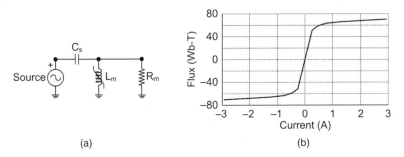

(a)   (b)

Figure 11.3 Case 1: Test circuit configuration. (a) Diagram of the test system. (b) Saturable inductance characteristic.

of such a number of files, the simulations are progressively made; for instance, if 200 runs are to be controlled from PCVP, the same file is run 4 times and the parameter is varied only 50 times within each run.

11.2.6 Case Studies

11.2.6.1 Case 1: An Illustrative Example

The basic circuit is shown in Figure 11.3a. The circuit consists of a 25 kV, 50 Hz power source, a capacitance of 0.1 μF, and a saturable inductance (see Figure 11.3b) paralleled with a resistance of 40 Ω. The shifting angle of the voltage source is −90°, being zero, both the initial capacitance voltage and the residual flux in the saturable inductance. The simulation is carried out without introducing any switching event. Using the above parameters, the resulting voltage in the inductor is that shown in Figure 11.4a; it can be easily identified as a periodic ferroresonant signal. Consider now that the value of the resistance is infinite; that is, assume a lossless circuit. Figure 11.4b shows the new time-domain response. Note, however, that the ferroresonant signal in Figure 11.4b cannot easily be characterized, since it is not possible to distinguish between a chaotic and a quasi-periodic mode. The 3D diagram, elaborated according to the

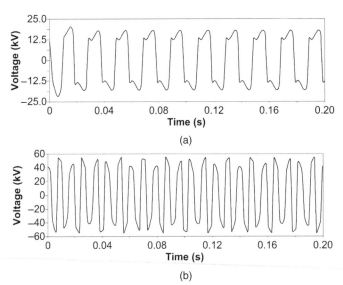

(a)

(b)

Figure 11.4 Case 1: Simulation results – Reactance voltage. (a) Lossy circuit, $R_m = 40\,k\Omega$ and $C_s = 0.1\,\mu F$. (b) Lossless circuit, $R_m = \infty$ and $C_s = 0.1\,\mu F$.

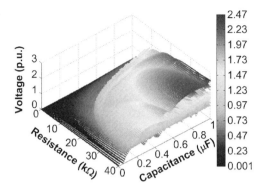

Figure 11.5 Case 1: 3D bifurcation diagram.

procedure summarized above and created by slowly varying C_s and R_m from 0 to 1 μF and 0 to 40 kΩ, respectively, is shown in Figure 11.5: even though ferroresonance is present in almost any C_s–R_m combination, there are two intervals (i.e. $C_s = 0 \div 0.2$ μF and $R_m \leq 10$ kΩ) in which the signal is completely damped. Note also that there is a surrounded area that remains below the nominal voltage. The 3D bifurcation diagram can be used to select design parameters for new systems, predict non-destructive zones, or analyse the impact on given equipment by indicating the range of values for which equipment failure could occur.

11.2.6.2 Case 2: Ferroresonant Behaviour of a Voltage Transformer
This study is based on a work presented in [10], and deals with a ferroresonance case that involves inductive voltage transformers (VTs) in a substation equipped with circuit breaker grading capacitors. Figure 11.6a,b show the model implemented in ATPDraw, based on the BCTRAN option, and relevant information for ferroresonance studies. Test system parameters are provided in Appendix B. A 400 kV bus feeds a transformer bank with no residual flux. The cable lengths are shown in the figure. All capacitances are initially discharged, the phase angle of the source is 0°, and the closing of the switch occurs at $t = 0.3$ s.

A catastrophic incident was reported when the three-phase circuit breaker (CB) was opened [10]; the operation left the grading capacitance connected in series to the nonlinear inductance of a transformer bank connected to the same line.

After a study aimed at analysing the influence of the value C_s, it was discovered that for a range of values the result could be much different [12]. The stray capacitance C_s and the grading capacitance C_g are recognized as major influence parameters in the ferroresonant system. In this example, the values of C_s and C_g are respectively 490 and 600 pF [10]. The 3D diagram depicted in Figure 11.7 was constructed using C_s and C_g as parameters. The z axis gives the pu value of the voltage at node N (phase B), and is referred to the source voltage. One can easily note that with most combinations of C_s and C_g, the resulting mode of ferroresonance will not exceed the 1 pu value, except for those values of C_s lower than 200 pF, an interval that can unfold chaotic mode with voltage values of up to 4 pu.

11.2.6.3 Case 3: Ferroresonance in a Five-Legged Core Transformer
This case is based on a series of studies conducted in the early 1990's and aimed at synthesizing experimental work and modelling of five-legged core transformers driven to a ferroresonance condition. Originally, the National Rural Electric Cooperative Association (NRECA, USA) funded a study in 1986 and 1987 to assess the problem on rural electric cooperative systems. Several reports were released describing overvoltage condition problems related to five-legged core transformers operating with the grounded-wye connection [27]. This situation induced

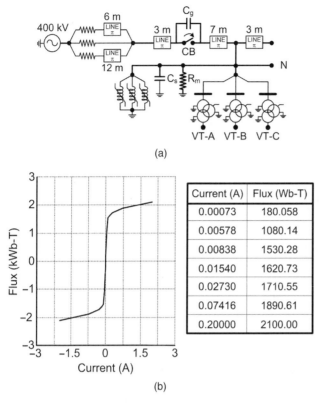

(a)

(b)

Figure 11.6 Case 2: Test system model. (a) Diagram of the test system. (b) Saturation curve of the voltage transformer.

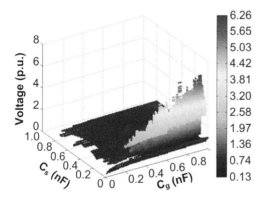

Figure 11.7 Case 2: 3D bifurcation diagram.

the need to develop a transformer model that would efficiently analyse the causes of the phenomenon [28]. After some validation work, the problem was categorized as a ferroresonance phenomenon [29]. An ATP model of a five-legged core transformer was later proposed in [30]. That study was aimed at analysing the ferroresonant response of the transformer to changes in the capacitance of the cable between the transformer and the source.

Figure 11.8a shows the transformer model, as implemented in ATPDraw. Figures 11.8b−e add information on saturable inductances. More information about the test system is provided in the Appendix B.

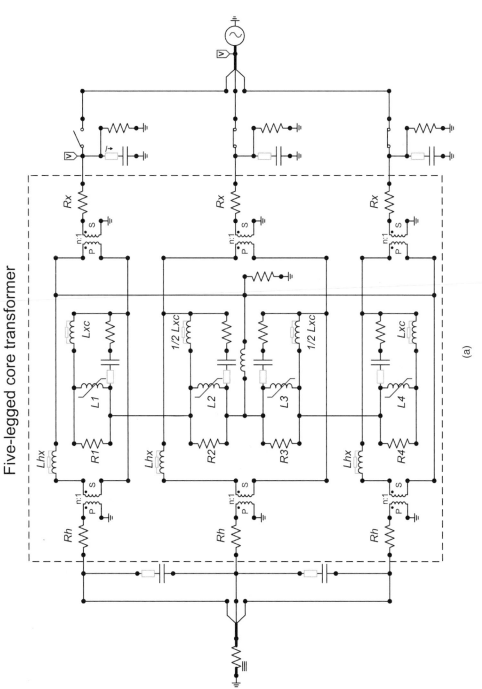

Figure 11.8 Case 3: Test system model. (a) ATPDraw implementation of the test system. (b) Saturation curve of the inductance L_1. (c) Saturation curve of the inductance L_2. (d) Saturation curve of the inductance L_3. (e) Saturation curve of the inductance L_4.

(a)

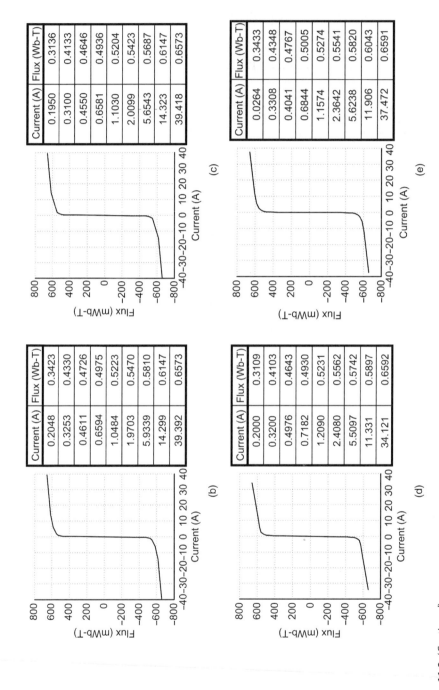

Current (A)	Flux (Wb-T)
0.2048	0.3423
0.3253	0.4330
0.4611	0.4726
0.6594	0.4975
1.0484	0.5223
1.9703	0.5470
5.9339	0.5810
14.299	0.6147
39.392	0.6573

(b)

Current (A)	Flux (Wb-T)
0.1950	0.3136
0.3100	0.4133
0.4550	0.4646
0.6581	0.4936
1.1030	0.5204
2.0099	0.5423
5.6543	0.5687
14.323	0.6147
39.418	0.6573

(c)

Current (A)	Flux (Wb-T)
0.2000	0.3109
0.3200	0.4103
0.4976	0.4643
0.7182	0.4930
1.2090	0.5231
2.4080	0.5562
5.5097	0.5742
11.331	0.5897
34.121	0.6592

(d)

Current (A)	Flux (Wb-T)
0.0264	0.3433
0.3308	0.4348
0.4041	0.4767
0.6844	0.5005
1.1574	0.5274
2.3642	0.5541
5.6238	0.5820
11.906	0.6043
37.472	0.6591

(e)

Figure 11.8 (Continued)

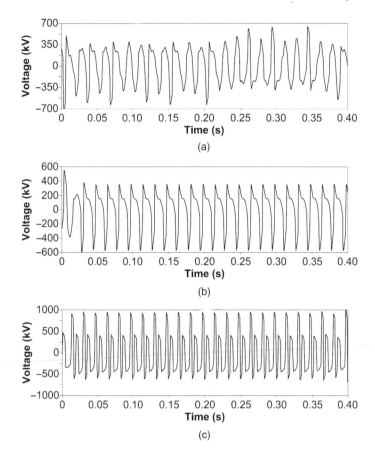

Figure 11.9 Case 3: Phase A – Some ferroresonant signals. (a) Shift angle = −90, Capacitance = 15 μF, Source voltage = 480 V. (b) Shift angle = −90, Capacitance = 10 μF, Source voltage = 440 V. (c) Shift angle = 0, Capacitance = 25 μF, Source voltage = 500 V.

This case analyses the effects in the ferroresonance final stage provoked by the cable capacitance, the phase shift angle variation, and the source voltage regulation (±10%). Some simulation results showing the ferroresonance state in phase A are presented in Figure 11.9. Since the ferroresonant situation starts along with the simulation, the switch between the source and the phase A is left open (see Figure 11.8a), and the initial capacitance voltage and the residual flux are set to zero. After running a series of simulations, it was possible to understand how the capacitance value, the switching angle, and the source voltage did impact the final state of the ferroresonance. Thus, the capacitance has been varied from 0 to 30 μF, the phase angle has been varied from −90° to 360°, and the source voltage has been varied from 432 to 528 V. All tests were carried out assuming a no-load condition (represented by means of a very high resistance).

An extensive parametric study has been resumed using the 3D bifurcation diagram. A pair of bifurcation diagrams have been obtained compressing each of them up to 90 000 simulations. Figure 11.10 shows 3D diagrams that result from the variation of the source voltage and the shift angle, respectively. From these plots, it is possible to easily locate those potentially destructive zones as well as zones where the oscillation peaks are harmless for equipment and protections (i.e. they are lower than 1 pu).

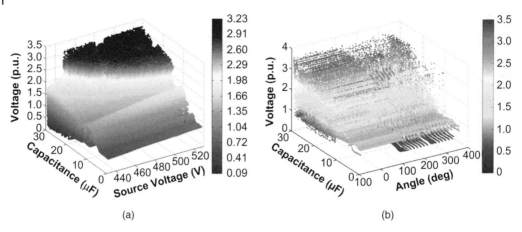

(a)   (b)

Figure 11.10 Case 3: 3D bifurcation diagrams. (a) Ferroresonant voltages as a function of the cable capacitance and the source voltage. (b) Ferroresonant voltages as a function of the cable capacitance and the shift angle.

11.2.7   Discussion

This section has presented the application of a MATLAB-ATP procedure for creating bifurcation diagrams that can characterize ferroresonant situations. 3D bifurcation diagrams provide a good understanding of the behaviour of a power system while being driven into a ferroresonant state. This option can also be used to establish the roadmap of a hypothetical ferroresonant situation, pointing out the safe zones in which the parameter under study should be selected.

The CPU time required by a single core system to carry out some of these analyses can be too long, mainly when a short or very short time step has to be used and a high number of runs is necessary to cover the full range of parameter values. The usage of a multicore installation where the runs are shared among the cores is a powerful alternative. Since the reduction of the simulation time is proportional to the number of cores, some studies can easily become affordable by using a few dozen cores. Some useful information about the cases studied in this section is provided in Table 11.1. In order to understand the advantage of this approach it can be mentioned that all the cases studied here required more than one day of single-CPU time to obtain all the information needed for creating the corresponding 3D bifurcation diagram.

11.3   Lightning Performance Analysis of Transmission Lines

11.3.1   Introduction

The lightning performance of an overhead line is, in general, measured by the number of flashovers per 100 km and year [31]. Transmission lines are usually shielded by ground wires.

Table 11.1 Characteristics of the test cases and simulation times – 50 cores.

Test case	Parameters	Tmax (s)	Time step (μs)	Runs	CPU time (s)
1	R_m: 0–40 kΩ; C_s: 0–1 μF	1	1	200 × 200	6 031
2	C_s: 0–1 ηF; C_g: 0–1 ηF	8	10	200 × 200	8 621
3	V: 432–528 V; C: 0–30 μF	1	1	300 × 300	35 558
	Angle: −90° to 360°; C: 0–30 μF	1	1	300 × 300	35 435

Since the level of overvoltages induced by strokes to ground is too low for transmission-level overhead lines, lightning failures can be caused by strokes to either a shield wire or a phase conductor. The Lightning Flashover Rate (LFOR) of a transmission line is estimated by adding the Backflashover Rate (BFOR) and the Shielding Failure Flashover Rate (SFFOR) [31, 32]. To obtain both quantities, an incidence model is required to discriminate strokes to shield wires from those to phase conductors and those to ground [31, 32].

Due to the random nature of lightning, the flashover rate calculation must be based on a statistical approach. The application of a Monte Carlo method is the usual solution for this type of studies [33]. However, the Monte Carlo method requires a very high number of runs to achieve accurate results. Besides, this is aggravated by the fact that the line model will be very complex and sophisticated [34]. An obvious solution to this disadvantage is the usage of a multicore installation: the distribution of the Monte Carlo runs among several cores can reduce the computing time in a ratio close to the number of cores.

This section presents a MATLAB-ATP procedure implemented for assessing the lightning performance of overhead transmission lines using parallel computing. The Monte Carlo procedure has been implemented in a MATLAB application that pre-calculates random values, distributes the ATP tasks between the various cores involved in lightning overvoltage calculations, and post-process simulation results [35]. The first two subsections summarize the characterization of lightning strokes and the modelling guidelines used for representing transmission lines in lightning overvoltage calculations. Only a summary is provided here as these aspects have already been covered, see Chapter 6 and references [33, 36]. The last subsections are dedicated to detailing the implementation of the parallel MATLAB-ATP procedure and its application to the analysis of an actual overhead transmission line.

11.3.2 Lightning Stroke Characterization

The most important aspects for a full characterization of a lightning flash are summarized below. For more details see [37].

Polarity. Most lightning flashes are of negative polarity. The incidence of positive flashes increases during winter, although their percentage exceeds 10% very rarely [37].

Multiplicity. Negative flashes can consist of multiple strokes, while positive flashes usually have a single stroke. Fewer than 10% of positive flashes have multiple pulses; see [37, 38].

Stroke waveform. A concave waveform, with no discontinuity at $t = 0$, is an accurate representation of a negative stroke. Several expressions have been proposed for such a form, the so-called Heidler model being one of the most widely used [39]. The waveform selected in this study for representing the stroke current is a concave waveform based on a double exponential, see Figure 11.11, and defined by the following expression:

$$i(t) = \begin{cases} i_1(t) & \text{when } i(t) \leq 0.999 I_p \\ i_1(t) + i_2(t) & \text{when } i(t) \geq 0.999 I_p \end{cases} \tag{11.1}$$

with

$$i_1(t) = I_p \left(1 - \exp\left(-\left(\frac{t - t_s}{t_f} \right)^n \right) \right) \qquad i_2(t) = I_p \left(\exp\left(-\left(\frac{t - t_w}{t_x} \right) \right) - 1 \right) \tag{11.2}$$

where I_p is the peak current, t_s is the start time of the waveform, $t_w = t$ when $i_1(t) = 0.999 I_p$, and

$$t_x = 1.44(t_h - t_f \cdot 6.7^{1/n}) \tag{11.3}$$

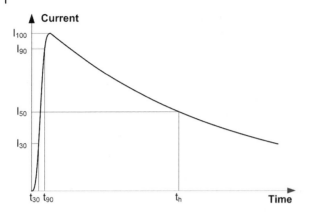

Figure 11.11 Parameters of a lightning stroke – Concave waveform.

The parameters used to define this waveform are the peak current magnitude, $I_{100} = I_p$, the rise time, t_f, and the tail time, t_h (i.e. the time interval between the start of the wave and the 50% of the peak current measured on tail). Parameter n is a correction factor of the wave front. In this study, parameter n is fitted to obtain $t_f = 1.67(t_{90} - t_{30})$, being t_{90} and t_{30} the instants at which the stroke current reaches the 90% and the 30% of the peak current value respectively; the resulting value is $n = 3.24$. The waveform shown in Figure 11.11 is similar to that proposed in standards and used by some authors; see, for instance, [31].

Stroke parameters. Although negative flashes have multiple strokes, only the first and the second strokes are of concern for transmission insulation levels. Actually, only an accurate knowledge of the parameters of the first negative stroke can be crucial for transmission lines with rated voltage 400 kV and above [40]. The peak current magnitudes of positive strokes are larger than those of negative polarity, and their front and tail times are far longer. In addition, they exhibit a seasonal variation: the number of positive flashes increases during winter [41, 42], when their statistical parameters are very different from those of negative flashes. The same conclusion is derived when comparing winter positive flashes to summer positive flashes, whose parameters are similar to those of negative strokes.

The statistical variation of the lightning stroke parameters can be approximated by a log-normal distribution, with the following probability density function [37]:

$$p(x) = \frac{1}{\sqrt{2\pi}x\sigma_{\ln x}} \exp\left[-\frac{1}{2}\left(\frac{\ln x - \ln x_m}{\sigma_{\ln x}}\right)^2\right] \tag{11.4}$$

where $\sigma_{\ln x}$ is the standard deviation of $\ln x$, and x_m is the median value of x.

A zero correlation coefficient between lightning stroke parameters is assumed here. See also references [33, 36].

11.3.3 Modelling for Lightning Overvoltage Calculations

Modelling guidelines for representing different parts of the transmission line involved in lightning overvoltage calculations are summarized in the following paragraphs. For more information on the guidelines, see references [33, 43–45]. See also Chapter 6.

1) Shield wires and phase conductors of the transmission line can be modelled by three or more spans at each side of the point of impact. A rigorous representation of each span should be based on a multiphase frequency-dependent untransposed distributed-parameter line model. For lightning overvoltage calculations, a constant parameter line model with parameters calculated at a high frequency (e.g. 500 kHz) can be accurate enough [33].

Figure 11.12 Linear corona model.

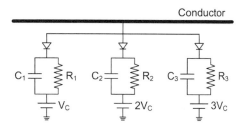

2) The line termination at each side of the above model can be represented by a long-enough line section to avoid reflections during the simulation time that could affect the calculated overvoltages around the point of impact. Only waves reflected at the towers closer to the point of impact should affect the voltages caused by the lightning stroke current.

3) Several models have been proposed to represent transmission line towers; they have been developed using a theoretical approach or based on an experimental work [46]. The simplest representation is a lossless distributed-parameter transmission line, characterized by a surge impedance and a travel time. For an introduction to tower modelling see [47].

4) Phase voltages at the instant at which the lightning stroke impacts the line have to be included in calculations; their values are deduced by randomly determining a phase reference angle.

5) Corona effect can impact the propagation of surges associated with lightning strokes by introducing a time delay to the wave front [47]. This time delay takes effect above the corona inception voltage and varies with surge magnitude. This variation can be expressed as a voltage-dependent capacitance added to the geometrical capacitance of the transmission line. In general, corona does not affect the tail of the surge. The circuit shown in Figure 11.12 has been used to model the dynamic capacitance region of the q/V curve in a piecewise linear fashion. However, this model has some limitations: (i) it is based on lumped elements and must be lumped at sufficiently small intervals along the line to minimize any error introduced by the discretization; (ii) it does not adequately address corona in a multiphase model. Corona effect is independent of the conductor size and geometry for high magnitude surges. For low magnitude surges, the effect will depend on conductor sizes and the corresponding corona inception voltages. Weather conditions have no significant impact on corona distortion. The approach proposed in [48] relies upon the observation that, for voltages substantially higher than the corona inception level, the time delay as a function of travel distance becomes linear and the steepness of the overvoltage is independent of the voltage value. The following relationship has been proposed in standards [49, 50]:

$$S = \frac{1}{\frac{1}{S_o} + A \cdot d} \tag{11.5}$$

where S_o is the original steepness of the overvoltage, S is the new steepness after the waveform travels for a distance d, and A is a constant. The constant A is a function of the line geometry and also depends on the surge polarity. Typical values are given in [49, 50]. Equation (11.5) can be used to estimate the variation with travel length of the steepness of lightning overvoltages that impact a substation [31].

6) Several models have been proposed to represent insulator strings in lightning studies. The most accurate representation relies on the application of the leader progression model [31, 51, 52], which can be used to account for non-standard lightning voltages. The approach used here is based on the voltage-time curves. These curves show the dependency of the peak voltage of the specific impulse shape on the time to breakdown; see Figure 2.9 [47].

These curves are experimentally determined for a specific gap or insulator string and may be represented with empirical equations, applicable only within the range of parameters covered experimentally [31]:

$$V_f = A + Bt^m \tag{11.6}$$

In practice, measurements can be affected by several factors: impulse front shape, front times of the applied standard lightning impulse, gap distance and geometry, polarity, internal impedance of the impulse generator (due to the predischarge currents in the gap).

7) An accurate representation of the footing impedance must be based on a nonlinear frequency-dependent circuit [47, 51, 53]. Since the information needed to derive such a model is not always available, very often a conservative approach is applied, and the impedance is modelled as a lumped nonlinear resistance, although this representation is not always adequate [47]. The value of this resistance can be approximated by the following expression [31, 49, 52]:

$$R_T = \frac{R_o}{\sqrt{1 + I/I_g}} \tag{11.7}$$

where R_o is the low-current and low-frequency footing resistance, I is the stroke current through the resistance, and I_g is the limiting current to initiate sufficient soil ionization, given by

$$I_g = \frac{E_0 \rho}{2\pi R_o^2} \tag{11.8}$$

where ρ is the soil resistivity (ohm-m) and E_0 is the soil ionization gradient (400 kV/m, [54]).

8) A lightning stroke is represented as a current source whose parameters, as well as its polarity and multiplicity, are randomly determined according to the distribution density functions recommended in the literature [31, 37]. See previous subsection.

11.3.4 Implementation of the Monte Carlo Procedure Using Parallel Computing

This subsection details the procedure implemented in MATLAB for calculating the LFOR of overhead transmission lines using parallel computing [35]. The main steps are basically those implemented in previous works [33, 36]. A summary of the procedure follows.

1) The values of random parameters are estimated. This step includes the calculation of the parameters of the lightning stroke (peak current, rise time, tail time, location of the vertical leader channel) and the phase conductor voltages.

2) The last step of a lightning stroke is determined by using the electrogeometric model; see Figure 11.13 [31]. The striking distance values are obtained from the following expressions:

$$r_c = \alpha \cdot I_p^\gamma \qquad r_g = \beta \cdot r_c \tag{11.9}$$

where α and γ are constants that depend on the object, β is a constant that depends on the electrogeometric model, and I_p is the stroke current peak [31, 32].

3) The previous step decides the final target of each lightning stroke: shield wire (tower or midspan point), phase conductors, ground.

4) Strokes to ground are discarded, while those to the line (either to a tower or to a midspan point) are edited to include the random parameters into the ATP file of the test line model. That is, once the point of impact of each lightning stroke is known, the ATP file that will simulate each stroke to the line is edited.

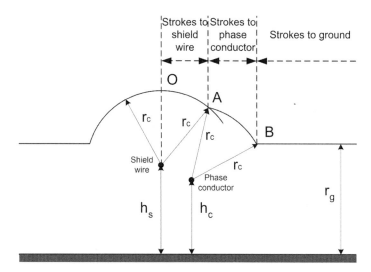

Figure 11.13 Application of the electrogeometric model.

5) The MATLAB application distributes the ATP files between the cores.
6) Overvoltage calculations are performed. The only difference between models for backflash and shielding failure simulations is the node to which the current source that represents the stroke must be connected. The result of each simulation (i.e. flashover, no flashover) is reported. In case of flashover, the individual simulation is stopped.
7) The MATLAB application post processes the result of each simulation to obtain the LFOR and any other distribution function of interest (e.g. the distribution of lightning stroke peak currents that caused flashover).

The convergence of the Monte Carlo method can be checked as the simulations progress or prior to any simulation by comparing the calculated probability density functions of all random parameters to the theoretical functions. The second option is applied here, so a high number of runs is selected to guarantee the convergence of the method prior to the simulation of any case.

Figure 11.14 shows a diagram of the procedure implemented in MATLAB. ATP capabilities are used to edit the line model template in which the current source that represents the stroke and the random parameters (i.e. lightning stroke parameters, phase conductor voltages) are embedded.

Table 11.2 lists the random values that can be generated by the application implemented for this study [33, 36]. In the present work, insulator string and footing resistance parameters are not random; instead, they are fully specified in the ATP overhead line template prior to the calculation of any random value.

11.3.5 Illustrative Example

11.3.5.1 Test Line
Figure 11.15 shows the tower design for the line tested in this study. It is an actual 400 kV line located in Northern Algeria. As shown in the figure, the line has two conductors per phase and one shield wire, whose characteristics are provided in Table 11.3.

11.3.5.2 Line and Lightning Stroke Parameters
A model of the test line was created using ATP capabilities and following the guidelines summarized above.

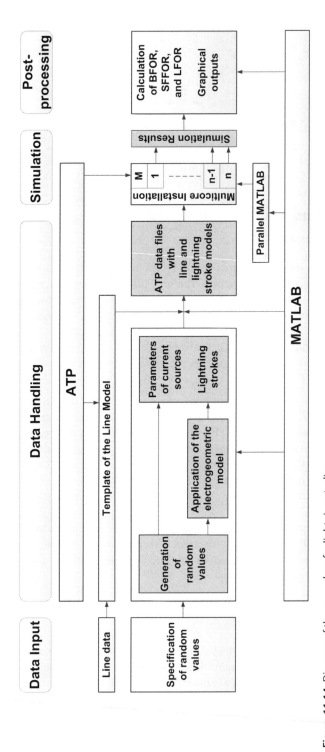

Figure 11.14 Diagram of the procedure for lightning studies.

Table 11.2 Probability density functions.

Parameter/variable	Probability density function
Ground coordinates of the lightning stroke channel	Uniform
Stroke parameters	Log-normal
Phase conductor reference angle	Uniform
Insulator string parameters	Weibull
Footing resistance	Normal

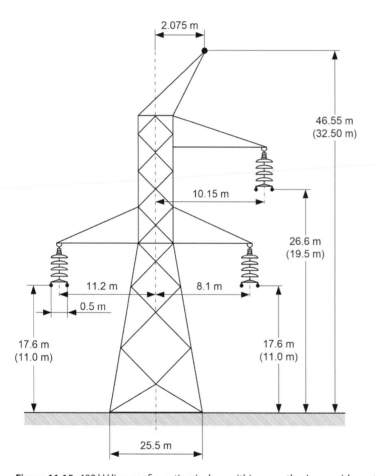

Figure 11.15 400 kV line configuration (values within parenthesis are midspan heights).

- The line is represented by five 390-m spans plus one 10-km section as line termination at each side of the point of impact. Each span is divided into thirteen 30-m sections (for corona effect modelling) represented by means of frequency-dependent distributed-parameter models.
- The towers are represented by means of the so-called multi-storey model proposed in [55]. See also [46, 47] for more details.
- Only single-stroke negative-polarity flashes are considered. A lightning stroke is represented by a concave waveform (see Figure 11.11), with parameter $n = 3.24$ in Eqs. (11.2) and (11.3).

Table 11.3 Transmission test line conductor characteristics.

	Type	Diameter (mm)	Resistance (Ω/km)
Phase conductors	Almelec $2 \times 570 \, \text{mm}^2$	31.05	0.0523
Shield wire	OPGW $185 \, \text{mm}^2$	19.84	0.185

- Footing impedances are represented as nonlinear resistances modelled according to Eq. (11.7) with $R_o = 20 \, \Omega$. The value of the soil resistivity is $100 \, \Omega$-m.
- Insulator strings are represented by means of a simple model (implemented in MODELS language) that duplicate a voltage-time curve (see Figure 2.9) The length of insulator strings is 5.6 m. Assuming the critical flashover voltage (CFO) of insulator strings for negative polarity strokes is estimated as [49]:

$$CFO^- = 700 \cdot d_s \tag{11.10}$$

where d_S is the striking distance of the insulator strings, the corresponding CFO^- is 3920 kV. The curve has been fitted to obtain breakdown at $2 \, \mu s$ with $1.58 \cdot CFO$ and breakdown at $3 \, \mu s$ with $1.36 \cdot CFO$ [47]. The resulting coefficients for Eq. (11.6) are $A = 2273.6E+3$, $B = 5448.8E+3$, and $m = -0.5$.
- Corona effect is incorporated into the line model by means of a simplified version of the circuit depicted in Figure 11.12, in which only the values of C_1, R_1, and V_i are to be specified. As mentioned above, each line span is divided into sections of 30 m, and the parameters for each section (i.e. V_c, R_1, C_1) are obtained according to the following equations:

$$V_c = 23.8 \, r \left(1 + \frac{0.67}{r^{0.4}} \right) \ln \left(\frac{2h}{r} \right) \quad \text{kV} \tag{11.11a}$$

$$R_1 = 60 \quad M\Omega \tag{11.11b}$$

$$C_1 = k_C \frac{1}{18 \ln \frac{2h}{r}} \quad \text{nF/m} \tag{11.11c}$$

where r is the phase equivalent radius, in cm, and h is the conductor height, in cm. K_c in Eq. (11.11c) is adjusted to obtain a propagation model as close to Eq. (11.5) as possible. An adequate value of K_c should be between 0.5 and 1.5; in this study $K_c = 0.6$.

The following probability density functions are used to obtain random values:

- *Lightning stroke.* Peak current magnitude, rise time and tail time are determined by assuming a log-normal distribution for all of them. Table 11.4 shows the values used for each parameter. All parameters are assumed independently distributed.
- *Stroke location.* It is randomly estimated by assuming a vertical path and a uniform ground distribution of the leader. Vertical channels are uniformly distributed in a surface with a single-line span length (i.e. 390 m) and within a 500 m distance at each line side (see Figure 11.16). The electrogeometric model is represented in (11.9) with $\alpha = 8$, $\gamma = 0.65$, and $\beta = 1$ [31].
- *Phase conductor voltages.* The reference angle is estimated by using a uniform distribution between $0°$ and $360°$.

Table 11.4 Statistical parameters of lightning strokes.

Parameter	x_m	ρ_{lnxm}
I_{100}, kA	34.0	0.740
t_f, μs	2.0	0.494
t_h, μs	77.5	0.577

Figure 11.16 Lightning strokes coordinates.

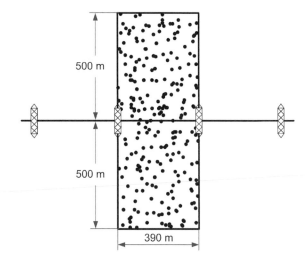

11.3.5.3 Simulation Results

The LFOR of the test line was estimated by generating up to 100 000 combinations of random numbers (i.e. those needed to characterize the lightning stroke and phase conductor voltages). Figure 11.16 shows the distribution of vertical channels within that surface after estimating the location of a few hundred channels.

Once the coordinates of the channels were known, the electrogeometric model was applied to obtain the strokes that would impact the line and discard those to ground. Actually, less than 9000 strokes did finally impact the line (see Table 11.5). Figure 11.17 provides the distribution of lightning strokes to the line, but distinguishing between those that impacted the shield wire from those that impacted a phase conductor. A summary of simulation results is presented in Table 11.5 and Figure 11.18. The figure provides the results obtained with two different models of the current source that represents the lightning stroke.

Although the percentage of strokes to phase conductors was not negligible, the SFFOR was zero in all studies. From the application of the electrogeometric model, the maximum peak current magnitude of strokes to phase conductors was always below 20 kA, and no stroke could cause flashover. The percentage of strokes to the shield wire was much higher; the flashovers caused by these strokes began with peak current values above 60 kA. Consequently, the distribution of stroke currents that caused flashover was that of strokes to the shield wire.

Given the surface within which the randomly generated lightning vertical channels were distributed and assuming that the flash ground density was $N_g = 1$ flash per km² and year, the generation of 100 000 vertical channels corresponds to an analysis of the test line during 256 410 years. Consider the case with a resistance of 400 Ω without including corona effect; since only 1820 strokes could cause flashovers, then the estimated LFOR for this test line is

Table 11.5 Lightning performance of the test line (100 000 runs, 50 cores).

Option	Number of simulations	Number of flashovers	LFOR (flashovers per 100 km and year)	Simulation time (s)
$R_s = 400\,\Omega$				
Without corona	8670	1820	1.820	679.1
With corona	8670	1731	1.731	698.3
		Ideal current source		
Without corona	8670	3075	3.075	628.1
With corona	8670	2872	2.872	673.2

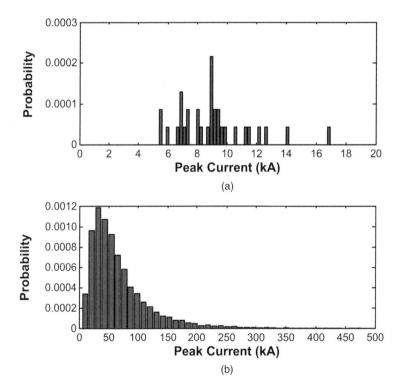

Figure 11.17 Distribution of lightning strokes to the line. (a) Stroke currents to phase conductors. (b) Stroke currents to the shield wire.

1.820 flashovers per 100 km and year. Remember that there is a proportionality with the flash ground density, so if this density was higher than 1, the estimated LFOR of the test line should be increased by the same ratio.

As expected, the number of flashovers increases with the value of the source resistance, and reaches its maximum value for an ideal source. One can observe that the addition of corona to the line model reduces the number of flashovers, and this effect increases with the parallel resistance of the source current that represents the stroke; with an ideal source current the final value of LFOR is about 6.5% lower than with corona effect, see Table 11.5.

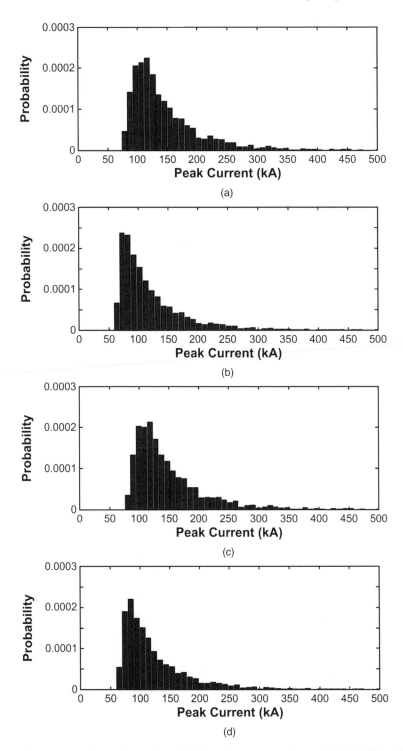

Figure 11.18 Simulation results. (a) LFOR without corona – $R_s = 400\,\Omega$. (b) LFOR without corona – Ideal current source. (c) LFOR with corona – $R_s = 400\,\Omega$. (d) LFOR with corona – Ideal current source.

The study was carried out with a multicore installation in which the ATP simulations were distributed between 50 cores. The total time required to carry out the entire study, including the time needed by the MATLAB application to pre-edit ATP files and post-process ATP results, and that needed by the ATP to calculate overvoltages (and decide whether the line will flashover or not) was less than 12 minutes in all case studies; see Table 11.5.

11.3.6 Discussion

A statistical analysis is the natural approach in calculations when some parameters are not well known or they are known with some uncertainties. This is a common situation in overvoltage calculations because both stress and strength are usually characterized by probability density functions. When the calculations have to be derived from the application of a time-domain software tool, such as ATP, a very long simulation time can be required to perform a statistical study based on the Monte Carlo method.

Although very powerful hardware platforms and very accurate software tools are currently available, simulation times of an order of hours will be usually needed. A solution for this drawback is the usage of a multicore environment. Parallel computing is a very simple solution for parametric and statistical studies that require very long simulation times since, in both cases, tasks can be distributed between several cores to significantly reduce CPU time. This section has proposed a MATLAB application to be used in combination with the ATP package to achieve this goal.

The transmission line model used for this study could have been implemented with more accurate models for some line components (i.e. insulator strings); therefore, the results derived from this study should be used with care.

A discussion about some aspects that must be accounted for to obtain the LFOR of an overhead transmission line and whose accuracy may affect the final results is presented below.

- The study has been carried out with some simplifications: several parts of the transmission line model have not been represented by the most accurate models (e.g. insulator strings, corona effect), voltages induced by electric and magnetic fields of lightning channels to shield wires and phase conductors are neglected, a vertical direction is assumed for the stroke leader when it approaches earth, only single-stroke negative-polarity flashes were assumed and the incidence model was based on the electrogeometric model, which is a very simple approach for representing a very complex physical phenomenon as lightning. See [33, 40, 56] for other aspects that could have been considered in this case study.
- Corona effect has been represented by means of a very simple model, which is not accurate enough for multiconductor line studies. The approach applied here was used to obtain a line model in which corona could introduce a steepness variation close to that recommended in Eq. (11.5).
- The BFOR is sensitive to the coefficient of correlation between the peak current magnitude and the rise time, although the SFFOR is not; consequently, the LFOR decreases as the coefficient of correlation increases; see [33, 40]. In other words, an accurate knowledge of this coefficient is an important issue for flashover rate calculations.
- Sensitivity studies can be very useful to analyse the influence of line and stroke parameters, and to determine what range of values might be of concern [33]. Although the number of parameters involved in lightning calculations is very high, only some of them can accurately be specified from the line geometry. For results presented in previous works [33], it is obvious that the flashover rate increases with the peak current magnitude and decreases with the rise time, while the influence of the footing resistance is not critical for low values of the soil resistivity and low values of R_o.

11.4 Optimum Design of a Hybrid HVDC Circuit Breaker

11.4.1 Introduction

It is widely accepted that the hybrid design presented in references [57–59] represent a significant advance towards a low-loss fast HVDC circuit breaker since that design eliminates on-state losses and provides a very fast operation under faulted conditions. The implementation in a time-domain simulation tool of a circuit breaker model based on that design has been presented in several works with different levels of complexity; see [60–63]. A parametric analysis of the transient response of this design was presented in reference [64]. A procedure for the optimum selection of parameters of the hybrid design was presented in [65]; the goal was to obtain a transient response of some circuit breaker variables within a specified range of values. The work proposed a MATLAB-ATP environment that combined the application of a genetic algorithm (GA) and the use of a multicore environment to obtain the optimum design of the hybrid HVDC circuit breaker. That work was based on a simplified representation of the hybrid circuit breaker taking into account one of the designs presented in the original patent. The optimization procedure applied here is that originally presented in [65] but using a more detailed model of the hybrid circuit breaker [66].

The design and principles of operation of the hybrid circuit breaker are summarized in the next subsection. Subsection 11.4.3 presents the ATP implementation of a detailed model and the ATP capabilities used for its implementation. Subsection 11.4.4 presents the test system used in this study and the modelling guidelines followed for representing the test system. A summary of para-metric analyses presented in previous works is provided in Subsection 11.4.5. The application of a GA for an optimum selection of circuit breaker parameters and its implementation in a multicore environment are detailed in Subsection 11.4.6. The main simulation results and a discussion of those results are presented in Subsections 11.4.7 and 11.4.8, respectively.

11.4.2 Design and Operation of the Hybrid HVDC Circuit Breaker

The hybrid circuit breaker design is based on the parallel connection of the *main branch*, which consists of the series connection of semiconductor-based sections with individual parallel arrester banks, and the *auxiliary branch*, which consists of a semiconductor-based load commutation switch (LCS) connected in series with a fast mechanical disconnector (FMD); see Figure 11.19 [57–59].

During normal operation, the current flows through the auxiliary branch; in case of fault, the current is transferred by the LCS to the (series-connected semiconductor cells of the) main branch, which will quickly open and transfer the current to the parallel arresters; see Figure 11.19. To protect the LCS from the voltage that builds up across the breaker, the FMD opens as soon as the auxiliary branch does not carry current. To protect the arrester banks from thermal overload, a residual current breaker (RCB) interrupts the residual arrester current and isolates the faulted system from the HVDC grid. The hybrid design incorporates a current limiting reactor (CLR) that can be required to obtain a maximum fault current below a specified value; see Figure 11.19. It is important to keep in mind that the operation of the circuit breaker during a fault requires a bidirectional main branch; that is, the current can reverse its flow during the fault condition with respect to the flow prior to the fault, as illustrated in the cases analysed with the test system presented in Subsection 11.4.4. See reference [60] for more details about the behaviour of this design during a fault condition.

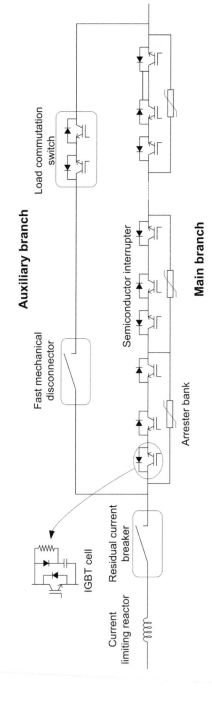

Figure 11.19 Hybrid HVDC circuit breaker.

11.4.3 ATP Implementation of the Hybrid HVDC Circuit Breaker

The model of the auxiliary branch, as shown in Figure 11.19, can consist of a single ideal switch with an antiparallel diode and a parallel snubber circuit to represent the LCS, and an ideal switch (without including arc dynamics) to represent the FMD. However, the representation of the main semiconductor-based branch (SCB) can be very complex due to the high number of series cells: several hundreds of IGBT models may be required to represent the entire main branch of an HVDC circuit breaker in cases of high-voltage operation (e.g. a voltage level equal or above 230 kV).

The breaker model has been implemented using a modular approach taking advantage of some ATP capabilities, as detailed below. ATP incorporates a capability (i.e. the DO loop) that can be used to quickly achieve a representation in which the same components are repeated as many times as required. In addition, some component parameters can automatically be obtained as some mathematical manipulation is possible at the time the DO loop is applied; for instance, the snubber parameters for each IGBT cell will depend on the number of cells to be included in the model (i.e. these parameter values can automatically be derived when creating the model of the main breaker path by means of a DO loop). Figure 11.20 illustrates the approach followed to implement a detailed representation of the main circuit breaker branch.

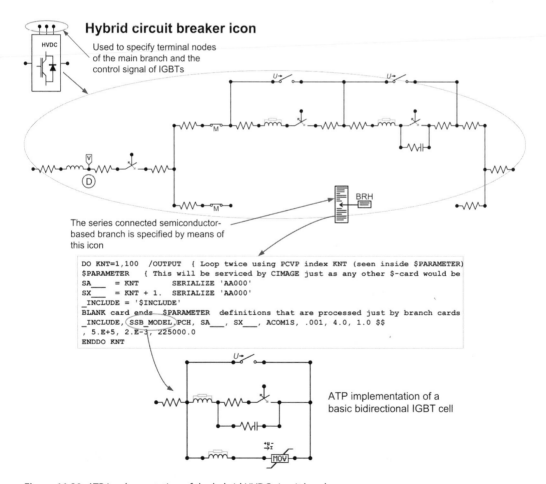

Figure 11.20 ATP implementation of the hybrid HVDC circuit breaker.

- Each bidirectional IGBT cell has been represented by means of an ideal controlled switch (Type 13) with a series resistor (that represents semiconductor losses), a series inductor (that represents parasitic inductances), and a parallel *RC* snubber circuit. The 'DO loop' capability is used to generate the main semiconductor-based branch, as shown in Figure 11.20. By using this approach, the number of IGBT cells to be included in the branch can be freely varied. If 100 bidirectional IGBT cells are required to represent the main branch, the code could be as follows (see Figure 11.20 and reference [66]):

```
DO KNT=1, 100 /OUTPUT
$PARAMETER
SW__A = KNT SERIALIZE 'SSB00'
SW__B = KNT + 1. SERIALIZE 'SSB00'
_INCLUDE = '$INCLUDE'
BLANK card
_INCLUDE, SSB_MODEL.PCH, SW__A, …
ENDDO KNT
```

Note that a very small resistance is used to separate contiguous IGBT cells, represented by the module 'SSB_MODEL'.
- Parallel arresters are modelled by means of a Type 92 nonlinear resistor, and are included within the model of an IGBT cell represented within the module 'SSB_MODEL'.

These capabilities can easily be used to represent a more complex and realistic semiconductor-based breaker. According to the design detailed in [59], parallel connection of IGBT modules can be required to increase current rating, while series connection may be used to increase voltage rating. In addition, either series or parallel connections are applied to increase reliability. Therefore, IGBT matrices can be used for high-power high-voltage applications.

The models of FMD and the RCB could be more complex since an arc model might be needed in each one. An arc voltage could help driving the commutation of the current from the mechanical switch to the main semiconductor-based branch. However, since it is assumed that both mechanical switches open with a zero or near-zero current, the arc dynamics have been disabled.

Mechanical disconnectors have been represented using the controlled Type 11 switch; this is due to the fact that the mechanical breaker will only be open when the current to be interrupted is zero (or below a specified very small value).

A parallel snubber circuit is not strictly required for any of these switches since the current across them is zero when they open.

The control strategies of the hybrid circuit breakers have been implemented using TACS (Transient Analysis of Control Systems) capabilities. Both arresters and snubber circuits in parallel to main-branch IGBT cells assure equal voltage distribution during current breaking.

11.4.4 Test System

The configuration of the test system is depicted in Figure 11.21a; it is based on the system analysed in reference [67]. The HVDC system links two High Voltage Alternating Current (HVAC) 400 kV systems and consists of two station converters plus an insulated cable. Each station is represented by a two-level dc voltage source with the corresponding equivalent impedances being the stations linked by means of two insulated cables of equal characteristics. Different

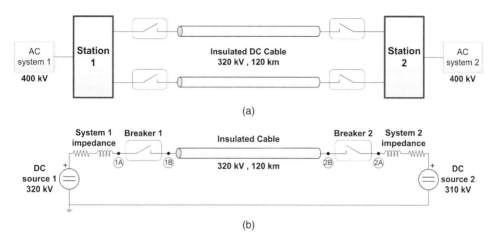

Figure 11.21 HVDC transmission test system configuration and simplified equivalent circuit. (a) Test system configuration. (b) Simplified equivalent circuit.

voltage values at each station allow some current flow between the two terminals. Figure 11.21b depicts the simplified model used in this study (see also [67]); the equivalent *RL* impedances of the two HVDC systems are the same. It is assumed the system is protected by hybrid circuit breakers at each station. Both pole-to-ground and pole-to-pole faults can be considered; due to the symmetry of the system, currents, and voltages for pole-to-ground faults and pole-to-pole faults will be very similar, so the transient response to the system in front of any type of fault will be analysed by means of the simplified system model shown in Figure 11.21b [67].

The models and parameter values used in this study are as follows (see also [65, 66]):

- *HVDC sources.* Each system side is represented by an ideal dc voltage source in series with its *RL* equivalent impedance, see Figure 11.21b. The rated value of source voltages is 320 kV, although voltage of source 2 is lower to allow a current flow from source 1 to source 2. The equivalent impedance is the same in both systems: $R = 8\,\Omega$, $L = 64\,\text{mH}$. Note that with these values, the highest prospective short-circuit current (i.e. that to be supplied from source 1) is 40 kA and the time constant of both system impedances (i.e. L/R) is 8 ms.
- *HVDC insulated cable.* It is a 320 kV 120 km single-core self-contained cable represented by a frequency-dependent distributed-parameter model. The cable configuration and parameters correspond to those of the cable used in [67]. The wave propagation velocity in the cable is close to 190 km/ms [64].
- *Hybrid HVDC circuit breaker.* The model implemented in this study is that presented in Figure 11.19 and Subsection 11.4.3. The LCS will open once the current exceeds a threshold value, I_{th}, while the commutation from the SCB to the arrester branch will occur with a fixed delay, t_d, with respect to the first current commutation (from the auxiliary branch to the SCB); this delay is meant to give time to the FMD for a full opening. The main branch of the circuit breaker model can incorporate any number of series IGBTs when using the ATP capabilities detailed in the previous subsection; in this study that number has been varied between 20 and 100.

11.4.5 Transient Response of the Hybrid Circuit Breaker

This subsection presents some simulation results aimed at illustrating the transient response of the hybrid circuit breaker working as current limiter and the effect that some circuit breaker parameters have on this transient response in front of a pole-to-ground fault in the insulated cable; see Figure 11.21b. The parameters whose effect is analysed are: (i) the *current threshold* I_{th} (i.e. the value used by the hybrid circuit breaker to start its operation); (ii) the *current commutation delay* t_d (i.e. the time period between the moment at which the fault current is transferred from the auxiliary branch to the SCB and the moment at which the current is transferred from the SCB to the arrester branch, see Figure 11.19); (iii) the inductance of the CLR, L_{CLR}; (iv) the arrester voltage, V_{arr}; (v) the fault location.

A detailed analysis of the transient response of the hybrid circuit breaker as a function of some system and breaker parameters using the same test system configuration that is analysed here was presented in [64]. A summary of the main conclusions is presented below.

1. *Currents.* The maximum fault current will always be supplied by the left-side system (i.e. System 1 in Figure 11.21) whose voltage is the highest one.
   - This maximum current value is reached when the fault is located at the left terminal of the cable (i.e. Node 1B in Figure 11.21b).
   - The peak of the fault current increases with the current threshold selected for detecting a fault current (i.e. the current peak with $I_{th} = 5\,\text{kA}$ is larger than with $I_{th} = 4\,\text{kA}$) and with the time delay between the first and the second current commutation within the breaker (i.e. the current peak with $t_d = 4\,\text{ms}$ is higher than with $t_d = 3\,\text{ms}$).
2. *Voltages.* The maximum voltages originated during the transient caused by a fault will always occur at node D (see Figure 11.20) of either Breaker 1 or Breaker 2.
   - The differences between the maximum voltages at node D obtained with different values of the current threshold I_{th} are not significant if all other parameters remain constant; that is, differences between maximum voltages at node D (of either Breaker 1 or Breaker 2) with $I_{th} = 4\,\text{kA}$ or with $I_{th} = 5\,\text{kA}$ are not too large.
   - The effect that the delay between current commutations has on the maximum voltage values when the other parameters remain the same is neither negligible nor too large.
   - The fault location (or the distance between the fault location and any substation) with which the maximum voltage is reached in one breaker depends on the cable length.

Figure 11.22 shows some voltages and currents obtained within the two circuit breakers (see Figure 11.21) and corresponding to faults located at both cable terminals.

All the results were derived from a circuit breaker model in which the main branch was represented by means of 50 series-connected semiconductor cells (see Figures 11.19 and 11.20). Note that the threshold current value used to detect a fault condition is different for each fault location. Each plot shows results derived with three different values of the time period between the two current commutations within each circuit breaker. In all cases, the fault condition occurs at $t = 2\,\text{ms}$. These results support those previously obtained with a simplified representation of the circuit breaker and presented in reference [64].

11.4.6   Implementation of a Parallel Genetic Algorithm

The application of a GA begins by defining a set of input data (i.e. the initial population), randomly selected from a search space. The outputs obtained will then allow calculation of the fitness of this initial population [68]. After the first evaluation, individuals considered fit are selected and become the parents for a new generation of individuals. Once the next generation is obtained, the process is repeated until a termination criterion is fulfilled or when a predefined

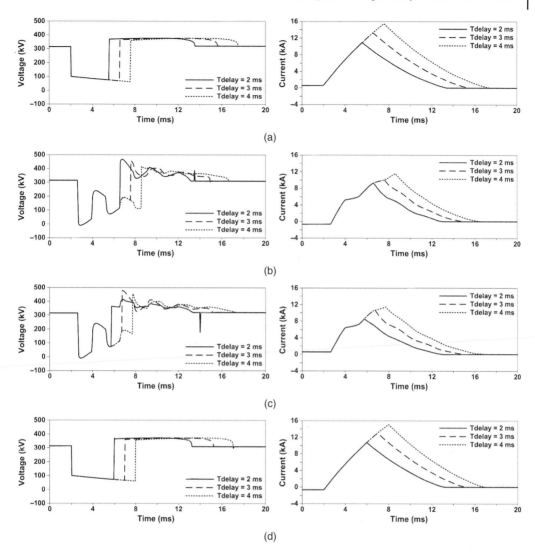

Figure 11.22 Simulation results – Main branch represented by means of 50 IGBT bidirectional cells. (a) Fault location: Left cable terminal – Breaker 1 variables (I_{th} = 5 kA; V_{arr} = 7.8 kV; L_{CLR} = 30 mH). (b) Fault location: Left cable terminal – Breaker 2 variables (I_{th} = 5 kA; V_{arr} = 7.8 kV; L_{CLR} = 30 mH). (c) Fault location: Right cable terminal – Breaker 1 variables (I_{th} = 3 kA; V_{arr} = 7.8 kV; L_{CLR} = 30 mH). (d) Fault location: Right cable terminal – Breaker 2 variables (I_{th} = 3 kA; V_{arr} = 7.8 kV; L_{CLR} = 30 mH).

number of generations has been accomplished. A GA for a particular problem must have the following components [69]:

1. a criterion to create the initial population (it can be random within a defined range);
2. an evaluation function to obtain the possible solutions before calculating its fitness;
3. a genetic operator to obtain the next generation;
4. a stopping criterion.

Several types of parallel GAs have been proposed: global single-population master-slave GAs (GPGAs), single-population fine-grained, multiple- population coarse-grained GAs [70]. GPGAs use a single population but the evaluations of the fitness function are distributed among different processors: the master processor stores the population and assigns a fraction of the population to each slave processor which evaluates and returns the fitness values. This type of GA is easy to implement, as there is no need to change the general guidelines of single-core GAs, and it can exhibit a significant processing speedup [71].

The solution of many engineering problems usually leads to a multi-objective optimization with constrains that must be fulfilled to obtain an optimum or near-to-optimum solution. A multi-objective GA can be seen as a procedure for solving the function being optimized while satisfying some constraints [72–75]. Multi-objective algorithms also imply having more than one solution, with each of the selected parameters coming from a constraint [75, 76].

This study uses the MATLAB-ATP environment previously proposed in [65, 66], and shown in Figure 11.23. It is aimed at performing an optimization procedure by means of a GA and using a multicore installation. According to this diagram, the evaluation of the GA is done in a MATLAB master platform, while the parallelization part is carried out using a multicore installation whose evaluation tasks are subdivided in MATLAB slave platforms [66]. ATP performs as a black box that is called by each slave processor to evaluate the data input from the GA evaluation and provides the respective outputs. ATP is also used to edit the test system template in which the set of random data created for each generation in the GA can easily be inserted. Therefore, it is crucial for the application to be able to dynamically insert new parameter values into the model without changing the rest of the test system information. Since a GA is an iterative process and a parallel GA is a collection of iterative processes running simultaneously, several data decks are sent to ATP every new generation. An easy way to carry out this task is to use $INCLUDE capability, which allows ATP to evaluate different data without changing the test system template.

11.4.7 Simulation Results

The main features of the multi-objective parallel GA implemented in this study may be summarized as follows [65]:

a) The search space is defined by means of the following value ranges: (i) the threshold current (I_{th}) can be any value between 3 and 6 kA; (ii) the current commutation delay (t_d) is assumed to vary between 2 and 4 ms; (iii) the limiting reactance (L_{LCR}) can vary from 1 to 40 mH; (iv) the total arrester voltage (V_{arr}) can be between 350 and 450 kV.
b) Any feasible solution should exhibit results with a maximum voltage equal or lower than 700 kV, a maximum current equal or lower than 15 kA, and a fault clearance time equal or shorter than 15 ms.
c) In order to obtain a feasible solution it is crucial to check the results derived from any fault position located between cable terminals. In this study, up to 13 fault positions uniformly located between the two cable terminals are considered. Therefore, the solution corresponding to any combination of the input parameters defined above is obtained by evaluating the 13 fault positions, and selecting only the maximum values of the output variables. Taking into account that the fault position with which the maximum current occurs is not the same position with which the maximum voltage is developed.

The general procedure may be summarized as follows [65]:

1) A set of random parameters (i.e. a population) is created taking into account the search area specified above.
2) Each combination of three input values (i.e. I_{th}, t_d, L_{LCR}) is evaluated. Current and voltage peaks, as well as the time period needed to clear the fault, are measured in the two circuit

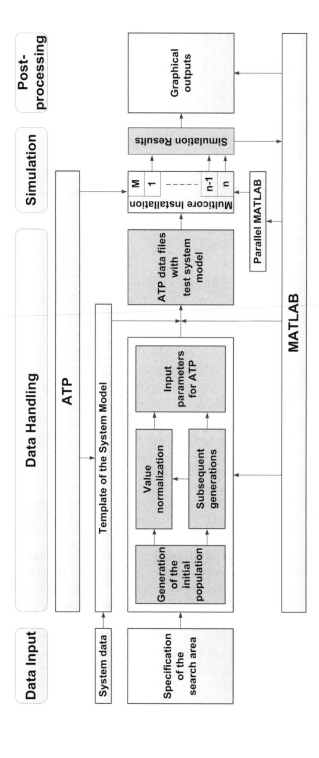

Figure 11.23 Diagram of the MATLAB-ATP interaction when using genetic algorithms.

Table 11.6 Simulation results – 50 IGBT cells.

Arrester voltage (kV)	Transient performance	Circuit breaker parameters
8.6	$7.7 \leq t_{fc} \leq 13\,\mathrm{ms}$, $V_{max} < 680\,\mathrm{kV}, 9 \leq I_{max} < 15\,\mathrm{kA}$	$3.5 \leq I_{th} \leq 5\,\mathrm{kA}, 2.1 \leq t_d \leq 4\,\mathrm{ms}$, $L_{CLR} \leq 40\,\mathrm{mH}$
8.4	$8 \leq t_{fc} \leq 14.2\,\mathrm{ms}$, $V_{max} < 668\,\mathrm{kV}, 9 \leq I_{max} < 15\,\mathrm{kA}$	$3.5 \leq I_{th} \leq 5\,\mathrm{kA}, 2.1 \leq t_d \leq 4\,\mathrm{ms}$, $L_{CLR} \leq 40\,\mathrm{mH}$
8.2	$9 \leq t_{fc} \leq 14.8\,\mathrm{ms}$, $V_{max} < 659\,\mathrm{kV}, 9 \leq I_{max} < 14.8\,\mathrm{kA}$	$3.5 \leq I_{th} \leq 5\,\mathrm{kA}, 2 \leq t_d \leq 4\,\mathrm{ms}$, $L_{CLR} \leq 40\,\mathrm{mH}$
8.0	$9.5 \leq t_{fc} \leq 14.9\,\mathrm{ms}$, $V_{max} < 646\,\mathrm{kV}, 9 \leq I_{max} < 14.8\,\mathrm{kA}$	$3.5 \leq I_{th} \leq 5\,\mathrm{kA}, 2 \leq t_d \leq 4\,\mathrm{ms}$, $L_{CLR} \leq 40\,\mathrm{mH}$
7.8	$9.8 \leq t_{fc} \leq 15\,\mathrm{ms}$, $V_{max} < 636\,\mathrm{kV}, 8.8 \leq I_{max} < 14.9\,\mathrm{kA}$	$3.5 \leq I_{th} \leq 5\,\mathrm{kA}, 2 \leq t_d \leq 4\,\mathrm{ms}$, $L_{CLR} \leq 40\,\mathrm{mH}$
7.6	$11.4 \leq t_{fc} \leq 14.9\,\mathrm{ms}$, $V_{max} < 622\,\mathrm{kV}, 8.7 \leq I_{max} < 14.9\,\mathrm{kA}$	$3.5 \leq I_{th} \leq 5.5\,\mathrm{kA}, 2 \leq t_d \leq 3.8\,\mathrm{ms}$, $4.2 \leq L_{CLR} \leq 40\,\mathrm{mH}$
7.4	$12 \leq t_{fc} \leq 14.9\,\mathrm{ms}$, $V_{max} < 614\,\mathrm{kV}, 8.7 \leq I_{max} < 14.6\,\mathrm{kA}$	$3.5 \leq I_{th} \leq 5.5\,\mathrm{kA}, 2 \leq t_d \leq 3.7\,\mathrm{ms}$, $4.2 \leq L_{CLR} \leq 38\,\mathrm{mH}$
7.2	$12 \leq t_{fc} \leq 14.0\,\mathrm{ms}$, $V_{max} < 598\,\mathrm{kV}, 8.8 \leq I_{max} < 14.6\,\mathrm{kA}$	$3.5 \leq I_{th} \leq 5.5\,\mathrm{kA}, 2 \leq t_d \leq 3.2\,\mathrm{ms}$, $4.2 \leq L_{CLR} \leq 38\,\mathrm{mH}$

breakers. To obtain the maximum current that corresponds to a combination of parameters, measuring the peak of the current through Breaker 1 when the fault is located at the left terminal of the cable suffices; however, to estimate the maximum voltage or the longer fault clearance time that can occur in any of the two circuit breakers it is necessary to obtain results with different fault locations. Only the combinations that fulfil the restrictions are processed as feasible solutions. This step is carried out for a single arrester voltage value, V_{arr}, which must be between the limits defined above.

3) A set of genetic operators (crossover and mutation) is applied to the feasible values.
4) The new set of values resulting for the previous step replaces the previous population. The process is repeated until the specified number of generations has been evaluated.

A total of 10 generations with 60 individuals per generation were simulated. Table 11.6 summarizes the results derived with all arrester voltages considered in this study when the main branch of the hybrid design was represented by means of 50 series IGBT cells. Figure 11.24 presents the results derived with some arrester voltages. The plot produced for each arrester voltage visualizes the surface derived from all the feasible solutions that satisfy the imposed constraints.

For arrester voltages below 7.2 kV, it was not possible to achieve the desired performance: the effect of decreasing the arrester voltage below 7.2 kV helps keep the maximum voltage values below the specified limit, but the effect in limiting the current is an undesirable delay of the moment at which the fault current will allow the opening of the RCB; that is, with values below 7.2 kV the circuit breaker cannot fully isolate the fault within the specified time period (15 ms). However, acceptable results (e.g. the peak voltage did not exceed the maximum allowed voltage, 700 kV) were obtained with arrester voltages above 8.6 kV.

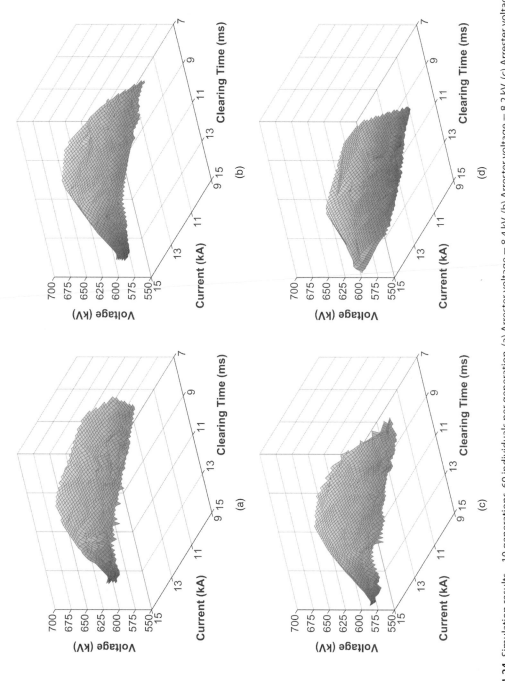

Figure 11.24 Simulation results – 10 generations, 60 individuals per generation. (a) Arrester voltage = 8.4 kV. (b) Arrester voltage = 8.2 kV. (c) Arrester voltage = 8.0 kV. (d) Arrester voltage = 7.8 kV.

11.4.8 Discussion

This study has presented the application of a multi-objective GA for optimizing the design of a hybrid HVDC circuit breaker operating as current limiter and represented by a detailed model of the main circuit breaker branch. The procedure has been implemented in a MATLAB-ATP environment for application with a multicore installation. The approach presented here uses a simplified model of the test system and some circuit breaker components (e.g. IGBT cells of the main breaker branch).

1. The maximum values for the current and voltage that the circuit breakers should withstand, as well as the maximum fault clearance time, were selected without using any field experience that would advise those values. However, in case of fixing different limits for those variables, the optimization procedure based on the GA would remain the same.
2. The effect of the threshold current and the current commutation delay is basically the same: they both result in a delayed limitation of the current and hence a higher current peak.
3. The computing time needed to solve the GAs for a given number of IGBTs cells has been analysed. Table 11.7 presents the times obtained with 60 cores and corresponding to a single arrester evaluation. The values provided in the table correspond to the results derived from the simulation of 10 generations and 60 individuals per generation. The main branch of each circuit breaker was represented by means of 50 series IGBT cells. Remember that the evaluation of each set of individuals involves the simulations corresponding to 13 fault locations; that is, in case of using a population of 60 individuals and performing the simulation of 10 generations, the number of ATP runs that have to be carried out is 7800 per arrester voltage value. The table lists results corresponding to 10 generations and a single arrester voltage value. It is obvious that the time required for each arrester voltage value depends directly on the number of IGBTs cells used for representing the main branch of the hybrid circuit breaker.
4. Table 11.8 provides the number of feasible solutions obtained with some arrester voltages for each generation and the cumulative quantity of solutions after a certain number of generations. From the results shown in the table, it is evident that the implemented procedure provides more than one single feasible solution (i.e. there can be several combinations of input parameters that fulfil the imposed constraints) per generation: the number of feasible solutions that were found after 10 generations with one arrester voltage varied between 506 (with 8.4 kV arresters) and 192 (with a 7.6 kV arrester). Those results can be used to decide about the convergence of the procedure. Since the computing time required to obtain a high enough number of feasible combinations depends not only on the number of cores used in computations but on the number of individuals per generation, the number of generations to be simulated will finally depend on the desired number of feasible combinations. For instance, assume that for the design of the hybrid circuit breaker a minimum of 200 combinations of feasible solutions per arrester voltage value is required; then, it is obvious that less than 10 generations will suffice with arrester voltages equal or above 7.8 kV.

Table 11.7 Genetic algorithm performance (60 cores – 10 generations – 60 individuals per generation).

Number of main-branch series cells	Simulation time per arrester voltage (s)
100	29 270
50	5 708
20	1 977

Table 11.8 Genetic algorithm performance.

Generation	Feasible solution per generation					Cumulative feasible solutions				
	Arrester voltage (kV)					Arrester voltage (kV)				
	7.6	7.8	8.0	8.2	8.4	7.6	7.8	8.0	8.2	8.4
1	3	2	49	50	47	3	2	49	50	47
2	3	2	54	49	48	6	4	103	99	95
3	10	7	50	48	48	16	11	153	147	143
4	12	12	46	51	54	28	23	199	198	197
5	17	26	50	52	47	45	49	249	250	244
6	20	34	47	46	51	65	83	296	296	295
7	24	48	49	47	51	89	131	345	343	346
8	32	56	51	56	52	121	187	396	399	398
9	36	52	43	56	52	157	239	439	455	450
10	35	47	33	58	56	192	286	472	513	506

Although a very complex model of the hybrid design has been used, the application of a multicore environment has allowed affordable computing times. This study has shown that the selection of the number of individuals for the application of the GA must be made wisely in order to save computing time; see also Discussion in [65].

Acknowledgement

All case studies included in this chapter were developed and implemented by the authors. However, some case studies used models or parameters provided by other authors. Thanks are due to B. Mork who provided the parameters of the transformer used in the third ferroresonance case study. Thanks are also due to S. Badoui who provided the line model used in the lightning study. The study on the hybrid HVDC circuit breaker was supported by the project ESPE, within the KIC InnoEnergy consortium. Finally, the authors want to express their gratitude to Dr. -Ing. M. Buehren for making his library for parallel processing available to MATLAB users.

References

1 Buehren, M. MATLAB Library for Parallel Processing on Multiple Cores. Copyright 2007. www.mathworks.com (accessed 03 July 2019).

2 Iravani, M.R., Chaudhary, A.K.S., Giewbrecht, W.J. et al. (2000). Modelling and analysis guidelines for slow transients: part III: the study of ferroresonance. *IEEE Transactions on Power Delivery* 15 (1): 255–265.

3 Ferracci, P. (1998). Ferroresonance, Cahier Technique no. 190. Groupe Schneider.

4 Hilborn, R. (2001). *Chaos and Nonlinear Dynamics: An Introduction for Scientists and Engineers*. New York (NY, USA): Oxford University Press.

5 Kieny, C. (1991). Application of the bifurcation theory in studying and understanding the global behavior of a ferroresonant electric power circuit. *IEEE Transactions on Power Delivery* 6 (2): 866–872.

6 Bodger, P.S., Irwin, G.D., Woodford, D.A., and Gole, A.M. (1996). Bifurcation route to chaos for a ferroresonant circuit using an electromagnetic transients program. *IEE Proceedings - Generation, Transmission and Distribution* 143 (3): 238–242.

7 Ben-Tal, A., Shein, D., and Zissu, S. (1999). Studying ferroresonance in actual power systems by bifurcation diagram. *Electric Power Systems Research* 49 (3): 175–183.

8 Jacobson, D.A.N., Lehn, P.W., and Menzies, R.W. (2002). Stability domain calculations of period-1 ferroresonance in a nonlinear resonant circuit. *IEEE Transaction on Power Delivery* 17 (3): 865–871.

9 Wörnle, F., Harrison, D.K., and Zhou, C. (2005). Analysis of a ferroresonant circuit using bifurcation theory and continuation techniques. *IEEE Transaction on Power Delivery* 20 (1): 191–196.

10 Escudero, M.V., Dudurych, I., and Redfern, M.A. (2010). Characterization of ferroresonant modes in HV substation with CB grading capacitor. *Electric Power Systems Research* 77 (1): 1506–1513.

11 Amar, B. and Dhifaoui, R. (2011). Study of the periodic ferroresonance in the electrical power networks by bifurcation diagrams. *Electrical Power and Energy Systems* 33: 61–85.

12 Corea-Araujo, J.A., González-Molina, F., Martínez, J.A. et al. (2014). Tools for characterization and assessment of ferroresonance using 3-D bifurcation diagrams. *IEEE Transactions on Power Delivery* 29 (6): 2543–2551.

13 Jazebi, S., Zirka, S.E., Lambert, M. et al. (2016). Duality derived transformer models for low-frequency electromagnetic transients – part I: topological models. *IEEE Transactions on Power Delivery* 31 (5): 2410–2419.

14 Martinez, J.A., Walling, R., Mork, B. et al. (2005). Parameter determination for modelling systems transients. Part III: transformers. *IEEE Transactions on Power Delivery* 20 (3): 2051–2062.

15 Mork, B.A., Gonzalez, F., Ishchenko, D. et al. (2007). Hybrid transformer model for transient simulation-part I: development and parameters. *IEEE Transactions on Power Delivery* 22 (1): 248–255.

16 de León, F., Gómez, P., Martinez-Velasco, J.A., and Rioual, M. (2009). Transformers. In: *Power System Transients. Parameter Determination* (ed. J.A. Martinez-Velasco), 177–249. Boca Raton (FL, USA): CRC Press.

17 Zirka, S.E., Moroz, Y.I., Arturi, C.M. et al. (2012). Topology-correct reversible transformer model. *IEEE Transactions on Power Delivery* 27 (4): 2037–2045.

18 Rezaei-Zare, A., Iravani, R., and Sanaye-Pasand, M. (2009). Impacts of transformer core hysteresis formation on stability domain of ferroresonance modes. *IEEE Transactions on Power Delivery* 24 (1): 177–186.

19 Rezaei-Zare, A., Iravani, R., Sanaye-Pasand, M. et al. (2008). An accurate hysteresis model for ferrorresonance analysis of a transformer. *IEEE Transactions on Power Delivery* 23 (3): 1448–1456.

20 Moses, P., Masoum, M.A.S., and Toliyat, H.A. (2011). Impact of hysteresis and magnetic couplings on the stability domain of ferrorresonance in asymmetric three-phase three-leg transformers. *IEEE Transactions on Power Delivery* 26 (2): 581–592.

21 Dommel, H.W. (1986). *EMTP Theory Book*. Portland (OR, USA): Bonneville Power Administration.

22 Milićević, K. and Emin, Z. (2009). Impact of initial conditions on the initiation of ferroresonance. *International Journal of Electrical Power and Energy Systems* 31: 146–152.

23 Milićević, K., Vinko, D., and Emin, Z. (2011). Identifying ferroresonance initiation for a range of initial conditions and parameters. *Nonlinear Dynamincs* 66: 755–762.

24 Corea-Araujo, J.A., Gonzalez-Molina, F., Martinez-Velasco, J.A. et al. (2013). An EMTP-based analysis of the switching shift angle effect during energization/de-energization in the final ferroresonance state. International Conference on Power Systems Transients (IPST) in Vancouver, Canada (July 2013).

25 Parker, T.S. and Chua, L.O. (1989). *Practical Numerical Algorithms for Chaotic Systems*. New York (NY, USA): Springer.

26 Guerra, G., Corea-Araujo, J.A., Martinez-Velasco, J.A. and González-Molina, F. (2015). Generation of bifurcation diagrams for ferroresonance characterization using parallel computing. European EMTP Users Group Meeting (EEUG) in Grenoble, France (September 2015).

27 Mairs, D.D., Stuehm, D.L., and Mork, B.A. (1989). Overvoltages on five-legged core transformers on rural electric systems. *IEEE Transactions on Industry Applications* 25 (2): 366–370.

28 Stuehm, D.L., Mork, B.A., and Mairs, D.D. (1989). Five-legged core transformer equivalent circuit. *IEEE Transactions on Power Delivery* 4 (3): 1786–1793.

29 Stuehm, D.L., Mork, B.A. and Mairs, D.D. (1988). Ferroresonance with three-phase five-legged core transformers. Minnesota Power Systems Conference in Minneapolis, USA (October 1988).

30 Mork, B.A. (1992). Ferroresonance and Chaos – Observation and Simulation of Ferroresonance in a Five Legged Core Distribution Transformer. Ph.D. Thesis. North Dakota State University.

31 Hileman, A.R. (1999). *Insulation Coordination for Power Systems*. New York (NY, USA): Marcel Dekker.

32 IEEE Std. 1243-1997. (1997). IEEE Guide for improving the lightning performance of transmission lines.

33 Martinez, J.A. and Castro-Aranda, F. (2005). Lightning performance analysis of overhead transmission lines using the EMTP. *IEEE Transactions on Power Delivery* 20 (3): 2200–2210.

34 Martínez-Velasco, J.A. and Castro-Aranda, F. (2010). Modelling of overhead transmission lines for lightning overvoltage calculations. *Ingeniare* 18 (1): 120–131.

35 Martínez-Velasco, J.A., Corea-Araujo, J., and Bedoui, S. (2018). Lightning performance analysis of transmission lines using the Monte Carlo method and parallel computing. *Ingeniare* 26 (3): 398–409.

36 Martínez-Velasco, J.A. and Castro-Aranda, F. (2008). EMTP implementation of a Monte Carlo method for lightning performance analysis of transmission lines. *Ingeniare* 16 (2): 169–180.

37 Chowdhuri, P., Anderson, J.G., Chisholm, W.A. et al. (2005). Parameters of lightning strokes: a review. *IEEE Transactions on Power Delivery* 20 (1): 346–358.

38 Anderson, R.B. and Eriksson, A.J. (1980). Lightning parameters for engineering applications. *Electra* 69: 65–102.

39 Heidler, F., Cvetic, J.M., and Stanic, B.V. (1999). Calculation of lightning current parameters. *IEEE Transactions on Power Delivery* 14 (2): 399–404.

40 Martinez, J.A. and Castro-Aranda, F. (2006). Lightning characterization for flashover rate calculation of overhead transmission lines. IEEE PES General Meeting in Montreal, Canada (June 2006).

41 Yokoyama, S., Miyake, K., Suzuki, T., and Kanao, S. (1990). Winter lightning on Japan sea coast – development of measuring system on progressing feature of lightning discharge. *IEEE Transactions on Power Delivery* 5 (3): 1418–1425.

42 Asakawa, A., Miyake, K., Yokoyama, S. et al. (1997). Two types of lightning discharge to a high stack on the coast of the sea of Japan in winter. *IEEE Transactions on Power Delivery* 12 (3): 1222–1231.

43 CIGRE WG 33–02. (1990). Guidelines for Representation of Network Elements when Calculating Transients. CIGRE Brochure 39.

44 IEEE (1996). Modelling guidelines for fast transients. *IEEE Transactions on Power Delivery* 11 (1): 493–506.

45 Gole, A.M., Martinez-Velasco, J.A. and Keri, A.J.F. (eds.) (1998). Modelling and Analysis of System Transients Using Digital Programs. IEEE PES Special Publication, Technical Report PES-TR7 (Formerly TP-133-0).

46 Martinez, J.A. and Castro-Aranda, F. (2005). Tower modelling for lightning analysis of overhead transmission lines. IEEE PES General Meeting in San Francisco, USA (June 2005).

47 Martinez-Velasco, J.A., Ramirez, A.I., and Dávila, M. (2009). Overhead lines. In: *Power System Transients. Parameter Determination* (ed. J.A. Martinez-Velasco), 17–135. Boca Raton (FL, USA): CRC Press.

48 Eriksson, A.J. and Weck, K.H. (1988). Simplified procedures for determining representative substation impinging lightning overvoltages. CIGRE Session in Paris, France.

49 IEC 60071-2. (2018). Insulation co-ordination, Part 2: Application guide, 4e.

50 IEEE Std 1313.2-1999. (1999). IEEE Guide for the Application of Insulation Coordination.

51 Pigini, A., Rizzi, G., Garbagnati, E. et al. (1989). Performance of large air gaps under lightning overvoltages: experimental study and analysis of accuracy of predetermination methods. *IEEE Transactions on Power Delivery* 4 (2): 1379–1392.

52 CIGRE WG 33-01. (1991). Guide to procedures for estimating the lightning performance of transmission lines. CIGRE Brochure 63.

53 Chisholm, W.A. and Janischewskyj, W. (1989). Lightning surge response of ground electrodes. *IEEE Transactions on Power Delivery* 14 (2): 1329–1337.

54 Mousa, A.M. (1994). The soil ionization gradient associated with discharge of high currents into concentrated electrodes. *IEEE Transactions on Power Delivery* 9 (3): 1669–1677.

55 Ishii, M., Kawamura, T., Kouno, T. et al. (1991). Multistory transmission tower model for lightning surge analysis. *IEEE Transactions on Power Delivery* 6 (3): 1327–1335.

56 Martinez, J.A. and Castro-Aranda, F. (2006). Influence of the stroke angle on the flashover rate of an overhead transmission line. IEEE PES General Meeting in Montreal, Canada (June 2006).

57 Häfner, J. and Jacobson, B. (2012). Device and method to break the current of a power transmission or distribution line and current limiting arrangement. Patent Application EP 2502248 A1.

58 Häfner, J. and Jacobson, B. (2011). Proactive hybrid HVDC breakers – A key innovation for reliable HVDC grids. CIGRE Bologna, Paper 0264.

59 Callavik, M., Blomberg, A., Häfner, J. and Jacobson, B. (2012). The hybrid HVDC breaker – An innovation breakthrough enabling reliable HVDC grids. ABB Grid Systems Technical Paper.

60 Martinez, J.A. and Magnusson, J. (2015). EMTP modelling of hybrid HVDC breakers. IEEE PES General Meeting in Denver, USA (July 2015).

61 Hassanpoor, A., Häfner, J. and Jacobson, B. (2014). Technical assessment of load commutation switch in hybrid HVDC breaker. International Power Electronics Conference (IPEC) in Hiroshima, Japan (May 2014).

62 Lin, W., Jovcic, D., Nguefeu, S., and Saad, H. (2016). Modelling of high-power hybrid DC circuit breaker for grid-level studies. *IET Power Electronics* 9 (2): 237–246.

63 Bucher, M.K. and Franck, C.M. (2016). Fault current interruption in multiterminal HVDC networks. *IEEE Transactions on Power Delivery* 31 (1): 87–95.

64 Martinez-Velasco, J.A. and Magnusson, J. (2017). Parametric analysis of the hybrid HVDC circuit breaker. *International Journal of Electrical Power and Energy Systems* 84: 284–295.

65 Corea-Araujo, J.A., Martinez-Velasco, J.A., and Magnusson, J. (2017). Optimum design of hybrid HVDC circuit breakers using a parallel genetic algorithm and a MATLAB-EMTP environment. *IET Generation, Transmission and Distribution* 11 (12): 2974–2982.

66 Corea-Araujo, J.A. and Martinez-Velasco, J.A. (2016). Application of ATP to optimization studies using genetic algorithms and parallel computing. European EMTP Users Group Meeting (EEUG) in Birmingham, UK (September 2016).

67 Saad, H., Denneriere, S., Mahseredjian, J. et al. (2014). Modular multilevel converter models for electromagnetic transients. *IEEE Transactions on Power Delivery* 9 (3): 1481–1489.

68 Mitchell, M. (1999). *An Introduction to Genetic Algorithms.* London (UK): MIT Press.

69 Michalewicz, Z. (1992). *Genetic Algorithms + Data Structures = Evolution Programs.* New York (NY, USA): Springer.

70 Altman, Y.M. (2014). *Accelerating MATLAB Performance: 1001 Tips to Speed Up MATLAB Programs.* Boca Raton (FL, USA): CRC Press.

71 Tylavsky, D.J., Bose, A., Alvarado, F. et al. (1992). Parallel processing in power systems computation. *IEEE Transactions on Power Systems* 7 (2): 629–637.

72 Yeh, E.C., Venkata, S.S. and Sumic, Z. (1995). Improved distribution system planning using computational evolution. IEEE Power Industry Computer Application Conference (PICA) in Salt Lake City, USA (May 1995).

73 Konfrst, Z. (2004). Parallel genetic algorithms: advances, computing trends, applications and perspectives. Presented at the 18th Parallel and Distributed Processing Symposium in Santa Fe, USA (April 2004).

74 de Toro Negro, F., Ortega, J., Ros, E. et al. (2004). PSFGA: parallel processing and evolutionary computation for multiobjective optimisation. *Parallel Computing* 30 (5): 721–739.

75 Farina, M., Deb, K., and Amato, P. (2004). Dynamic multiobjective optimization problems: test cases, approximations, and applications. *IEEE Transactions on Evolutionary Computation* 8 (5): 425–442.

76 Cámara, M., Ortega, J., and de Toro, F. (2009). A single front genetic algorithm for parallel multi-objective optimization in dynamic environments. *Neurocomputing* 72 ((16): 3570–3579.

A Characteristics of the Multicore Installation

The main characteristics of the multicore installation used in this chapter are listed below:

- Number of servers: 4
- Model: Fujitsu PRIMERGY CX 250 S1
- Processor: 2 Intel Xeon E5-2660 (8 Cores, clock frequency = 2.2–3 GHz)
- Hard disc memory: 500 GB
- RAM memory: 128 GB
- Communication: 2x Port Gigabit Ethernet LAN.

B Test System Parameters for Ferroresonance Studies

Case 2 – Main Parameters

Power source (ideal). Phase-to-phase rms voltage = 400 kV; Frequency = 50 Hz.
Cables. Represented as constant-parameters PI models, with lengths between 3 and 12 m.
Circuit breaker. Represented by means of a grading capacitance, C_g, with variable value.
Voltage transformers (VTs):

- Single-phase three-winding transformers.
- Ratings: 400/0.1/0.033 kV, 0.1/0.1/0.05 kVA.
- Short-circuit test values: $Z_{sh,1w\text{-}2w} = 13.46\%$, $Z_{sh,1w\text{-}3w} = 20\%$, $Z_{sh,2w\text{-}3w} = 20\%$.
- No load test losses: $R_m = 182\,\text{M}\Omega$.
- Saturation curve: See Figure 11.6.

Stray capacitance (C_s). Variable value.

Case 3 – Main Parameters

Power source. Phase-to-phase rms voltage = 480 V; Frequency = 60 Hz.

Three-phase transformer:

- YNy0 Three-phase two-winding transformer, grounded through a 2.4 Ω impedance.
- Rated voltages = 12 470/7200Yn − 480/277Yn V.
- Rated power = 75 kVA.
- High voltage side data: $R_h = 4.7\,\Omega$, $L_{hx} = 0.118\,\Omega$.
- Low voltage side data: $R_x = 0.0224\,\Omega$, $L_{xc} = 0.08\,\Omega$.
- Saturation curves: See Figure 11.8.
- Core resistances: $R_1 = 411\,\Omega$, $R_2 = 337\,\Omega$, $R_3 = 372\,\Omega$, $R_4 = 395.6\,\Omega$.

To Probe Further

ATP files corresponding to all cases studied in this chapter are available in the companion web-site of this book. Readers are encouraged to run the files and check the results presented here. The files are edited for use in a single-CPU computer and do not include optimization and parallel computing features.

Index

a

active filter 423, 425–428, 441–449, 462
adjustable-speed drive 129, 210, 276, 333, 334, 338, 447, 456
admittance matrix 14, 27, 37, 44, 90, 98, 101, 112, 118, 121
aiming time 234, 235, 238
air region 15, 56
Alternative Transients Program (ATP) 1–8, 15, 23, 26–28, 30, 45, 46, 61, 64–66, 76, 85, 87, 89, 92, 97–101, 107–136, 139–201, 203, 204, 216–218, 224, 230, 241, 243, 246, 247, 253, 267, 275–279, 281, 282, 284–287, 298, 302, 309, 323, 328, 334–345, 349, 355, 365, 367, 368, 376, 377, 382, 405–407, 411, 419–426, 428, 431, 434, 440, 441, 447, 453, 471–474, 476, 477, 482, 484, 490–493, 508, 513, 514, 531, 539–575
analog-to-digital converter (ADC) 486
angular position 55, 282, 326
angular speed 55, 113, 282, 283, 302, 314, 318, 319, 321, 322, 327, 328, 371, 373, 374
arcing period 60
arcing time 60
ARMA *see* auto-regressive moving-average
armature relay 309
armature winding 57, 279, 281
armour 24, 25, 27
arrester energy 247, 259, 261
arrester lead 197, 210, 261
arrester selection 204, 256
asymmetrical fault current 477
asymmetrical short-circuit current 32, 58–60, 567
asynchronous machine 75, 102, 276
ATP *see* Alternative Transients Program

b

backflash 205, 555
backflashover 214, 253, 259
backflashover rate (BFOR) 244, 255, 551, 562
back-to-back capacitor bank 157, 158, 160, 162, 163
BCTRAN (routine) 37, 111, 112, 116, 123, 224, 225, 474, 545
Bergeron model 110, 224, 233
bifurcation diagram 539–550
bitmap file format 120, 122, 127
black-box model 30, 38, 42, 59–61, 485, 488
bonding 24–26, 207
brute force approach 97, 101, 119

ATP Control Center 3, 107
ATPDesigner 3, 107
ATPDraw 3–7, 107, 112, 120, 122–125, 139, 140, 145, 147, 148, 153, 157, 158, 164, 166, 169, 171, 172, 174, 176, 177, 180, 183, 185, 189, 192, 194, 199, 200, 216–220, 224, 230–235, 237–239, 241, 247, 248, 258, 259, 261, 263, 264, 266, 269, 275, 277, 278, 284, 288, 291, 292, 296, 298, 300, 306, 308, 310, 311, 314, 315, 318, 320, 325, 326, 350–254, 356, 358, 360, 362, 363, 365, 367–369, 372, 377, 382, 385, 405, 406, 408, 411, 414, 417, 419, 421, 423, 426, 431, 436, 437, 439, 441, 442, 444–446, 449, 450, 462, 480, 482, 493, 494, 496, 499, 500, 503, 505, 514, 515, 518, 520, 522, 524, 525, 528, 545–547
automatic initialization 97, 100, 113, 275, 281, 286, 287, 310, 311, 313, 315, 328, 371
auto-regressive moving-average (ARMA) 110

Transient Analysis of Power Systems: A Practical Approach, First Edition. Edited by Juan A. Martinez-Velasco.
© 2020 John Wiley & Sons Ltd. Published 2020 by John Wiley & Sons Ltd.
Companion Website: www.wiley.com/go/martinez/power_systems

built-in model 7, 107, 110, 112, 216, 217, 253, 275, 287, 339, 344, 349, 365, 406, 408, 439
bus duct 214, 263
bushing 209, 213–215, 475
buswork 209, 212

c

cable 3, 23–28, 109, 110, 116, 123, 133, 143, 144, 168, 184, 187, 190, 203–205, 207, 208, 216, 217, 221–226, 229, 230, 298, 315, 317, 318, 338, 439, 472–473, 489, 540, 541, 545, 546, 549, 550, 566–570, 572, 579
CABLE CONSTANTS (routine) 23, 110, 111, 116, 224, 241
cable energization 205
CABLE PARAMETERS (routine) 23, 110, 116
capacitance 13–15, 19, 27–29, 31, 35, 37, 41, 42, 44, 45, 56, 58, 77, 79–81, 84, 91–93, 112, 141, 142, 145, 148, 150, 166, 168, 169, 171, 172, 174–176, 178, 180, 184, 185, 187, 190, 193, 194, 204, 209, 210, 212, 213, 215, 219, 221, 222, 224, 225, 278, 282, 303, 334, 336, 343, 365, 368, 396, 407, 439, 441, 442, 473–476, 479, 482, 541, 544–546, 549, 550, 553, 579, 580
capacitance matrix 15, 28
capacitive coupling 31, 32, 41, 56, 475
capacitive divider 478
capacitive voltage transformer 8, 478–479, 482, 483
capacitor bank 11, 130–131, 145, 155, 157–163, 165, 205, 207, 209, 227, 240, 242–244, 306, 314–316, 323, 326, 393, 431, 432, 434, 436, 437
capacitor bank energizing 205, 227, 242
capacitor switching 30, 125, 158, 238, 425
case-peak method 234
Cassie model 61, 62
cathode spot region 62
Cauer equivalent circuit 36
CCVT *see* coupling capacitor voltage transformer
CFO *see* critical flashover voltage
chaotic mode 540, 542, 545

chopping 61, 63, 78, 168, 172, 174, 175, 205, 208, 209, 476
CIGRE 42, 119, 206, 214, 246, 248, 434, 439
CIGRE load model 434, 439
circuit breaker 8, 58–66, 92, 97, 132, 155, 162, 168, 170–172, 174, 192, 193, 195, 208, 209, 213, 215, 218, 222, 225, 227, 229, 234, 236, 347, 471, 475, 476, 498, 503, 506, 507, 509, 510, 514–518, 526, 527, 529–532, 539, 545, 563–565, 567–568, 572, 574
circuit breaker model 59–66, 475–476, 510, 516, 563, 567, 568
closed-loop interaction 471, 486, 487
closed-loop method 487
closing operation 59, 64–66, 230, 232–235, 239, 263, 267, 518
closing time 59, 64–66, 171, 184, 209, 234, 236, 238
coercive current 34, 35
compact grounding system 18
companion circuit 78–86, 88, 118, 146, 188, 198
companion equivalent 79, 198
compensating reactor 478
compensation-based method 87, 113, 283, 284
compensation method 87, 88, 90, 117, 118
Computer Business Equipment Manufacturers Association (CBEMA) curve 126, 456
COMTRADE 125
concave waveform 246, 551, 552, 557
concentric design 29
conditioned energy resource 348
conductance 13, 28, 39, 40, 60–62, 64–66, 85, 89, 117, 118, 121, 177, 282, 337, 342, 405
constant-parameter model 102, 259
continuous operating voltage 204, 256
continuous power-frequency voltage 203
control
  control block 96, 100, 113, 119, 383, 384
  control system 3, 6, 76, 96–97, 100–102, 108–113, 117, 119, 121, 140, 177, 275, 277, 282, 285, 286, 323, 338, 339, 341, 343–345, 405, 409, 475, 488, 566
  control system dynamics 119–120, 129, 405
controlled switch 110, 111, 209, 340, 350, 367, 385, 396, 476, 511–513, 518, 566
convolution 14, 110

core design   29, 32, 292, 299, 303, 307
core model   32–34
corona
  corona effect   12–15, 207, 212, 259, 553, 557–560, 562
  corona-inception voltage   12, 15, 22, 212, 553
  corona model   212, 553
counterpoise   18–20, 211
coupling capacitor voltage transformer (CCVT)   476, 478–480, 484, 487, 496, 497, 501, 502
critical damping adjustment (CDA)   101, 338
critical flashover voltage (CFO)   21, 23, 558
crossarm   16, 17
cross-bonding   25, 26, 207
cross saturation   49
current chopping   78, 168, 172–175, 208, 476
current commutation delay   568, 570, 574
current injection method   191, 192, 194
current interruption   59–60, 92, 103, 170, 175, 180, 193, 195, 225, 511, 512
current limiting fuse (CLF)   511, 512, 520, 521
current limiting reactor (CLR)   142, 155, 563
current-mode control   348
current threshold   568
current transformer (CT)
  CT burden   477
  CT burden resistance   477
  CT core loss resistance   477
  CT flux-current curve   178
  CT hysteresis loop   476
  CT hysteresis representation   477
  CT magnetizing branch   476
  CT magnetizing inductance   477
  CT model   215, 477, 480
  CT nonlinear magnetization characteristic   476
  CT remanence   477
  CT remanent flux   477, 478
  CT saturation   477, 480, 484
  CT saturation flux density   477
current zero   60–63, 174, 192, 193, 208, 335, 340, 349, 475, 476, 511
custom-made
  custom-made model   5–8, 109, 117, 122, 136, 216, 217, 224, 241, 248, 293, 318, 344, 406, 414, 421, 424, 470, 472, 514, 515, 537, 539

custom-made module   107, 216, 218, 288, 310, 314, 405–419, 426–428, 432, 441, 450, 462
Custom Power   1, 7, 107, 333, 336, 345, 347, 419, 425, 426, 462

d

damper winding   52
damping
  damping coefficient   55, 282, 325
  damping factor   93, 369, 373
  damping resistor   94, 205, 479
DATA BASE MODULE   5, 6, 117, 122, 216, 217, 406–411, 423, 441, 445
data conversion procedure   2, 5, 284–286
data display   126–127
data formatting   127, 128
data management   125–126
data modularization   406, 408
data module   5, 128, 407
data processing   127
data sorting by class   406, 409
data symbol replacement   5, 128
d-axis equivalent circuit   48, 53
d-axis subtransient reactance   289, 292, 299, 307, 317
d-axis synchronous reactance   289, 292, 299, 303, 307, 317
d-axis transient reactance   289, 292, 299, 303, 307, 317
de-energization transient   150–155
deep-rotor bar effect   49
detection   426, 486, 508, 514
detector   485, 486, 498
dielectric failure   60, 155, 168, 171, 276, 539
dielectric recovery period   60
dielectric strength   15, 21, 23, 60, 64, 170, 174, 263, 512
dielectric stress   38
digital fault recorder (DFR)   485, 488
diode   95, 111, 334–337, 339, 340, 342, 349–351, 354, 356, 357, 359–362, 365, 366, 369, 441, 445, 485, 565
dip *see* voltage dip
discharge   17, 20–22, 27, 60, 133, 154, 164–167, 169, 180, 195, 197, 205, 213, 215, 216, 256–259, 261

disconnector operation   206, 215, 261–263, 266

disconnector switching   206

discretization   49, 337, 553

discretized circuit   118, 146, 152

distance measuring unit   498, 503

distance relay   485, 486, 489, 491–493, 495, 497, 498, 503

distortion   4, 8, 15, 128, 189, 212, 215, 333, 343, 344, 347, 419, 422–424, 432, 437, 447, 449, 482, 487, 496, 539, 553

distributed energy resource (DER)   334, 337–349, 472

distributed-parameter component   7, 140, 187–201, 285

distributed-parameter model   12, 15, 19, 20, 28, 39, 41, 54, 57, 110, 140, 144, 187, 188, 196, 208, 211, 213, 224, 227, 229, 241, 473, 557, 567

distribution feeder protection   508

distribution system   129, 130, 217, 219–225, 333, 346–347, 373, 388, 390, 393, 396, 425, 426, 447–465, 472, 508–531, 539

Dommel scheme   78, 100, 117, 118

doubly-fed induction generator (DFIG)   309

doubly-fed induction machine (DFIM)   309, 310, 312, 323

dual active bridge (DAB)   388, 392

duality   32, 33, 36, 37, 112, 137, 222, 541

duality model   32

dynamic arc representation   59, 61–62, 170, 510

dynamic average model   334–335, 339, 367

dynamic model   15

dynamic state-space model   491

dynamic voltage restorer (DVR)   347, 427, 462, 464, 465

e

eddy currents   29–32, 34–36, 38, 56, 57, 208, 276, 279, 286, 474, 475

electrogeometric model   246, 247, 253, 554, 555, 558, 559, 562

electromagnetic torque   50, 54, 55, 281–283, 310, 311, 319, 324, 325, 328, 371, 374, 492, 494, 495, 501, 502

electromagnetic transient   3, 6, 11, 12, 23, 24, 34, 38, 75–103, 107, 112, 139–201, 206, 472, 541

Electromagnetic Transients Program (EMTP)   6, 76, 96, 99, 101, 107, 118, 125, 276, 281, 425, 426, 486, 487, 541

electromechanical distance relay   491–497

electromechanical MHO distance relay   491

electromechanical relay   478, 484, 485, 490, 491, 493

electromechanical transient   3, 75, 100, 107, 302

electronically-interfaced DER system   347, 348

electronic relay   485

EMTP *see* Electromagnetic Transients Program

energization
  energization of a back-to-back capacitor bank   157–158, 160–163
  energization of a capacitor bank   155–157, 159–161
  energization of a nonlinear reactance   181–184
  energization of lines and cables   187–191, 227
  energization transient   145–150

energy storage   333, 347, 425, 462

equivalent circuit   14, 15, 19, 28, 32, 33, 35–37, 41, 42, 44, 45, 48, 52, 53, 56–58, 75, 87, 89, 91, 103, 113, 130, 133, 134, 192, 193, 210, 214, 221, 280, 283–287, 361, 476, 477, 567

exact-PI circuit model   473

excitation controller   296

excitation loss   37, 112, 474

expulsion fuse   511, 512, 521

extended grounding system   17, 18

f

factorization   86, 118, 337

fast front
  fast-front lightning overvoltage   205
  fast-front overvoltage   12, 203, 205, 206, 210–215, 243–270
  fast-front surge   30, 275, 276
  fast-front switching overvoltage   205
  fast-front transient   11–13, 30, 31, 38, 39, 41–43, 45, 112, 196, 203, 213, 215, 243, 257, 258, 261–263, 267–270, 276

fast mechanical disconnector (FMD)   563, 565–567

fault
  fault clearing   191–195, 204, 205, 301, 308, 449, 454, 462, 491
  fault current   27, 192, 193, 450, 451, 457, 458, 477, 482, 509–514, 521–523, 563, 568, 572
  fault overvoltage   204, 205
feasible solution   570, 572, 574, 575
ferroresonance
  ferroresonance characterization   539–550
  ferroresonance mode   540
  ferroresonance suppression circuit (FSC)   478, 479, 482–484
filter model   207, 441, 445, 464, 498
five-legged core   545–550
five-legged stacked-core   29
five-legged wound-core   29
fixed-point miscellaneous data   140, 141
FIX SOURCE option   298
flexible AC transmission systems (FACTS)   1, 7, 97, 107, 129, 333, 346, 382
floating subnetwork   142
flux linkage   46, 47, 49, 51, 52, 56, 123, 178, 181, 183
footing impedance   12, 554, 558
foreign model   136, 406
FORTRAN statement   113, 114
Foster equivalent circuit   35
Fourier analysis   119, 411, 427, 428, 434, 447
frequency-dependent   12, 13, 15, 26–29, 33–36, 42, 44, 56, 57, 69, 98, 102, 110–112, 116, 119, 129, 144, 208, 210, 211, 213, 216, 217, 229, 233, 234, 246, 247, 259, 343, 422, 424, 434, 439, 472–474, 496, 552, 554, 557, 567
frequency-dependent distributed-parameter line model   208
frequency-domain simulation   4, 109, 110, 128, 337, 422, 423
frequency-domain solution technique   118–119
frequency range   2, 11, 12, 14–17, 20, 27, 28, 31, 36, 38, 42, 44, 45, 77, 101, 112, 139, 143, 203, 206–208, 210, 216, 275, 284, 339, 422, 473, 474
frequency relay   475, 486, 490

FREQUENCY SCAN (FS)   119, 123, 230, 423
Fuse   471, 508, 511–514, 520, 521, 524–527, 529–531

g
gapless metal oxide surge arrester   133, 209
gas-filled circuit breaker   61–62, 170
gas insulated substation (GIS)   203, 206, 214–215, 261–270
gate turn-off thyristors (GTOs)   334, 335, 341, 350, 353–355, 358, 384
Gaussian distribution   21, 209
gearbox   323–325, 328
generation of harmonic waveform   429, 430
generation of random numbers   253
generator
  generator-infinite bus system   298
  generator protection   508
  generator synchronization   45
  generic model   485
genetic algorithm (GA)   563, 568–571, 574, 575
GIFU (option)   355–362, 365
governor   125, 296, 304, 475
graphical output   107, 127
graphical user interface (GUI)   2, 3, 6, 76, 107, 108, 120–125, 140, 406
grey-box model   30
ground
  ground electrode   17–20
  ground fault factor   205
  ground rod   18, 20
grounded primary transformer winding   513–514
grounding
  grounding impedance   13, 17–19, 207, 210, 211, 246, 318, 449, 450, 459
  grounding model   17, 20, 245, 251, 256
  grounding system   17, 18, 215, 399, 425
group option   417–419, 441, 445
GTO *see* gate turn-off thyristors
GTPPLOT   3, 107
GUI *see* graphical user interface
guideline   2, 6–8, 12, 23, 28, 31, 77, 203, 204, 207–210, 214, 216, 246, 259, 276, 284, 287, 288, 334, 338, 367, 369, 384, 393, 396, 424, 425, 441, 471–476, 513, 539–541, 551, 552, 555, 563, 570

h

harmonic   3, 4, 8, 30, 31, 97–98, 102, 103, 109, 118, 119, 128, 129, 143, 144, 168, 185, 207, 286, 333, 337, 338, 343, 344, 347, 382, 384, 405, 419, 422–447, 449, 462, 479, 539

harmonic current compensation   428, 441–447

HARMONIC FREQUENCY SCAN (HFS)   405, 423, 425, 434–441

harmonic interaction   30

harmonic-neutralized voltage-source converter (HN-VSC)   384

harmonic resonance   333, 431–434

harmonics analysis   424, 426–447

Heidler model   246, 247, 258, 551

hemisphere electrode   20, 21

HFS *see* HARMONIC FREQUENCY SCAN

high-frequency

high-frequency model   17–20, 37–39, 55–58

high-frequency transformer model   42, 475

high-frequency transient   27, 29, 31, 37–46, 52, 217, 257, 285, 341

high-voltage direct current (HVDC)   96–98, 129, 333, 345, 346, 539, 563–575

history term   79, 89, 118, 121, 188, 198, 199, 283

HVDC *see* high-voltage direct current

HVDC system   566, 567

hybrid HVDC circuit breaker   563–575

hybrid model   32, 37, 123, 125, 222, 227, 293

HYSDAT (routine)   112, 116, 178

hysteresis   29, 34, 35, 55, 89, 90, 112, 116, 117, 175, 178, 225, 344, 441, 444, 447, 474, 476, 477, 541

HYSTERESIS HEVIA (routine)   112, 117, 178

i

ideal line   80, 81, 188, 190, 192

ideal model   335, 515

ideal switch   59–61, 65, 66, 77, 78, 90, 142, 166, 172, 208, 209, 263, 335, 336, 362, 377, 475, 510, 565

ideal transformer   77, 111, 112, 324, 396, 474

ideal voltage source   207, 350, 351, 362, 369, 434

IEC   7, 21, 206, 256, 449, 456

IEC standard   21

IEEE

IEEE FLASH   16

IEEE standard   21

IGBT   334–336, 341, 361, 362, 364, 374, 441–444, 565–567, 569, 572, 574

impedance relay   477, 491, 495, 496, 503, 512, 515

implementation   14, 61, 100, 114, 118, 147–148, 152, 153, 169, 171, 174, 177, 180, 185, 189, 194, 199, 200, 204, 217, 219, 220, 224, 230–235, 237, 239, 241, 246–248, 253, 256–259, 261, 264, 266, 269, 276, 288, 291, 292, 296, 300, 304, 308, 310, 315, 318, 320, 323, 326, 342, 346, 350–354, 356, 358, 360–363, 365, 367–369, 372, 373, 376–378, 383–386, 389, 405, 412, 414–418, 421, 430, 431, 436, 439, 441, 442, 444, 446, 453, 471, 480, 482, 486, 491, 493, 498–500, 505, 515, 518, 520, 522, 525, 528, 540, 542, 547, 551, 554, 563, 565, 568–570

inadvertent energization   45

INCLUDE   4, 406, 407, 566, 570

incremental inductance   49, 53, 54, 116, 178

induced voltage   14, 26, 205

inductance   13, 15, 19, 27, 28, 30, 32–36, 38, 41, 45, 47, 49, 50, 52–54, 56, 77, 79, 80, 83, 86, 89, 90, 92–94, 99, 102, 111, 112, 117, 141, 142, 144, 145, 154, 162, 168, 169, 171, 172, 175–180, 184, 185, 190, 192, 193, 209, 213, 222, 276, 282, 310, 315, 317, 321, 334–336, 350, 365, 369, 373, 376, 377, 379, 390, 392, 396, 473–477, 479, 493, 541, 544–547, 566, 568

induction machine   7, 53, 113, 125, 276, 277, 281, 284, 285, 309–328

induction machine equations   46–51

Information Technology Industry Council (ITI)   456, 457, 459

initialization   97–102, 113, 115, 118, 119, 121, 129, 275, 277, 279, 281, 286, 287, 311, 313, 315, 328, 338, 339, 342, 368, 369, 371, 376–378, 408, 423–425

Initialization with Harmonics (IwH)   102

input data file   6, 113, 117, 125, 139–141, 144, 147, 152, 382

inrush current   34, 157, 168, 181–183, 204, 209

instrument transformer 8, 184, 213, 219, 471, 476–484, 494–496, 503, 515, 518

insulated cable 23–28, 110, 116, 133, 222, 229, 317, 473, 566–568

insulation 3, 4, 12, 21, 22, 24, 26–28, 30, 38, 56, 65, 75, 135, 175, 184, 195, 203, 205, 212, 214, 229, 262, 275, 336, 539

insulation permittivity 27

insulator 12, 13, 16, 22, 23, 59, 198, 205, 207, 210, 211, 213, 214, 247, 250–253, 259

insulator string 16, 21, 212, 217, 244–246, 251, 253, 255, 256, 259, 553–555, 557, 558, 562

integer miscellaneous data 140, 141

integration method 23, 212, 336, 343

interactivity 5

interconnection protection 508–509, 513, 514, 529

intercontact spark voltage 267

intercontact voltage 267

inter-electrode region 62

interface 2, 3, 56, 76, 96, 97, 102, 109, 113, 116, 120, 121, 125, 140, 275, 276, 278, 280–282, 284, 286, 287, 323, 334, 337, 342–344, 347–349, 384, 406, 423, 449, 462, 486, 488, 503

interharmonic 419

interleaved design 29

internal model 38, 55

interpolation 90, 101, 338, 377

interpolation error 143

interruption

    interruption of an inductive current 169, 171, 173, 174, 177, 180–181

    interruption of a small inductive current 170, 180

inter-turn insulation failure 275

inter-turn voltage 276, 278

inter-winding capacitance 474, 476

inverse Fourier transform 14

island 314, 508, 513, 514, 529, 531

isotropic rotor 49, 52

ITI curve 456, 457

j

JMarti model 110, 111, 116, 230, 232, 233

JMARTI SETUP (routine) 110, 116

k

Kirchhoff's current law 85

l

Laplace antitransform 166

Laplace transform 145, 151, 156, 157, 159, 165, 176, 193

lattice diagram 16

leader progression model (LPM) 23, 212, 246, 252, 553

leakage inductance 32, 33, 35, 44, 45, 392, 396, 477, 479

let-thru current 512

lightning

    lightning flash 12, 251, 551

    lightning flashover rate (LFOR) 59, 216, 244, 245, 251, 255, 256, 551, 554, 555, 559–562

    lightning overvoltage 4, 7, 16, 21, 128, 130, 187, 196–201, 205, 206, 211, 212, 217, 243, 244, 246, 248, 251, 551–554

    lightning performance analysis 539, 550–562

    lightning return stroke current 247

    lightning stroke 1, 16, 19, 21, 75, 77, 133, 135, 196, 197, 199, 203, 211, 212, 214, 246–248, 250–252, 259, 261, 551–560

    lightning stroke characterization 551–552

linear system 78, 98, 121, 187, 191, 337, 424, 425

LINE CONSTANTS (routine) 15–16, 110, 111, 116, 210, 213

line energization 65, 208, 225, 227–238, 240

line model 12, 14–16, 27, 39, 102, 110, 111, 123, 129, 193, 208, 214, 220, 229–233, 245–247, 249, 256, 278, 294, 427, 472, 473, 498, 551, 552, 554, 555, 558, 560, 562, 575

LINE MODEL FREQUENCY SCAN 123, 230

line shielding 247

line-to-ground system voltage 205

load commutation switch (LCS) 563, 565, 567

load-flow solution 98–100

load rejection 12, 45, 203–206, 217, 276, 287–298
log-normal distribution 51, 214, 552, 558
longitudinal
    longitudinal insulation 203, 205
    longitudinal overvoltage 204
    longitudinal temporary overvoltage 203
loss
    loss of parallel 514
    loss of synchronism 276, 514
lossless line 16, 17, 77, 80–82, 84, 211, 215, 263, 269
lossless tower model 17
lossy insulation 24
low-current model 63, 207
low-current resistance 20
low-frequency low-current model 207
low-frequency model 17–18, 31–32, 285
low-frequency overvoltage 12, 216–225
low-frequency transient 11, 31–37, 46–55, 112, 113, 275, 276, 287, 288, 323, 339, 474
low pass filter 257, 486, 497, 498
lumped mass 279
lumped-parameter component 82, 117
lumped-parameter element 77, 143
lumped-parameter model 19, 38, 40, 41, 57, 110, 155, 208, 293, 473

m
machine-based DER system 347
machine coil 57, 278
machine winding 55, 56, 286
magnetic circuit 32, 33, 382, 384
magnetic flux 27, 30, 35, 36, 57
manual initialization 286, 311, 313
MATLAB/Simulink 539
matrix model 474
matrix representation 31, 37, 116, 474
maximum system operating voltage 203
Mayr model 61, 62
mechanical disconnector 563, 566
mechanical parameter 280, 284–287, 306, 317, 318, 325, 328
mechanical system equation 50–51, 54–55
mechanical torque 50, 55, 113, 282, 314, 316, 317, 319, 328, 371, 374
melting mechanism 511, 512

melting period 511
metal oxide surge arrester (MOSA) 133, 195, 209, 213
method of characteristics 76
MHO distance relay 486, 491, 492, 495
microprocessor-based relay 8
misoperation 419, 476, 480, 482, 484, 488
mitigation 217, 427, 447, 449, 462–465
mitigation technique 405, 419, 422–426
modelling 2, 5–8, 11–66, 76–78, 100, 102, 109–117, 123, 125, 140, 142, 155, 177, 178, 181, 187, 203, 204, 206–217, 221, 243, 246, 259, 275, 276, 287, 288, 333, 334, 336, 338, 341, 342, 344, 347, 365, 367, 369, 384–385, 393–396, 419, 422–425, 441, 471–476, 479, 484–513, 539–541, 545, 551–554, 557, 563
modelling guideline 6–8, 12, 31, 203, 204, 207–216, 246, 259, 276, 287, 288, 334, 369, 393–396, 424, 441, 471–476, 539–541, 551, 552, 563
MODELS 6, 97, 107–108, 111–117, 120, 122–123, 125, 136, 216–218, 234, 263–264, 275, 279, 286, 339, 342, 344, 367, 375–378, 406, 411–417, 423, 426, 428–429, 432, 459, 491, 493–494, 503
modified-augmented-nodal analysis (MANA) 101, 103
modular validation 489, 490
moment of inertia 50, 55, 282, 284, 288, 289, 292, 299, 303, 307, 310, 315, 317, 321, 325, 373, 494
Monte Carlo method 251, 253, 255, 551, 555, 562
Monte Carlo procedure 251, 253, 254, 551, 554
Monte Carlo simulation 128
MOSA *see* metal oxide surge arrester
motor 34, 46, 50, 51, 58, 203, 206, 210, 213, 261, 276, 286, 309–315, 317–319, 321, 336, 369, 371, 373, 425, 447, 462
motor startup 317, 374
multi-conductor line 13, 16, 56, 278
multi-conductor transmission line (MTL) 13–16, 39
multicore environment 8, 539, 540, 542–544, 563, 575
multicore installation 540, 543, 550, 551, 562, 570, 574

multi-machine system   298

multi-mass shaft system   55

multi-objective algorithm   570

multi-objective genetic algorithm   570

multiphase harmonic load flow (MHLF)   102

multiple run   5, 128, 132, 216–218, 234, 253, 459

multi-storey model   17, 557

mutual-damping coefficient   55, 325

mutual inductance matrix   47

n

natural frequency   143, 146, 151, 152, 154, 156, 159, 160, 162, 164, 169, 172, 175, 209, 306

natural resonant frequency   302

negative polarity   251, 252, 261, 263, 551, 552, 557, 558, 562

network equivalent   207, 210, 230, 318, 496

network reduction   91

Newton-Raphson method   88, 89, 102

nodal admittance equation   78, 85–87, 98, 405

nodal analysis   337

nodal conductance matrix   85, 118, 121

NODA SETUP (routine)   110, 116

noise   419, 422

nominal-PI circuit model   473

nonlinear

nonlinear circuit   176–178, 195–201, 336, 540, 541

nonlinear component   3, 5, 7, 97, 117–119, 176–178, 405, 425, 474

nonlinear load   424, 431, 432, 434, 435, 437, 442, 445, 448, 449

nonlinear magnetizing inductance   49

nonlinear reactance   178–181, 184

non-uniform line   14

normal distribution   21, 214, 236, 238, 251, 252, 552, 558

Norton companion equivalent   79, 80, 82

Norton equivalent   103, 117

notch   419, 422, 498

numerical

numerical distance relay   491, 497–508

numerical error   93

numerical integration   78, 336

numerical oscillation   76, 78, 91, 92, 94, 95, 97, 101, 118, 142, 171, 172, 337, 340, 341, 345, 349, 350, 365, 367, 401, 410

numerical relay   484–487, 490

numerical solution technique   38

o

off-line simulation   76

on-line simulation   76, 128

open-circuit d-axis subtransient time constant   289, 292, 299, 307, 317

open-circuit d-axis transient time constant   289, 299, 303, 307

open-circuit q-axis subtransient time constant   289, 299, 307, 317

open-circuit q-axis transient time constant   289, 299, 303, 307

opening   64, 78, 94, 136, 141, 148, 150, 153, 162, 164, 166, 169, 171, 172, 194, 222, 229, 233, 262, 264, 265, 266, 337, 359, 471, 476, 512, 515, 567, 572

opening operation   59–64, 150, 155, 168, 191, 208, 209, 242, 263, 509, 510, 518, 531

open-loop control   368, 373, 376

open-loop interaction   487

open-loop method   487

open system   5

optimization   120, 124, 563, 570, 574, 580

optimum allocation   130–132

optimum design   563–575

optimum selection   539, 563

oscillatory   64, 66, 150, 154, 156, 160, 166, 203, 205, 419, 422, 424, 496

out-of-step synchronization   279

overcurrent   8, 75, 396, 474, 524

overcurrent relay   484, 485, 509–510, 514–518, 529

overdamped   146, 150, 151, 154

overhang region   56, 57, 578

overhead line   12–24, 28, 86, 110, 116, 133, 144, 199, 205, 210, 212, 229, 230, 256, 290, 291, 318, 321, 427, 449, 450, 459, 473, 489, 550, 551, 555

overvoltage   7, 11–12, 203–270

overvoltage condition   545

overvoltage magnitude   7, 204, 234

p

pancake design   29

parallel computing   539–575

parallel damping   92, 164, 185, 350, 365
parallel genetic algorithm   568–570
parallel resonance   132–133, 221, 479
parallel RLC circuit   140, 145, 150–155, 168, 175
PARAMETER   5, 117, 132, 218, 409, 410, 439, 566
parametric study   4, 5, 11, 128, 132, 133, 216, 218, 221, 244, 246, 251, 252, 261, 312, 317, 542, 543, 549
Park's transformation   52, 285
passive filter   425–427, 432, 437–439, 441
PCC *see* point of common coupling
PCPlot   3, 107
PCVP *see* POCKECT CALCULATOR VARIES PARAMETERS
permittivity   23, 26–28
phase angle   142, 160, 184, 190, 232, 261, 263, 306, 309, 346, 348, 350, 351, 368, 382–386, 411, 419, 451, 453, 455, 464, 476, 480, 497, 503, 545, 549
phase-case method   234
phase conductor   12, 15–17, 27, 198, 205, 210, 229, 245–248, 251, 252, 255, 261, 262, 291, 551, 552, 554, 555, 557–560, 562
phase-locked-loop (PLL)   344
phase opposition   204
phase-to-ground fault   132, 204, 205, 218, 236, 318, 322, 349, 399, 450, 451, 453, 464, 496, 503, 506, 507, 515, 518, 520, 521
phase-to-ground overvoltage   204, 234, 238
phase-to-phase overvoltage   205
phasor solution   90, 99, 118, 119, 144, 405, 434
photovoltaic (PV)   347, 508
physical arc model   59
physical model   22–23
pickup signal   514, 515
piecewise linear inductance   35, 89, 90, 118, 474
piecewise linear representation   49, 54, 87, 89–90, 118
piecewise linear resistance   89, 90, 111
PI equivalent   207, 221, 224, 541
PI model   12, 33, 40, 123, 218, 473
pipe-type cable   25, 110
pitch control   323
PlotXY   3, 6, 107

POCKECT CALCULATOR VARIES PARAMETERS (PCVP)   5, 120, 131, 132, 217, 218, 543, 544
Poincaré map   540
point of common coupling (PCC)   333, 431, 432, 435, 436, 437, 438, 441, 442, 445, 449, 508
polarity   21, 23, 47, 62, 205, 212, 214, 251, 252, 261, 263, 264, 356–359, 551–554, 557, 558, 562
pole span   66, 209
positive polarity   47, 261, 356, 358
post-arc current   60, 62, 64, 170
postprocessor   3, 4, 107, 108, 125–128, 143
power coefficient   323, 324, 326
power component   1–3, 5–7, 11–66, 75–77, 87, 123, 129, 135, 187, 196, 206, 275, 405, 424
power electronics   6, 7, 94, 101, 102, 107, 117, 128, 129, 131, 140, 207, 333–404, 424, 425, 455, 462
power electronics system   333, 334, 338–345, 349–366, 425
power flow reversal   398, 399
power-frequency model   44, 207
power-frequency overvoltage   12, 203, 204, 207, 214
power-frequency recovery voltage (PFRV)   60
power-frequency voltage   12, 203, 214
power-frequency waveshape   203
power quality   1, 4, 7, 125, 130, 347, 386, 396, 405, 419–427, 447, 455, 508, 511
power supply   150, 187, 207, 210, 343, 347, 386, 419, 426, 456, 529
power supply model   207, 210
power system component   12, 76–78, 123–125, 204, 207, 210, 216, 300, 337, 419, 423, 427, 531
power system protection   488
power system transient   1, 75, 476
prediction   30, 280, 283, 287, 298, 310, 311, 333, 335, 342, 343, 462, 541
prediction-based method   113, 284
pre-insertion   66, 209, 225, 234, 236, 238, 239
prestrike   59, 64, 65, 208, 263, 269
prestrike model   66
pre-striking   65, 73
principle of superposition   191

probability density function   251–253, 552, 555, 557, 558, 562

probability distribution   27, 64–66, 203, 227, 236, 238, 251, 253

propagation constant matrix   14

propagation time   23, 135, 144, 188–190

propagation velocity   16, 27, 28, 81, 187, 190, 194, 199, 201, 214, 215, 263, 269, 567

protection   1, 3, 6–8, 17, 75, 76, 129, 130, 132, 136, 187, 195, 196, 198, 213, 218, 256, 298, 300, 302, 336, 382, 384, 425, 449, 462, 471–532, 549

protective device   7, 8, 302, 471, 472, 478, 508–515, 524–531

protective margin   135

protective relay   1, 8, 96, 97, 107, 406, 471, 478, 488, 514

proximity effect   18, 24, 28, 29, 39

pseudo non-linear component   117, 177

pulse width modulation (PWM)   344, 348, 390, 393

PV *see* photovoltaic

PWM *see* pulse width modulation

q

q-axis equivalent circuit   48, 53, 285

q-axis subtransient reactance   289, 292, 299, 307, 317

q-axis synchronous reactance   289, 292, 299, 303, 307, 317

q-axis transient reactance   289, 292, 299, 303, 307, 317

qd0 induction machine model   47–48

qd0 synchronous machine model   52

quasi-periodic mode   540, 544

quenching   61, 63, 64, 66

r

random parameter   554, 555, 570

range of frequencies   2, 6, 11, 31, 48, 75, 77, 187, 206, 208, 213, 214, 275, 473

rational function   44, 116

rational transfer function   96, 119

real-time simulation

real-time simulation platform   76

real-time simulation tool   76

recloser   221, 508, 510–514, 518–520, 522–525, 527, 529–532

reclosing

reclosing operation   64, 162, 230, 234, 236, 239, 503, 511, 518, 520, 524, 531

reclosing overvoltage   205, 234, 236

reclosing signal   503

reference frame   14, 47, 48, 52, 284, 285, 393

reflection coefficient   189, 190

reignition   60–63, 168, 169, 174, 205, 512

relay

relay model   117, 471, 514, 515

relay modelling   484–508

relay signal   507, 532

relay technology   471, 472, 485

relief wave   195, 196, 198

remanent flux   35, 81, 474, 477, 478

renewable energy integration   333

residual current breaker (RCB)   563, 566, 572

residual flux   34, 50, 54, 541, 544, 545, 549

resistance   15, 17–21, 25–28, 32, 34–37, 40, 44, 47, 49, 52, 56, 59–61, 77, 79, 83, 88–90, 92–94, 111, 113, 116, 117, 141–143, 153, 154, 157, 160, 162, 164, 166, 168, 169, 172, 176–178, 180, 181, 184, 185, 187–191, 193, 195, 197–199, 209, 211, 213, 215, 218, 221, 229, 238, 244–246, 248, 250–253, 255, 256, 259, 261, 262, 267, 278, 280, 284, 285, 287, 289, 291, 292, 299, 303, 307, 312, 314, 317, 333, 335, 336, 341, 350, 357, 359, 361, 369, 377, 385, 390, 396, 410, 432, 434, 441, 473, 474, 476, 477, 480, 482, 489, 490, 493, 496, 512, 541, 544, 549, 554, 555, 557–560, 562, 566, 580

resonance   3, 4, 28, 30, 31, 45, 107, 118, 129, 130, 132, 133, 168, 184, 185, 203, 204, 207, 216–218, 221, 276, 302, 306, 343, 422–424, 426, 431, 432, 479, 540

restrike   60, 63, 155, 170, 205, 206, 242–244, 263

re-triangularization   91, 117, 177

rise of the recovery voltage (RRRV)   192–195

rod electrode   20, 21

rotating machine   2, 6, 7, 45–58, 97, 110, 112–114, 120, 125, 129, 140, 217, 275–328, 333, 338, 339, 343, 344, 347, 422, 424

rotating reference   47, 55, 284

rotor
  rotor angle stability   75, 298
  rotor inductance matrix   52
  rotor magnetic axis   48
round-rotor induction machine   50

S
sag *see* dip/voltage dip
saliency factor   52–54
salient-pole machine   52
SATURA (routine)   112, 116, 123, 178, 222, 477
saturable inductance   77, 89, 187, 222, 544, 546
saturable reactance   168, 207, 209
saturable transformer component (STC)   32, 111, 112, 123, 474
saturable transformer model   112, 474, 482
saturated magnetizing inductance   32, 52
saturation   29–34, 42, 49, 50, 52–54, 77, 112, 116, 123, 178, 179, 183, 207, 209, 215, 222, 223, 240, 276, 281, 285, 287, 292, 299, 306, 474, 476–478, 480, 482, 484, 546, 547, 580
screen   24–28
search space   568, 570
secondary injection   489
sectionalizer   221, 508, 512–514, 520, 522–525, 527
self-contained cable   24, 567
self-damping coefficient   55
self-excitation   281, 306, 308, 309
self-inductance matrix   47, 52, 222
semiconducting layer   26–28
semiconducting material   27
semiconducting screen   24, 27, 28
SEMLYEN SETUP (routine)   110, 116
sensitive load   343, 347, 456–458, 462
sensitivity curve   427
separation distance effect   196–198, 200
separation effect   133, 210
series-compensated system   306
series coupling transformer   382, 385, 386
series damping   92, 93, 350
series impedance matrix   13, 15
series RLC circuit   145–151, 155, 156, 159, 166
sheath   24, 25, 27, 28, 56, 62–64
shell core   29

shielding failure   205, 244, 247, 251, 253, 255, 259, 551, 555
shielding failure flashover rate (SFFOR)   244, 551, 559, 562
shield wire   5, 12, 205, 210, 229, 244–248, 250, 251, 253, 255, 259, 261, 290, 291, 551, 554, 555, 558–560, 562
short-circuit current   32, 58, 567
short-circuited rotor   47, 277, 313
shunt admittance matrix   27
shunt coupling transformer   382
shunt reactor   90, 130, 132, 133, 204, 207, 217, 218
simulation
  simulation time   2, 4, 8, 11, 35, 95, 128, 136, 142, 143, 213, 259, 335, 336, 338, 340, 539, 550, 553, 560, 562, 574
  simulation tool   1–4, 6, 8, 76, 101, 107, 108, 136, 216, 337, 422, 423, 426, 471, 472, 484–486, 539, 540, 563
single-conductor line model   16
single-core self-contained cable   24, 567
single-end bonding   25, 26
single-input single-output model   44, 58
single-phase inverter   361–362, 364
single-phase lossless line   77, 80–82
single-phase transformer model   32–36
single-phase transmission line (STL)   39, 41
skin effect   13, 24, 28–30, 35, 49, 56, 210, 215, 230, 473
slip   113, 125, 310, 312, 314–316, 318, 326, 328
slot region   56
slow-front overvoltage   12, 203, 205, 206, 208, 225–243
slow-front surge   275
slow-front transient   15, 31, 112, 210, 229
snapshot   102, 119
snubber circuit   94, 95, 142, 171, 172, 336, 337, 343, 349, 350, 354, 365, 367, 369, 373, 377, 409, 410, 441, 522, 565, 566
soil ionization   17, 19–21, 245, 246, 255, 554
soil resistivity   19, 20, 244, 246, 252, 554, 558, 562
solid-state converter   339
solid-state technology   333
solid state transformer (SST)   7, 367, 386–399, 426

solution method   3, 75, 76, 90–91, 96, 100, 101, 103, 109, 113, 118–121, 130, 275–284, 287, 337–338, 345, 405

solution technique   1, 6, 75–103, 117–119, 139, 158, 337

source data   140–142

source model   99, 217, 258, 423, 475

space charge sheath   62, 63

space vector modulation (SVM)   393

spark dynamics   215

speed control   296, 300, 304

spring constant   282, 307

squirrel-cage induction machine (SCIM)   309–313, 323, 325

SSR *see* subsynchronous resonance

standard   21, 22, 30, 33, 36, 37, 65, 107, 120, 122, 178, 203, 212, 221, 222, 230, 235, 236, 238, 239, 252, 256, 258, 281, 287, 344, 388, 419, 455, 478, 509, 541, 552–554

standard lightning impulse   21, 554

startup   284, 309–314, 317–319, 371, 374

static electronic relay   485

static model   15, 485

static relay   485, 489, 490

static synchronous compensator (STATCOM)   346, 382, 383, 385, 386

static synchronous series compensator (SSSC)   346, 382–385

static var compensator (SVC)   151, 367, 373, 375, 376, 378, 379, 402, 425

statistical method   135, 209

statistical study   4, 128, 136, 216, 217, 230, 232–236, 246, 539, 562

statistical switch   65–66, 128, 129, 209, 230, 423

statistical switching   209, 230

statistics switch   65, 236

statistics switching   236

stator   46, 47, 49, 51, 52, 56, 281, 282, 288, 300, 310–313, 315, 317, 321, 371, 373

stator self-inductance matrix   52

steady-state calculation   3, 12, 99, 102, 110, 132, 230, 302

steady-state initialization   97–103, 121, 129, 277, 279, 286, 338, 342, 368, 376, 423, 424

steady-state model   485, 503

steady-state solution   76, 97–100, 102, 111, 118, 119, 139, 141, 143, 144, 148, 152, 158, 185, 191, 209, 338, 405, 424

steep-fronted voltage surge   38

step-down transformer   411, 431, 478, 479, 482

step-up transformer   213, 291–293, 298, 299, 303, 306, 307

stray capacitance   44, 45, 77, 92, 112, 142, 213, 221, 343, 479, 541, 545, 580

streamer   22, 23

striking distance   245, 247, 554, 558

stroke

  stroke parameter   199, 255, 552, 555, 557, 562

  stroke waveform   199–201, 246, 262, 551

structural models   485

study zone   2, 221, 541

subharmonic mode   540

subsidence transient   479, 480, 484, 496

substation   130, 133, 184, 192–196, 203, 205, 207, 211–214, 219, 222, 225, 240, 241, 243, 315, 318, 321, 396, 435, 437, 438, 449, 450, 513, 514, 524, 529–532, 539, 545, 553, 568

substation busbar model   207

subsynchronous resonance (SSR)   3, 45, 107, 129, 130, 276, 279, 287, 302, 306, 308, 309

subtransient impedance   210

superposition   81, 88, 91, 185, 342

superposition principle   191

supporting routine   3–5, 15, 23, 27, 28, 37, 109–112, 116–117, 129, 178, 217, 222, 406, 473, 474

surge arrester

  surge arrester protection   195–196

  surge arrester selection   256

surge impedance   15–19, 27, 28, 42, 45, 57, 81, 86, 116, 133, 144, 146, 151, 156, 159, 175, 188, 190, 192–194, 197–199, 210, 211, 213, 215, 246, 251, 263, 267, 553

susceptance matrix   15

switch   59–61, 65, 66, 77, 78, 86, 90, 91, 94, 111, 113, 114, 116–118, 125, 140, 141, 144–148, 150–154, 157–166, 168, 169, 172, 183, 184, 187–192, 194, 199, 203, 204, 208, 209, 211, 217, 221, 224, 227, 234, 236, 238, 242–244, 261–267, 269, 315, 333–336, 339–342, 347, 349–354,

switch (*contd.*)
  357, 359–361, 382, 385, 406, 409–411,
  442, 450, 475, 476, 496, 510–513, 518,
  541, 545, 549, 563, 565, 566
switch data 65, 140, 141
switchgear 206, 208–209, 261, 348
switching
  switching model 334, 335, 339, 367
  switching overvoltage 7, 12, 64–66, 205,
    206, 227, 230, 236–238
  switching transient 35, 140, 143, 155–164,
    208, 225, 333, 336, 472
symmetrical fault current 478
symmetrical short-circuit current 32, 58–60,
  567
synchronization 45, 204, 279, 390
synchronous machine 7, 46, 51–55, 75, 102,
  111, 113, 129, 143, 276–281, 284, 285,
  287–309, 344
synthetisation 14, 36, 44, 58, 208, 545
systematic switch 65, 111, 237, 238, 476
systematic switching 234–236
system grounding 184, 204, 219
system zone model 229

t

TACS *see* Transient Analysis of Control Systems
TACS 69 97, 136, 217, 406
tank model 36, 37
temporary overvoltage (TOV) 12, 168, 203,
  204, 206, 207, 244, 256, 288
terminal model 35, 38, 41–45, 55, 58, 112
The Output Processor (TOP) 3, 6, 107,
  125–128
thermal breakdown 170
thermal failure 60
Thevenin equivalent 87–89, 91, 113, 117, 176,
  210, 280, 283, 284, 287, 342
three-legged stacked-core 29
three-phase asynchronous machine 276
three-phase controlled rectifier 361, 367–369,
  408–410
three-phase induction machine 7, 46–48, 113,
  277, 309–328
three-phase line-commutated rectifier
  362–365
three-phase self-contained cable 24–25

three-phase synchronous machine 7, 46, 51,
  52, 129, 276–281, 284, 287–309, 344
three-phase transformer model 36–37
thyristor 334, 335, 337, 339–341, 349, 350,
  352, 354, 356, 362, 367–370, 373, 377,
  379, 409–411
time constant 19, 60, 62, 95, 143, 146, 148,
  151, 152, 154, 166, 285, 289, 292, 299,
  303, 307, 317, 339, 567
time-controlled switch 111
time-current characteristic 509–511, 515,
  518, 529
time-domain simulation 4, 15, 77, 100, 128,
  216, 334, 337, 343, 422–424, 432, 471,
  484–486, 540–542, 563
time-domain solution 6, 76, 110, 117, 130,
  156, 157, 193, 337, 405
time step 15, 79, 85, 88, 89, 91, 93–97, 101,
  113, 116, 118, 120, 139, 141–144, 147,
  152, 156, 158, 160, 176, 177, 208, 209,
  213, 267, 280, 281, 283, 284, 335–337,
  339–345, 349, 350, 354, 365, 376, 377,
  405, 543, 550
time-to-breakdown 21–23
time-varying resistance 60, 61, 88, 111, 116
tip speed ratio 323, 324
TOP *see* The Output Processor
topologically-correct model 33, 222
topological model 29, 32, 33, 37, 44–45, 177,
  222
torque amplification 306
total clearing curve 511
total harmonic distortion (THD) 428, 432,
  434, 447
TOV *see* temporary overvoltage
tower 12, 13, 15–19, 207, 210–212, 214, 229,
  244, 246, 248, 251, 252, 259, 261, 262,
  290, 324, 325, 553–555, 557
TPBIG 3, 4, 6, 107–117, 120, 122–124, 129,
  136, 423
TPPLOT 3, 107
transformer
  transformer coil 39
  transformer energization 30, 34, 204
  transformer inrush current 34, 204
  transformer model 28–45, 110, 112, 116,
    120, 123–125, 207, 213, 215, 221, 224,
    241, 300, 464, 471, 473–477, 479, 480,
    482, 496, 515, 541, 546

transformer winding   33, 37–41, 45, 46, 56, 184, 221, 474, 475, 513, 541

transient analysis   1, 3, 13, 41, 75–103, 111, 112, 139, 187–201, 230, 257, 275, 302, 339, 422–424, 471, 473, 476, 539, 566

Transient Analysis of Control Systems (TACS) 96, 97, 100, 102, 111–116, 119, 120, 125, 129, 136, 148, 217, 275, 279, 286, 293, 324–326, 328, 339, 344, 361, 367–369, 396, 406, 411, 415, 419, 423, 441, 445, 491, 512, 514, 515, 566

transient model   257, 485, 489, 490

transient network analyser (TNA)   75

transient overvoltage   38, 64, 227, 229, 267, 509

transient recovery voltage (TRV)   60, 92, 145, 170, 173, 174, 191–195, 206, 222, 242, 512

transients analysis   1–8, 78–95

transient stability   45, 129, 278, 298–302, 305

transmission line
    transmission line equations   13–16
    transmission line grounding   17–21
    transmission line reclosure   276
    transmission line tower   16–17

transmission system   12, 97, 110, 302, 333, 345, 346, 385, 449, 529

transposed line   232–234

transverse electromagnetic mode (TEM)   16, 56

trapezoidal rule   76, 78–80, 82, 91, 96, 101, 118, 119, 280, 337, 405

trapped charge   98, 102, 111, 162–165, 205, 229, 233, 236, 263, 265, 270

travel time   16–19, 81, 86, 88, 89, 143, 192, 198, 211, 215, 268, 553

triplex core   29, 222, 240, 292, 299, 303, 307

tripping signal   59–61, 485

true non-linear component   32, 110, 114, 117

TRV *see* transient recovery voltage

turbine blade   55, 285, 323, 324, 326, 328

turn-off characteristic   335

turn-on characteristic   335

two-winding single-phase transformer   32, 33

Type 11   114, 339, 340, 349, 350, 361, 411, 566

Type 12   111, 339, 349

Type 13   111, 339, 340, 349, 350, 361, 362, 566

Type 56   113, 114, 125, 276, 277

Type 58   111, 114, 125, 133, 276–281, 286, 344

Type 59   111, 113, 114, 277–281, 286, 287, 289, 344

Type 92   111, 114, 566

Type 93   111, 114, 178

Type 94   5, 111, 116, 117, 177, 342, 406

Type 96   111, 116, 178, 480

Type 98   111, 178, 180

u

UM *see* universal machine

uncontrolled rectifier   361, 369, 417, 418, 420, 421, 427

undervoltage condition   459

unfaulted phase   204, 236, 399, 513

ungrounded capacitor bank   205

ungrounded primary transformer winding 513

unified power flow controller (UPFC)   346, 367, 382, 384, 385, 387, 399

unified power quality conditioner (UPQC) 347

uniform distribution   209, 558

uniformly-distributed parameter   14

uninterruptible power supply (UPS)   347, 419, 424, 425

universal machine (UM)   7, 46, 98, 111, 113, 276, 277, 281–284, 287, 369
    UM-8 DC machine   277
    UM-6 single-phase induction machine   277
    UM-3 three-phase squirrel-cage induction machine   277
    UM-1 three-phase synchronous machine 277
    UM-4 three-phase wound-rotor induction machine   277

untransposed line   33, 229, 232, 472, 473

user-defined component   108

user-defined module   411, 441, 445, 456

v

vacuum circuit breaker   61–64

validation   7, 140, 268, 333, 423, 479, 487–491, 546

velocity of propagation   15, 211, 213

very fast-front overvoltage (VFFO)   203, 206, 214–215

very fast-front transient (VFFT)   11, 13, 31,
    41–43, 45, 203, 215, 261–263, 266–268
VFFO *see* very fast-front overvoltage
VFFT *see* very fast-front transient
voltage angle jump   455
voltage balance   388, 392
voltage control   304, 383, 386, 390, 393, 396,
    426
voltage-controlled switch   110, 111
voltage-dependent load   427, 457
voltage dip
    voltage dip analysis   449, 453, 454, 457, 458,
        460
    voltage dip characterisation   453–462
    voltage dip compensation   464
    voltage dip duration   454–455, 459, 462
    voltage dip magnitude   427, 453–454
    voltage dip measurement   449–453, 457,
        461
    voltage dip mitigation   462–465
voltage drop   60, 209, 318, 335–337, 340, 349,
    447, 507, 531
voltage equations   47, 51, 52, 80, 81, 145, 156,
    159
voltage escalation   168, 170–173, 175, 177
voltage magnification   242, 243
voltage-mode control   348
voltage oriented control (VOC)   390
voltage-source converter (VSC)   334, 336, 346,
    347, 382–384, 462

voltage-time curve   21–22, 211, 212, 553, 558
voltage-tolerance curve   453, 456, 459
voltage transfer   38, 41, 42, 44, 45, 215
voltage transformer (VT)   8, 207, 441, 476,
    478–479, 482, 483, 491, 496, 545, 546

W
wave propagation   187, 567
WECS *see* wind energy conversion system
Weibull distribution   21, 252
white-box model   30
wind energy conversion system (WECS)   323,
    328, 349
wind gust   325–328
winding design   29, 33
winding model   32–34, 58, 208
winding resistance   32, 33, 35, 37, 474, 477
wind speed   323–328
WINDSYN   120, 125, 276, 277, 284, 286

X
XFORMER model   112, 116

Z
zero-sequence equivalent circuit   48
zero-sequence parameters   207, 472, 473
zero-sequence reactance   289, 292, 299, 303,
    307, 317
ZNO FITTER (routine)   117